Size-Dependent Effects in Materials for Environmental Protection and Energy Application

Size-Dependent Effects in Materials for Environmental Protection and Energy Application

Editors

Radostina Stoyanova
Ivelina Georgieva
Albena Bachvarova-Nedelcheva

MDPI

Basel • Beijing • Wuhan • Barcelona • Belgrade • Novi Sad • Cluj • Manchester

Editors
Radostina Stoyanova
Institute of General and Inorganic Chemistry
Bulgarian Academy of Sciences
Sofia
Bulgaria

Ivelina Georgieva
Institute of General and Inorganic Chemistry
Bulgarian Academy of Sciences
Sofia
Bulgaria

Albena Bachvarova-Nedelcheva
Institute of General and Inorganic Chemistry
Bulgarian Academy of Sciences
Sofia
Bulgaria

Editorial Office
MDPI
St. Alban-Anlage 66
4052 Basel, Switzerland

This is a reprint of articles from the Special Issue published online in the open access journal *Materials* (ISSN 1996-1944) (available at: https://www.mdpi.com/journal/materials/special_issues/XI603PZFQ8).

For citation purposes, cite each article independently as indicated on the article page online and as indicated below:

Lastname, A.A.; Lastname, B.B. Article Title. *Journal Name* **Year**, *Volume Number*, Page Range.

ISBN 978-3-7258-0762-8 (Hbk)
ISBN 978-3-7258-0761-1 (PDF)
doi.org/10.3390/books978-3-7258-0761-1

Contents

Preface

This Special Issue includes selected articles from the "*Fourth Workshop on Size-Dependent Effect in Materials for Environmental Protection and Energy Application (SizeMat4)*" published by MDPI with the financial support of TwinTeam Project Д01-92/06.2022, "European Network on Materials for Clean Technologies".

The concept of size-dependent material properties becomes dominant in the materials science community. Understanding the size-dependent properties of materials is the most challenging issue in advanced material science. To a great extent, this is a result of the technological requirements concerning the development of materials with controlled properties, as well as of the recent progress in materials science, nanotechnology, and computational chemistry.

The aim of this Special Issue was to discuss the dimensional effects of materials for environmental protection and clean energy production as an innovative approach for the development of innovative materials with improved properties. The Special Issue comprises, but is not limited to, the following three main classes of advanced inorganic materials that form the basis of modern technologies: (1) materials and thin films for environmental protection; (2) materials for clean energy storage; and (3) ceramics/bioceramics and glasses for better living (with applications in optics, molecular electronics, and medicine).

Radostina Stoyanova, Ivelina Georgieva, and Albena Bachvarova-Nedelcheva
Editors

materials

MDPI

Article

The Beneficial Impact of Mineral Content in Spent-Coffee-Ground-Derived Hard Carbon on Sodium-Ion Storage

Sonya Harizanova, Ivan Uzunov, Lyubomir Aleksandrov, Maria Shipochka, Ivanka Spassova and Mariya Kalapsazova *

Institute of General and Inorganic Chemistry, Bulgarian Academy of Sciences, 1113 Sofia, Bulgaria; sonya@svr.igic.bas.bg (S.H.); uzunov_iv@svr.igic.bas.bg (I.U.); lubomir@svr.igic.bas.bg (L.A.); shipochka@svr.igic.bas.bg (M.S.); ispasova@svr.igic.bas.bg (I.S.)
* Correspondence: maria_l_k@svr.igic.bas.bg

Abstract: The key technological implementation of sodium-ion batteries is converting biomass-derived hard carbons into effective anode materials. This becomes feasible if appropriate knowledge of the relations between the structure of carbonized biomass products, the mineral ash content in them, and Na storage properties is gained. In this study, we examine the simultaneous impact of the ash phase composition and carbon structure on the Na storage properties of hard carbons derived from spent coffee grounds (SCGs). The carbon structure is modified using the pre-carbonization of SCGs at 750 °C, followed by annealing at 1100 °C in an Ar atmosphere. Two variants of the pre-carbonization procedure are adopted: the pre-carbonization of SCGs in a fixed bed and CO_2 flow. For the sake of comparison, the pre-carbonized products are chemically treated to remove the ash content. The Na storage performance of SCG-derived carbons is examined in model two and three Na-ion cells. It was found that ash-containing carbons outperformed the ash-free analogs with respect to cycling stability, Coulombic efficiency, and rate capability. The enhanced performance is explained in terms of the modification of the carbon surface by ash phases (mainly albite) and its interaction with the electrolyte, which is monitored by ex situ XPS.

Keywords: hard carbon; spent coffee grounds; sodium-ion storage; mineral content; albite

Citation: Harizanova, S.; Uzunov, I.; Aleksandrov, L.; Shipochka, M.; Spassova, I.; Kalapsazova, M. The Beneficial Impact of Mineral Content in Spent-Coffee-Ground-Derived Hard Carbon on Sodium-Ion Storage. *Materials* **2024**, *17*, 1016. https://doi.org/10.3390/ma17051016

Academic Editors: Alessandro Dell'Era and Christian M. Julien

Received: 29 December 2023
Revised: 19 February 2024
Accepted: 20 February 2024
Published: 22 February 2024

1. Introduction

The pursuit of climate neutrality is the main task of the large-scale implementation of renewable energy [1]. Since renewable energy is dependent on meteorological conditions, devices are needed to store and send the generated energy when needed [2]. Rechargeable lithium-ion batteries (LIBs) appear attractive due to their high energy density and good cycling stability. Nevertheless, LIBs could not meet the requirements for large-scale stationary energy storage systems due to the limited and uneven distribution of lithium resources, as well as the rapidly increasing price. For this reason, it is necessary to develop non-lithium energy storage technologies that can replace LIBs for large-scale applications. Although several forms of energy storage devices have been commercialized, the development of new ones that are more cost-effective and more sustainable than existing ones is imperative [1,2]. One of the most attractive alternatives among emerging technologies for large-scale energy storage is that of sodium-ion batteries (SIBs) [1,3,4]. Although SIBs are on the cusp of commercialization, the key to their technological implementation is finding an appropriate anode material.

In recent years, many anode materials have been tried for SIBs, among which non-graphitized carbons (amorphous carbons, soft carbons, and hard carbons) stand out due to their low cost, structural stability, and ease of industrial production. Currently, most amorphous carbons are industrially synthesized from coal petrochemical products [5], but the growing demand for the sustainability and recycling of materials draws attention to agricultural waste and so-called waste biomass like avocado peels [6], peanut shells [7], and corn straw pith [8]. Among the leading sources of bio-waste is coffee [9], with Indonesia as the largest coffee producer [10]. The statistics of the International Coffee Organization show that for 2020/2021, the world consumption of coffee was almost 10 tons, with the largest amount in Europe—about 33% [11]. Each ton of green coffee produces around 650 kg of spent coffee grounds (SCGs) [9,12], and these large amounts of waste require strategies to be found and developed for its reuse. Coffee waste is used for the production of enzymes, bioactive components, and biofuels [9], and in recent years, it has also attracted attention as a raw material for the production of electrode materials for SIBs [13,14]. SCGs are rich in many organic compounds and, during pyrolysis, yield hard carbons (HCs) whose structure and properties depend on the carbonization procedure. The potential of SCGs as an anode for SIBs was first demonstrated by Gao et al., where a high specific capacity and unsatisfactory rate capability were found [15]. To improve the electrochemical performance of SCGs, the combination of pyrolysis with chemical treatment has been applied [16,17]. The most frequently used chemical reagents are H_3PO_4 and/or HCl, which ensure the hydrolysis of the pretreated biomass [18]. As an activation agent, KOH is utilized with the aim of enlarging the pore size of coffee-derived carbons [19,20]. Hydrothermal carbonization (HTC) is another step used to enhance the electrochemical performance of carbons [21]. The final temperature of carbonization also affects the Na-storage properties [17]. The most widely applied carbonization process for obtaining hard carbon is one-step pyrolysis at temperatures between 900 and 1600 °C [7,8]. Although most studies indicate that carbonized temperatures between 800 and 970 °C are sufficient to form electrochemically active SCG-derived carbons [15,19,20], some studies propose higher temperatures (such as 2000 °C) as most suitable [22]. These discrepancies come from the fact that one-step and two-step pre-carbonization procedures are used. Some research groups have found that the two-step heat treatment enhances the electrochemical performance of biomass-derived hard carbon [23,24]. According to Zhang et al., mineral impurities (coming from ash content in pinecone biomass, such as Mg, P, K, S, Ca, and Si) may occupy some active sites, as a result of which the insertion of sodium ions becomes difficult [25]. They found out that the pre-pyrolyzed samples at 500 °C after treatment with KOH and HCl and additional carbonization at 1400 °C showed a better cycling performance in comparison with unwashed samples, i.e., 328 and 299 mAh g^{-1} after 100 cycles, respectively [25]. All the aforementioned chemical pre-treatments aim to modify the carbon content, but they are usually non-clean techniques and make biomass-derived carbons more expensive.

In addition to the organic components, the SCGs contain a relatively low number of inorganic components composed of potassium, magnesium, phosphorus, calcium, etc. (i.e., around 2 wt.%) [26–28]. During the pyrolysis of SCGs, the inorganic components are transformed, forming a variety of ash phases, namely potassium magnesium silicate, magnesium oxide, dicalcium silicate, pyrophyllite ($Al_2Si_4O_{10}(OH)_2$), etc. [29–31]. The impact of ash phases on the electrochemical properties of biomass-derived hard carbon is a subject of intensive study nowadays, but a unified picture is missed [32–34]. Most of the research groups highlight a strong requirement for the removal of mineral ash from biomass (before or after carbonization) in order to improve the Na storage properties of hard carbon [32,35]. However, some studies report an augmentation in the Na^+ diffusion kinetics due to the inclusion of ash-forming elements in the graphitic layers [36]. The inconsistency in these reported data is a consequence of the complexity and variety of the ash phase composition, which, in its turn, depends on the nature of biomass sources. Understanding the role of the ash phase content is also important from the point of view of avoiding the use of expensive and pollutant chemical pre-treatment procedures during the production of carbonaceous electrodes from bio-waste.

In this study, we examine the simultaneous impact of the ash phase composition and carbon structure on the Na storage properties of hard carbons derived from spent coffee grounds (SCGs). The carbon structure is modified using the pre-carbonization of SCGs at 750 °C, followed by annealing at 1100 °C in an Ar atmosphere. Two variants of the pre-carbonization procedure are adopted: the pre-carbonization of SCGs in a fixed bed and CO_2 flow. For the sake of comparison, the pre-carbonized products are chemically treated to remove the ash content. The Na storage performance of SCG-derived carbons is examined in model two and three Na-ion cells. The electrode–electrolyte interaction after the electrochemical reaction is monitored by ex situ XPS.

2. Materials and Methods

2.1. Synthesis Procedure

The spent coffee grounds (SCGs) were obtained from leftovers of freshly ground coffee beans boiled in a coffee machine. The wet waste was dried at room temperature. The hard carbons were synthesized by the pyrolysis of dried waste following two temperature steps (Figure 1)—the coffee waste was first pre-carbonized at 750 °C and then annealed at 1100 °C in an Ar flow. The pre-carbonization of SCGs was performed in the following two ways: pyrolysis in a fixed bed (i.e., pyrolysis in an atmosphere of evolved gasses and lack of oxygen) and pyrolysis in a CO_2 flow. Two types of pre-carbonized samples were denoted as TC and PC, respectively. After pyrolysis, the solid residues were divided into two parts, one of which was chemically treated to remove the ash content, while the other part remained untreated. The chemical treatment involves the successive boiling of TC and PC samples in HCl, HNO_3, and NaOH for 3 h under reflux conditions. After each treatment, the samples were washed with deionized water until pH = 7 and dried. All chemically treated samples (TC and PC) were signified with an additional letter Z (i.e., TCZ and PCZ). The final temperature treatment was carried out at 1100 °C for 1 h, at a constant Ar flow of 0.3 ln min^{-1}. The tube furnace equipped with a gas supply system (Nabertherm, Lilienthal, Germany) was used for this procedure. After treatment at 1100 °C, the products were slowly cooled down to an ambient temperature in an Ar atmosphere. These carbonaceous materials were ground in a planetary ball-milling system (Pulverisette 6, Fritsch, Idar-Oberstein, Germany) for 3 h, and a fraction with a 0.1 mm size was collected and used for characterization. As a result of these procedures, the pre-carbonized samples at 750 °C are denoted as TC-8, TCZ-8, PC-8, and PCZ-8, while the pyrolyzed samples at 1100 °C are TC-11, TCZ-11, PC-11, and PCZ-11, respectively.

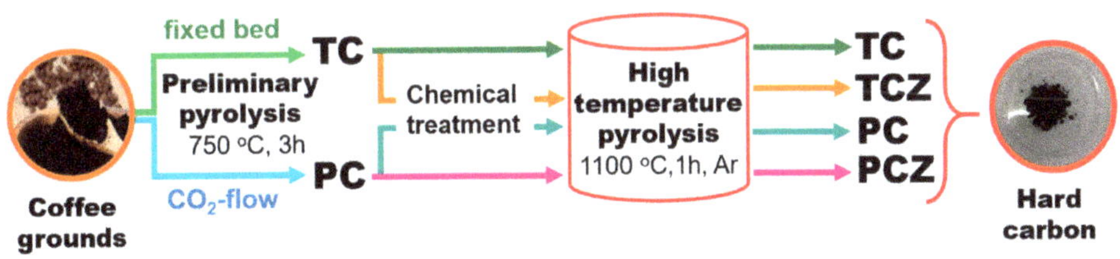

Figure 1. Schematic representation of the preparation of hard carbon from spent coffee.

2.2. Characterization Methods

The moisture and ash content of the air-dried SCGs were determined in accordance with TAPPI T-550 and T-211 standards, respectively [37].

Differential thermal analysis with thermogravimetry (DTA/TGA) of the SCGs and carbonaceous materials was performed with a Seteram Labsysis Evo 1600 instrument (SETARAM Company, Caluire, France) in the temperature range from room temperature up to 1000 °C at a heating rate of 5 K min^{-1} in oxygen or argon atmosphere.

The X-ray powder diffraction patterns of ash and carbonaceous materials were collected on the diffractometer Bruker D8 Advance (Bruker Corporation, Karlsruhe, Germany) with Cu Kα radiation.

The morphology of hard carbons was monitored by scanning electron microscope (SEM, JEOL JSM 6390, Tokyo, Japan) equipped with an energy-dispersive X-ray spectrometer (EDS, AZtec, Oxford Instruments, Abingdon, UK).

The elemental analysis (determination of C, H, N, S) was performed on a Vario Macro Cube (Elementar Analyzensysteme GmbH, Langenselbold, Germany). Oxygen content is determined by the difference.

Textural characterization of the carbon materials was carried out by low-temperature nitrogen adsorption at −196 °C using a Quantachrome NOVA 1200e (AntonPaar Quanta Tech Inc., Boynton Beach, FL, USA) instrument. Before the experiments, the samples were outgassed under a vacuum overnight. The adsorption–desorption isotherms are used for the evaluation of the specific surface area (by the BET method) and the pore volumes (according to the Gurvitch rule). The pore-size distributions were made by DFT (using the slit pore NLDFT equilibrium model).

The film composition and electronic structure were investigated by X-ray photoelectron spectroscopy (XPS). The measurements were carried out on an AXIS Supra electron-spectrometer (Kratos Analitycal Ltd., Manchester, UK) using monochromatic AlKα radiation with a photon energy of 1486.6 eV and a charge neutralization system. The binding energies (BEs) were determined with an accuracy of ±0.1 eV. The chemical composition in the depth of the films was determined by monitoring the areas and binding energies of C 1s, O 1s, N 1s, Ca 2p, Si 2p, Mg 2p, and K 2p photoelectron peaks. Using the commercial data-processing software of Kratos Analytical Ltd. (ESCApe™ 1.2.0.1325, Kratos Analytical A Shimadzu, Stretford, UK), the concentrations of the different chemical elements (in atomic %) were calculated by normalizing the areas of the photoelectron peaks to their relative sensitivity factors.

2.3. Electrode Preparation and Electrochemical Characterization

The electrodes were prepared by mixing 80% hard carbons with 10% carbon black (Super C65, TiMCAL Ltd., Bodio, Switzerland) and 10% carboxymethyl cellulose (CMC, Merck, St. Louis, MO, USA) in ultra-pure water. The mixture was homogenized in a planetary centrifugal mixer (Thinky Corporation, Tokyo, Japan) for 15 min at 2000 rpm. The obtained slurry was cast on conducted carbon-coated aluminum foil using a Doctor Blade film coater (Proceq SA, Scherzenbach, Switzerland), followed by vacuum drying at 80 °C overnight. The disk electrodes with a diameter of 10 mm were cut, pressed, and dried at 120 °C under a vacuum. The mass loading of active material varied between 2.8 mg and 3.2 mg.

All electrochemical experiments were carried out on a Biologic VMP-3e battery cycler (BioLogic, Seyssinet-Pariset, France) in a climatic chamber at 20 ± 1 °C. The three-electrode Swagelok-type cells were used for cyclic voltammetry tests, and two-electrode Swagelok-type cells were used for the rate capability and cycling stability tests. The model cells were mounted in a glovebox (MB-Unilab Pro SP (1500/780) with H_2O and O_2 content <0.1 ppm, MBraun, Garching, Germany). Clean sodium metal (Sigma Aldrich, St. Louis, MO, USA) was used as a counter and reference electrode. The electrolyte at 1 M $NaPF_6$ in PC (both from Sigma Aldrich, St. Louis, MO, USA) was soaked in glass microfiber separators (Whatman GF/D). To ensure the accuracy of the measurement, each electrochemical experiment was repeated at least twice.

For ex situ XPS experiments, the cells were dissembled in a glovebox, and the recovered electrodes were further analyzed.

3. Results

3.1. Formation of Carbons from SCGs

The elemental analysis data of pristine and carbonized SCGs are listed in Table 1. Pristine SCGs are rich in carbon, oxygen, and hydrogen elements, which is typical for coffee waste [38]. In addition, a low amount of nitrogen and traces of sulfur were also observed. These elements are usually associated not only with the organic components but also with ash–mineral components [29]. In SCG, the ash-forming elements (determined by EDS analysis) are Si, Mg, K, Al, and Ca (see the discussion below).

Table 1. Elemental analysis of pristine SCGs and carbons after low-temperature annealing (750 °C) and high-temperature pyrolysis (1100 °C).

Sample	C, Mass %	H, Mass %	N, Mass %	S, Mass %	O, Difference	O/C
SCGs	49.91	7.41	2.83	0.16	39.66	0.79
TC-8	77.74	2.19	4.27	0.24	15.56	0.20
TC-11	79.75	0.85	1.02	0.18	19.05	0.24
TCZ-8	57.13	2.25	3.68	0.19	36.75	0.64
TCZ-11	80.93	1.41	1.75	0.17	15.74	0.19
PC-8	74.82	2.19	3.70	0.21	19.08	0.26
PC-11	80.58	0.94	0.46	0.15	17.87	0.22
PCZ-8	51.09	2.54	3.94	0.19	42.24	0.83
PCZ-11	78.42	1.43	1.70	0.15	18.30	0.23

Because of the higher amount of oxygen and hydrogen, SCGs are easily decomposed in both an oxidizing and inert atmosphere (Figure 2). In oxygen flow, the DTA curve shows one endo process at around 77 °C, followed by strong exo processes at around 287, 389, and 450 °C. All thermal processes are accompanied by a corresponding mass loss; for up to 150 °C, the mass loss is 10 m%, and between 200 and 900 °C the mass loss reaches about 86 m.%. (For better resolution, the TG curve is presented in the form of the first derivative, inset Figure 2). Based on previous studies on the thermal behavior of coffee waste [39–41], the low-temperature endo process can be assigned to the evaporation of moisture and volatile compounds. The moisture calculated from TG analysis coincides well with that determined from TAPPI standards (Table S1), thus confirming once again the origin of the DTA/TG peak at 75 °C. The next exo processes at 287 and 389 °C can be attributed to the hemicellulose and cellulose degradation process, while the last exothermic process at 450 °C comes from the thermal decomposition of lignin. The higher thermal stability of the lignin than that of hemicellulose and cellulose is a result of its chemical composition: the lignin is richer in aromatic constituents and poorer in oxygen groups [42]. At 900 °C, the solid residue determined from TG analysis is 4 m.%. This value matches the ash content determined by the TAPPI T-211 standard [37] (i.e., ash content of 2.1 m.%, Table S1). According to the XRD analysis (Figure S1), SCG-derived ash consists of a phase mixture between albite calcian low ($Na_{1.0–0.9}Ca_{0.1–0.3}Al_{1.0–1.1}Si_{3.0–2.9}O_8$), MgO, K_2SO_4 and $Ca_5(PO_4)_3(OH,Cl,F)$ (Figure S1). A similar phase composition has been reported by Nieto et al. [21] in raw spent coffee grounds as follows: K_2SO_4, MgO, $Ca_5(PO_4)_3(OH,Cl,F)$, and $CaCO_3$.

When the thermal process takes place in an inert Ar atmosphere, the low-temperature endothermic effect (i.e., peak at ~61 °C and a mass loss of 11 m.%) is preserved, while only two exothermic effects appear at around 290 and 330 °C. The endothermic effect corresponds to moisture evaporation, while exothermic effects reflect the pyrolysis of hemicellulose and cellulose components [43,44]. The exothermic effect at around 707 °C can be explained by a release of some functional groups in cellulose residue [44]. At 900 °C, the residual mass of SCGs is 17.61%, which is higher than that determined in oxygen flow (i.e., of 4.06 wt.%).

Based on the chemical analysis and thermal properties of SCGs, we adopted a synthetic procedure including the pre-carbonization of SCGs at 750 °C, followed by the chemical treatment of pre-carbonized samples and, finally, the annealing of samples at 1100 °C. When pre-carbonization was performed in a fixed bed and in CO_2 flow, both the oxygen and hydrogen content decreased drastically, while the N-content appeared unchanged (Table 1). It is interesting that the O and H amounts did not show any dependence on the pre-carbonization at 750 °C. The chemical treatment with HCl, HNO_3, and NaOH yields a complete dissolution of ash containing albite and MgO phases (Table 1) [45,46]. This causes an increase in the oxygen content, thus indicating the procedure of side reactions between pre-carbonized samples and chemical reagents.

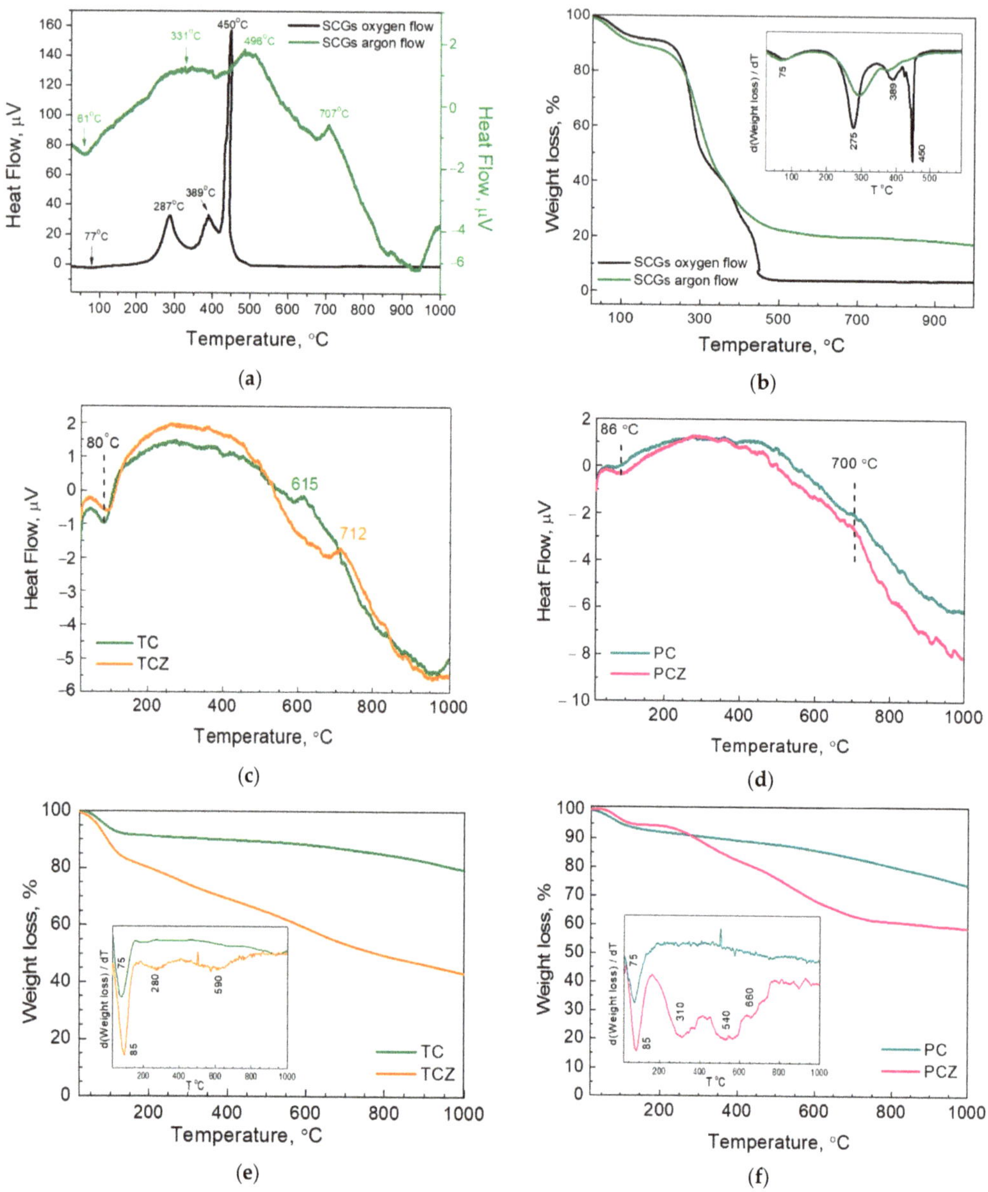

Figure 2. DTA (**a**) and TG (**b**) curves of SCGs in oxygen and argon atmospheres. DTA (**c**,**d**) and TG (**e**,**f**) curves of pre-carbonized samples, TC-8, TCZ-8, PC-8, and PCZ-8 samples in an argon atmosphere. The insets show the first derivative of the corresponding TG curves.

The annealing of pre-carbonized TC-8 and PC-8 at 1100 °C leads to a further decrease in the H content, but the O content seems unchanged. The same behavior in H and O content after annealing was observed by Lin et al. [47]: the decrease in hydrogen content was associated with the oxidation of alkyl groups, while the increase in the oxygen content was a result of the formation of C=O groups. In addition to the H and O content, the nitrogen content was also reduced after annealing at 1100 °C. In comparison with pre-carbonized TC-8 and PC-8, the annealing of chemically treated TCZ-8 and PCZ-8 samples produces a drastic decrease in the O content (more than two times), as a result of which this content becomes nearly the same for all types of samples. Moreover, after annealing the H and N content for ash-free PCZ-8 and TCZ-8 samples, these are also comparable to that of ash-containing ones.

The changes in the chemical composition of pre-carbonized samples with and without ash were further monitored by DTA and TG experiments (Figure 2). Up to 150 °C, all samples lost about 10 w.% due to the evaporation of moisture and volatile matters. On the other hand, this implies that the pre-carbonized samples easily absorb moisture, which is insensitive to their ash content. The availability of mineral ash in SCG-derived samples has an effect on their pyrolysis. Between 150 and 1000 °C, the pre-carbonized TC-8 and PC-8 samples lose an additional 10–15 wt.% without the release of measurable DTA heat. At 1000 °C, the total yield after pyrolysis is around 80–85%. For ash-free pre-carbonized samples, the weight loss between 150 and 1000 °C is more than two times that of the ash-containing samples (Table S1). The close inspection of the TG curve of TCZ-8 signifies that the loss takes place at two stages, 280 and 590 °C, respectively, while for the PCZ-8, at least three stages can be distinguished at 310, 540, and 660 °C. The enhanced mass loss of the ash-free samples correlates with their increased O content after chemical treatment (Table 1). It is of importance that the total yield after pyrolysis reaches around 50 wt.% irrespective of the observed differences in TG curves of TCZ-8 and PCZ-8.

The structure of SCG-derived samples is assessed by means of X-ray diffraction. Figure 3a,b compares the XRD patterns of the pre-carbonized and pyrolyzed samples. The XRD patterns of all samples display two broad diffraction peaks at approximately $2\theta = 23°$ and 44°, which correspond to the reflections of the (002) and (100)/(101) graphite planes. It is worth mentioning that these XRD features are typical for disordered carbons obtained from biomass [6,13,47]. The (002) reflection expressing the distance between carbon layers has a magnitude of 3.74 Å for all pre-carbonized samples. By increasing the pyrolyzing temperature from 750 to 1000 °C, the *d*-spacing shows a tendency to increase, reaching a value of 3.81–3.88 Å. This means that interlayer space adopts a similar magnitude that is insensitive to the method of pyrolysis. It is of importance that d-spacing is of benefit to the insertion and extraction of Na- ions since it is larger than the required minimum of 3.70 Å [48]. For the ash-containing carbons TC-11 and PC-11, additional narrow peaks were observed, which correspond to the contamination of mineral ash containing albite calcian low, MgO, and K_2SO_4. The detection of these crystalline phases at 1100 °C serves as a sign for their formation as secondary products during the pyrolysis of organic components of SCGs; because of the relatively low melting temperature (i.e., 1100–1120 °C), the albite easily crystallizes in the carbonized product. This is a consequence of the high amount of the Si-element in pristine SCGs [29,49]. It is worth mentioning that biomass rich in Si includes coffee waste, rice husks, wood, etc., while soybean husks, grapevine waste, and sunflower pellets and stalks contain a low content of Si [50,51].

The functional groups formed during the carbonization of SCGs are accessed by means of XPS analysis (Figures 3 and S2). The C 1s spectrum of pre-carbonized samples with and without ash consists of an intensive peak at 284.8 eV, which is attributed to the sp^2-hybridized graphitic carbon [52,53]. Superimposed to the peak at 284.8 eV, three less intensive peaks at (286.1 ± 0.2) eV, (287.4 ± 0.2) eV and (288.8 ± 0.2) eV can be distinguished. Although the peak at 286.1 eV can be assigned to C atoms bonded to oxygen in hydroxyl/epoxide groups (C–O), the carbonyl and carboxyl groups (C=O and O–C=O) give rise to the peaks at 287.4 eV and 288.8 [46,52]. The carboxyl groups appear only for the ash-free samples TCZ-8 and PCZ-8. Supporting this assignment of the C 1s signals, the O 1s spectra of pre-carbonized samples can be deconvoluted for at least three peaks at 529.0, 531.5, and 533.0 eV. The peak at 529.0 eV is adequately resolved for the ash-containing TC-8 and PC-8 and can be assigned to the O atoms bonded to ash elements such as Si, Mg, and K. The peak at 531.5 eV comes, most probably, from the double-bonded O atoms in esters and/or acids, while the single-bonded O atoms in ketons and/or ethers are responsible for the peak at 533.0 eV [54,55]. The functional C–O and C=O groups remain stable even after pyrolysis at 1100 °C. Moreover, the functional groups seem insensitive towards the method of carbonization, as well as towards ash removal.

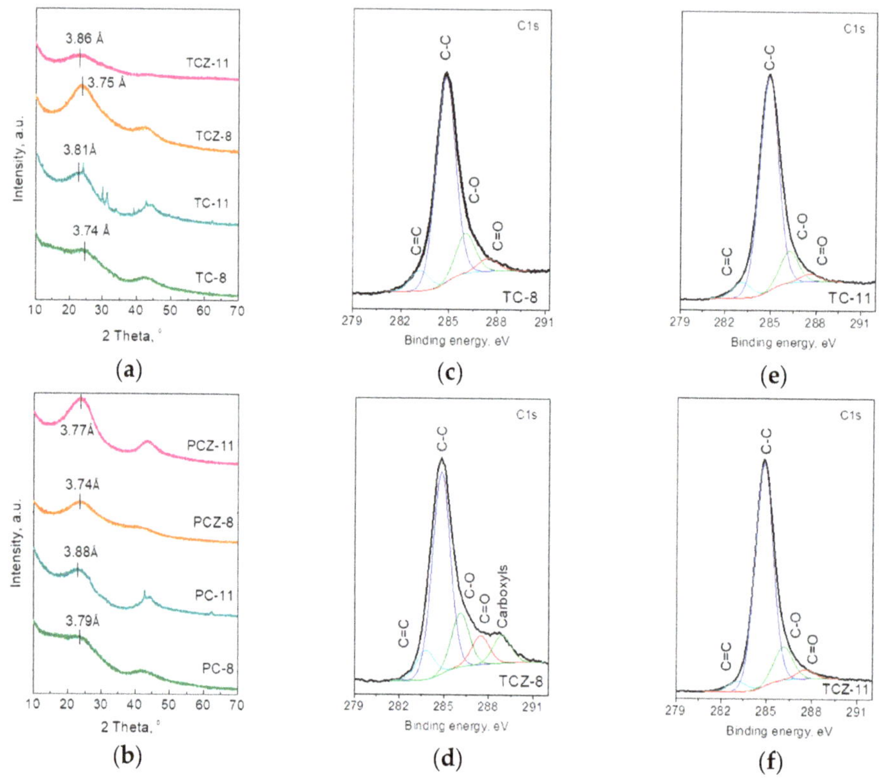

Figure 3. XRD patterns of pyrolyzed SCGs in (**a**) a fixed bed with and without ash, TC-8 and TCZ-8, and their analogs after high-temperature pyrolysis, TC-11 and TCZ-11; (**b**) CO2 flow SCGs with and without ash PC-8 and PCZ-8, and their analogs after high-temperature pyrolysis, PC-11 and PCZ-11. XPS spectra in the energy regions of C 1s for pyrolyzed SCGs obtained in a fixed bed with and without ash (**c**,**d**) and their high-temperature analogs (**e**,**f**).

The only parameter that is changed after carbonization is the elements' concentration. Figure 4 compares the element concentration determined by XPS and EDS analysis. These techniques are used in a complementary way to gain access to the distribution of elements along the depth of the carbonized samples at 1100 °C; the XPS method permits the identification and quantification of elements located on the outermost surface layers (i.e., up to 5 nm), while the EDS provides information on elements located inside the thicker layers (more than 500 nm). The data disclose that the oxygen is inhomogeneously distributed along the depth of particles, especially for ash-containing TC-11 and PC-11: the oxygen content is higher on the surface than the inside of the volume. For the ash-free samples, the oxygen is more or less homogeneously distributed. Furthermore, the amount of ash elements determined by XPS is higher than those evaluated by EDS. (The XPS spectra of Si 2p are given in Figure S3). This is a mark that mineral ash accumulates mainly on the surface of carbonized products, this being insensitive towards the method of pre-carbonization. In analogy with ash elements, nitrogen is also exclusively located on the surface.

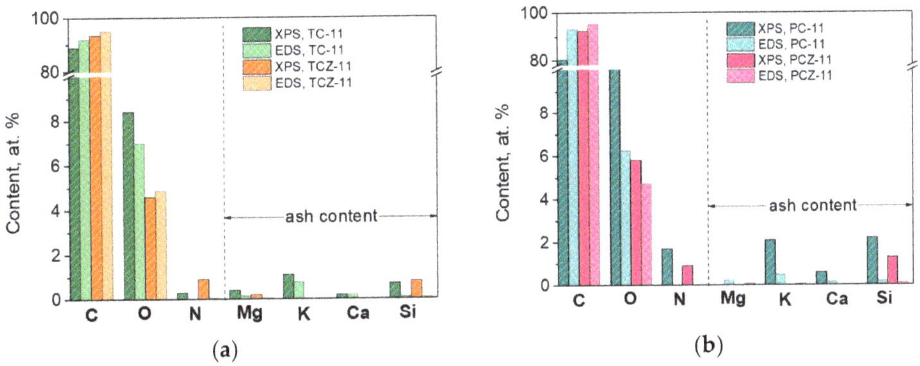

Figure 4. The concentration of elements determined by XPS and EDS analysis obtained after high-temperature pyrolysis and being pre-carbonized in a fixed bed (**a**) and CO2 flow (**b**).

The morphology of SCG-derived carbons was monitored by scanning electron microscopy, and the corresponding SEM images are shown in Figure S4. For ash-containing TC-11, the morphology consists of plate-like particles packed in columns, the aggregate sizes being around 1.5–2.5 μm. After ash removal, plate-like aggregates retain, in general, their sizes, but the degree of aggregate packing seems to diminish. In comparison to TC-11, the PC-11 displays plate-like aggregates with a more inhomogeneous size distribution spreading from 0.5 to 3.0 μm. The removal of ash from PC-11 does not have a significant effect on the carbon morphology.

The texture properties of SCG-derived carbons enable us to differentiate with respect to the method of carbonization and ash removal (Figures 5 and S5). The sample TC-8 obtained in a fixed bed has a low specific surface area of 17 $m^2\ g^{-1}$, while after ash removal, the specific surface area increases more than 10 times (i.e., up to 409 $m^2\ g^{-1}$) (Table 2). Irrespective of the observed change in the specific surface area, both TC-8 and TCZ-8 possess isotherms classified as I-type (Figure S5a,b, insets), which is characteristic of microporous materials. The pore size distribution and NLDFT calculation show that besides micropores, both carbons have mesopores with sizes between 5 and 7 nm. After the high-temperature pyrolysis, the specific surface area additionally increases, especially for the ash-free sample TCZ-11 (Table 2). Even in this case, TCZ-11 still exhibits an isotherm of the I-type, and the ratio between micro- and mesopore volumes Vmi/Vmes remains nearly unchanged (Figure 5b, inset). In comparison with the ash-free sample, the ash-containing sample shows a different behavior after high-temperature pyrolysis: the isotherm is changed from the I- to the H4-type [56,57], which is associated with the occurrence of narrow slit pores. The H4-type isotherm is often observed for carbons obtained from lignin-rich biomass [6,47]. The specific surface area of TC-11 also increases, but the ratio Vmi/Vmes decreases (Table 2), thus indicating the formation of more mesopores (Figure 5a).

In comparison with TC-8 and TCZ-8, the samples obtained in CO_2 flow (i.e., PC-8 and PCZ-8) exhibit isotherms belonging to the H3-type (Figure S5c, inset). These isotherms with a specific hysteresis loop are typical for materials with non-rigid aggregates of plate-like particles characterized by unlimited adsorption at high p/po [56,57]. After high-temperature pyrolysis at 1100 °C, the isotherm of PC-11 is transformed from the H3 to the II-type (Figure 5c), thus revealing a formation of a nonporous or macroporous-carbonized sample. The NLDFT calculation reveals a broad pore size distribution—from ~2.5 to ~33 nm (Figure 5c)—with a large quantity of mesopores (Table 2). Contrary to PC-11, the transformation of the isotherm for PCZ-11 is from H3 to H4 (Figure S5d, inset). The hysteresis loop is associated with narrow slip-shaped pores.

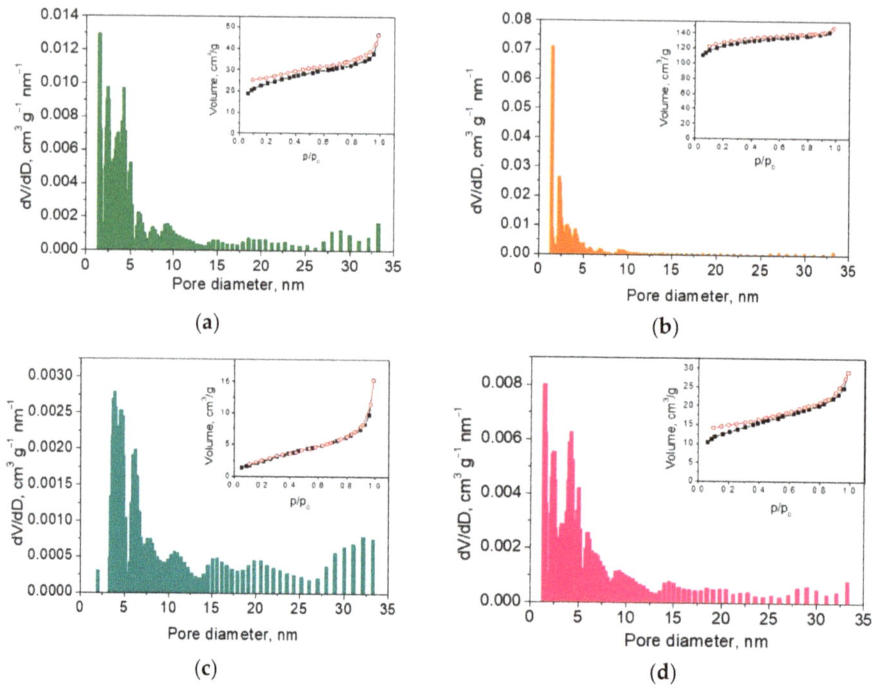

Figure 5. NLDFT pore size distribution of (**a**) TC-11, (**b**) TCZ-11, (**c**) PC-11, and (**d**) PCZ-11. The corresponding nitrogen adsorption–desorption isotherms are shown as insets.

Table 2. BET specific surface area, total pore volume, average pore size, micropore volume, and volume ratio between the micro- and mesopores.

Samples	S, $m^2\ g^{-1}$	V, $cm^3\ g^{-1}$	Dav, nm	Vmi, $cm^3\ g^{-1}$	Vmi/Vmes
TC-8	17	0.008	2.00	0.006	3
TC-11	86	0.07	1.9	0.02	0.4
TCZ-8	409	0.20	1.9	0.14	2.33
TCZ-11	471	0.23	1.7	0.16	2.3
PC-8	10	0.001	3.2	-	-
PC-11	10	0.02	2.9	0.006	0.31
PCZ-8	8	0.008	3.2	0.001	0.14
PCZ-11	48	0.05	1.9	0.03	1.5

In general, the comparison between the ash-containing and ash-free analogs (i.e., TC-11 and TCZ-11, as well as PC-11 and PCZ-11) shows that the specific surface area increases after the ash removal, while the amount of mesopores is significantly greater for ash-containing samples (Table 2, Figures 5 and S5).

3.2. Sodium Storage Performance of SCG-Derived Carbons

The capability of SCG-derived carbons to store Na is evaluated by CV measurements at a potential range of 0.05–2.0 V and a scanning rate varying from 0.01 mV s^{-1} to 100 mV s^{-1} (Figure S6). For all the samples, the first cathodic scan at a rate of 0.01 mV s^{-1} displays a broad wave between 0.9 and 0.3 V, which is not restored during the reverse anodic scan. This wave is assigned to the decomposition of the electrolyte, leading to the formation of the solid electrolyte interphase (SEI) on the electrode surface [35,36,48]. The calculated capacitance corresponding to the first cathodic scan between 2.0 and 0.3 V is shown in

Figure S6b. This comparison provides evidence that PC samples exhibit higher irreversible capacitance than TC analogs. This means that a thicker SEI is formed on the surface of PC samples. The thicker SEI for the PC samples could be related to their broader pore size distribution [58] (Figure S6). In addition, the effect of the ash content is small. On the other hand, this can serve as an indirect sign that the specific surface area is not a leading factor for the conditioning of SEI: the ash-free samples have the highest specific surface area (Table 2), but the biggest capacitance is determined for the ash-containing sample PC-11. This is in agreement with previous studies, where the contributions of a specific surface area and pore size distribution in the SEI formation have been demonstrated [59].

After the first irreversible cathodic scan, the CV curves adopt stable profiles, where sharp redox peaks below 0.1 V are clearly observed at slow scanning rates (Figures 6 and S7). These peaks can be associated with the reversible intercalation of Na^+ between graphitic layers and their further adsorption in the micropores [60]. By increasing the scan rate from 0.01 mV s^{-1} to 1 mV s^{-1}, the very broad redox wave centered at around 0.25 V becomes dominant at the expense of the sharp redox peaks. This wave comes from the adsorption-induced Na^+ storage on the pores and defect sites of hard carbons [48]. At the highest scan rate (i.e., 100 mV s^{-1}), the CV profiles resemble those of capacitive reactions (Figure S7). The established dependence of the Na^+ intercalation and adsorption reactions on the scan rate corroborates with previous data on the Na^+ adsorption energy and diffusion barrier: Na^+ adsorption proceeds faster than Na^+ intercalation [61]. The important finding is that all of the SCG-derived carbons have the same CV features, thus revealing a constancy in the Na storage mechanism irrespective of the carbonization procedure, as well as the ash content.

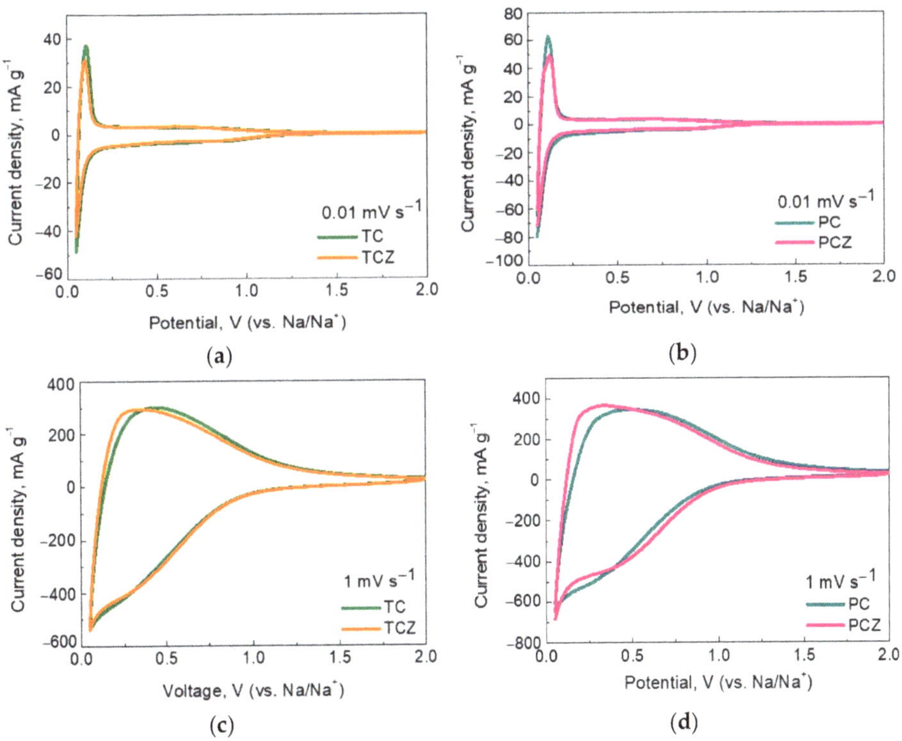

Figure 6. CV curves of SCG-derived carbon at 0.01 mV s^{-1} (**a**,**b**) and 1 mV s^{-1} (**c**,**d**).

The galvanostatic experiments further support the main features of the Na storage mechanism. The SEI on carbons is formed after the first cycle, the measure of which is the first irreversible capacity. The highest magnitude of the first irreversible capacity reaches the PC samples (208 and 147 mAh g^{-1} corresponding to ICE 49% and 60% for PC-11 and PCZ-11, respectively), thus confirming the growth of a thicker SEI on them in comparison with the TC samples (119 and 145 mAh g^{-1} corresponding to ICE 56% and 52% for TC-11 and TCZ-11, respectively). This trend is valid irrespective of the ash content. It is noticeable that the first irreversible capacity follows the order already established from CV data (Figures 6 and S7). The comparison of the galvanostatic and CV data evidence that the ash content slightly affects the SEI formation, while the manner of pore size distribution is a main factor contributing to SEI's growth.

The second cycle enables an evaluation of the reversible capacity delivered by SCG-derived carbons (Figure 7c,d). To avoid sodium plating and reactivity at low potentials in the sodium-half cell [62,63], the low voltage limit is set to 0.05 V. These data show that TC samples with and without ash possess around a 110 mAh g^{-1} capacity due to the Na^+ adsorption, while the Na^+ intercalation gives rise to a capacity of around 70 mAh g^{-1}, respectively. In general, PC samples deliver high capacity in comparison with TC samples.

The cycling stability and rate capability are the next parameters that enable the differentiation of samples with respect to the ash content (Figure 7e,f). At a lower current load (5 mA g^{-1}), ash-free TCZ-11 delivers a slightly higher capacity than that of the ash-containing analog. However, Coulombic efficiency is significantly better for ash-containing TC-11, especially after the first five cycles. By increasing the current load from 5 to 500 mA g^{-1}, the capacity of TC-11 gradually becomes higher than that of TCZ-11, thus disclosing the better rate capability of the ash-containing sample. For a whole rate-capability test, the Coulombic efficiency for TC-11 remains more stable than that of TCZ-11. In comparison with TC, the PC samples display the same features: ash-containing PC-11 has a better rate capability and Coulombic efficiency than the ash-free analog PCZ-11. All these data give evidence for the better electrochemical performance of the ash-containing samples than their ash-free analogs. The improved performance could be correlated with the surface modification of the carbonized materials through an accumulation of the mineral ash (namely low albite calcian, MgO, and K_2SO_4). This mechanism is different from that established by Zhang et al., where the ash-containing samples display worse cycling stability due to the occupancy of active Na storage sites by mineral impurities [25]. However, the occurrence of impurity potassium in hard carbon obtained from coconut endocarp leads to facilitating Na^+ diffusion by expanding the interlayer spacing of the graphitic layers [36]. This suggests that the impact of the mineral ash on carbon performance depends not only on the phase composition but also on the manner of ash distribution.

Figure 8 compares the cycling stability of ash-containing TC-11 and PC-11 under a current load of 50 mA g^{-1}. Despite the electrochemical inactivity of mineral ash, the comparison shows that TC-11 outperforms PC-11. This implies that the Na storage performance depends both on the carbon structure (which is a result of the carbonization procedure) and ash phases.

Taking into account that most studies provide information on the specific capacity of deeply discharged carbons (i.e., the low voltage limit of 0.01 V or 0.002 V), we performed additional electrochemical tests on TC-11 by decreasing the low voltage limit from 0.05 V to 0.01 V and keeping the upper voltage limit to 2.0 V. Figure 8b shows the charge/discharge curves for "softly" and "deeply" discharged TC-11 after the 1st and 5th cycles under 50 mA g^{-1}. By decreasing the low voltage limit, both the discharge and charge capacities increased dramatically, but the irreversible capacity remained the same (i.e., 113 mAh g^{-1}). This resulted in an increase in the initial Coulombic efficiency (ICE): from 37 to 56%. After further cycling, the Coulombic efficiency (CE) tended to be 99–100% without any loss of capacity. For the 20th cycle, the specific capacities for "softly" and "deeply" discharged TC-11 were 75 and 168 mAh g^{-1}, respectively. To demonstrate the electrochemical performance of ash-containing carbons, Table 3 compares our data with previously reported ones for coffee-derived electrode materials using different preparation methods. The comparison reveals that ash-containing TC-11 displays electrochemical parameters that are quite comparable to carbons undergoing additional chemical treatment. This means that mineral ash has a beneficial effect on the Na storage performance of SCG-derived carbons.

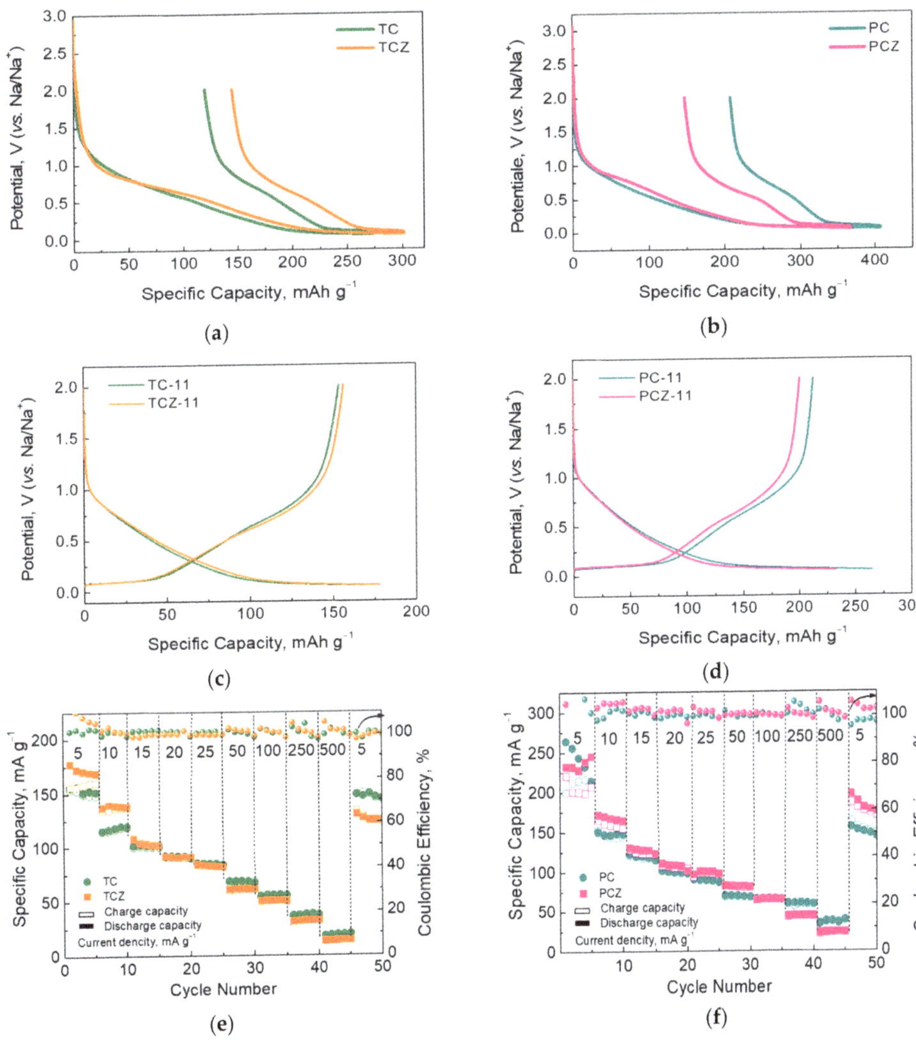

Figure 7. Initial discharge and reversible charge capacity of high-temperature samples with and without ash obtained in a fixed bed, TC and TCZ (**a**) and in CO_2-flow, PC and PCZ (**b**); original charge–discharge curves for the second cycle of TC/Z (**c**) and PC/Z (**d**); rate capability TC/Z (**e**) and PC/Z (**f**).

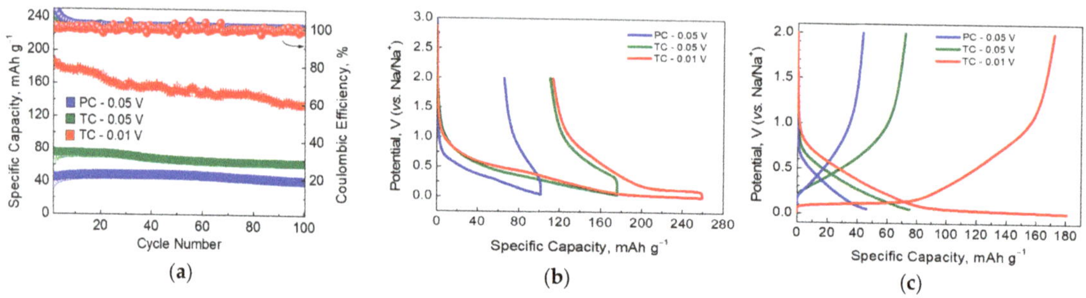

Figure 8. (**a**) Cycling stability at a current load of 50 mA g^{-1} for ash-containing high-temperature samples. (**b**) Charge/discharge curves for ash-containing SCG-derived carbons during the first and (**c**) fifth cycles.

Table 3. Comparison of the electrochemical parameters of coffee-derived carbons.

Bio-Waste	Synthesis Procedure	Temperature Obtained	Current Density	Potential Range, (vs. Na/Na⁺)	ICE, %	First Cycle Reversible Capacity, mAh/g	Ref.
Spent coffee grounds	(1) Hydrothermal synthesis (HtS); (2) Chemical treatment after HtS; (3) One-step pyrolysis in Ar-flow.	700 °C	50 mA/g	0.01–3.00 V	55	198	[15]
Coffee grounds (CGs)	(1) Pre-treatment of CGs with KOH for 24 h; (2) One-step carbonization in N_2.	900 °C	50 mA/g	0.01–3.00 V	64 39	190 (untreated) 144 (treated with KOH 1:1)	[19]
Coffee grounds (CGs)	(1) Pre-treatment of CGs with HCl for 3 days; (2) One-step pyrolysis in Ar-flow.	970 °C	C/5 (1C = 300 mAh/g)	0.02–2.2 V	~39	~130 (CMC binder)	[20]
Spent Coffee Grounds	One-step carbonization in N_2. (1) Hydrothermal pretreatment (HTC); (2) One-step pyrolysis in N_2.	1200 °C	C/15 (1C = 372 mAh/g)	0.002–2.0 V	34 32	199 (untreated) 219 (HTC pre-treated)	[21]
Waste beverage coffee (WBC)	Two-step pyrolysis (first step at 800 °C in N_2 and the second in Ar).	1600 °C 2000 °C 2400 °C 2600 °C	10 mA/g	0.01–2.0 V	56 90 75 64	~168 ~270 ~225 ~194	[22]
Our samples—spent coffee grounds	Two step pyrolysis in a fixed bed (TC) at 750 °C and 1100 °C in Ar flow.	1100 °C	50 mA/g	0.05–2.0 V 0.01–2.0 V	37 56	65 146	

3.3. *Ex Situ XPS Analysis*

To rationalize the performance of carbon-based electrodes, ex situ XPS was undertaken with the aim of assessing the decomposition products forming the SEI. In this aspect, XPS spectra in the binding-energy region of C, Na, F, and P elements were of primary interest (Figure 9). After the electrochemical reaction, the C 1s spectra of all electrodes were dominated by the signals due to C–O, C=O, and O–C=O groups. In addition, the characteristic signals due to the C-F and C-F_2 groups also appeared, especially for TCZ-11. The detection of these surface products can be regarded as a result of the interaction of the electrode surface with the electrolyte, leading to the deposition of C–O, C=O, and O–C=O and C-F groups. For the sake of better comparison, the calculated ratios (C–O)-to-(C–C) and (C=O)-to-(C–C) for the electrodes and pristine samples are given in Figure 10a. The comparison shows that the C–O and C=O groups increase dramatically after the electrochemical reaction, especially for the TC samples. This means that the electrolyte–electrode interaction leads to the deposition of C–O and C=O groups at higher amounts for the TC-11 and TCZ-11 electrodes in comparison with PC-electrodes.

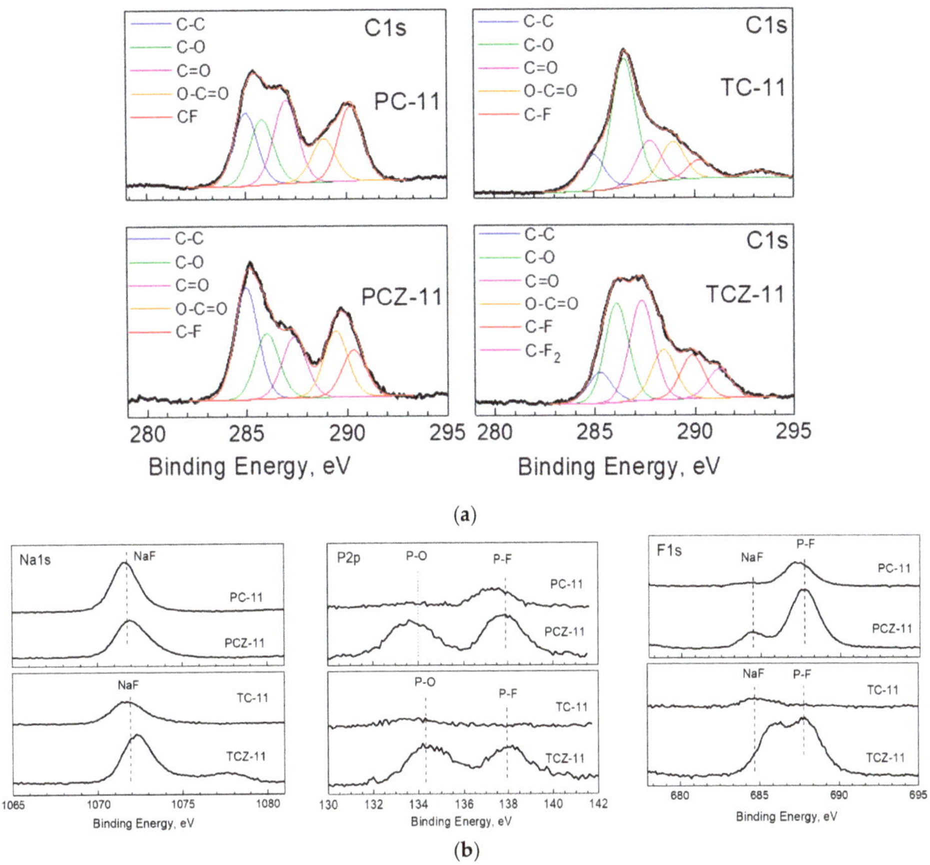

Figure 9. XPS spectra in the energy regions of C 1s of the cycled electrodes for pyrolyzed samples with and without ash obtained in a fixed bed and CO_2 flow (**a**), as well as Na 1s, P 2p, and F 1s (**b**).

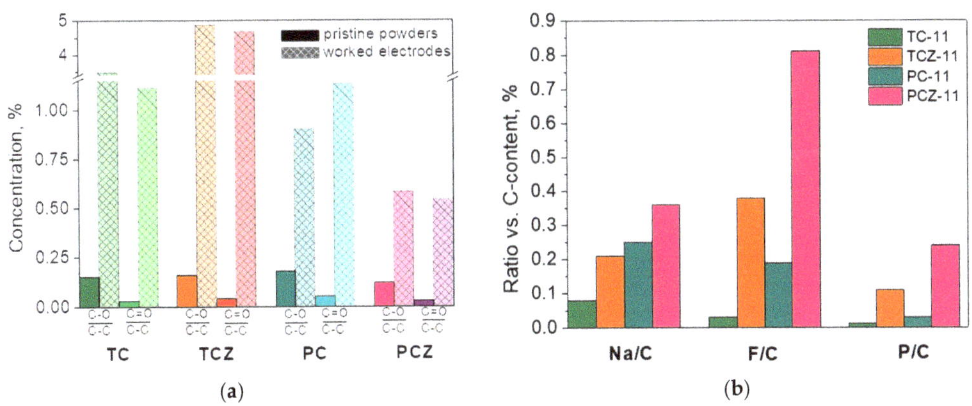

Figure 10. Calculated ratio (C–O)-to-(C–C) and (C=O)-to-(C–C) for the electrodes and pristine samples (**a**). The amount of Na, F, and P elements relative to carbon content (**b**).

While the C 1 spectra provide information on the products coming from the electrolyte solvent decomposition (i.e., PC molecules), electrolyte salt, $NaPF_6$, and decomposition on the electrode surface is evaluated by Na 1s, F 1s and P 2p spectra (Figure 9b). In the energy region of Na 1s, one broad signal centered at 1072 eV is observed for all carbon electrodes. The binding energies of Na 1s correspond to Na atoms in NaF [64]. The Na 1s signal assignment is supported by the F 1s spectra, where a signal centered at 684.5 eV is clearly resolved, and it is associated with the F atom in NaF [65]. In addition, the F 1 spectra display a second signal at 687.6 eV, which is attributed to F atoms in $Na_xPF_yO_z$ [65]. In agreement with F 1, the P 2p spectra consist of two overlapping signals centered at 133.7 and 137.7 eV. The signal with a lower binding energy could be assigned to P_2O_5 and/or $Na_xPF_yO_z$, while the signal with a higher binding energy comes from Na_xPF_y. It is worth mentioning that the ash-free samples PCZ-11 and TCZ-11 are covered predominately with $Na_xPF_yO_z$ and Na_xPF_y products, while NaF is the main surface product for the ash-containing samples.

The next characteristics for electrodes determined by ex situ XPS analysis are the amounts of Na, F, and P elements (Figure 10b). The comparison discloses that the surface of PC samples is richer on Na than that of the TC samples. The removal of ash from both TC and PC samples facilitates Na deposition. While the changes in the Na content between TC and PC samples are relatively moderate (between 1.5 and 2 times), the changes in the F and P content are dramatic (more than 3–5 times). As in the case of Na, the surface of PC samples concentrates more F and P elements than that of the TC samples. Furthermore, the F and P elements are predominantly deposed when ash is removed from the carbons. A close inspection of Figure 10 shows that the ash phases significantly reduce the decomposition of $NaPF_6$ salt.

Based on ex situ XPS analysis, one can conclude that the SEI of carbon electrodes is composed of organic and inorganic products coming mainly from the decomposition of the electrolyte solvent (i.e., propylene carbonate) and electrolyte salt (i.e., $NaPF_6$), with the ratio between them being dependent on the method of carbonization and the ash content. The surface of carbons obtained in a fixed bed is richer on organic decomposition products, while inorganic decomposition products are deposed on the surface of carbons obtained in CO_2 flow. In comparison with organic products, it appears that inorganic products contribute to a higher extent to the formation of SEI, as a result of which PC samples form a thicker SEI (Figure 10b). Ash-free carbons form thicker SEI than that ash-containing analogs. This implies that ash phases, mainly albite, stimulate the decomposition of the solvent electrolyte molecules and their deposition on the electrode surface. Along with this, ash phases prevent electrolyte salt decomposition.

4. Conclusions

This study demonstrates that complementary to the carbon structure and texture, mineral ash phases and their distribution have an impact on the Na storage properties of SCG-derived carbons. Because of the higher amount of Si and Al in SCGs, the carbonized products contain mineral ash composed mainly of low albite calcian, MgO, and K_2SO_4. The mineral ash is mainly located on the outmost surface of carbonized products.

The Na storage proceeds the same mechanism for ash-containing and ash-free carbons; for up to 0.1 V, there is Na^+ adsorption, while below 0.1 V, both Na^+ intercalation and/or Na^+ nanopore filling occurs. However, ash-containing carbons outperform ash-free analogs with respect to cycling stability, Coulombic efficiency, and rate capability. At a current load of 50 mA g^{-1}, the "softly" and "deeply" discharged carbons (i.e., up to 0.05 and 0.01 V) deliver a reversible capacity of 75 and 168 mAh g^{-1}. The better performance of ash-containing carbons can be explained by the electrode–electrolyte interaction. The SEI of ash-containing carbons is mainly composed of organic products resulting from electrolyte solvent decomposition. The electrolyte salt decomposition is suppressed by the presence of the ash phases. The SEI richer in inorganic decomposition products is preferentially formed on carbons obtained in a CO_2 flow.

In general, the presence of mineral ash and its distribution on the carbon surface helps to improve the performance of SCG-derived carbons. This study reveals that there is no need to utilize additional and costly procedures for the removal of mineral ash.

Supplementary Materials: The following supporting information can be downloaded at: https://www.mdpi.com/article/10.3390/ma17051016/s1, Table S1: Moisture and ash content in the SCGs and carbon yields after pyrolysis; Figure S1: XRD pattern of the SCG-derived ash at 525 °C; Figure S2: XPS spectra in the energy regions of C1s for pyrolyzed SCGs obtained in CO_2 flow with and without ash (a,b) and their high-temperature analogs (c,d); Figure S3: XPS spectra in the energy regions of O 1s and Si 2p for pyrolyzed SCGs in a fixed bed (a,b) and in CO_2 flow (c,d), respectively; Figure S4: SEM images of (a) TC-11, (b) TCZ-11, (c) PC-11, and (d) PCZ-11 obtained after high-temperature pyrolysis; Figure S5: NLDFT pore size distribution of the (a) TC-8, (b) TCZ-8, (c) PC-8, (d) PCZ-8. The corresponding nitrogen adsorption–desorption isotherms are shown as insets; Figure S6: CV curves during the first cathodic cycle at 0.01 mV s^{-1} scan rate (a) and the calculated capacitance in the potential window 2.0–0.3 V (b); Figure S7: CV curves of SCG-derived carbon at 100 mV s^{-1}.

Author Contributions: Conceptualization, M.K. and S.H.; methodology, I.U., M.K. and S.H.; validation, S.H. and M.K.; investigation, S.H., I.U., L.A., M.S. and I.S.; resources, M.K.; data curation, S.H., I.U., L.A., M.S., I.S. and M.K.; writing—original draft preparation, M.K.; writing—review and editing, M.K.; visualization, S.H., I.U., L.A., M.S. and I.S.; supervision, M.K.; project administration, M.K.; funding acquisition, M.K. and S.H. All authors have read and agreed to the published version of the manuscript.

Funding: This research was funded by the Bulgarian National Science Fund, grant number KΠ-06-H69/7 from 15 December 2022.

Institutional Review Board Statement: Not applicable.

Informed Consent Statement: Not applicable.

Data Availability Statement: Data are contained within the article and Supplementary Materials.

Acknowledgments: The authors acknowledge the TwinTeam project D01-272 "European Network on Materials for Clean Technologies" for providing the opportunity to present the results at the SizeMat 4 conference, as well as for the financial publication support. The authors are sincerely grateful to Radostina Stoyanova for her significant contribution to the quality of this article with valuable and well-aimed comments and remarks during the discussion.

Conflicts of Interest: The authors declare no conflicts of interest.

References

1. Hirsh, H.S.; Li, Y.; Tan, D.H.S.; Zhang, M.; Zhao, E.; Meng, Y.S. Sodium-Ion Batteries Paving the Way for Grid Energy Storage. *Adv. Energy Mater.* **2020**, *10*, 2001274. [CrossRef]
2. Simons, S.; Schmitt, J.; Tom, B.; Bao, H.; Pettinato, B.; Pechulis, M. Chapter 10-Advanced concepts. In *Thermal, Mechanical, and Hybrid Chemical Energy Storage System*; Brun, K., Allison, T., Dennis, R., Eds.; Academic Press, Elsevier Inc.: Cambridge, MA, USA, 2021; pp. 569–596.
3. Pan, H.; Hu, Y.-S.; Chen, L. Room-temperature stationary sodium-ion batteries for large-scale electric energy storage. *Energy Environ. Sci.* **2013**, *6*, 2338–2360. [CrossRef]
4. Hwang, J.-Y.; Myung, S.-T.; Sun, Y.-K. Sodium-ion batteries: Present and future. *Chem. Soc. Rev.* **2017**, *46*, 3529–3614. [CrossRef]
5. Li, H.Q.; He, X.J.; Wu, T.T.; Jin, B.; Yang, L.; Qiu, J. Synthesis, modification strategies and applications of coal-based carbon materials. *Fuel Process. Technol.* **2022**, *230*, 107203. [CrossRef]
6. Genier, F.S.; Pathreeker, S.; Schuarca, R.L.; Islam, M.; Hosein, I.D. Hard Carbon Derived from Avocado Peels as a High-Capacity, High-Performance Anode Material for Sodium-Ion Batteries. *ECS Adv.* **2022**, *1*, 030502. [CrossRef]
7. Dou, X.; Hasa, I.; Hekmatfar, M.; Diemant, T.; Behm, R.J.; Buchholz, D.; Passerini, S. Pectin, Hemicellulose, or Lignin? Impact of the Biowaste Source on the Performance of Hard Carbons for Sodium Ion Batteries. *ChemSusChem* **2017**, *10*, 2668–2676. [CrossRef]
8. Zhu, Y.-E.; Gu, H.; Chen, Y.-N.; Yang, D.; Wei, J.; Zhou, Z. Hard carbon derived from corn straw piths as anode materials for sodium ion batteries. *Ionics* **2018**, *24*, 1075–1081. [CrossRef]
9. Janissen, B.; Huynh, T. Chemical composition and value-adding applications of coffee industry byproducts: A review. *Resour. Conserv. Recycl.* **2018**, *128*, 110–117. [CrossRef]

10. Khusyaeri, H.; Pratiwi, D.; Kurniawan, H.A.; Nurohmah, A.R.; Yudha, C.S.; Purwanto, A. Synthesis of High-Performance Hard Carbon from Waste Coffee Ground as Sodium Ion Battery Anode Material: A Review. *Mater. Sci. Forum* **2021**, *1044*, 25–39. [CrossRef]
11. International Coffee Organization. Available online: https://www.ico.org/trade_statistics.asp?section=Statistics (accessed on 13 November 2023).
12. Forcina, A.; Petrillo, A.; Travaglioni, M.; di Chiara, S.; De Felice, F. A comparative life cycle assessment of different spent coffee ground reuse strategies and a sensitivity analysis for verifying the environmental convenience based on the location of sites. *J. Clean. Prod.* **2023**, *385*, 135727. [CrossRef]
13. Rios, C.d.M.S.; Simonin, L.; Geyer, A.d.; Ghimbeu, C.M.; Dupont, C. Unraveling the Properties of Biomass-Derived Hard Carbons upon Thermal Treatment for a Practical Application in Na-Ion Batteries. *Energies* **2020**, *13*, 3513. [CrossRef]
14. Krikstolaityte, V.; Joshua, O.E.Y.; Veksha, A.; Wai, N.; Lisak, G.; Lim, T.M. Conversion of Spent Coffee Beans to Electrode Material for Vanadium Redox Flow Batteries. *Batteries* **2018**, *4*, 56. [CrossRef]
15. Gao, G.; Cheong, L.-Z.; Wang, D.; Shen, C. Pyrolytic carbon derived from spent coffee grounds as anode for sodium-ion batteries. *Carbon Resour. Convers.* **2018**, *1*, 104–108. [CrossRef]
16. Liu, X.; Zhang, S.; Wen, X.; Chen, X.; Wen, Y.; Shi, X.; Mijowska, E. High yield conversion of biowaste coffee grounds into hierarchical porous carbon for superior capacitive energy storage. *Sci. Rep.* **2020**, *10*, 3518. [CrossRef] [PubMed]
17. Navarro-Suárez, A.M.; Saurel, D.; Sánchez-Fontecoba, P.; Castillo-Martínez, E.; Carretero-González, J.; Rojo, T. Temperature effect on the synthesis of lignin-derived carbons for electrochemical energy storage applications. *J. Power Sources* **2018**, *397*, 296–306. [CrossRef]
18. Wang, C.-C.; Su, W.-L. Understanding acid pretreatment of lotus leaves to prepare hard carbons as anodes for sodium ion batteries. *J. Surf. Coat.* **2021**, *415*, 127125. [CrossRef]
19. Chiang, P.-H.; Liu, S.-F.; Hung, Y.-H.; Tseng, H.; Guo, C.-H.; Chen, H.-Y. Coffee-Ground-Derived Nanoporous Carbon Anodes for Sodium-Ion Batteries with High Rate Performance and Cyclic Stability. *Energy Fuels* **2020**, *34*, 7666–7675. [CrossRef]
20. Darjazi, H.; Staffolani, A.; Sbrascini, L.; Bottoni, L.; Tossici, R.; Nobili, F. Sustainable Anodes for Lithium- and Sodium-Ion Batteries Based on Coffee Ground-Derived Hard Carbon and Green Binders. *Energies* **2020**, *13*, 6216. [CrossRef]
21. Nieto, N.; Porte, J.; Saurel, D.; Djuandhi, L.; Sharma, N.; Lopez-Urionabarrenechea, A.; Palomares, V.; Rojo, T. Use of Hydrothermal Carbonization to Improve the Performance of Biowaste-Derived Hard Carbons in Sodium Ion-Batteries. *ChemSusChem* **2023**, *16*, e202301053. [CrossRef]
22. Lee, M.E.; Kwak, H.W.; Jin, H.-J.; Yun, Y.S. Waste Beverage Coffee-Induced Hard Carbon Granules for SodiumIon Batteries. *ACS Sustain. Chem. Eng.* **2019**, *7*, 12734–12740. [CrossRef]
23. Stevens, D.A.; Dahn, J.R. High Capacity Anode Materials for Rechargeable Sodium-Ion Batteries. *J. Electrochem. Soc.* **2000**, *147*, 1271–1273. [CrossRef]
24. Alvin, S.; Yoon, D.; Chandra, C.; Susanti, R.F.; Chang, W.; Ryu, C.; Kim, J. Extended flat voltage profile of hard carbon synthesized using a two-step carbonization approach as an anode in sodium ion batteries. *J. Power Sources* **2019**, *430*, 157–168. [CrossRef]
25. Zhang, T.; Mao, J.; Liu, X.; Xuan, M.; Bi, K.; Zhang, X.L.; Hu, J.; Fan, J.; Chen, S.; Shao, G. Pinecone biomass-derived hard carbon anodes for high-performance sodium-ion batteries. *RSC Adv.* **2017**, *7*, 41504–41511. [CrossRef]
26. Rios, C.d.M.S.; Simone, V.; Simonin, L.; Martinet, S.; Dupont, C. Biochars from various biomass types as precursors for hard carbon anodes in sodium-ion batteries. *Biomass Bioenergy* **2018**, *117*, 32–37. [CrossRef]
27. Ballesteros, L.F.; Teixeira, J.A.; Mussatto, S.I. Chemical, Functional, and Structural Properties of Spent Coffee Grounds and Coffee Silverskin. *Food Bioprocess Technol.* **2014**, *7*, 3493–3503. [CrossRef]
28. Campos-Vega, R.; Loarca-Piña, G.; Vergara-Castañeda, H.A.; Oomah, B.D. Spent coffee grounds: A review on current research and future prospects. *Trends Food Sci. Technol.* **2015**, *45*, 24–36. [CrossRef]
29. Mussatto, S.I.; Ballesteros, L.F.; Martins, S.; Teixeira, J.A. Extraction of antioxidant phenolic compounds from spent coffee grounds. *Sep. Purif. Technol.* **2011**, *83*, 173–179. [CrossRef]
30. Colantoni, A.; Paris, E.; Bianchini, L.; Ferri, S.; Marcantonio, V.; Carnevale, M.; Palma, A.; Civitarese, V.; Gallucci, F. Spent coffee ground characterization, pelletization test and emissions assessment in the combustion process. *Sci. Rep.* **2021**, *11*, 5119. [CrossRef]
31. Eliche-Quesada, D.; Martínez-García, C.; Martínez-Cartas, M.L.; Cotes-Palomino, M.T.; Pérez-Villarejo, L.; Cruz-Pérez, N.; Corpas-Iglesias, F.A. The use of different forms of waste in the manufacture of ceramic bricks. *Appl. Clay Sci.* **2011**, *52*, 270–276. [CrossRef]
32. Boström, D.; Skoglund, N.; Grimm, A.; Boman, C.; Öhman, M.; Broström, M.; Backman, R. Ash Transformation Chemistry during Combustion of Biomass. *Energy Fuels* **2012**, *26*, 85–93. [CrossRef]
33. Susanti, R.F.; Alvin, S.; Kim, J. Toward high-performance hard carbon as an anode for sodium-ion batteries: Demineralization of biomass as a critical step. *J. Ind. Eng. Chem.* **2020**, *91*, 317–329. [CrossRef]
34. Rios, C.d.M.S.; Simonin, C.M.; Ghimbeu, L.; Vaulot, C.; Perez, D.d.S.; Dupont, C. Impact of the biomass precursor composition in the hard carbon properties and performance for application in a Na-ion battery. *Fuel Process. Technol.* **2022**, *231*, 107223. [CrossRef]
35. Beda, A.; Le Meins, J.-M.; Taberna, P.-L.; Simon, P.; Ghimbeu, C.M. Impact of biomass inorganic impurities on hard carbon properties and performance in Na-ion batteries. *Sustain. Mater. Technol.* **2020**, *26*, e00227. [CrossRef]

36. Wu, F.; Liu, L.; Yuan, Y.; Li, Y.; Bai, Y.; Li, T.; Lu, J.; Wu, C. Expanding Interlayer Spacing of Hard Carbon by Natural K+ Doping to Boost Na-Ion Storage. *ACS Appl. Mater. Interfaces* **2018**, *10*, 27030–27038. [CrossRef]
37. Technical Association of the Pulp and Paper Industry (TAPPI) Standard. Available online: https://www.tappi.org/ (accessed on 30 January 2023).
38. Pujol, D.; Liu, C.; Gomincho, J.; Olivella, M.À.; Fiol, N.; Villaescusa, I.; Pereira, H. The chemical composition of exhausted coffee waste. *Ind. Crops Prod.* **2013**, *50*, 423–429. [CrossRef]
39. Garcia, E.; Ejim, I.F.; Liu, H. Thermogravimetric analysis of co-combustion of a bituminous coal and coffee industry by-products. *Thermochim. Acta* **2022**, *715*, 179296. [CrossRef]
40. Chen, J.; Liu, J.; He, Y.; Huang, L.; Sun, S.; Sun, J.; Chang, K.; Kuo, J.; Huang, S.; Ning, X. Investigation of co-combustion characteristics of sewage sludge and coffee grounds mixtures using thermogravimetric analysis coupled to artificial neural networks modeling. *Bioresour. Technol.* **2017**, *225*, 234–245. [CrossRef]
41. Zarrinbakhsh, M.; Wang, T.; Rodriguez-Uribe, A.; Misra, M.; Mohanty, A.K. Characterization of wastes and coproducts from the coffee industry for composite material production. *BioResources* **2016**, *11*, 7637–7653. [CrossRef]
42. Wang, G.; Dai, Y.; Yang, H.; Xiong, Q.; Wang, K.; Zhou, J.; Li, Y.; Wang, S. A Review of Recent Advances in Biomass Pyrolysis. *Energy Fuels* **2020**, *34*, 15557–15578. [CrossRef]
43. Yang, H.; Yan, R.; Chen, H.; Ho Lee, D.; Zheng, C. Characteristic of hemicellulose, cellulose and lignin pyrolysis. *Fuel* **2007**, *86*, 1781–1788. [CrossRef]
44. Wacheter, I.; Rantuch, P.; Drienovsky, M.; Martinka, J.; Haz, A.; Stefko, T. Determining the activation energy of spent coffee grounds by the thermogravimetric analysis. *J. Univ. Chem. Technol. Metall.* **2022**, *57*, 1006–1018.
45. Tsuzuki, Y.; Kadota, S.; Takashima, I. Dissolution process of albite and albite glass in acid solutions at 47 °C. *Chem. Geol.* **1985**, *49*, 127–140. [CrossRef]
46. Gruber, C.; Kutuzov, I.; Ganor, J. The combined effect of temperature and pH on albite dissolution rate under far-from-equilibrium conditions. *Geochim. Cosmochim. Acta* **2016**, *186*, 154–167. [CrossRef]
47. Lin, X.; Liu, Y.; Tan, H.; Zhang, B. Advanced lignin-derived hard carbon for Na-ion batteries and a comparison with Li and K ion storage. *Carbon* **2020**, *157*, 316–323. [CrossRef]
48. Cao, Y.; Xiao, L.; Sushko, M.L.; Wang, W.; Schwenzer, B.; Xiao, J.; Nie, Z.; Saraf, L.V.; Yang, Z.; Liu, J. Sodium Ion Insertion in Hollow Carbon Nanowires for Battery Applications. *Nano Lett.* **2012**, *12*, 3783–3787. [CrossRef]
49. Vassilev, S.V.; Vassileva, C.G.; Song, Y.-C.; Li, W.-Y.; Feng, J. Ash contents and ash-forming elements of biomass and their significance for solid biofuel combustion. *Fuel* **2017**, *208*, 377–409. [CrossRef]
50. Vassilev, S.V.; Baxter, D.; Andersen, L.K.; Vassileva, C.G. An overview of the composition and application of biomass ash. Part 1. Phase–Mineral and chemical composition and classification. *Fuel* **2013**, *105*, 40–76. [CrossRef]
51. Jordan, C.A.; Akay, G. Speciation and distribution of alkali, alkali earth metals and major ash forming elements during gasification of fuel cane bagasse. *Fuel* **2012**, *91*, 253–263. [CrossRef]
52. Ganguly, A.; Sharma, S.; Papakonstantinou, P.; Hamilton, J. Probing the Thermal Deoxygenation of Graphene Oxide Using High-Resolution In Situ X-ray-Based Spectroscopies. *J. Phys. Chem. C* **2011**, *115*, 17009–17019. [CrossRef]
53. Reiche, S.; Blume, R.; Zhao, X.C.; Su, D.; Kunkes, E.; Behrens, M.; Schlögl, R. Reactivity of mesoporous carbon against water–An in-situ XPS study. *Carbon* **2014**, *77*, 175–183. [CrossRef]
54. Kundu, S.; Wang, Y.M.; Xia, W.; Muhler, M. Thermal stability and reducibility of oxygen-containing functional groups on multiwalled carbon nanotube surfaces: A quantitative highresolution XPS and TPD/TPR study. *J. Phys. Chem. C* **2008**, *112*, 16869–16878. [CrossRef]
55. Clark, D.T.; Dilks, A. ESCA applied to polymers. 23. RF glow discharge modification of polymers in pure oxygen and helium oxygen mixtures. *J. Polym. Sci. Pol. Chem.* **1979**, *17*, 957–976. [CrossRef]
56. Thommes, M.; Kaneko, K.; Neimark, A.; Olivier, J.P.; Rodriguez-Reinoso, F.; Rouquerol, J.; Sing, K.S.W. Physisorption of gases, with special reference to the evaluation of surface area and pore size distribution (IUPAC Technical Report). *Pure Appl. Chem.* **2015**, *87*, 1051–1069. [CrossRef]
57. Shields, J.E.; Lowell, S.; Thomas, M.A.; Thommes, M. *Characterization of Porous Solids and Powders: Surface Area, Pore Size and Density*; Kluwer Academic Publisher: Boston, MA, USA, 2004; pp. 43–45.
58. Yang, J.; Wang, X.; Dai, W.; Lian, X.; Cui, X.; Zhang, W.; Zhang, K.; Lin, M.; Zou, R.; Loh, K.P.; et al. From Micropores to Ultra-micropores inside Hard Carbon: Toward Enhanced Capacity in Room-/Low-Temperature Sodium-Ion Storage. *Nano-Micro Lett.* **2021**, *13*, 98. [CrossRef]
59. Zheng, Y.; Lu, Y.; Qi, X.; Wang, Y.; Mu, L.; Li, Y.; Ma, Q.; Li, J.; Hu, Y.-S. Superior electrochemical performance of sodium-ion full-cell using poplar wood derived hard carbon anode. *Energy Storage Mater.* **2019**, *18*, 269–279. [CrossRef]
60. Alvin, S.; Yoon, D.; Chandra, C.; Cahyadi, H.S.; Park, J.-H.; Chang, W.; Chung, K.Y.; Kim, J. Revealing sodium ion storage mechanism in hard carbon. *Carbon* **2019**, *145*, 67–81. [CrossRef]
61. Yin, X.; Lu, Z.; Wang, J.; Feng, X.; Roy, S.; Liu, X.; Yang, Y.; Zhao, Y.; Zhang, J. Enabling Fast Na+ Transfer Kinetics in the Whole-Voltage-Region of Hard-Carbon Anodes for Ultrahigh-Rate Sodium Storage. *Adv. Mater.* **2022**, *34*, 2109282. [CrossRef] [PubMed]

62. Hijazi, H.; Ye, Z.; Zhang, L.; Deshmukh, J.; Johnson, M.B.; Dahn, J.R.; Metzger, M. Impact of Sodium Metal Plating on Cycling Performance of Layered Oxide/Hard Carbon Sodium-ion Pouch Cells with Different Voltage Cut-offs. *J. Electrochem. Soc.* **2023**, *170*, 070512. [CrossRef]
63. Pfeifer, K.; Arnold, S.; Becherer, J.; Das, C.; Maibach, J.; Ehrenberg, H.; Dsoke, S. Can Metallic Sodium Electrodes Affect the Electrochemistry of Sodium-Ion Batteries? Reactivity Issues and Perspectives. *ChemSusChem* **2019**, *12*, 3312–3319. [CrossRef]
64. Brisson, P.-Y.; Darmstadt, H.; Fafard, M.; Adnot, A.; Servant, G.; Soucy, G. X-ray photoelectron spectroscopy study of sodium reactions in carbon cathode blocks of aluminium oxide reduction cells. *Carbon* **2006**, *44*, 1438–1447. [CrossRef]
65. Kalapsazova, M.; Zhecheva, E.; Tyuliev, G.; Nihtianova, D.; Mihaylov, L.; Stoyanova, R. Effects of the Particle Size Distribution and the Electrolyte Salt on the Intercalation Properties of $P3\text{-}Na_{2/3}Ni_{1/2}Mn_{1/2}O_2$. *J. Phys. Chem. C* **2017**, *121*, 5931–5940. [CrossRef]

materials MDPI

Article

Water Splitting Reaction Mechanism on Transition Metal (Fe-Cu) Sulphide and Selenide Clusters—A DFT Study

Ellie Uzunova *, Ivelina Georgieva and Tsvetan Zahariev

Institute of General and Inorganic Chemistry, Bulgarian Academy of Sciences, 1113 Sofia, Bulgaria; ivelina@svr.igic.bas.bg (I.G.); tzahariev@svr.igic.bas.bg (T.Z.)
* Correspondence: ellie@svr.igic.bas.bg

Abstract: The tetracarbonyl complexes of transition metal chalcogenides $M_2X_2(CO)_4$, where M = Fe, Co, Ni, Cu and X = S, Se, are examined by density functional theory (DFT). The M_2X_2 core is cyclic with either planar or non-planar geometry. As a sulfide, it is present in natural enzymes and has a selective redox capacity. The reduced forms of the selenide and sulfide complexes are relevant to the hydrogen evolution reaction (HER) and they provide different positions of hydride ligand binding: (i) at a chalcogenide site, (ii) at a particular cation site and (iii) in a midway position forming equal bonds to both cation sites. The full pathway of water decomposition to molecular hydrogen and oxygen is traced by transition state theory. The iron and cobalt complexes, cobalt selenide, in particular, provide lower energy barriers in HER as compared to the nickel and copper complexes. In the oxygen evolution reaction (OER), cobalt and iron selenide tetracarbonyls provide a low energy barrier via OOH* intermediate. All of the intermediate species possess favorable excitation transitions in the visible light spectrum, as evidenced by TD-DFT calculations and they allow photoactivation. In conclusion, cobalt and iron selenide tetracarbonyl complexes emerge as promising photocatalysts in water splitting.

Keywords: chalcogenides; ab initio methods; DFT; artificial photosynthesis; transition metal sulfides; selenides; carbonyl complexes

Citation: Uzunova, E.; Georgieva, I.; Zahariev, T. Water Splitting Reaction Mechanism on Transition Metal (Fe-Cu) Sulphide and Selenide Clusters—A DFT Study. *Materials* **2024**, *17*, 56. https://doi.org/10.3390/ma17010056

Academic Editor: Gennady L. Gutsev

Received: 15 November 2023
Revised: 16 December 2023
Accepted: 20 December 2023
Published: 22 December 2023

1. Introduction

Transition metal chalcogenides as clusters, layers or coordination compounds with various ligands possess the unique property to accept, store and donate electrons to substrates. Coordination of cyanide, carbonyl or more complex protein-like ligands, strongly influences the electron distribution within these clusters and enables their application in redox catalysis, including water splitting and carbon dioxide reduction. Water dissociation to hydrogen and oxygen, $2H_2O(g) \rightarrow 2H_2(g) + O_2(g)$, is a strongly endothermic process with an enthalpy of 483.7 kJ mol^{-1} at 298 K [1]. In addition to the high endothermic effect, reaction barriers certainly add to the overall energy needed to split water. The electrochemical route requires a four-electron transfer for the cathode reaction delivering hydrogen $4H^+ + 4e^- \rightarrow 2H_2(g)$, and the anode reaction delivering oxygen, $2H_2O \rightarrow 2O_2(g) + 4H^+ + 4e^-$. Numerous studies were focused on the optimization of the electrochemical reaction, its pH dependence and the well-known overpotential in the oxygen evolution reaction (OER) [2–7]. The OER and the O–O bond formation chemistry was explored using a variety of materials: theoretical studies on cobalt oxide clusters and on photosystem II, which includes a manganese complex [3,4], and heterogeneous metallic electrocatalysts [5,6]. The importance of an oxygenated intermediate after $[H^+,e^-]$ removal was outlined [3–8]. The research efforts have targeted the analogues of natural enzymes, used in photosynthesis, in order to obtain efficient photocatalysts and overcome the high reaction barriers [9–13]. The photocatalysts can either be applied directly as electrode materials, or used as bulk materials in a photoelectrochemical cell. Theoretical studies help in preliminary studies for discerning promising photocatalytic materials and in the elucidation of the reaction mechanism.

As structural analogs of ferredoxin and hydrogenase enzymes, the sulfides of iron and nickel, with carbonyl or halogen ligands, have been the subject of experimental and theoretical studies [10–15]. Despite the promising results on hydrogen evolution by photoactivated $Fe_2S_2(CO)_6$ complexes [10], other transition metal chalcogenide complexes received much less attention. The present study simulates enzyme analogs of the transition metals, Fe, Co, Ni and Cu, which form binuclear clusters M_2X_2 (M = Fe, Co, Ni, Cu and X = S, Se) and they are coordinated by carbonyl ligands so as to form symmetric tetracarbonyl complexes. The electronic structure of these complexes is examined by density functional theory and the reaction of water splitting is traced by transition state theory. The catalytic pathway with the possibility of photoactivation of the two half-reactions: Hydrogen Evolution Reaction (HER) and Oxygen Evolution Reaction (OER) are traced by time-dependent density functional theory (TD-DFT).

2. Materials and Methods

All calculations were performed with the B3LYP functional [16–19], which includes local and non-local terms as implemented in the Gaussian 16 package [20]. The standard 6-311+G(2df) basis set with diffuse and polarization functions was employed, which consists of the McLean–Chandler (12s, 9p) $\rightarrow$ (621111,52111) basis sets for second-row atoms and the Wachters–Hay all-electron basis set for the first transition row, using the scaling factors of Raghavachari and Trucks [21–25]. In their ground states, selenium-containing clusters were reoptimized using the QZVP basis set [26,27], but no significant change in bond lengths (within 1.6%) or relative energies (within 0.9%) occurred. The differences obtained with the LanL2DZ basis [28–30] were even smaller, differing by only 0.7% regarding bond lengths and by 0.9% regarding relative energies. The selection of the density functional and basis set was based on calculations of the diatomic molecules and the diiron disulfide hexacarbonyl complex from previous studies [31–33], where different density functionals were compared, as there are sufficiently accurate experimental data for these compounds. Proton and electron affinities are calculated as the energy required to attach a proton or electron, respectively. For proton–electron couples a subsequent proton and electron attachment are calculated.

The spin-unrestricted formalism was applied and calculations in the broken symmetry (BS) approach were performed, which consists of the localization of the opposite spins on different parts of the molecule to give a mono-determinant representation of the spin exchange interactions, thus reducing the symmetry of the space and spin wavefunctions with respect to that of the nuclear framework [31–34]. The synchronous transit-guided quasi-Newton (STQN) method was used for the transition state optimizations [35,36]. Intrinsic reaction coordinate (IRC) calculations were performed to confirm the transition state structures and to evaluate activation energies [37,38]. Reaction studies using water as a solvent were performed using the Polarizable Continuum Model [39] (PCM). Time-dependent (TD) DFT was used [40,41] to determine the excitation energies of the ground state cluster complexes, the reaction intermediates and oscillator strengths. Dispersion effects were taken into account for the ground states and the reaction intermediates by using the formula of Grimme with Becke–Johnson damping [42]. The bond populations and charge distributions were examined by using natural orbitals and natural bond orbital (NBO) analysis [43,44].

3. Results and Discussion

3.1. Structure and Bonding of the Tetracarbonyl Complexes of Metal Disulfides and Diselenides, $M_2X_2(CO)_4$

The tetracarbonyl complexes of cobalt, iron and nickel disulfides possess two conformations of the core M_2X_2: rhombic and planar; see Figure 1. The global energy minimum structures of all $M_2S_2(CO)_4$ complexes (M = Fe, Co, Ni, Cu) have a rhombic non-planar core M_2X_2. The energy gap between non-planar and planar configurations is 67 kJ mol^{-1} for $Fe_2S_2(CO)_4$ and 88 kJ mol^{-1} for $Co_2S_2(CO)_4$. It is much smaller for $Ni_2S_2(CO)_4$, 21 kJ mol^{-1}. The selenide complexes $M_2Se_2(CO)_4$ (M = Fe, Co, Cu) form only a non-planar rhombic core M_2Se_2, whereas $Ni_2Se_2(CO)_4$ is found as a planar and non-planar structure, the planar being the global minimum and the rhombic one lying by 39 kJ mol^{-1} higher in energy. All of the sulfides and selenides with rhombic non-planar structures contain an S-S or Se-Se bond.

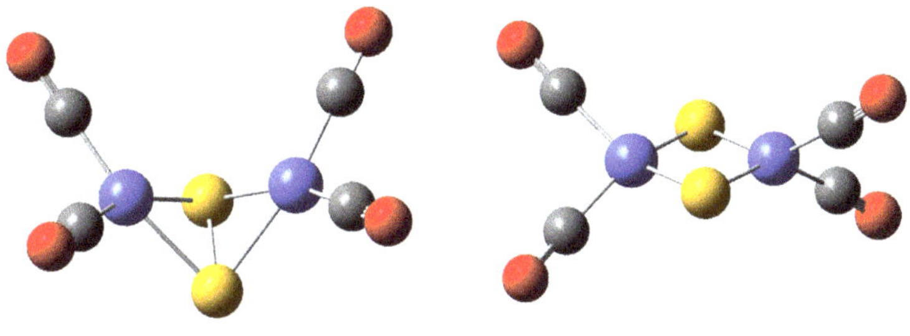

(**a**) (**b**)

Figure 1. (**a**) The global minimum of $Co_2S_2(CO)_4$ in rhombic non-planar configuration. (**b**) $Co_2S_2(CO)_4$ with a planar core. Legend: Co atoms are blue, sulfur atoms—yellow; carbon atoms—grey; oxygen atoms—red.

According to our results, the geometry flexibility of the metal disulfide and diselenide core matters to the reactivity of these complexes by providing a favorable orientation towards substrate molecules. Thus, water adsorption proves to be an exothermic process; however, the adsorption energy depends on the core configuration: on the planar clusters, the heat of adsorption is 25–30 kJ mol^{-1}, while on the non-planar clusters, it is weaker, at 8–13 kJ mol^{-1}. Further, in the subsequent reaction steps (dissociation, HER, OER), the active site M_2X_2 may present variable deviation from planarity.

For all $M_2X_2(CO)_4$ complexes, which have two conformations (planar and non-planar), the M-X bond lengths in the planar core are shorter than in the non-planar rhombic core, as shown in Figure 2. The Fe-S and Co-S bond lengths variation is much smaller as compared to the Ni-S or Ni-Se bond lengths difference. For $M_2X_2(CO)_4$ complexes studied, the Co-S/Se bond length is the shortest one and this result corresponds to the strong Co-S bond found in the diatomic molecule [45]. For the global minimum structures of $M_2X_2(CO)_4$, the bond lengths increase as follows: Co-S < Fe-S < Ni-S < Cu-S < Co-Se < Fe-Se < Cu-Se < Ni-Se < Zn-S. While zinc sulfide exists in the solid state and forms clusters without ligands, the filled 3d shell does not allow interaction with a strong electron donor as the carbonyl groups. The Zn-CO bonds reach 2.183 Å, while in the other complexes, they do not exceed 1.95 Å. The Zn-S bonds are also lengthened to 2.541 Å. Thus, $Zn_2S_2(CO)_4$ is not examined further in the present study.

The calculated proton affinities of the sulfide complexes $M_2S_2(CO)_4$ complexes are higher than the proton affinities of the corresponding selenide complexes Table 1. The chalcogenide complexes of Cu and Ni have markedly higher proton affinities as compared to the chalcogenides of Fe and Co, but subsequent electron addition with the formation of proton–electron couples [H^+,e^-] indicates lower affinities for $Ni_2Se_2(CO)_4$, $Cu_2S_2(CO)_4$ and particularly for $Cu_2Se_2(CO)_4$. The protons are always attached at the chalcogenide center (S, Se), while the hydride ligand formed upon neutralization of the positive charge (H^+) can either remain located at the chalcogenide center or bind at the metal cation sites. Usually, the hydride ligand is centered between the metal cations and forms equal M-H-M bonds, and this was experimentally proven in diiron disulfide complexes, but configurations with a single M-H bond by binding predominantly at one cation site are also possible [9,31–33]. The values of the proton–electron affinities for different sites allow us to discern the stability of configurations with hydride ligands, as shown in Figure 3. In iron sulfides and selenides, protonation occurs at a chalcogenide site, and subsequent reduction shifts the hydride ligand to a single stable position with equal Fe-H bond lengths, as shown in Figure 3c.

In nickel and copper sulfides and selenides, the hydride ligand remains bonded at the chalcogenide site. Only cobalt centers provide several stable sites for the coordination of a hydride ligand: S-H (Se-H) and Co-H-Co for the disulfides and diselenides, as well as a stable site Co-H, available only in the diselenide complex, as shown in Figure 3a. The role of hydride ligands is crucial for the hydrogen evolution reaction and for the redox capacity in other reactions such as carbon dioxide reduction.

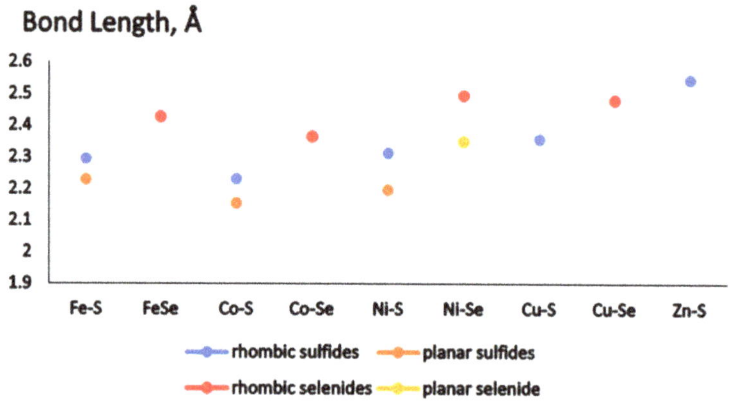

Figure 2. Metal cation-to-chalcogenide bond lengths for $M_2X_2(CO)_4$ complexes, M = Fe, Co, Ni, Cu, Zn; X = S, Se.

Table 1. Calculated proton affinities (PA, kJ mol^{-1}) and [H^+,e^-] affinities (kJ mol^{-1}) for [M2-X2] tetracarbonyls, M = Fe, Co, Ni, Cu and X = S, Se.

Cluster/Binding	PA	[H^+,e^-]
$[Fe_2S_2H]^+(CO)_4$; S-H	651	
$[Fe_2HS_2](CO)_4$; Fe-H-Fe		229
$[Fe_2Se_2H]^+(CO)_4$; Se-H	639	
$[Fe_2HSe_2](CO)_4$; Fe-H-Fe		228
$[Co_2S_2H]^+(CO)_4$; $[Co_2S_2H](CO)_4$; S-H	767	224
$[Co_2HS_2](CO)_4$; Co-H-Co		177
$[Co_2Se_2H]^+(CO)_4$; $[Co_2Se_2H](CO)_4$; Se-H	748	191
$[Co_2HSe_2](CO)_4$; Co-H-Co		179
$[Co_2HSe_2](CO)_4$; Co-H		201
$[Ni_2S_2H]^+(CO)_4$; $[Ni_2S_2H](CO)_4$; S-H	935	181
$[Ni_2Se_2H]^+(CO)_4$; $[Ni_2Se_2H](CO)_4$; Se-H	907	168
$[Cu_2S_2H]^+(CO)_4$; $[Cu_2S_2H](CO)_4$; S-H	968	134
$[Cu_2Se_2H]^+(CO)_4$; $[Cu_2Se_2H](CO)_4$; Se-H	936	81

The rhombic core efficiently redistributes the positive charges induced by the binding of protons, substrates, or electron density from electron-donor ligands. The bare chalcogenide rhombic clusters M_2X_2 with M = Fe, Co, Ni, Cu and X = S, Se, are structural analogs of the corresponding oxide clusters, with planar configuration in their global minima, stabilized by antiferromagnetic coupling [46–49]. They readily coordinate electron-donor ligands, e.g., water molecules, halogen ligands and carbonyl groups [15,46,47]. Up to six carbonyl groups can be attached, with three at each cation site [10–14,50]. The loss of a carbonyl ligand from $Fe_2S_2(CO)_6$ requires 154 kJ mol^{-1} and, similarly, for $Co_2S_2(CO)_6$

the energy needed is 151 kJ mol^{-1}, but the selenide complexes have low stability as hexacarbonyls: our calculations indicate that $Co_2Se_2(CO)_6$ loses a carbonyl ligand by only 12 kJ mol^{-1}. The tetracarbonyl complexes proved stable for both the sulfide and the selenide complexes—the loss of a carbonyl ligand from $Co_2Se_2(CO)_4$ requires 197 kJ mol^{-1}. It is thus useful to compare the electronic structure properties of the hexacarbonyl and tetracarbonyl complexes of cobalt, which form the shortest Co-S and Co-Se bonds. In their global minima, the hexacarbonyl and tetracarbonyl complexes contain a non-planar Co_2X_2 core with S-S or Se-Se bond, denoted as $Co_2(X_2)$, as shown in Figure 1a, Table 2. The metal cation-to-carbon bonds are of typical lengths within 1.75–1.82 Å. Natural orbital analysis reveals that in disulfides, cobalt valence orbital occupancies vary in the frame Co 4s(0.43–0.46) 3d(8.32–8.49) 4p(0.80–0.90) and this applies for both the tetracarbonyl and hexacarbonyl complexes, $Co_2(S_2)(CO)_6$ and $Co_2(S_2)(CO)_4$. The electron density on cobalt centers is thus significantly increased, as compared to the $3d^7$ electron configuration of Co(II). The 4p orbital population indicates 4s3d4p hybridization on cobalt. The sulfide and selenide centers also act as ligands, but the population of the sulfur valence levels is slightly higher than it is on selenium, namely S 3s(1.75) 3p(3.90–4.30) and Se 4s(1.75)4p(3.70–3.91). Cobalt centers in diselenide complexes increase their local valence orbital population, respectively, to Co 4s(0.45–0.47) 3d(8.4–8.6)4p(1.15–1.23). In the presence of a hydride ligand, the H 1s orbital population is between 0.83–1.01 and corresponds to a hydrogen atom, 1s(1), but with a partial positive charge, or minor negative for 1s(1.01) in $[Co_2HS_2](CO)_6$, Table 2. Among the cobalt disulfide complexes, the HOMO(SOMO)-LUMO gaps are higher for the hexacarbonyls than they are for the tetracarbonyls. The attachment of a hydride ligand increases the HOMO(SOMO)-LUMO gap in the hexacarbonyls, but in tetracarbonyls, the binding of a hydride ligand at a chalcogenide center (S, Se) always lowers the HOMO(SOMO)-LUMO energy gap. When the hydride ligand binds to the cobalt centers, the SOMO-LUMO gap in the tetracarbonyls strongly increases. In $Co_2(Se_2)(CO)_4$, the HOMO-LUMO energy gap is higher than in the disulfide complexes. The binding of hydride in a midway position between cobalt centers, Figure 2, increases the SOMO-LUMO gap in $[Co_2HS_2](CO)_6$ by 0.26 eV relative to the $Co_2(S_2)(CO)_6$, while in the corresponding tetracarbonyl, $[(Co_2H)S_2](CO)_4$, it decreases by 0.21 eV relative to the $Co_2(S_2)(CO)_4$ complex. The formation of a single Co-H bond, which is observed only in selenide complexes, e.g., $[Co(Co\text{-}H)Se_2](CO)_4$, increases the SOMO-LUMO gap by only 0.04 eV relative to $Co_2(Se_2)(CO)_4$, but this is the highest energy gap observed among the sulfide and selenide complexes with or without hydride ligand.

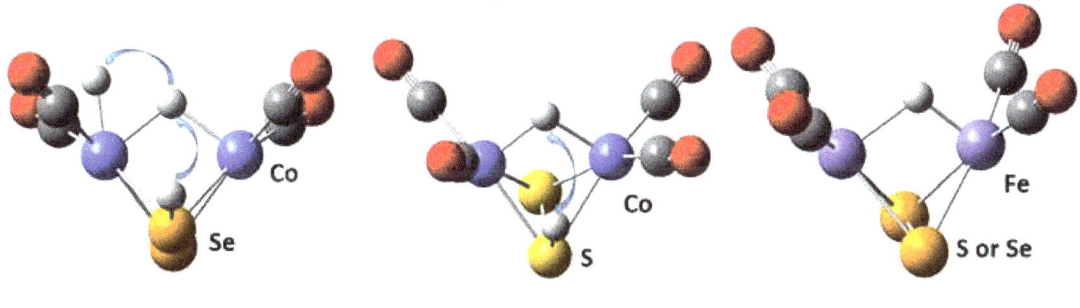

(**a**) (**b**) (**c**)

Figure 3. (**a**) The three distinct positions of hydride in $Co_2Se_2(CO)_4$. (**b**) The two distinct positions of hydride in $Co_2S_2(CO)_4$. (**c**) The single position of hydride in $Fe_2S_2(CO)_4$ valid also for $Fe_2Se_2(CO)_4$. Legend as in Figure 1.

Table 2. HOMO(SOMO)-LUMO (H-L) energy gaps (eV) and electron distribution on the hydride ligand, in cobalt tetracarbonyl and hexacarbonyl complexes calculated by natural population analysis.

Hexacarbonyl Complexes	H-L	H 1s	Tetracarbonyl Complexes	H-L	H 1s	Tetracarbonyl Complexes	H-L	H 1s
$Co_2(S_2)(CO)_6$	2.84		$Co_2(S_2)(CO)_4$	2.83		$Co_2(Se_2)(CO)_4$	3.29	
$[Co_2S_2H](CO)_6$	2.91	0.89	$[Co_2S_2H](CO)_4$	1.67	0.87	$[Co_2Se_2H](CO)_4$	1.56	0.91
$[Co_2HS_2](CO)_6$	3.10	1.01	$[(Co_2H)S_2](CO)_4$	2.62	0.86	$[(Co_2H)Se_2](CO)_4$	2.31	0.85
						$[Co(Co\text{-}H)Se_2](CO)_4$	3.33	0.83

According to the TD-DFT calculations, the protonated and reduced cobalt complexes possess intense light-absorption bands in the visible part of the spectrum, as shown in Table 3. Typically, the bands are blue shifted for the selenide complexes when analogous conformations are compared. The TD-DFT calculated highest intensity bands in the UV-Vis spectra correspond to multiple vertical electron excitations within the Co_2X_2 core, with dominant Co $\rightarrow$ X and Co $\rightarrow$ H transitions, or metal to ligand charge transfer bands, MLCT. Though the hydride-bonded complexes are powerful reducing agents even without photoactivation, the presence of intense bands in the visible region allows excitation and further enhancement of reactivity towards inert molecules such as CO_2.

Table 3. TD-DFT results for protonated and reduced (H^+,e^-) cobalt sulfides and selenides. The most intense light absorption bands listed.

Complex	Light Absorption Bands, nm	Oscillator Strength
$[Co_2(S_2H)]^+(CO)_4$	703; 910	0.0012; 0.0130
$[Co_2S\text{-}SH](CO)_4$	689; 824	0.0060; 0.0010
$[Co\text{-}H\text{-}CoS_2](CO)_4$	750; 803	0.0020; 0.0040
$[Co_2(Se_2H)]^+(CO)_4$	519; 753	0.0028; 0.0036
$[Co_2Se\text{-}SeH](CO)_4$	628; 749	0.0011; 0.0017
$[Co\text{-}H\text{-}CoSe_2](CO)_4$	748; 1610	0.0006; 0.0250
$[Co(Co\text{-}H)Se_2](CO)_4$	546; 639	0.0017; 0.0024

3.2. *The Hydrogen Evolution Reaction (HER) on Tetracarbonyl Complexes of Metal Disulfides and Diselenides, $M_2X_2(CO)_4$*

The reaction path in water dissociation with hydrogen evolution includes an intermediate step of breaking an H-OH bond with the formation of an S-H or Se-H bond and a bridging hydroxyl group, as shown in Figure 4.

In a photoactivated reaction, the following elementary steps are followed:

$$[M_2X_2](CO)_4 + H_2O + h\nu \Rightarrow [M_2X_2H^*](CO)_4 + OH^* \Rightarrow [M_2(OH)X_2H](CO)_4 \Rightarrow [M_2OX_2](CO)_4 + H_2 \quad (1)$$

The reaction mechanism was traced for the global minima of the complexes, which correspond to either a diamagnetic (*d*) singlet or antiferromagnetic (*afm*) singlet ground states. The chalcogenide complexes of cobalt and iron are presented in Figure 4. Dicobalt diselenide, $Co_2Se_2(CO)_4$ (*d*) and diiron disulfide, $Fe_2S_2(CO)_4$ (*afm*) provide a more favorable energy path in the first reaction step of water dissociation, but in the following step of dihydrogen formation, the corresponding energy barriers are with the highest values. The reverse case is observed with $Fe_2Se_2(CO)_4$ (*afm*), and though it reaches the highest energy barrier in the first reaction step, in the next step of dihydrogen formation, it provides the lowest energy path. Overall, it may be concluded that $Co_2Se_2(CO)_4$, (*d*), $Co_2S_2(CO)_4$ (*afm*), and $Fe_2Se_2(CO)_4$ (*afm*) perform best: for $Co_2Se_2(CO)_4$, the first reaction barrier is 107 kJ mol^{-1} and the second reaction barrier is 135 kJ mol^{-1}, whereas the lowest value for the second step of hydrogen formation is for $Fe_2Se_2(CO)_4$, 75 kJ mol^{-1}, preceded by a barrier of 174 kJ mol^{-1}. $Co_2S_2(CO)_4$ stays between these values—the first barrier at 167 kJ mol^{-1} and the second barrier at 88 kJ mol^{-1}. The chalcogenide complexes of the remaining elements—nickel and copper perform worse in the water dissociation and HER, as shown in Figure 5. The lowest energy barrier for the first step is 167 kJ mol^{-1} for $Ni_2Se_2(CO)_4$, followed by 369 kJ mol^{-1} for dihydrogen formation. Among these complexes, $Cu_2Se_2(CO)_4$ performs best, with a first reaction step barrier of 175 kJ mol^{-1} and a second barrier of 305 kJ mol^{-1}. Though the reaction barriers for the second reaction step look prohibitively high, the role of the first step of water dissociation is important, as pointed out in studies on carbon dioxide trapping and activation [51,52]. It provides surface hydroxyl or sulfonyl groups, which are able to activate the CO_2 molecule and promote in this way carboxyl or carbonate formation.

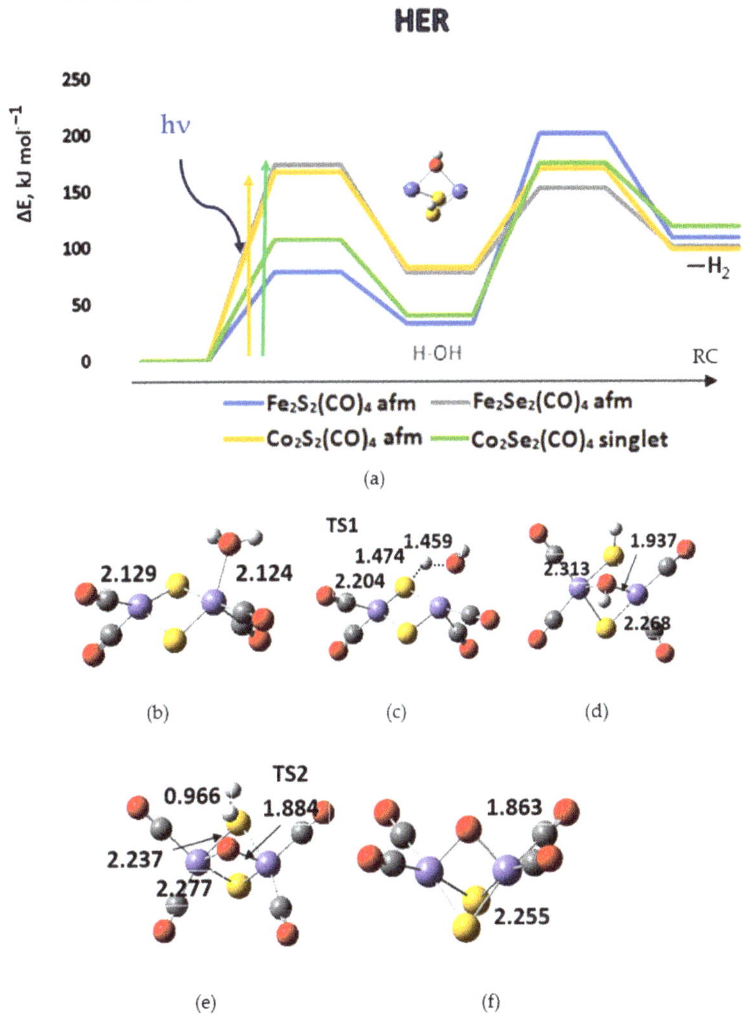

Figure 4. (**a**) The reaction path of water dissociation and hydrogen evolution (HER) on different tetracarbonyl complexes of iron and cobalt. TS1 corresponds to the reaction barrier of primary dissociation, and TS2 to dihydrogen formation. afm denotes antiferromagnetic singlet ground states. ΔE is the energy difference relative to the ground state complexes; RC—reaction coordinate. The excitation energies of representative most intense singlet-singlet transitions induced by light absorption are denoted by vertical arrows. (**b**) Structure of the water adsorption complex on $Fe_2S_2(CO)_4$; (**c**) structure of TS1; (**d**) dissociated water on $Fe_2S_2(CO)_4$; (**e**) structure of TS2; (**f**) structure of the product with bridging oxygen after hydrogen desorption. Cartesian coordinates of intermediate species are presented in Supporting Information.

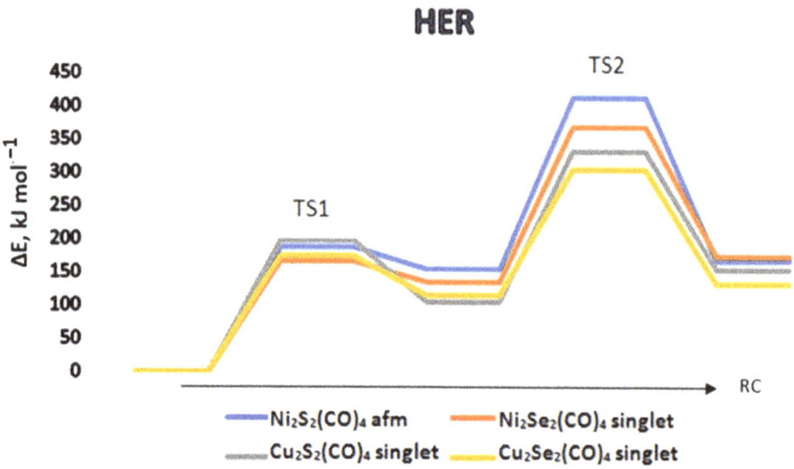

Figure 5. The reaction path of water dissociation and hydrogen evolution (HER) on different tetracarbonyl complexes of nickel and copper. TS1 corresponds to the reaction barrier of primary dissociation, and TS2 to dihydrogen formation. afm denotes antiferromagnetic singlet ground states. ΔE is the energy difference relative to the ground state complexes; RC—reaction coordinate.

In addition, the triplet potential energy surfaces of the complexes $M_2X_2(CO)_4$ were also examined as all of them have stable triplet states (local minima); see Supporting Information (SI). All of the triplet states are higher in energy than the singlet ground states (by 47 ÷ 152 kJ mol^{-1}) and the energy gaps between the triplet and singlet state minima are presented in Table S1 in SI. The calculated triplet HER pathway of chalcogenide complexes of cobalt and iron in Figure S1 (SI) showed that the lowest energy barrier in the first step is provided by $Co_2Se_2(CO)_4$, but in the following step of dihydrogen formation, it reaches the highest energy value of 213 kJ mol^{-1}. This triplet state was found to be 68 kJ mol^{-1} higher in energy than the singlet ground state. The triplet state reaction path does not provide a lowering of the second reaction barrier, which is rate determining. The triplet state reaction path for nickel and iron chalcogenide complexes is highly unfavorable: the lowest energy barrier for the rate-determining step is 670 kJ mo^{-1} for $Ni_2Se_2(CO)_4$, see Figure S2 in SI.

The activation barriers in HER differ largely depending on the type of system studied [53–56] and the calculated values for TS1 varied in the range of 40–210 kJ mol^{-1} for $(MO_2)_n$ clusters (M = Ti, Zr, Hf, n = 1–3) [56]. The barriers for TS2 were not much larger, ranging from 63 to 210 kJ mol^{-1}. A systematic dependence on the type and size of the clusters was not reported. Our results on the chalcogenides of cobalt and iron fall within a similar range, 68–174 kJ mol^{-1} for TS1 and 88–213 kJ mol^{-1} for TS2 and we also observe a lack of systematic change depending on the composition of the complexes.

3.3. The Oxygen Evolution Reaction (OER) on Tetracarbonyl Complexes of Metal Disulfides and Diselenides, $M_2X_2(CO)_4$

Previous studies outlined the role of pH and oxygenated reaction intermediates after proton–electron removal [4–8,53–55]. The oxygen evolution reaction on the chalcogenide complexes also proved to be pH dependent and in acidic media it proceeds via protonation and a peroxo intermediate OOH*, which is the more favorable route, Equation (2). Here again, the lowest energy barriers of 91 and 93 kJ mol^{-1} are obtained for the selenide complexes, $Fe_2Se_2(CO)_4$, and $Co_2Se_2(CO)_4$, followed by the disulfides of iron and cobalt, $Fe_2S_2(CO)_4$ and $Co_2S_2(CO)_4$ with barriers of 114 and 109 kJ mol^{-1}, as shown in Figure 6.

The OER reaction begins with an oxygen-bridged complex, which comes out from the HER reaction.

$$[M_2OX_2](CO)_4 + O[H^+,e^-] + h\nu \Rightarrow [M_2(OOH)X_2](CO)_4 \Rightarrow [M_2HX_2](CO)_4 + O_2 \quad (2)$$

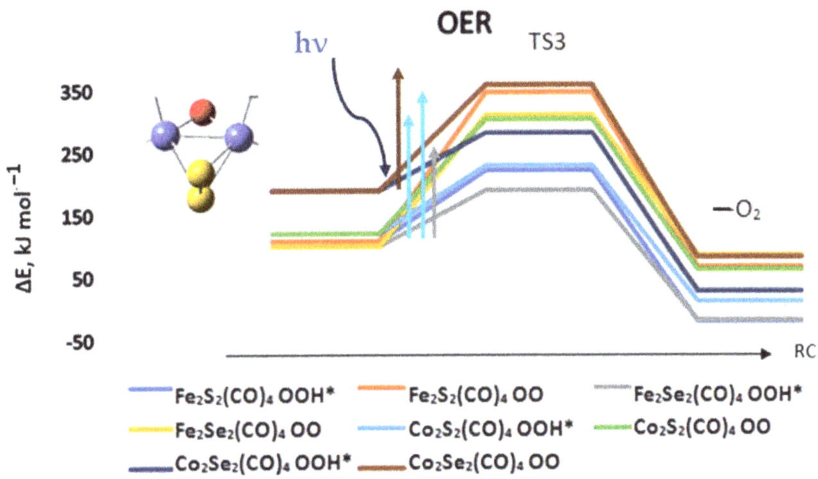

Figure 6. The reaction path of oxygen formation and oxygen evolution for cobalt and iron chalcogenide complexes. ΔE is the energy difference relative to the ground state complexes; RC—reaction coordinate. The global minima of the complexes were used, as denoted in Figure 4. The excitation energies of representative most intense transitions induced by light absorption are denoted by vertical arrows.

The reaction barriers for OER are thus not prohibitively high, but the presence of favorable light absorption bands would certainly allow a photocatalytic pathway. TD-DFT calculations indeed indicate the presence of intense bands in the visible spectrum for the dioxygen intermediates and for the peroxo intermediates, Table 4. They are slightly blue-shifted, as compared to the reduced forms of the corresponding complexes. The energy provided by light absorption is sufficient to provide activation relevant to the reaction barriers of OER via OOH*, and for the high-lying barriers of dioxygen formation. Among the most favorable energy pathways for OER, as reported in the literature, is that on a molecular cubane complex, and a reaction barrier of 84 kJ mol^{-1} was experimentally determined [8]. Theoretical modeling with small cobalt oxide clusters provided an accurate estimate of this barrier and the reported calculated value is 97 kJ mol^{-1} [4]. The cobalt and iron chalcogenides thus provide comparable reaction barriers, according to our results, via the peroxo intermediate OOH*, as shown in Figure 6. The OER can be started from the triplet state of the complexes, but the resulting activation barriers are higher by 87–150 kJ mol^{-1}; see Figure S3 in SI.

In a pH-neutral or alkaline solvent, $Co_2Se_2(CO)_4$ and $Co_2S_2(CO)_4$ provide the lowest energy path for dioxygen formation with barriers of 170 and 183 kJ mol^{-1}, respectively. The reaction pathway in this case is the following Equation (3):

$$[M_2OX_2](CO)_4 + 2OH^* + h\nu \Rightarrow [M_2(OO)X_2](CO)_4 + [H^+,e^-]\text{-}OH \Rightarrow [M_2X_2](CO)_4 + H_2O + O_2 \quad (3)$$

Both reaction paths include proton–electron transfer. The light absorption bands correspond to electron excitations within the M_2X_2 core, M → X of MLCT character, but they include the bonded dioxygen species with transitions O → M, which correspond to ligand-to-metal charge transfer (LMCT). This is another proof of the great capacity of the

$M_2X_2(CO)_4$ complexes to redistribute electron density. The transitions are of the type triplet to triplet for Equation (3) and doublet to doublet for Equation (2).

Table 4. TD-DFT results for OER intermediates of selected sulfide and selenide tetracarbonyl complexes. The most intense light absorption bands are listed.

Complex	Light Absorption Bands, nm	Oscillator Strength
$Co_2S_2(CO)_4$; O-O	651	0.0014
$Co_2Se_2(CO)_4$; O-O	582	0.0091
$Fe_2Se_2(CO)_4$; O-O	807	0.0131
$Cu_2Se_2(CO)_4$; O-O	430	0.0094
$Co_2S_2(CO)_4$; OOH*	507; 573	0.0023; 0.0027
$Co_2Se_2(CO)_4$; OOH*	531	0.0014
$Fe_2Se_2(CO)_4$; OOH*	798	0.0052
$Cu_2Se_2(CO)_4$; OOH*	590	0.0130
$Ni_2Se_2(CO)_4$; OOH*	558	0.0023

For the complexes of nickel and copper, the role of acidity is not pronounced, as shown in Figure 7. The lowest reaction barrier is indeed for OOH* formation on $Cu_2Se_2(CO)_4$ and it is 156 kJ mol^{-1}, followed by OOH* formation on $Ni_2S_2(CO)_4$ with a barrier of 169 kJ mol^{-1}. The formation of dioxygen on $Ni_2S_2(CO)_4$ does not change the barrier significantly—it goes up to 185 kJ mol^{-1}. On $Ni_2Se_2(CO)_4$ the barrier heights for OOH* and OO formation are reversed: dioxygen formation requires 203 kJ mol^{-1}, while the pathway via OOH* intermediate goes through a slightly higher barrier of 211 kJ mol^{-1}. The OER can be started from the triplet state of the complexes, but the resulting activation barriers are higher by 157–250 kJ mol^{-1} higher; see Figure S3 in SI.

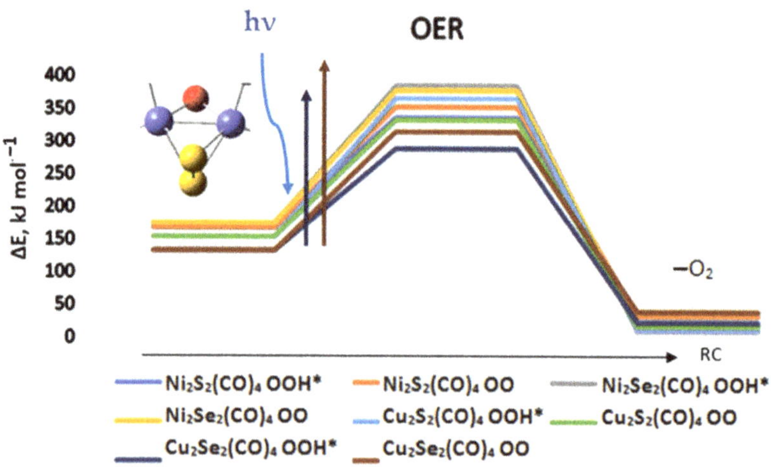

Figure 7. The reaction path of oxygen formation and oxygen evolution for nickel and copper chalcogenide complexes. ΔE is the energy difference relative to the ground state complexes; RC—reaction coordinate. The global minima complexes were used, as denoted in Figure 5. The excitation energies of representative most intense transitions induced by light absorption are denoted by vertical arrows.

4. Conclusions

The chalcogenide tetracarbonyl complexes of the 3d transition metal elements (Fe-Cu) follow a pathway with similar intermediates in the reaction of water splitting, with low energy barriers for the singlet pathway, and the presence of visible light-absorption bands favor photoactivation. Though only sulfides are direct structural analogs of natural enzymes, selenides have similar proton affinities, proton–electron affinities, and light absorption bands and may outperform sulfides in the OER reaction. Cobalt and iron sulfides and selenides perform better than the corresponding complexes of nickel and copper for both the HER and OER reactions. Protonation affects positively the energy barriers for OER in the case of cobalt and iron chalcogenide complexes, but the effect is weaker for the nickel and copper analogs. The hydride intermediates, relevant to hydrogen evolution, and also the oxidized intermediates possess favorable light absorption bands in the visible spectrum. They allow photoactivation in the complexes, for which the reaction barriers are high.

Supplementary Materials: The following supporting information can be downloaded at: https://www.mdpi.com/article/10.3390/ma17010056/s1, Cartesian coordinates of the complexes presented in Figures S1, S3, S4b,c,d,e,f. HER/OER reaction path on the triplet potential energy surface, as Figures S1–S3 and singlet to triplet excitation energies as Table S1.

Author Contributions: Conceptualization, E.U. and I.G.; methodology, E.U., I.G. and T.Z.; validation, E.U., I.G. and T.Z.; investigation, E.U., I.G. and T.Z.; writing—original draft preparation, E.U.; writing—review and editing, E.U., I.G. and T.Z.; visualization, E.U., I.G. and T.Z. All authors have read and agreed to the published version of the manuscript.

Funding: This research received funding by TwinTeam project Д01-272 "European Network on Materials for Clean Technologies" for providing the opportunity to present the results at the SizeMat4 conference, as well as for the publication financial support and also thanks to the funding by the Bulgarian National Science Fund of Bulgarian Ministry of Education and Science, Grant КП-06-Н59/6 (2021), project (PhotoMetalMod).

Data Availability Statement: Data are contained within the article and Supplementary Materials.

Acknowledgments: The authors thank the TwinTeam Project Д01-272 "European Network on Materials for Clean Technologies", funded by the Ministry of Education and Science under the National Program "European Scientific Networks" for providing the opportunity to present the results at the SizeMat4 conference, as well as for the publication financial support. The authors acknowledge the financial support of the Bulgarian National Science Fund of the Bulgarian Ministry of Education and Science, Grant КП-06-Н59/6 (2021), project (PhotoMetalMod). This work was supported by the European Regional Development Fund within the Operational Programme "Science and Education for Smart Growth 2014–2020" under the Project CoE "National Center of Mechatronics and Clean Technologies" (BG05M2OP001-1.001-0008) (for supplying a license for program package Gaussian16). The authors also acknowledge the access provided to the e-infrastructure of the NCHDC—part of the Bulgarian National Roadmap on RIs, with financial support through Grant No D01-168/28.07.2022.

Conflicts of Interest: The authors declare no conflict of interest.

References

1. Chase, M.W., Jr. *NIST-JANAF Thermochemical Tables*, 4th ed.; American Institute of Physics: College Park, MD, USA, 1998.
2. Song, J.; Wei, C.; Huang, Z.-F.; Liu, C.; Zeng, L.; Wang, X.; Xu, Z. A review on fundamentals for designing oxygen evolution electrocatalysts. *Chem. Soc. Rev.* **2020**, *49*, 2196–2214. [CrossRef] [PubMed]
3. Greife, P.; Schönborn, M.; Capone, M.; Assunção, R.; Narzi, D.; Guidoni, L.; Dau, H. The electron–proton bottleneck of photosynthetic oxygen evolution. *Nature* **2023**, *617*, 623–628. [CrossRef] [PubMed]
4. Li, X.; Siegbahn, P.E.M. Water Oxidation Mechanism for Synthetic Co–Oxides with Small Nuclearity. *J. Am. Chem. Soc.* **2013**, *135*, 13804–13813. [CrossRef]
5. Birss, V.I.; Damjanovic, A.; Hudson, P.G. Oxygen Evolution at Platinum Electrodes in Alkaline Solutions: II. Mechanism of the Reaction. *J. Electrochem. Soc.* **1986**, *133*, 1621. [CrossRef]
6. Conway, B.E.; Liu, T.C. Characterization of electrocatalysis in the oxygen evolution reaction at platinum by evaluation of behavior of surface intermediate states at the oxide film. *Langmuir* **1990**, *6*, 268–276. [CrossRef]
7. Gerencsér, L.; Dau, H. Water Oxidation by Photosystem II: H_2O–D_2O Exchange and the Influence of pH Support Formation of an Intermediate by Removal of a Proton before Dioxygen Creation. *Biochemistry* **2010**, *49*, 10098–10106. [CrossRef]
8. McCool, N.S.; Robinson, D.M.; Sheats, J.E.; Dismukes, C. A Co_4O_4 "Cubane" Water Oxidation Catalyst Inspired by Photosynthesis. *J. Am. Chem. Soc.* **2011**, *133*, 11446–11449. [CrossRef]

9. Appel, A.M.; Bercaw, J.E.; Bocarsly, A.B.; Dobbek, H.; DuBois, D.L.; Dupuis, M.; Ferry, J.G.; Fujita, E.; Hille, R.; Kenis, P.J.A.; et al. Frontiers, Opportunities, and Challenges in Biochemical and Chemical Catalysis of CO_2 Fixation. *Chem. Rev.* **2013**, *113*, 6621–6658. [CrossRef]
10. Nann, T.; Ibrahim, S.K.; Woi, P.-M.; Xu, S.; Ziegler, J.; Pickett, C.J. Water Splitting by Visible Light: A Nanophotocathode for Hydrogen Production. *Angew. Chem. Int. Ed.* **2010**, *49*, 1574–1577. [CrossRef]
11. Dubois, M.R.; Dubois, D.L. Development of Molecular Electrocatalysts for CO_2 Reduction and H_2 Production/Oxidation. *Acc. Chem. Res.* **2009**, *42*, 1974–1982. [CrossRef]
12. Gloaguen, F.; Rauchfuss, T.B. Small molecule mimics of hydrogenases: Hydrides and redox. *Chem. Soc. Rev.* **2009**, *38*, 100–108. [CrossRef]
13. Tard, C.; Pickett, C.J. Structural and Functional Analogues of the Active Sites of the [Fe]-, [NiFe]-, and [FeFe]-Hydrogenases. *Chem. Rev.* **2009**, *109*, 2245–2274. [CrossRef] [PubMed]
14. Tard, C.; Liu, X.; Ibrahim, S.K.; Bruschi, M.; De Gioia, L.; Davies, S.C.; Yang, X.; Wang, L.-S.; Sawers, G.; Pickett, C.J. Synthesis of the H-cluster framework of iron-only hydrogenase. *Nature* **2005**, *433*, 610–613. [CrossRef] [PubMed]
15. Varley, J.B.; Hansen, H.A.; Ammitzbøll, N.L.; Grabow, L.C.; Peterson, A.A.; Rossmeisl, J.; Nørskov, J.K. Ni–Fe–S Cubanes in CO_2 Reduction Electrocatalysis: A DFT Study. *ACS Catal.* **2013**, *3*, 2640–2643. [CrossRef]
16. Becke, A.D. Density-functional thermochemistry. III. The role of exact exchange. *J. Chem. Phys.* **1993**, *98*, 5648–5652. [CrossRef]
17. Becke, A.D. Density-functional thermochemistry. IV. A new dynamical correlation functional and implications for exact-exchange mixing. *J. Chem. Phys.* **1996**, *104*, 1040–1046. [CrossRef]
18. Lee, C.; Yang, W.; Parr, R.G. Development of the Colle-Salvetti correlation-energy formula into a functional of the electron density. *Phys. Rev. B* **1988**, *37*, 785–789. [CrossRef]
19. Miehlich, B.; Savin, A.; Stoll, H.; Preuss, H. Results obtained with the correlation-energy density functionals of Becke and Lee, Yang and Parr. *Chem. Phys. Lett.* **1989**, *157*, 200–206. [CrossRef]
20. Frisch, M.J.; Trucks, G.W.; Schlegel, H.B.; Scuseria, G.E.; Robb, M.A.; Cheeseman, J.R.; Scalmani, G.; Barone, V.; Petersson, G.A.; Nakatsuji, H.; et al. *Gaussian 16, Revision C.01*; Gaussian, Inc.: Wallingford, CT, USA, 2016.
21. McLean, A.D.; Chandler, G.S. Contracted Gaussian basis sets for molecular calculations. I. Second row atoms, Z = 11–18. *J. Chem. Phys.* **1980**, *72*, 5639–5648. [CrossRef]
22. Krishnan, R.; Binkley, J.S.; Seeger, R.; Pople, J.A.J. Self-consistent molecular orbital methods. XX. A basis set for correlated wave functions. *Chem. Phys.* **1980**, *72*, 650–654. [CrossRef]
23. Wachters, A.J.H. Gaussian basis set for molecular wavefunctions containing third-row atoms. *J. Chem. Phys.* **1970**, *52*, 1033–1036. [CrossRef]
24. Hay, P.J. Gaussian basis sets for molecular calculations—Representation of 3D orbitals in transition-metal atoms. *J. Chem. Phys.* **1977**, *66*, 4377–4384. [CrossRef]
25. Raghavachari, K.; Trucks, G.W. Highly correlated systems: Excitation energies of first row transition metals Sc-Cu. *J. Chem. Phys.* **1989**, *91*, 1062–1065. [CrossRef]
26. Weigend, F.; Ahlrichs, R. Balanced basis sets of split valence, triple zeta valence and quadruple zeta valence quality for H to Rn: Design and assessment of accuracy. *Phys. Chem. Chem. Phys.* **2005**, *7*, 3297–3305. [CrossRef] [PubMed]
27. Weigend, F. Accurate Coulomb-fitting basis sets for H to Rn. *Phys. Chem. Chem. Phys.* **2006**, *8*, 1057–1065. [CrossRef] [PubMed]
28. Hay, P.J.; Wadt, W.R. Ab initio effective core potentials for molecular calculations—Potentials for the transition-metal atoms Sc to Hg. *J. Chem. Phys.* **1985**, *82*, 270–283. [CrossRef]
29. Wadt, W.R.; Hay, P.J. Ab initio effective core potentials for molecular calculations—Potentials for main group elements Na to Bi. *J. Chem. Phys.* **1985**, *82*, 284–298. [CrossRef]
30. Hay, P.J.; Wadt, W.R. Ab initio effective core potentials for molecular calculations—Potentials for K to Au including the outermost core orbitals. *J. Chem. Phys.* **1985**, *82*, 299–310. [CrossRef]
31. Uzunova, E.L.; Mikosch, H. Electronic, Magnetic Structure and Water Splitting Reactivity of the Iron-Sulfur dimers and their Hexacarbonyl Complexes: A Density Functional Study. *J. Chem. Phys.* **2014**, *141*, 044307. [CrossRef]
32. Uzunova, E.L. Pathways of selective catalytic CO_2 two-step reduction on di-iron, di-cobalt and iron-cobalt disulfide carbonyls—An electronic structure study. *Catal. Sci. Technol.* **2019**, *9*, 1039–1047. [CrossRef]
33. Uzunova, E.L. Cation binding of Li(I), Na(I) and Zn(II) to cobalt and iron sulphide clusters—Electronic structure study. *Phys. Chem. Chem. Phys.* **2022**, *24*, 20228–20238. [CrossRef] [PubMed]
34. Lovell, T.; Himo, F.; Han, W.-G.; Noodleman, L. Density functional methods applied to metalloenzymes. *Coord. Chem. Rev.* **2003**, *238–239*, 211–232. [CrossRef]
35. Halgren, T.A.; Lipscomb, W.N. The Synchronous Transit Method for Determining Reaction Pathways and Locating Transition States. *Chem. Phys. Lett.* **1977**, *49*, 225–232. [CrossRef]
36. Peng, C.; Ayala, P.Y.; Schlegel, H.B.; Frisch, M.J. Using redundant internal coordinates to optimize equilibrium geometries and transition states. *J. Comp. Chem.* **1996**, *17*, 49–56. [CrossRef]
37. Fukui, K. The path of chemical-reactions—The IRC approach. *Acc. Chem. Res.* **1981**, *14*, 363–368. [CrossRef]
38. Hratchian, H.P.; Schlegel, H.B. Accurate reaction paths using a Hessian based predictor-corrector integrator. *J. Chem. Phys.* **2004**, *120*, 9918–9924. [CrossRef] [PubMed]

39. Tomasi, J.; Mennucci, B.; Cammi, R. Quantum Mechanical Continuum Solvation Models. *Chem. Rev.* **2005**, *105*, 2999–3093. [CrossRef] [PubMed]
40. Bauernschmitt, R.; Ahlrichs, R. Treatment of electronic excitations within the adiabatic approximation of time dependent density functional theory. *Chem. Phys. Lett.* **1996**, *256*, 454–464. [CrossRef]
41. Furche, F.; Ahlrichs, R. Adiabatic time-dependent density functional methods for excited state properties. *J. Chem. Phys.* **2002**, *117*, 7433–7447. [CrossRef]
42. Grimme, S.; Ehrlich, S.; Goerigk, L. Effect of the damping function in dispersion corrected density functional theory. *J. Comput. Chem.* **2011**, *32*, 1456–1465. [CrossRef]
43. Reed, A.E.; Curtiss, L.A.; Weinhold, F. Intermolecular interactions from a natural bond orbital, donor-acceptor viewpoint. *Chem. Rev.* **1988**, *88*, 899–926. [CrossRef]
44. Weinhold, F.; Carpenter, J.E. *The Structure of Small Molecules and Ions*; Plenum: New York, NY, USA, 1988.
45. Flory, M.A.; McLamarrah, S.K.; Ziurys, L.M. High-resolution spectroscopy of CoS (X4Δi): Examining 3d transition-metal sulfide bonds. *J. Chem. Phys.* **2005**, *123*, 164312. [CrossRef] [PubMed]
46. Haider, S.; Di Tommaso, D.; de Leeuw, N.H. Density functional theory simulations of the structure, stability and dynamics of iron sulphide clusters in water. *Phys. Chem. Chem. Phys.* **2013**, *15*, 4310–4319. [CrossRef] [PubMed]
47. Terranova, U.; de Leeuw, N.H. Aqueous Fe_2S_2 cluster: Structure, magnetic coupling, and hydration behaviour from Hubbard U density functional theory. *Phys. Chem. Chem. Phys.* **2014**, *16*, 13426–13433. [CrossRef]
48. Staemmler, V.; Reinhardt, P.; Allouti, F.; Alikhani, M.E. A theoretical study of the electronic structure of the Co_2O_2 molecule. *Chem. Phys.* **2008**, *349*, 83–90. [CrossRef]
49. Gutsev, G.L.; Weatherford, C.A.; Jena, P.; Johnson, E.; Ramachandran, B.R. Competition between surface chemisorption and cage formation in $Fe_{12}O_{12}$ clusters. *Chem. Phys. Lett.* **2013**, *556*, 211–216. [CrossRef]
50. Kunkely, H.; Vogler, A. Photoreactivity of $Fe_2S_2(CO)_6$ originating from dσ* metal-to-ligand charge transfer excitation. *J. Organomet. Chem.* **1998**, *568*, 291–293. [CrossRef]
51. Rybakov, A.A.; Trubnikov, D.N.; Larin, A.V. The role of water in the catalytic CO_2 binding by alkaline earth Y faujasites. *Microporous Mesoporous Mater.* **2022**, *343*, 112125. [CrossRef]
52. Uzunova, E.L.; Seriani, N.; Mikosch, H. CO_2 Conversion to Methanol on Cu(I) Oxide Nanolayers and Clusters: Electronic Structure Insight into the Reaction Mechanism. *Phys. Chem. Chem. Phys.* **2015**, *17*, 11088–11094. [CrossRef]
53. Gust, D.; Thomas, A.; Moore, T.A.; Ana, L.; Moore, A.L. Solar Fuels via Artificial Photosynthesis. *Acc. Chem. Res.* **2009**, *42*, 1890–1898. [CrossRef]
54. Wang, N.; Wang, M.; Liu, J.; Jin, K.; Chen, L.; Sun, L. Preparation, Facile Deprotonation, and Rapid H/D Exchange of the μ-Hydride Diiron Model Complexes of the [FeFe]-Hydrogenase Containing a Pendant Amine in a Chelating Diphosphine Ligand. *Inorg.Chem.* **2009**, *48*, 11551–11558. [CrossRef] [PubMed]
55. Camara, J.M.; Rauchfuss, T.B. Mild Redox Complementation Enables H_2 Activation by [FeFe]-Hydrogenase Models. *J. Am. Chem. Soc.* **2011**, *133*, 8098–8101. [CrossRef]
56. Fang, Z.; Dixon, D.A. Computational Study of H_2 and O_2 Production from Water Splitting by Small $(MO_2)_n$ Clusters (M = Ti, Zr, Hf). *J. Phys. Chem. A* **2013**, *117*, 3539–3555. [CrossRef] [PubMed]

materials

MDPI

Article

Corrosion-Resistive ZrO_2 Barrier Films on Selected Zn-Based Alloys

Irina Stambolova [1,*], Daniela Stoyanova [1], Maria Shipochka [1], Nelly Boshkova [2], Silviya Simeonova [3], Nikolay Grozev [3], Georgi Avdeev [2], Ognian Dimitrov [4] and Nikolai Boshkov [2]

[1] Institute of General and Inorganic Chemistry, Bulgarian Academy of Sciences, "Acad. G. Bonchev" St. Bl. 11, 1113 Sofia, Bulgaria; dsto@svr.igic.bas.bg (D.S.); shipochka@svr.igic.bas.bg (M.S.)

[2] Institute of Physical Chemistry "R. Kaishev", Bulgarian Academy of Sciences, 1113 Sofia, Bulgaria; nelly.boshkova@ipc.bas.bg (N.B.); g_avdeev@ipc.bas.bg (G.A.); nboshkov@ipc.bas.bg (N.B.)

[3] Faculty of Chemistry and Pharmacy, Sofia University, 1164 Sofia, Bulgaria; fhsss@chem.uni-sofia.bg (S.S.); fhng@chem.uni-sofia.bg (N.G.)

[4] Institute of Electrochemistry and Energy Systems, Bulgarian Academy of Sciences, "Acad. G. Bonchev" St. Bl. 10, 1113 Sofia, Bulgaria; ognian.dimitrov@iees.bas.bg

* Correspondence: stambolova@yahoo.com

Abstract: This work presents the enhanced corrosion resistance of newly developed two-layer composite coatings deposited on low-carbon steel: electrodeposited zinc alloy coatings (Zn–Ni with 10 wt.% Ni (ZN) or Zn–Co with 3 wt.% Co (ZC), respectively) and a top ZrO_2 sol–gel layer. Surface morphology peculiarities and anti-corrosion characteristics were examined using scanning electron microscopy (SEM), energy dispersive spectroscopy (EDX), atomic force microscopy (AFM), water contact angle (WCA) measurements, X-ray diffraction (XRD), X-ray photoelectron spectroscopy (XPS) analyses, potentiodynamic polarization (PDP) curves, corrosion potential (Ecorr), polarization resistance (Rp) measurements (for a prolonged period of 25 days) and open-circuit potential (OCP). The results were compared with the corrosion peculiarities of usual zinc coating. The zirconia top coatings in both systems were amorphous and dense, possessing hydrophobic nature. The experimental data revealed an increased corrosion resistance and protective ability of the ZC system in comparison to that of ZN due to its smooth, homogeneous surface and the presence of poorly crystallized oxides (ZnO and Co_3O_4), both later playing the role of a barrier for corrosive agents.

Keywords: zirconium oxide films; zinc alloy coatings; surface morphology; corrosion resistance; structure

Citation: Stambolova, I.; Stoyanova, D.; Shipochka, M.; Boshkova, N.; Simeonova, S.; Grozev, N.; Avdeev, G.; Dimitrov, O.; Boshkov, N. Corrosion-Resistive ZrO_2 Barrier Films on Selected Zn-Based Alloys. *Materials* **2023**, *16*, 7673. https://doi.org/10.3390/ma16247673

Academic Editor: Juan José Santana-Rodríguez

Received: 13 November 2023
Revised: 6 December 2023
Accepted: 8 December 2023
Published: 15 December 2023

1. Introduction

Zirconium oxide is an important ceramic material used in many branches of industry: optics, biomedicine, etc. It has a wide band gap and a very high thermal expansion coefficient, compatible with many bulk metals. ZrO_2 also exhibits high chemical stability, wear resistance and hardness, which are the prerequisites for effective corrosion protection [1,2]. Zirconia-based coatings can be prepared using various physical and chemical techniques, such as physical vapor deposition (PVD) [3], magnetron sputtering [4], chemical vapor deposition (CVD) [5], plasma spray [6], hydrothermal process [7] and sol–gel process [8].

Chemical deposition techniques are widely used due to the low cost of the equipment, low process temperature and the possibility of obtaining films on large areas for the easy control of the films' stoichiometry, the degree of crystallization, etc.

In the current literature, the data devoted to the investigation of anti-corrosion properties of chemically deposited ZrO_2 on galvanized steel are very scarce. Recently, only a handful of authors have successfully applied the spray method for the deposition of ZrO_2 coatings on some hot-dip galvanized low-carbon steel types (the latter containing Al between 5 and 55 wt.%) with the thickness of the hot-dip (zinc) coating within the range of 13–30 μm. Zirconium acetylacetonate–water or ethanol–water solutions were used as the precursor. Samples with deposited ZrO_2 layers were tested in 0.5 M NaCl solution, and they showed good anti-corrosion performance compared to pure galvanized steel substrates [9–12].

The sol–gel method is a low-temperature method, and it is financially beneficial, enabling the possibility of obtaining a wide spectrum of oxide coatings. It also offers good adhesion to metallic surfaces, which is a very important advantage in the preparation of corrosion-resistive coatings [13]. The sol–gel method is an interesting, feasible alternative for the preparation of corrosion-protective layers on metallic substrates with various shapes and sizes, which make them suitable for industrial applications. Nevertheless, in the available literature to date, the data regarding investigations devoted to systems composed of sol–gel ZrO_2 coatings deposited on steel and galvanized steel are not observable.

The corrosion of metals is a very common worldwide problem, arising from their interaction with the environment and leading to great financial, material, environmental and sometimes also human losses [14]. It is well known that iron corrodes, forming porous layers of corrosion products (red rust) and thus enabling a relatively easy access of the corrosive agents deep inside the bulk materials. For this reason, the corrosion resistance of metallic materials needs to be improved. As is well known, galvanic zinc and zinc alloy plating seems to be a widespread method for protection of low-carbon steels. The anti-corrosive resistance of zinc concerning steel (Fe) is due to the potential difference between these metals, leading to anodic dissolution of Zn. The latter plays a sacrificial role and dissolves first in the case of corrosion attack, leading to the appearance of corrosion products with low solubility value and composition, which depends on the composition of the corrosive medium; for example, zinc hydroxide chloride appears in the case of a 5% NaCl medium. These corrosion products create an additional protective layer with a barrier characteristic, inhibiting the penetration of an aggressive medium deep inside [15–17]. One possible disadvantage of this method could be the insufficient protection of the steel substrate only from zinc corrosion products in very aggressive media. In order to further protect zinc-coated steel against corrosion, surface modification is generally adopted. The good corrosion resistance of galvanized steel could be improved by alloying Zn with selected metals, such as Co, Ni, Mn, Fe, etc., which generally leads to higher polarization resistance values in chlorine-containing medium compared to a case where ordinary Zn is used as a substrate [16]. Another approach for enhanced corrosion protection is the application of some inhibitors [18].

In the present paper, we aim to prolong the lifetime of low-carbon steel in chlorine-containing media by means of newly developed systems composed of selected Zn alloys (Zn modified with 10 wt.% Ni or 3 wt.% Co) applied as sub-layers, the latter being covered by sol–gel ZrO_2 films. We also test the behavior of these new systems for a prolonged period of time, aiming to check their suitability for some potential industrial applications.

2. Materials and Methods

2.1. Deposition of the Samples

Low-carbon steel plates (Metalsnab, Sofia, Bulgaria) with dimensions of 30 mm × 10 mm × 1 mm and the following elemental composition balance (wt.%) (C 0.05–0.12; S $\leq$ 0.04; P $\leq$ 0.35; Mn 0.25–0.5; Cr $\leq$ 0.1; Si $\leq$ 0.03; Ni $\leq$ 0.3; Cu $\leq$ 0.3; As $\leq$ 0.08; Fe) were taken as substrates. Two kinds of systems were realized and tested:

- System ZN: Zn–Ni (10 wt.%)—sub-layer; ZrO_2—top layer;
- System ZC: Zn–Co (3 wt.%)—sub-layer; ZrO_2—top layer.

System ZN was prepared using the following procedures: First, the alloy coating Zn–Ni (10 wt.%) was electrodeposited in an experimental cell via circulation from an electrolyte having a composition $NiSO_4 \cdot 7H_2O$:$NiCl_2 \cdot 6H_2O$:$ZnCl_2$:β-alanine—100:100:30:10 (g/L). All the constituents were purchased from Valerus Ltd., Sofia, Bulgaria. The pH value was 4. The electrodeposition process was held using a cathodic current density of 2 A/dm^2, a temperature of 40 °C and applying non-soluble Ti-Pt networks as anodes [16]. Afterward, the zirconia-based top films were prepared using zirconium butoxide—$Zr(OC_4H_9)_4$ (Sigma

Aldrich, St. Louis, MO, USA; CAS No.: 1071-76-7, 80 wt.% in 1-butanol, molecular weight: 383.68). The Zr precursor was diluted by adding isopropanol and some amounts of acetyl acetone (Sigma Aldrich, CAS No.: 123-54-6, ≥99%, molecular weight: 100.12) and acetic acid (Sigma Aldrich, CAS No.: 64-19-7, purity 99.8%, molecular weight: 60.05) as complexating agents in order to obtain 0.2 M solution. The next step was to add nitric acid to prevent hydrolysis (Sigma Aldrich, CAS No.: 7697-37-2, purity 99.8%, molecular weight: 63.01) and polyethylene glycol (PEG 400; Sigma Aldrich, CAS No.: 25322-68-3). The final solution was stirred for 2 h until the appearance of a yellowish transparent color. The low-carbon steel samples, previously galvanized using Zn–Ni (10 wt.%) alloy, were dipped into the zirconium precursor solution. The samples were dried up at room temperature and subsequently heated at 100 °C for 30 min. The deposition and drying processes of ZrO_2 coating were repeated 3 times. The final thermal treatment was at 380 °C for 1 h.

The System ZC was prepared using similar procedures: firstly, the alloy coating Zn–Co (3 wt.%) was electrodeposited from an electrolytic solution, which contains $ZnSO_4{\cdot}7H_2O{:}CoSO_4{\cdot}7H_2O{:}NH_4Cl{:}H_3BO_3$—100:120:30:25 g/L (Valerus Ltd., Sofia, Bulgaria). The pH value of the electrolyte was ~ 4.0, and the cathodic current density was 2 A/dm^2. In addition, soluble zinc anodes (Valerus, Sofia, Bulgaria) and additives ZC-1 (wetting agent—20 mL/L, IPC-BAS, Sofia, Bulgaria) and ZC-2 (brightener—2 mL/L, IPC-BAS, Sofia, Bulgaria) were used [16]. The low-carbon steel samples, previously galvanized using Zn–Co alloy, were immersed into the zirconium precursor solution, following the abovementioned dip coating and treatment procedures.

For comparison, an ordinary zinc coating was obtained from solution having a composition of 150 g/L $ZnSO_4{\cdot}7H_2O$, 30 g/L NH_4Cl and 30 g/L H_3BO_3 under the following conditions: pH 4.5–5.0, cathodic current density 2 A/dm^2, soluble zinc anodes and 2 additives: wetting agent (AZ1—40 mL/L, IPC-BAS, Sofia, Bulgaria) and brightener (AZ2—10 mL/L, IPC-BAS, Sofia, Bulgaria) [16]. The thickness of all investigated samples was about 11–12 μm.

2.2. Investigations with SEM/EDX Methods

The SEM/EDX investigations were carried out via scanning electron microscopy unit ZEISS Evo 10 (Oxford Instruments, Oxford, UK) in a high vacuum mode using secondary electron imaging and 25 keV accelerating voltage. The elemental composition of the samples was studied with energy dispersive spectroscopy (EDS) probe Oxford Ultim Max 40 and the results were compiled using the AZtec Live—Advanced software, version 6.1. The EDX analyses of the samples were realized in four different points.

2.3. AFM Studies and Hydrophobicity Measurements

Atomic Force microscope NanoScopeV system with Nanoscope software, Bruker Ltd., Mannheim, Germany, (operating in a tapping mode in air at room temperature, Cantilever force—40 N/m; resonance frequency of 300 kHz) was applied. The scanning rate was 1 Hz. The roughness analysis enables us to obtain the parameter Ra, the latter being an arithmetic average of the absolute values of the surface height deviations, obtained based on the mean plane, while Rq is the root-mean-squared average of height deviations, received from the mean image data. The water contact angles (WCAs) were evaluated with Ramé-Hart automated goniometer model 290 with DROP image advanced v2.4 (Succasunna, NJ, USA) at ambient temperature. Water drops having a volume within the 2–5 μL range were created and deposited with Ramé-Hart automatic dispensing system. The contact angles were measured from 10 consecutive drops positioned at random sample locations. The values of the mean angle and the mean error were obtained based on them. The contact angle values allow us to determine the wettability of the investigated surfaces [19].

2.4. Chemical and Phase Compositions

The X-ray photoelectron spectroscopy (XPS) was used for the identification of the chemical composition and electronic structure of the films. The measurements were carried out on AXIS Supra electron-spectrometer (Kratos Analitycal Ltd., Manchester, UK) by the application of achromatic AlKα (photon energy of 1486.6 eV) having charge neutralization system. The binding energies (BEs) were calculated with an accuracy of ±0.1 eV, using the C1s line at 284.6 eV (due to adsorbed hydrocarbons). The areas and binding energies of C1s, O1s, Zn2p and Zr3d photoelectron peaks were monitored to evaluate the chemical composition of the films. Using the commercial data processing software ESCApe™, version 1.2.0.1325 of Kratos Analytical Ltd., Manchester, UK, the concentrations of the different chemical elements (in atomic %) were rated. The sample surfaces were studied after etching. It was performed on VG ESCALAB II electron spectrometer with Ar+ ions with 3 keV of energy, a current density of 10 $\mu A/cm^2$ and an etching rate of 2 nm/min.

2.5. XRD Analyses

The X-ray diffraction analysis allows us to define the phase composition of the samples by using an X-ray diffractometer (CuKα radiation; generator voltage 40 kV), equipped with a PW1830 generator and a PW1050 goniometer manufactured by Philips. Data were obtained within the angular range of 5–90° 2-theta with a step of 0.05° 2-theta and exposure of 3 s. The phase analysis was performed using the HighScore Plus 3.0 program, Inorganic Crystal Structure Database (ICSD) and Powder Diffraction File™ (PDF-2 2023).

2.6. Electrochemical Tests

Classical widespread electrochemical methods have been used with the aim to determine the anticorrosion properties of the sample systems and those of ordinary zinc. These are the potentiodynamic (PDP) polarization method (anodic and cathodic curves, potential range varying between −1.2 V and 0 V relative to SCE, scan rate 1 mV/s) as well as "open-circuit potential (OCP)"and polarization resistance (Rp) measurements. The latter was realized by applying a "Corrovit" unit ("Tacussel", Villeurbanne, France). These methods and their peculiarities were applied by the authors previously for other test metallic/alloy objects [16]. The investigations were carried out after 15 min stabilization of the corrosion potential in a three-electrode experimental cell having a volume of 300 mL. Saturated calomel electrode was the reference electrode and platinum wire was the counter electrode. Potentiostat Model VersaStat 4 PAR, AMETEK, Oak Ridge, TN, USA was applied to realize the PDP and OCP investigations.

2.7. Data Reproducibility and Corrosive Medium

Five specimens of every sample type were studied electrochemically in a model corrosive test medium (5% NaCl solution).

3. Results

3.1. SEM and EDX Measurements

The SEM photograph of Zn–Ni alloy coating reveals homogeneous fine-grained surface morphology, while the Zn–Co alloy possesses a smooth surface, covered with some micro-crack zones of different shapes and widths (Figure 1). The EDX analyses were performed on the previously etched samples (the procedure is described in Section 2.4). Figures 2 and 3 reveal that the surfaces of the ZN and ZC systems repeat in detail the morphological features of the corresponding sub-layer. The results, obtained via SEM analyses, clearly show that the nature of the zinc alloy sub-layer significantly influences the morphology of ZrO_2 top-layer, which is a result most probably arising from the different thermal expansion coefficients and the surface roughness of Fe, the Zn–Co and Zn–Ni alloys, as well as that of ZrO_2.

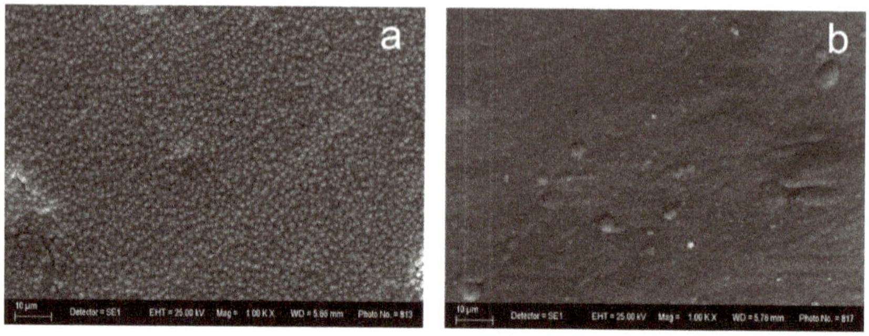

Figure 1. SEM analyses of the (**a**) Zn–Ni and (**b**) Zn–Co sub-layers (magnification 1000).

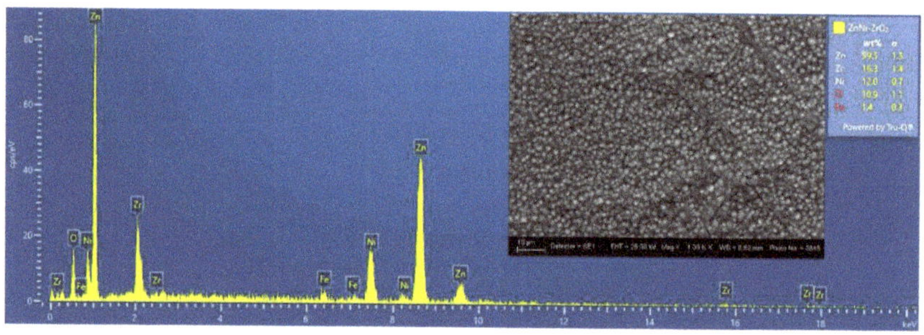

Figure 2. EDX/SEM analyses of the System ZN (magnification 1000).

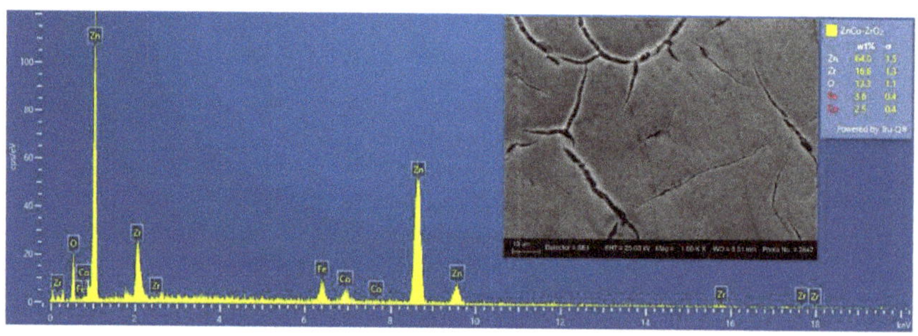

Figure 3. EDX/SEM analyses of the System ZC (magnification 1000).

3.2. *AFM Analyses and Surface Hydrophobicity*

Figure 4 represents the topography of the low-carbon steel (LCS) and galvanized steel (LCS/Zn) substrates for comparison to those of the ZN and ZC systems. The roughness values and corresponding water contact angles are shown in Table 1. The AFM images of Zn–Ni galvanized steel reveal a rough and grained surface, similar to those of steel and galvanized steel substrates (Figure 5). Figure 6 shows the dense topography without any visible pores, cracks, etc., of the System ZN after thermal treatment. The coating also has no visible detached parts from the substrate, which proves its good adhesion. The average roughness Ra is 105 nm, while Rq is 131 nm.

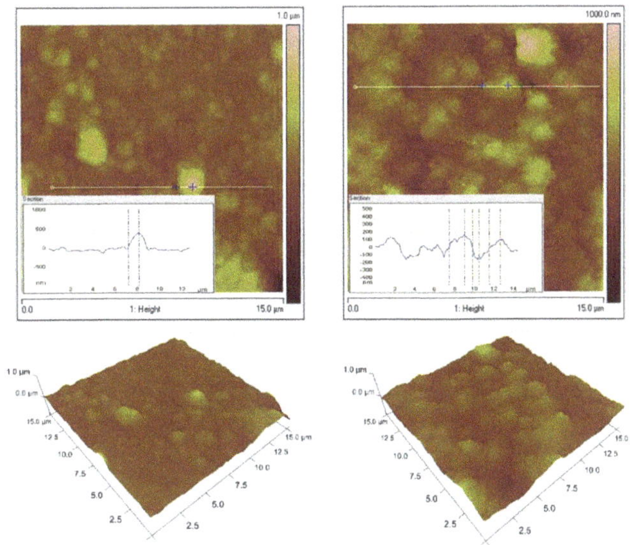

Figure 4. AFM 2D and 3D images of LCS (**left**); LCS/Zn (**right**).

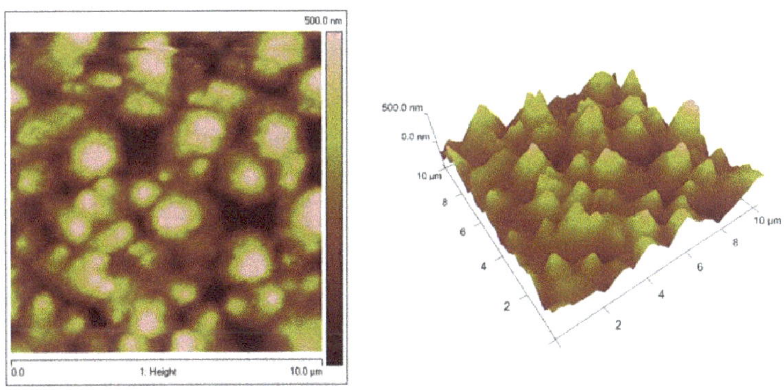

Figure 5. AFM (2D and 3D) images of Zn–Ni substrate.

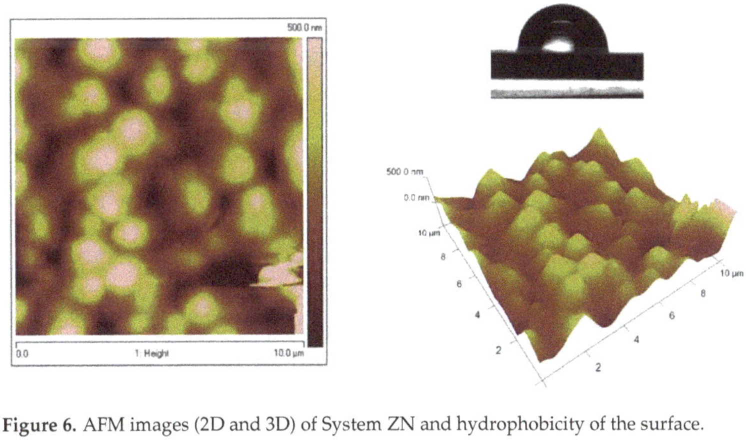

Figure 6. AFM images (2D and 3D) of System ZN and hydrophobicity of the surface.

The surfaces of Zn–Co substrate and the System ZC shows significantly smoother surface, which are demonstrated in Figures 7 and 8, respectively. These data confirm the conclusions, based on the results of the SEM analyses, that the type of the previously electrodeposited zinc alloy coating is responsible for the final morphology of the top ZrO_2 film. Table 1 represents the WCAs for System ZN, System ZC and different types of low carbon steel substrates.

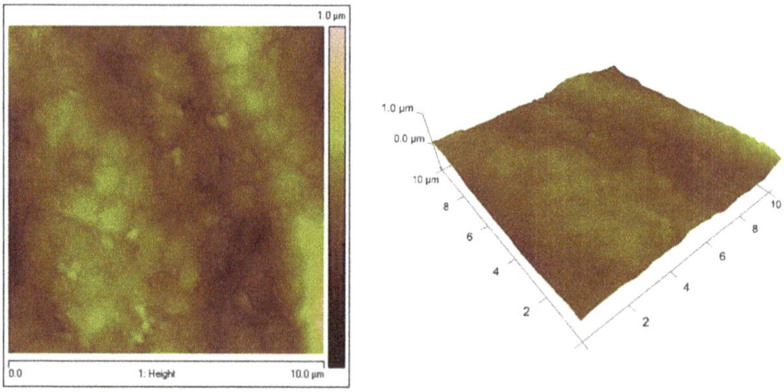

Figure 7. AFM images (2D and 3D) of Zn–Co substrate.

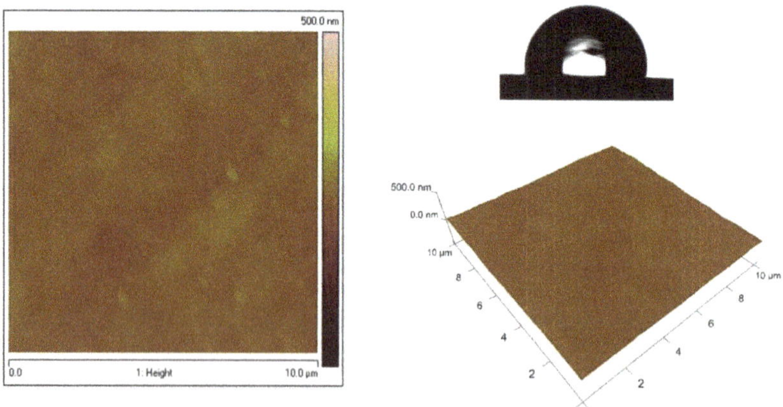

Figure 8. AFM images (2D and 3D) of System ZC and hydrophobicity of the surface.

Table 1. Water contact angles for System ZN, System ZC and different types of low-carbon steel substrates (LCSs).

Sample	LCS	LCS /Zn	LCS /Zn–Ni	LCS /Zn–Co	LCS/Zn–Ni-ZrO_2 (ZN)	LCS/Zn–Co-ZrO_2 (ZC)
WCA (°)	92	122	97	95	93	90
Rp (nm)	59	69	68	117	131	9.5
Ra (nm)	46	56	53	92	105	7.4

The low-carbon steel substrate shows a typical hydrophobic surface character having a WCA value of 92°, while the steel, coated with electrodeposited Zn coating, has a much higher WCA value of $122 \pm 2°$. The modification of Zn with Ni or Co leads to a significantly decreased hydrophobicity. The deposition of ZrO_2 coating upon the Zn alloys and subsequent heat treatment maintains the hydrophobic nature of the surface of the final

ZN and ZC systems. The WCA values of the ZN and ZC systems are close to those of the corresponding sub-layers.

3.3. XRD Analyses

The System ZN containing nickel has a significantly simple phase composition (Figure 9). The co-deposition of nickel and zinc and subsequent annealing results in the formation of an intermetallic compound of composition Ni_2Zn_{11}. Any peaks, belonging to the ZrO_2 crystallographic phases, have not been identified, most probably due to the amorphous nature of the top inorganic coating. A uniform coating was formed on the surface of the System ZC, in which the diffraction peaks of the metallic zinc were wider than those normally obtained (Figure 10). This is a sign that most likely some of the cobalt was included in its lattice. On the other hand, the coating has a complex phase composition, which is associated with a partial oxidation of the surface. This leads to some additional formation of zinc oxide (ZnO) and cobalt oxide (Co_3O_4), which possess a low degree of crystallization. Some weakly expressed peaks between 2-theta 35 and 65° probably belong to a non-identified phase as a result of the interaction between the zirconia layer and the substrate constituents.

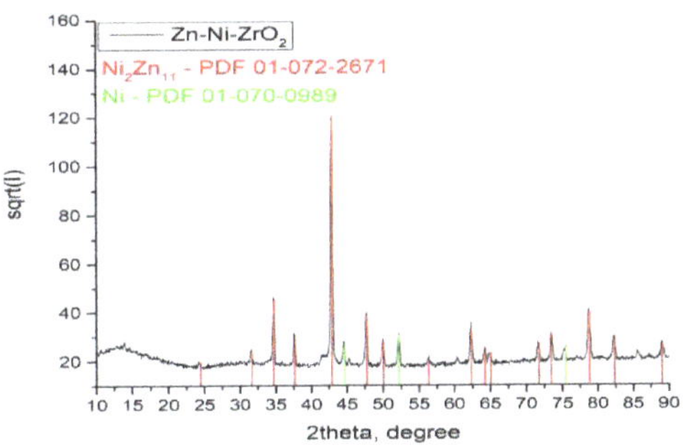

Figure 9. XRD of System ZN.

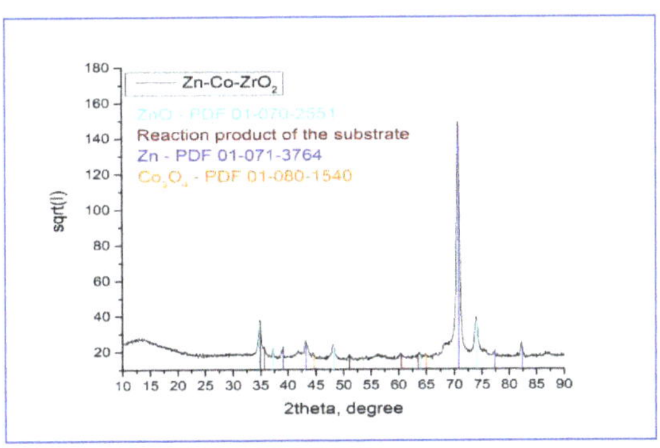

Figure 10. XRD of System ZC.

3.4. XPS Analyses

The XPS analysis represents that the chemical elements Zn, Zr, C, O and Na are registered on the surface of the layers. The Zr3d peaks (presented in Figure 11a,b) are intensive with spin–orbit splitting between $Zr3d_{5/2}$ and $Zr3d_{3/2}$ from 2.4 eV and this shows that zirconia exists in the form of two phases: ZrO_2 (182.2 eV) and ZrOH (183.6 eV) [20]. The shape of spectra and the binding energy of zinc at 1021.9 eV strongly suggest a 2^+ oxidation state (Figure 11c). For a more precise determination of the state of zinc, the modified Auger parameter was also calculated, which is 2009.2 eV, and it corresponds to

ZnO. The oxygen spectra are broad and asymmetric. A Lorentzian–Gaussian curve fitting into several components was applied in the oxygen spectra (Figure 12). The binding energy at ~529.8 eV corresponds to the O–Zr bond, while that at ~531.4 eV is attributed to O in the Zn crystal lattice. The last two components refer to the oxygen–carbon bond at ~532.7 eV and to the oxygen in the water molecule at ~534.0 eV (only for the ZN sample). The carbon spectra indicate that there is a layer of adventitious carbon contamination on the surface of the samples. Most samples, exposed to the atmosphere, have adventitious carbon on the surface. This statement is also proven by the components of the decomposed spectra of carbon, which are attributed to C–C, C–O–C and O–C=O bonds. Small amounts of sodium (0.7 at %) for ZN sample were registered, probably due to contamination.

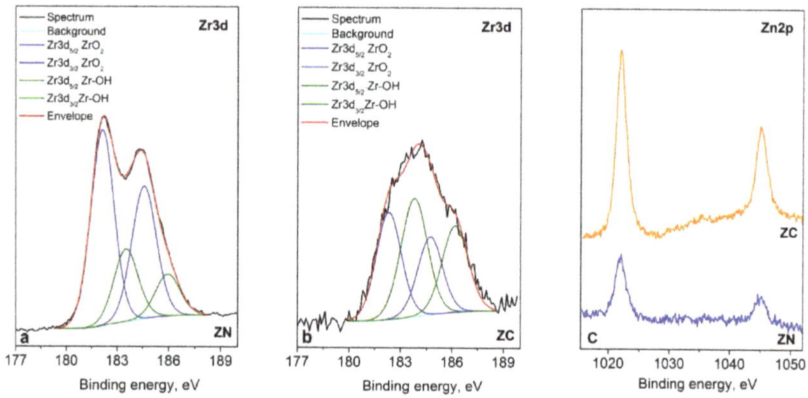

Figure 11. Deconvolution of Zr3d (**a**) ZN sample and (**b**) ZC sample and (**c**) Zn2p core level spectra of the systems.

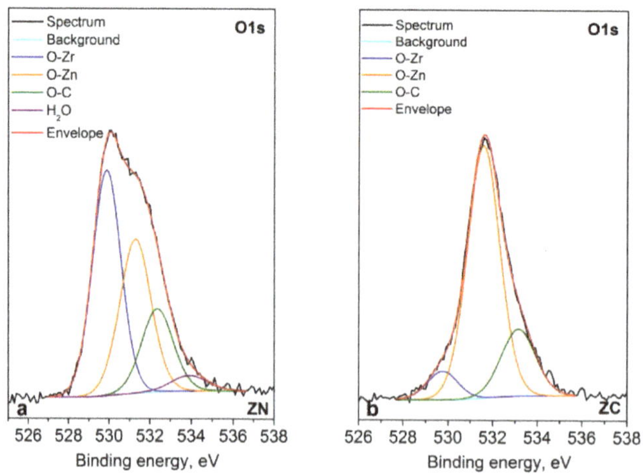

Figure 12. Deconvolution of O1s core level spectra of (**a**) ZN and (**b**) ZC systems.

3.5. Electrochemical Tests

3.5.1. Measurement of the Polarization Resistance

The polarization resistance in 5% NaCl solution, measured during a time period of 25 days for the ZC (Zn–Co/ZrO_2), ZN (Zn–Ni/ZrO_2) and the ordinary zinc coating systems can be observed in Figure 13. According to the figure, the *Rp* values of the ordinary zinc are the lowest ones (~800–900 ohms.cm^2) and they are very close in value. System ZC demonstrates an almost increasing tendency (except for the 10th day) and the highest peak of *Rp* values on the 20th day (~15,000 ohms.cm^2). This system reaches the polarization resistance value of about 13,000 ohms.cm^2 at the end of the test. System ZN represents the highest *Rp* value at the 10th day (~9500 ohms.cm^2), which is very close to the value of the same parameter of System ZC. The *Rp* values of System ZN decrease until the 20th day, after which some separate "red points" appear on its surface, indicating the presence of the so-called "red rust", i.e., the corrosion process has reached the low-carbon steel substrate and the latter has begun to dissolve as a result of the corrosion attack. Bearing in mind this behavior, the *Rp* measurement of the system is stopped. Contrary to this, both System ZC and the ordinary zinc do not show any "red rust" zones, which is a sign of their better protective characteristics during the test period in a corrosive environment.

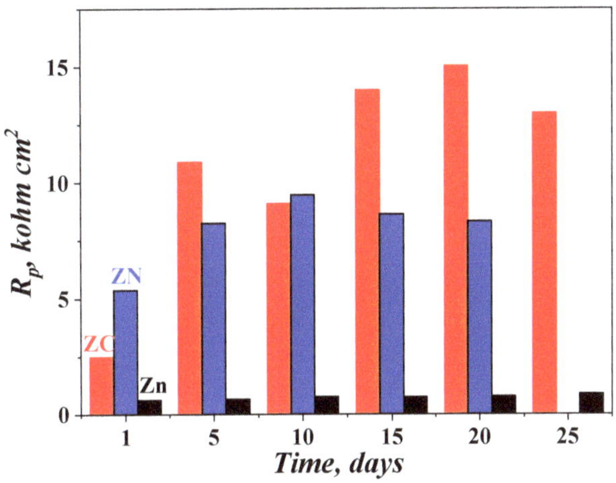

Figure 13. Polarization resistance of the investigated samples in 5% NaCl solution.

3.5.2. Potentiodynamic Polarization Curves of Fresh Samples

The PDP curves of the investigated usual zinc coating and both systems are shown in Figure 14. It can be observed that the corrosion potential of the ordinary zinc coating is manifested at a more negative value, compared to that of System ZC and System ZN. In addition, Zn represents a curve, which is very short in the anodic branch, compared to both systems. This observation means that this metal lasts a shorter time period at external anodic polarization in that medium. The anodic curves of System ZC and System ZN are very similar in their course, although some differences appear—System ZC shows a zone with lower current density values, for example, in the potential interval between −0.7 and 0.4 V. It can also be registered that the anodic part of the PDP curve of System ZN seems to be longer than that of System ZC. However, the anodic curve of System ZN is positioned at higher current densities after the potentials of −0.3 V, i.e., the anodic process is realized more intensively.

Table 2 presents the most important parameters of the PDP curves: corrosion potential E_{corr} and corrosion current density I_{corr}. The results indicate that the zinc sample has the most negative potential value compared to both other systems, which means that the latter are more noble. This can be expected when keeping in mind the nature of the alloying elements and the final surface oxide layer. It can also be concluded that the corrosion current densities of the systems are much lower compared to that of zinc, i.e., their corrosion resistance and protective ability are greater.

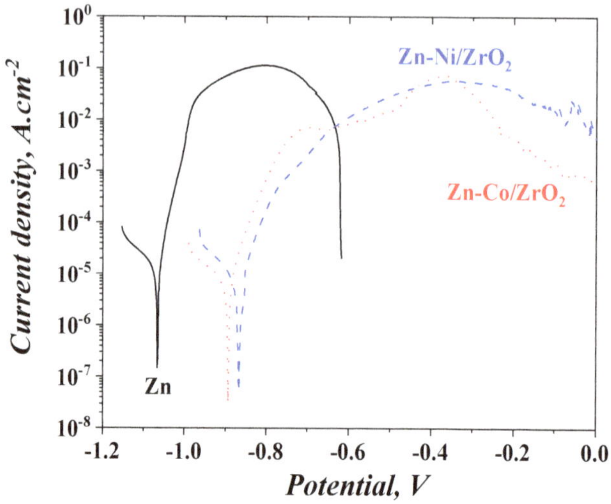

Figure 14. Potentiodynamic polarization curves of the fresh samples in 5% NaCl solution (vs. SCE).

Table 2. Electrochemical parameters of the PDP curves from Figure 14.

Sample	E_{corr}, V	I_{corr}, A.cm^{-2}
Zn	−1.065	1.8×10^{-5}
Zn–Co/ZrO_2	−0.895	6.5×10^{-6}
Zn–Ni/ZrO_2	−0.868	7.6×10^{-6}

3.5.3. Long Period Test (25 Days in Test Medium) of the Samples Presented by Potentiodynamic Polarization Curves

The corrosion characterization of the samples has also been carried out through the application of curves after the 25-day immersion in the corrosion medium (Figure 15 and Table 3). The corrosion potential (E_{corr}) of both systems ZC and ZN are additionally strongly shifted in a positive direction with about 140 mV, i.e., they become more noble. E_{corr} of the pure zinc is also positively shifted with approximately 40 mV (compare Figures 14 and 16). The reason for this result is most probably the changes in sample surface as a result of the corrosive treatment. As it is well known, the corrosion potential depends on the surface oxidation state and on the appearance of different corrosion products, and it can be expected that some differences will appear in its value. Compared to the results from Table 2, it can be concluded that the I_{corr} values of ZN and ZC systems become about four times lower, while that of the ordinary zinc remains the same. In the case of the ordinary zinc, the presence of some newly appeared corrosive products might affect the potential shift. When both systems are under corrosion attack, some other corrosion products are obtained on the surface, influencing the corrosion parameters. Another important reason could be the presence of cracks, pores or some surface non-homogeneities, which could ensure a faster access of the corrosive medium deep inside.

Table 3. Electrochemical parameters of the PDP curves from Figure 15.

Sample	E_{corr}, V	I_{corr}, A.cm^{-2}
Zn	−1.029	1.8×10^{-5}
Zn–Co/ZrO_2	−0.763	1.6×10^{-6}
Zn–Ni/ZrO_2	−0.748	1.7×10^{-6}

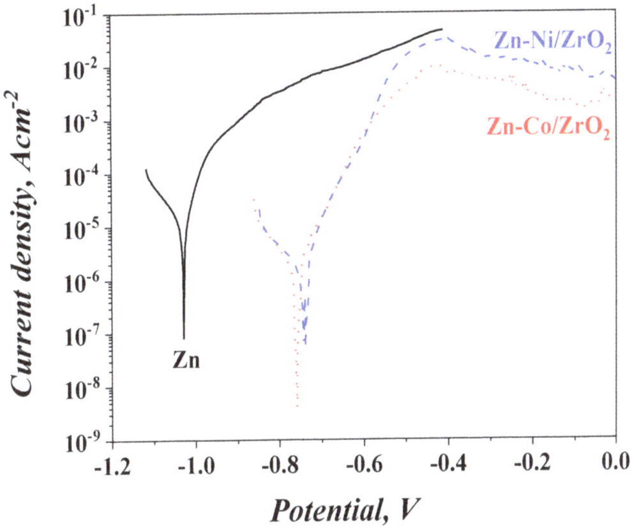

Figure 15. Potentiodynamic polarization curves of the investigated samples after 25 days of continuous immersion in 5% NaCl solution (vs. SCE).

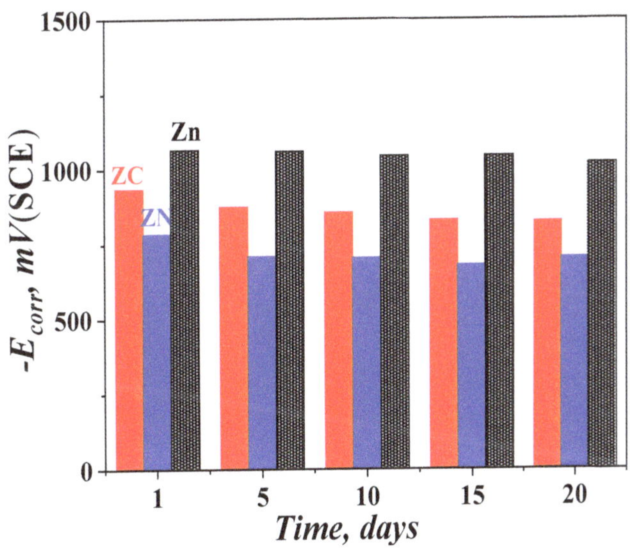

Figure 16. Open-circuit potential (OCP) values of the investigated samples after 25 days of immersion in 5% NaCl solution (vs. SCE).

The results proved that the influence of the surface oxide layer and of the subsequently newly appeared corrosion products positively affect the corrosion behavior and protective ability of the investigated systems in that test medium.

3.5.4. Open-Circuit Potential (OCP)

The results for the OCP values of the tested objects are shown in Figure 16. It is well visible that the values of the ordinary zinc are the most negative ones in comparison to those of ZC and ZN systems, which means that this coating plays a sacrificial role and will dissolve earlier, ensuring an appearance of zinc-based corrosion products. Contrary to this, the OCP values of both systems are more positive (noble), most probably due to the appearance of protective barrier layers of corrosive products. It could be supposed that the appearance of "red rust" for System ZN results from the presence of cracks and surface defects on the layer, leading to a faster corrosion attack deep inside.

4. Discussion

The electrochemical data, obtained during the experiments conducted with the application of both zinc-based alloys as sub-layers, show enhanced anticorrosion characteristics of the systems. For example, ZN and ZC demonstrate a positive potential shift of the corrosion potential (E_{corr}) compared to that of the usual zinc, which is proof for their better-expressed "nobility". This potential shift is observed both for the fresh samples and for the samples after prolonged treatment in the corrosion medium.

A similar tendency can be observed in view of the "open-circuit potential" measurements—the potential value of the zinc is more negative, compared to both systems during the whole investigation period of time. In addition, the corrosion current density values of ZN and ZC taken from the PDP curves are lower compared to that of the pure zinc.

Potentiodynamic curves of ZN and ZC are about 400 mV longer than that of the zinc, i.e., these systems last longer under conditions of external anodic polarization.

The polarization resistance measurements during a 25-day time period confirm in general the results from the accelerated tests. However, bearing in mind the obtained results, System ZN seems to be insufficiently resistant. One possible reason could be the presence of more cracks and pores, as well as a greater surface inhomogeneity, which makes the penetration of the chloride ions deep inside easier, leading to the appearance of some "red rust" zones and the deterioration of the protective characteristics.

Compared to the other investigations on this topic, which are scarce, the following conclusions could be summarized: (i) the other authors use spray pyrolysis to obtain the ZrO_2 layers onto aluminized steel with different content; and (ii) the surface morphology, for example, is different due to the characteristics of the applied method. Bearing all these peculiarities in mind, it seems to be very difficult to compare the results obtained for all systems. However, based on the experimental results, it can be concluded that System ZC demonstrates very good protective characteristics and it can be successfully applied for the protection of low-carbon steel in chloride-containing media.

It is known that several physical–chemical parameters can affect the anti-corrosion behavior of the barrier oxide coatings: surface morphology, hydrophobicity, the presence of specific crystalline phases and/or the presence of amorphous phase, the presence of defects, pores, etc.

The surface morphology of the zirconia layers in our case is influenced by the type of Zn contained in the sub-layer: the modification with Co of galvanized steel leads to the formation of smooth films. On the other hand, the presence of nickel-doped galvanized steel promotes the formation of a finely grained ZrO_2 surface. These results have also been proven by the estimated values of the average surface roughness (105 nm vs. 75 nm for ZN and ZC, respectively). Keeping in mind the obtained experimental results, it seems that the surface characteristics of the corrosion-resistant coatings do not have a well-expressed relationship with the corrosion stability of the investigated systems.

The phase composition and/or the presence of an amorphous phase of ZrO_2 coating is another important factor, influencing its corrosive protection properties. The XRD analyses have not registered any crystallographic phases of zirconia, so the ZrO_2 films obtained in both systems possess an amorphous nature. These findings were proven through the studies of other researchers. According to the results obtained in [21,22], the amorphous phase leads to a lower conductivity (ionic and electronic), thus retarding the electrochemical reactions on the surface. It has been proven that the good anti-corrosion properties of the ZrO_2 films are due to their high density and partially amorphous character [23]. In the case of ZC system, the appearance of new phases, consisting of low crystalline ZnO and Co_3O_4

oxides, can additionally improve the protective characteristics of the system in comparison to the System ZN. It could be assumed that the presence of these oxides can impede the penetration of the corrosion agents, thus slowing down the rate of the corrosion process due to their barrier properties.

Further, the surface hydrophobicity is also responsible for the high anti-corrosion efficiency of the systems. As a rule, it inhibits the corrosion process by limiting the access of surface corrosive species (water, halide ions, etc.) and interaction. Subsequently, the destruction of the metal oxide films becomes slower. Jothi et al. have proven that hydrophobicity (122°) of the surface improves the corrosion resistance. They have obtained polyurethane-Pd-ZrO_2 nanocomposites (in different combinations), possessing good anti-corrosion performance in a 3.5% NaCl medium [24]. On the other hand, our research group has revealed the absence of a direct correlation between the high contact angle of water and better anti-corrosion protection [25]. Huang and co-authors [26] also found that superhydrophobic TiO_2 nanotubes possess low corrosion resistance.

In our case, the lower contact angle of the ZN and ZC systems, compared to that of Zn-coated low-carbon steel, does not lead to the deterioration of the corrosion characteristics. It can be assumed that this is due to the presence of micro- or nano-pores, which are connected in subchannels. These subchannels enable the penetration of the aggressive agents (Cl anions) through the ZrO_2 coating. The zirconia layer has a densely amorphous structure, suggesting that some of these channels are also narrower than those of galvanized steel. This induces a limitation of the areas, subjected to a corrosive medium. Similar suppositions have been suggested by Yu et al., which can explain the anti-corrosion properties of a three-layer composite coating (containing TiO_2 and ZnO layers) [27]. It could be summarized that the synergism between the abovementioned physicochemical features of the surface ensures the good corrosion-protective properties of the systems presented in this article.

5. Conclusions

The results presented in the article have established that the newly obtained systems based on low-carbon steel covered with Zn–Co or Zn–Ni alloy (as sub-layer) and an amorphous hydrophobic ZrO_2 sol–gel layer (as the top layer) showed generally improved protective characteristics, compared to those of the ordinary zinc in a model medium with chlorine ions. The System ZC possesses a lower corrosion current density value ($1.6.10^{-6}$ $A.cm^{-2}$) and improved polarization resistance (13,000 $ohms.cm^2$) to those of ordinary zinc and System ZN, which is due to the formation of poorly crystallized oxides (ZnO and Co_3O_4, both of which play a role as a barrier toward corrosive agents).

The polarization resistance of both systems at the end of the 25-day prolonged test time period is much higher, compared to that of the usual zinc; however, some "red rust" zones appear on ZN (most probably due to the presence of cracks), which is a sign of lower protective characteristics. Contrary to this, System ZC reveals great potential for industrial applications, owing to both the good protective characteristics and the advantages of the sol–gel method: good adhesion to the metal substrates and a possibility to deposit layers on large surface areas with unique shape.

Author Contributions: Conceptualization, I.S., D.S., N.B. (Nelly Boshkova) and N.B. (Nikolai Boshkov); Methodology, S.S., N.G., G.A. and M.S.; Software, O.D., G.A., S.S. and N.G.; Validation, N.B. (Nikolai Boshkov); Formal analysis, M.S.; O.D.; Investigation, N.B. (Nelly Boshkova), N.B. (Nikolai Boshkov), D.S. and I.S.; Writing—original draft, I.S., D.S. and N.B. (Nikolai Boshkov); Writing—review and editing, I.S., D.S. and N.B. (Nikolai Boshkov); Visualization, O.D., S.S., M.S. and G.A.; Supervision, I.S., D.S. and N.B. (Nikolai Boshkov); Project administration, N.B. (Nelly Boshkova) and N.B. (Nikolai Boshkov). All authors have read and agreed to the published version of the manuscript.

Funding: The authors express their gratitude to the project European Regional Development Fund within the OP Science and Education for Smart Growth 2014–2020, Project CoE: National Centre for Mechatronics and Clean Technologies, No. BG05M2OP001-1.001-0008 and the "Scientific Investigations Fund", Bulgaria, project number KP-06-N37/16 (КП-06-Н37/16), "New environmentally friendly one and multi-coatings for corrosion protection of structural materials with wide application" for the financial support and for the possibility to publish the obtained results.

Institutional Review Board Statement: Not applicable.

Informed Consent Statement: Not applicable.

Data Availability Statement: Data are contained within the article.

Acknowledgments: The authors express their gratitude for support of the contract "Environmentally friendly anticorrosive hybrid coatings" is within the Non-currency Equivalent Exchange Bilateral Cooperation between the Bulgarian Academy of Sciences and the Serbian Academy of Sciences and Fine Arts.

Conflicts of Interest: The authors declare no conflict of interest.

References

1. Lovchinov, K.; Gergova, R.; Alexieva, G. Structural, Morphological and Optical Properties of Nanostructured ZrO_2 Films Obtained by an Electrochemical Process at Different Deposition Temperature. *Coatings* **2022**, *12*, 972. [CrossRef]
2. Stambolova, I.; Dimitrov, O.; Vassilev, S.; Yordanov, S.; Blaskov, V.; Boshkov, N.; Shipochka, M. Preparation of newly developed CeO_2/ZrO_2 multilayers: Efffect of tthe treatment temperature on the structrure and corrosion performance of stainless steel. *J. Alloys Compd.* **2019**, *806*, 1357–1367. [CrossRef]
3. Shen, Z.; Liu, Z.; Mu, R.; He, L.; Liu, G. Y–Er–ZrO_2 thermal barrier coatings by EB-PVD: Thermal conductivity, thermal shock life and failure mechanism. *Appl. Surf. Sci. Adv.* **2021**, *3*, 100043. [CrossRef]
4. Rezek, J.; Vlček, J.; Houška, J.; Čapek, J.; Baroch, P. Enhancement of the deposition rate in reactive mid-frequency ac magnetron sputtering of hard and optically transparent ZrO_2 films. *Surf. Coat. Technol.* **2018**, *336*, 54–60. [CrossRef]
5. Bjormander, C. CVD deposition and characterization of coloured Al_2O_3/ZrO_2 multilayers. *Surf. Coat. Technol.* **2006**, *201*, 4032–4036. [CrossRef]
6. Mehar, S.; Sapate, S.G.; Vashishta, N.; Rathod, A.; Bagde, P. Tribological performance of plasma sprayed Al_2O_3-TiO_2-ZrO_2 ceramic coating. *Mater. Today Proc.* **2021**, *45*, 4737–4741. [CrossRef]
7. Garg, N.; Bera, S.; Mangamma, G.; Mittal, V.K.; Krishnan, R.; Velmurugan, S. Study of Fe_2O_3-ZrO_2 interface of ZrO_2 coating grown by hydrothermal process on stainless steel. *Surf. Coat. Technol.* **2014**, *258*, 597–604. [CrossRef]
8. Dimitrov, O.; Stambolova, I.; Vassilev, S.; Lazarova, K.; Simeonova, S. Surface and Optical Properties of Gd-Doped ZrO_2 Nano Films. *Mater. Proc.* **2021**, *4*, 4. [CrossRef]
9. Lopez-Ibanez, R.; Martin, F.; Ramos-Barrado, J.R.; Brucker, F.; Leinen, D. Oxide barrier coatings on steel strip by spray pyrolysis. *Surf. Coat. Technol.* **2004**, *188–189*, 675–683. [CrossRef]
10. Romero-Pareja, R.; Lopez-Ibanez, R.; Martin, F.; Ramos-Barrado, J.R.; Leinen, D. Corrosion behaviour of zirconia barrier coatings on galvanized steel. *Surf. Coat. Technol.* **2006**, *200*, 6606–6610. [CrossRef]
11. Romero-Pareja, R.; Martin, F.; Ramos-Barrado, J.R.; Leinen, D. Study of different inorganic oxide thin films as barrier coatings against the corrosion of galvanized steel. *Surf. Coat. Technol.* **2010**, *204*, 2060–2063. [CrossRef]
12. Lopez-Ibanez, R.; Martin, F.; Ramos-Barrado, J.R.; Leinen, D. Large area zirconia coatings on galvanized steel sheet. *Surf. Coat. Technol.* **2008**, *202*, 2408–2412. [CrossRef]
13. Ha, S.; Lee, H.; Lee, W.-Y.; Jang, B.; Kwon, H.-J.; Kim, K.; Jang, J. Effect of Annealing Environment on the Performance of Sol–Gel-Processed ZrO_2 RRAM. *Electronics* **2019**, *8*, 947. [CrossRef]
14. Koch, G.; Brongers, M.; Thomson, N.; Virmani, Y.; Payer, J. *Corrosion Cost and Preventive Strategies in the United States*; NACE International: Washington, DC, USA, 2016.
15. Zhao, G.; Zhang, W.; Zhao, M. Investigation of Metal Coated D32 Steel Corrosion in Marine Environment. *Intern. J. Electrochem. Sci.* **2022**, *17*, 220134. [CrossRef]
16. Boshkov, N.; Petrov, K.; Vitkova, S.; Nemska, S.; Raichevski, G. Composition of the corrosion products of galvanic coatings Zn-Co and their influence on the protective ability. *Surf. Coat. Technol.* **2002**, *157*, 2–3. [CrossRef]
17. Gao, F.; Mu, J.; Bi, Z.; Wang, S.; Li, Z. Recent advances of polyaniline composites in anticorrosive coatings: A review. *Prog. Org. Coat.* **2021**, *151*, 106071. [CrossRef]
18. Raviprabha, K.; Bhat, R.S. Corrosion inhibition of mild steel in 0.5 M HCL by substituted 1,3,4-oxadiazole. *Egypt. J. Pet.* **2023**, *32*, 1–10. [CrossRef]
19. Adamson, A.W.; Gast, A.P. *Physical Chemistry of Surfaces*, 6th ed.; John Wiley & Sons: New York, NY, USA, 1997; pp. 347–380.

20. Gondal, M.A.; Fasasi, T.A.; Baig, U.; Mekki, A. Effects of Oxidizing Media on the Composition, Morphology and Optical Properties of Colloidal Zirconium Oxide Nanoparticles Synthesized via Pulsed Laser Ablation in Liquid Technique. *J. Nanosci. Nanotechnol.* **2018**, *18*, 4030–4039. [CrossRef]
21. Kadhum, A.H.; Mohamad, A.B.; Hammed, L.A.; Al-Amiery, A.A.; San, N.H.; Musa, A.Y. Inhibition of mild steel corrosion in hydrochloric acid solution by new cumarin. *Materials* **2014**, *7*, 4335–4348. [CrossRef]
22. Ghasemi, T.; Shahrabi, A.A.; Oskuie, H.H.; Sanjabi, S. Effect of heat treatment on corrosion properties of sol–gel titania–ceria nanocomposite coating. *J. Alloys Compd.* **2010**, *504*, 237–242. [CrossRef]
23. Holgado, J.P.; Pérez-Sánchez, M.; Yubero, F.; Espinós, J.P.; González-Elipe, A.R. Corrosion resistant ZrO_2 thin films prepared at room temperature by ion beam chemical vapor deposition. *Surf. Coat. Technol.* **2002**, *151–152*, 449–453. [CrossRef]
24. Jothi, K.J.; Balachandran, S.K.; Mohanraj, K.; Prakash, N.; Subhasri, A.; Santhana Gopala Krishnan, P.; Palanivel, K. Fabrications of hybrid Polyurethane-Pd doped ZrO_2 smart carriers for self-healing high corrosion protective coatings. *Environ. Res.* **2022**, *211*, 113095. [CrossRef] [PubMed]
25. Stambolova, I.; Stoyanova, D.; Shipochka, M.; Boshkova, N.; Eliyas, A.; Simeonova, S.; Grozev, N.; Boshkov, N. Surface Morphological and Chemical Features of Anticorrosion ZrO_2–TiO_2 Coatings: Impact of Zirconium Precursor. *Coatings* **2021**, *11*, 703. [CrossRef]
26. Huang, Q.; Yang, Y.; Hu, R.; Lin, C.; Sun, L.; Vogler, E.A. Reduced platelet adhesion and improved corrosion resistance of superhydrophobic TiO_2-nanotube-coated 316L stainless steel. *Colloids Surf. B* **2015**, *125*, 134–141. [CrossRef] [PubMed]
27. Yu, D.; Tian, J.; Dai, J.; Wang, X. Corrosion resistance of three layer superhydrophobic composite coating on carbon steel in seawater. *Electrochim. Acta* **2013**, *97*, 409–419. [CrossRef]

materials

Article

Naphthalene Monoimides with *Peri*-Annulated Disulfide Bridge—Synthesis and Electrochemical Redox Activity

Monika Mutovska [1], Natali Simeonova [1], Stanimir Stoyanov [1], Yulian Zagranyarski [1,*], Silva Stanchovska [2] and Delyana Marinova [2,*]

[1] Faculty of Chemistry and Pharmacy, Sofia University "St. Kliment Ohridski", 1164 Sofia, Bulgaria; ohmgm@chem.uni-sofia.bg (M.M.); natali.n.simeonova@gmail.com (N.S.); sstoyanov@chem.uni-sofia.bg (S.S.)
[2] Institute of General and Inorganic Chemistry, Bulgarian Academy of Sciences, 1113 Sofia, Bulgaria; stanchovska@svr.igic.bas.bg
* Correspondence: ohjz@chem.uni-sofia.bg (Y.Z.); manasieva@svr.igic.bas.bg (D.M.)

Abstract: Nowadays, organosulfur compounds provide new options in the development of full organic ion batteries. However, many drawbacks (such as kinetics limitations during the reversible oxidation of disulfides with cleavage of S–S bond, as well as solubility in non-aqueous electrolytes) make their commercialization difficult. Herein, a new concept for the design of organosulfur compounds with regulated redox properties and limited solubility is proposed. As a proof-of-concept, we designed *peri*-disulfo-substituted 1,8-naphthalimide derivatives, in which the alkyl chain length and halogen substituents (Cl or Br) at positions 3 and 6 are varied. The compounds were synthesized by an originally developed procedure starting from tetrahalonaphthalic anhydride via nucleophilic substitution at both *peri*-positions in the respective imide. Using ionic liquid electrolyte, it was found that the new *peri*-dithiolo-1,8-naphthalimides can participate in *n*- and *p*-type redox reactions at about 2.0 V and above 4.0 V vs. Li/Li^+, respectively. The redox potentials are sensitive mainly to whether Cl or Br substituents are available in the molecule architecture, while the alkyl chain length determines the kinetics of the redox reactions. Among all compounds, the chloro-substituted compound with the shorter alkyl chain displays the best kinetics for both low- and high-voltage redox reactions.

Keywords: organosulfur electrode materials; *n*- and *p*-type redox reactions; dithiole heterocycle annulation; 1,8-naphthalimide derivatives

Citation: Mutovska, M.; Simeonova, N.; Stoyanov, S.; Zagranyarski, Y.; Stanchovska, S.; Marinova, D. Naphthalene Monoimides with *Peri*-Annulated Disulfide Bridge—Synthesis and Electrochemical Redox Activity. *Materials* **2023**, *16*, 7471. https://doi.org/10.3390/ma16237471

Academic Editor: Wolfgang Linert

Received: 7 November 2023
Revised: 25 November 2023
Accepted: 28 November 2023
Published: 1 December 2023

Correction Statement: This article has been republished with a minor change. The change does not affect the scientific content of the article and further details are available within the backmatter of the website version of this article.

1. Introduction

Lithium-ion batteries (LiIBs) are nowadays indispensable devices for mobile energy storage [1,2]. They function thanks to the reversible redox reactions of Li^+ ions with the two electrodes [1–3]. Although LiIBs offer the highest energy density, they are in disconformity with strong environmental requirements since the key ingredient of electrode materials is the toxic and expensive Co element [4,5]. The cobalt content in widely used batteries (i.e., those worked with layered $LiCoO_2$ and $LiCo_xNi_yMnzO_2$ electrodes) varies in the range of 0.05–0.37 kg Co per kW h [6]. Depending on how the spent batteries are disposed, Co can leach out from batteries as ions, which are easily accumulated in natural waters, lands, plants and crops, thus provoking problems for human health and all living organisms [7]. The toxicity of Co also encompasses the release of oxide nanomaterials (such as layered $LiCoO_2$ and $LiCo_xNi_yMn_zO_2$ electrodes), which become dangerous for humans and wildlife if they are breathed [7].

In searching for greener electrodes without compromising their energy density, redox organic materials emerge as an alternative to conventional inorganic materials due to their greater variety of compositions and redox reaction mechanisms, a finer structure design and easier accommodation of structural stresses occurring during the redox reactions, a weaker dependence of the redox properties from the ionic sizes, and charges of the metal ions participating as oxidizers or reducers [8,9].

Among organic materials, organopolysulfide compounds can be singled out because of their unique redox properties [10–12]. The reversible redox reaction of Li^+ with organopolysulfide compounds has been shown to proceed with cleavage and subsequent formation of the sulfur–sulfur (S–S) bond: RS–SR + $2Li^+$ + $2e^-$ ↔ 2RSLi [13]. Several types of organic groups bonded to sulfur bridge have been examined, starting from aromatic molecules reaching to polymers [14,15]. Irrespective of the variety of organosulfur compositions (i.e., in terms of the linkage of the sulfur with organic groups and the number of sulfur chain [10]), the two-electron redox reaction of Li^+ with the (S–S) bond enables one to reach a high specific capacity, but at the expense of the slow kinetics. Furthermore, the organic electrode materials suffer from their relatively high dissolution in the electrolyte, as well as their instability during electrochemical reaction [13]. There are two approaches to overcoming these drawbacks: (i) functionalization or doping of organosulfur compounds; and (ii) finding a suitable electrolyte. Through the functionalization of phenyl disulfide on CNT, both kinetics and reversibility of the redox reaction are significantly enhanced [16]. The addition of *N*-heterocycles into organosulfur compounds enables a better control on the redox properties; for example, after the substitution of two carbons with N atoms in dipyridyl disulfide, the discharge voltage increases from 2.2 to 2.45 V together with the improvement of the cycling stability [17]. The most utilized electrolytes encompass liquids ones, such as solutions of the given lithium salt (predominately $LiClO_4$ or LiTFSI) into carbonate-based solvent [13].

Although the redox reaction of the polysulfide moieties proceeds with the participation of Li^+ ions (i.e., *n*-type redox reaction), the organosulfur compounds display a *p*-type redox reactivity, according to which the organic fragment is first oxidized and is then bound to electrolyte anion. The main representatives of the *p*-type compounds are the thioethers [13]. In general, the *p*-type redox reaction is usually accomplished at higher redox potentials than that of the *n*-type. Because of the different mechanisms, both *n*- and *p*-type reactions are characterized with different kinetics and capacity. However, the main drawback concerning the solubility of organosulfur compounds in electrolytes remains for *p*- and *n*-type compounds. Therefore, the state-of-the-art studies have mainly focused on finding organosulfur compounds with improved properties.

Among the classes of functional organic compounds, the 1,8-naphthalen monoimides (NMIs) and diimides (NDIs) have attracted substantial scientific interest due to their excellent chemo- and photostability, accessibility, and relatively easy derivatization capabilities. Core-substituted NDIs have found numerous applications in organic electronics and electrochemistry, for example, as field effect transistors [18] and new cathode materials for LiIBs [19]. Typically, they undergo reversible redox reactions at $E_{1/2} = -1.10$ V and $E_{1/2} = -1.51$ V vs. Fc/Fc+ (or $E_{1/2} = 2.32$ V and $E_{1/2} = 1.91$ V expressing in Li/Li^+ scale) [20]. The presence of electron-withdrawing imide fragments in the NMI structure allows for straightforward substitution at the *peri*-positions (4 and 5) with various nucleophiles, thus forming donor–π–acceptor type systems with interesting electronic properties due to the intramolecular charge transfer (ICT) in their molecules. The push–pull effect leads to a drastic decrease in the HOMO-LUMO transition energy, manifested in a strong bathochromic shift of the absorption and fluorescence bands with respect to the unsubstituted NMIs. The desired photophysical properties allowed for their successful application as OLEDs [21], optical sensors [22], fluorescent cellular imaging agents, and DNA targeting binders [23], etc. Concerning electrochemical energy storage, the organic charge transfer complexes have been proposed as a new approach for improving the electrical conductivity and dissolution of redox-active organic compounds [24].

This study aims to propose a new concept for design of organosulfur compounds with regulated redox properties and limited solubility. As a proof-of-concept, we designed *peri*-disulfo-substituted 1,8-naphthalimide derivatives, in which the alkyl chain length and halogen substituents (Cl or Br) at positions 3 and 6 are varied. Thus, designed *peri*-dithiolo annulated 1,8-naphthalimides are compared with previously reported organosulfur compounds (Figure 1). It is worth mentioning that the redox properties of complexes of iron with naphthalene monoimide derivatives of *peri*-substituted dichalcogenides have been already examined in aqueous solutions [25]. As far as we know, there are no data on the redox properties of *peri*-dithiolo-1,8-naphthalimide compounds in non-aqueous lithium electrolytes. Therefore, the first part of the study is focused on the synthesis of these compounds. The second part of the study covers their redox properties, which are examined in model Li-ion cells using ionic liquid as an electrolyte. The used methodology allows us to monitor the effects of the halogen substituents and alkyl chain length on the redox properties of the new *peri*-dithiolo-1,8-naphthalimides.

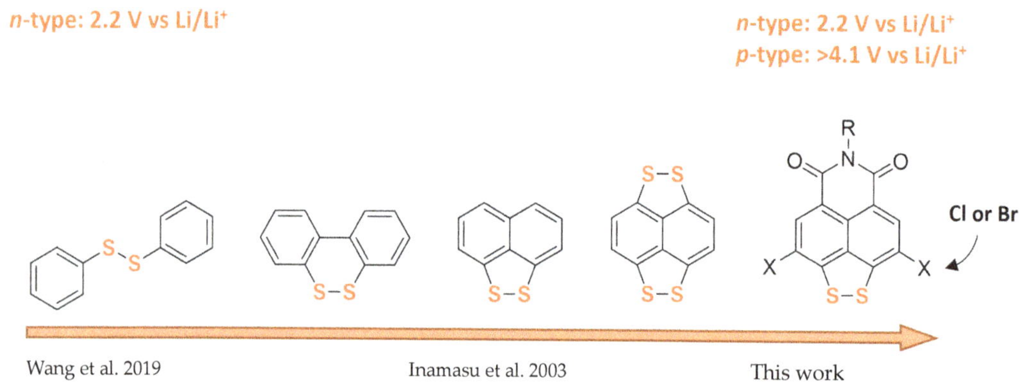

Figure 1. Evolution of disulfide cathode materials [17,26].

2. Materials and Methods

All starting materials (excepting the synthesized by us) and solvents were commercially available and used without additional purification (Merck (Darmstadt, Germany), Fisher Scientific (Hampton, NH, USA), Fluorochem (Glossop, UK), Sigma-Aldrich (St. Louis, MO, USA)). Thin-layer chromatography was used to monitor the progress of all reactions (Macherey-Nagel F 254 silica gel sheet, Macherey-Nagel, Duren, Germany) using appropriate mixture of solvents as eluent (described for each compound in the synthetic procedures). Column chromatography on silica gel (Macherey Nagel, 0.063 mm–0.200 mm) was used for purification. NMR spectra were recorded on a Bruker Avance 500 MHz instrument (Bruker, Karlsruhe, Germany). Spectra (^{1}H, ^{13}C{1H}) were referenced to appropriate residual solvent signals ($CDCl_3$, $C_2D_2Cl_4$). Elemental analyses were carried out on a Leco CHNS-932 (Leco Europe, Geleen, The Netherlands).

2.1. Synthesis

2.1.1. Synthesis of (*N*-octyl)-3,4,5,6-tetrahalo-1,8-naphthalimides **3** and **4**

General procedure: The corresponding 1,8-naphthalic anhydride **1** or **2** (50.0 mmol) was suspended in a mixture of 150 mL NMP and 150 mL acetic acid. After *n*-octylamine (1.5 eq, 75.0 mmol, 9.70 g) was added, the resulting mixture was stirred for 1 h at 110 °C then cooled down slowly to room temperature and afterwards poured into ice. The formed precipitate was filtered, washed thoroughly with water, and dried. The crude product was purified by column chromatography on silica (hexane/dichloromethane as eluent) to afford the target compounds as a slightly yellowish solid.

Synthesis of (*N*-octyl)-3,4,5,6-tetrachloro-1,8-naphthalimide **3**

Yield after column chromatography 20.80 g (93%).

^{1}H-NMR (δ (ppm), $CDCl_3$): 0.87 (t, 3H, CH_3, J = 7.0 Hz); 1.23–1.42 (m, 10H, CH_2); 1.69 (p, 2H, CH_2, J = 7.5 Hz); 4.11–4.14 (m, 2H); 8.65 (s, 2H).

^{13}C-NMR (δ (ppm), $CDCl_3$): 14.23; 22.77; 27.16; 28.04; 29.31; 29.31; 29.40; 31.39; 41.15; 122.43; 127.75; 128.96; 132.94; 135.37; 137.14; 161.96.

Anal. calcd. $C_{20}H_{19}Cl_4NO_2$: C, 53.72; H, 4.28; N, 3.13; Found: C, 53.55; H, 4.01; N, 3.27.

Synthesis of (*N*-octyl)-3,4,5,6-tetrabromo-1,8-naphthalimide **4**

Yield after column chromatography 28.12 g (90%).

^{1}H-NMR (δ (ppm), $CDCl_3$): 0.87 (t, 3H, CH_3, J = 6.9 Hz); 1.27–1.41 (m, 10H, CH_2); 1.69 (p, 2H, CH_2, J = 7.5 Hz); 4.10–4.13 (m, 2H); 8.78 (s, 2H).

^{13}C-NMR (δ (ppm), $CDCl_3$): 14.23; 22.77; 27.15; 28.04; 29.31; 29.40; 31.39; 41.12; 122.91; 127.87; 129.59; 131.38; 132.73; 135.78; 162.03.

Anal. calcd. $C_{20}H_{19}Br_4NO_2$: C, 38.44; H, 3.06; N, 2.24; Found: C, 38.31; H, 2.99; N, 1.95.

2.1.2. Synthesis of (*N*-butyl)-3,4,5,6-tetrahalo-1,8-naphthalimides **5** and **6**

General procedure: The corresponding 1,8-naphthalic anhydride **1** or **2** (50.0 mmol) was suspended in a mixture of 200 mL NMP and 200 mL acetic acid. After *n*-butylamine (1.5 eq, 75.0 mmol, 5.49 g) was added the resulting mixture was stirred for 2 h at 110 °C then cooled down slowly to room temperature and afterwards poured into ice. The formed precipitate was filtered, washed thoroughly with water, and dried. The crude product was purified by column chromatography on silica (hexane/dichloromethane as eluent) to afford the target compounds as pale gray solid.

Synthesis of (*N*-butyl)-3,4,5,6-tetrachloro-1,8-naphthalimide **5**

Yield after column chromatography 17.79 g (91%).

^{1}H-NMR (δ (ppm), $CDCl_3$): 0.97 (t, 3H, CH_3, J = 7.4 Hz); 1.42 (h, 2H, CH_2, J = 7.4 Hz); 1.69 (p, 2H, CH_2, J = 7.6 Hz); 4.12–4.15 (m, 2H); 8.65 (s, 2H).

^{13}C-NMR (δ (ppm), $CDCl_3$): 13.91; 20.41; 30.10; 40.89; 122.42; 127.74; 128.95; 132.94; 135.38; 137.14; 161.97.

Anal. calcd. $C_{16}H_{11}Cl_4NO_2$: C, 49.14; H, 2.84; N, 3.58; Found: C, 48.88; H, 2.61; N, 3.29.

Synthesis of (*N*-butyl)-3,4,5,6-tetrabromo-1,8-naphthalimide **6**

Yield after column chromatography 25.03 g (88%).

^{1}H-NMR (δ (ppm), $CDCl_3$): 0.97 (t, 3H, CH_3, J = 7.3 Hz); 1.42 (h, 2H, CH_2, J = 7.4 Hz); 1.68 (p, 2H, CH_2, J = 7.6 Hz); 4.12–4.15 (m, 2H); 8.79 (s, 2H).

^{13}C-NMR (δ (ppm), $CDCl_3$): 13.91; 20.41; 30.12; 40.87; 122.91; 127.88; 129.61; 131.38; 132.75; 135.79; 162.06.

Anal. calcd. $C_{16}H_{11}Br_4NO_2$: C, 33.78; H, 1.95; N, 2.46; Found: C, 33.98; H, 1.63; N, 2.45.

2.1.3. Synthesis of 6-Alkyl-3,9-dihalo-5*H*-[1,2]dithiolo[3′,4′,5′:4,5]naphtho[1,8-cd]pyridine-5,7(6*H*)-diones **7–10**

General procedure: A mixture of the corresponding 3,4,5,6-tetrahalo-1,8-naphthalimide (10.0 mmol) and sulfur (2 eq, 40.0 mmol, 1.28 g) in 50 mL NMP was stirred for 4–5 h at 175 °C. The mixture was slowly cooled down to room temperature and poured into ice. The formed precipitate was filtered, washed thoroughly with water, and dried. The crude product was purified via column chromatography on silica (hexane/dichloromethane as eluent) or recrystallized from dioxane to afford the target compounds as a yellow solid.

Synthesis of 3,9-Dichloro-6-octyl-5*H*-[1,2]dithiolo[3′,4′,5′:4,5]naphtho[1,8-cd]-pyridine-5,7(6*H*)-dione **SCl8**

Yield after column chromatography 4.10 g (93%).

^{1}H-NMR (δ (ppm), $CDCl_3$): 0.87 (t, 3H, CH_3, J = 6.8 Hz); 1.27–1.42 (m, 2H, CH_2); 1.68 (p, 2H, CH_2, J = 7.5 Hz); 4.11–4.14 (m, 2H, CH_2); 8.32 (s, 2H).

^{13}C-NMR (δ (ppm), $CDCl_3$): 14.24; 22.78; 27.25; 28.13; 29.36; 29.47; 31.95; 40.97; 119.67; 123.28; 126.47; 131.98; 134.59; 150.34; 162.12.

Anal. calcd. $C_{20}H_{19}Cl_2NO_2S_2$: C, 54.55; H, 4.35; N, 3.18; Found: C, 54.27; H, 4.18; N, 3.35.

Synthesis of 3,9-Dibromo-6-octyl-5*H*-[1,2]dithiolo[3′,4′,5′:4,5]naphtho[1,8-cd]-pyridine-5,7(6*H*)-dione **SBr8**

Yield after column chromatography 4.76 g (90%).

^{1}H-NMR (δ (ppm), $CDCl_3$): 0.87 (t, 3H, CH_3, J = 6.8 Hz); 1.23–1.42 (m, 2H, CH_2); 1.68 (p, 2H, CH_2, J = 7.5 Hz); 4.11–4.14 (m, 2H, CH_2); 8.43 (s, 2H).

^{13}C-NMR (δ (ppm), $CDCl_3$): 14.24; 22.79; 27.25; 28.13; 29.36; 29.47; 31.96; 40.96; 110.55; 119.45; 127.12; 133.39; 134.50; 153.49; 161.98.

Anal. calcd. $C_{20}H_{19}Br_2NO_2S_2$: C, 45.38; H, 3.62; N, 2.65; Found: C, 45.14; H, 3.88; N, 2.89.

Synthesis of 6-Butyl-3,9-dichloro-5*H*-[1,2]dithiolo[3′,4′,5′:4,5]naphtho[1,8-cd]-pyridine-5,7(6*H*)-dione **SCl4**

Yield after column chromatography 3.65 g (95%).

^{1}H-NMR (δ (ppm), $CDCl_3$): 0.98 (t, 3H, CH_3, J = 7.3 Hz); 1.42 (h, 2H, CH_2, J = 7.3 Hz); 1.67 (p, 2H, CH_2, J = 7.7 Hz); 4.13 (t, 2H, J = 7.5 Hz); 8.33 (s, 2H).

^{13}C-NMR (δ (ppm), $CDCl_3$): 13.85; 20.26; 29.90; 40.46; 119.20; 122.96; 126.21; 131.74; 134.38; 150.34; 161.89.

Anal. calcd. $C_{16}H_{11}Cl_2NO_2S_2$: C, 50.01; H, 2.89; N, 3.64; Found: C, 49.78; H, 2.99; N, 3.85.

Synthesis of 3,9-Dibromo-6-butyl-5*H*-[1,2]dithiolo[3′,4′,5′:4,5]naphtho[1,8-cd]-pyridine-5,7(6*H*)-dione **SBr4**

Yield after recrystallization 4.45 g (94%).

^{1}H-NMR (δ (ppm), $CDCl_3$): 0.97 (t, 3H, CH_3, J = 7.3 Hz); 1.42 (h, 2H, CH_2, J = 7.4 Hz); 1.67 (p, 2H, CH_2, J = 7.9 Hz); 4.13 (t, 2H, J = 7.6 Hz); 8.48 (s, 2H).

^{13}C-NMR (δ (ppm), $CDCl_3$): 14.85; 21.26; 30.91; 41.45; 111.29; 120.01; 127.93; 134.26; 135.31; 154.55; 162.80.

Anal. calcd. $C_{16}H_{11}Br_2NO_2S_2$: C, 40.61; H, 2.34; N, 2.96; Found: C, 40.38; H, 2.11; N, 3.17.

2.2. Electrochemical Characterization

The cycling voltammetry experiments were carried out using Swagelok-type three-electrode cells. The positive electrode consists of a mixture between *peri*-substituted disulfides (**SCl4**, **SCl8**, **SBr4**, or **SBr8**), 10% Super C65 (TIMCAL), and 10% sodium carboxymethyl cellulose (CMC) in a ratio 80-10-10 wt%. The slurry was spread on an aluminum foil, double side carbon-coated, and dried overnight at 80 °C. Disks with a diameter of 10 mm were cut and additionally dried under vacuum. The active mass loaded on Al collectors was about 2–3 mg. The electrolyte for the electrochemical tests was a 1 M LiTFSI-$Pyr_{1,3}$FSI (Lithium bis (trifluoromethanesulfonyl) imide in *N*-methyl, propyl pyrrolidinium bis (fluorosulfonyl) imide) 1:9 by ratio). This electrolyte is chosen over the conventional carbonate-based electrolyte (i.e., 1 M $LiPF_6$ in EC/DMC) due to the insolubility of the organosulfur compounds.

The assembly of the cells was carried out in an MB-Unilab glovebox model Pro SP (1500/780), with a low content of moisture and oxygen (under 0.1 ppm). The electrochemical reactions were performed in potentiostatic mode on a multi-channel potentiostat/galvanostat Biologic VMP-3e including impedance meter. The voltage window for cycling of model lithium half-cells was 1.5–5.0 V and scan rates were between 100 mV/s and 1 mV/s.

3. Results and Discussion

3.1. Design Concept

Despite being one of the most promising electrode materials and the subject of enormous scientific interest, disulfide compounds suffer from several unsolved problems: (i) Poor recombination (oxidation) of dithiolate ions back to disulfide due to particle diffusion or rotation in the cyclic compounds; (ii) the LUMO levels are usually too high, and the molecule does not undergo cleavage of the two-electron sulfide bond but rather a reversible one-electron reduction [27]; (iii) problems with the solubility of the electrode material [28,29].

To tackle these limitations, we designed the bridged *peri*-disilfo-substituted 1,8-naphthalimide core structure shown on Figure 2. The main advantage of this architecture is the fixed dichalcogenide bridge, offering a high degree of recombination and hence good electrochemical reversibility. Furthermore, the dithiole cycle is aromatic, which makes its reconstruction even more favorable. Another benefit is the presence of an imide acceptor fragment, perfectly positioned *peri* to the disulfide bridge, which should lead to a drastic decrease in the LUMO levels and hence a significantly easier progress of the reduction process. Additionally, it is possible to fine-tune the physical and mechanical properties (solubility, crystallization ability, thermo- and chemostability, etc.) by varying the type and length of the alkyl chain at one hand, as well as the electron-accepting properties of the naphthalimide core due to the presence of different halogen atoms in positions 3 and 6 on the other.

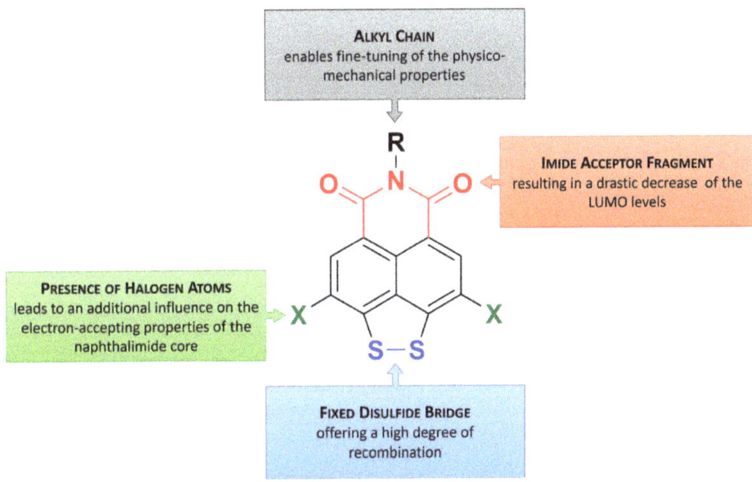

Figure 2. Design of target compounds.

Based on the design concept, the following target compounds were planned for synthesis (Figure 3).

3.2. Synthesis

Tetrahaloanhydrides **1** and **2** were identified as suitable starting compounds and were prepared according to a previously published procedure [25]. The first step was the synthesis of the corresponding naphthalimides **3**–**6** (Scheme 1). The higher nucleophilicity of the aliphatic amines and the good solubility of the tetrahaloanhydrides allowed the imidization to be carried out at relatively low temperatures compared to those previously described [25,30]. The imidization of **1** and **2** with *n*-octylamine was carried out in a mixture of *N*-methylpyrrolidone (NMP) and acetic acid at 110 °C for 1 h. After workup and purification via column chromatography, we isolated 3,4,5,6-tetrachloro-1,8-naphthalimide **3** and 3,4,5,6-tetrabromo-1,8-naphthalimide **4** with excellent yields (93% and 90%, respectively). Both *N*-(*n*-octyl) substituted imides are very soluble in organic solvents, making them suitable as starting compounds for the formation of a dithiole ring with the participation of the activated halogen atoms at positions 4 and 5.

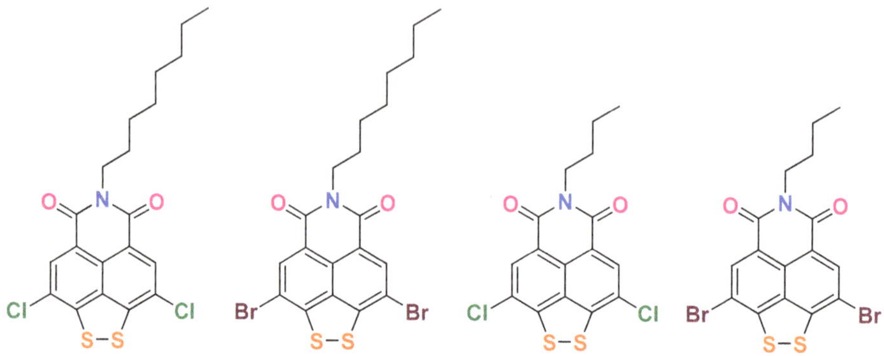

SCl8 **SBr8** **SCl4** **SBr4**

Figure 3. Structures of target compounds.

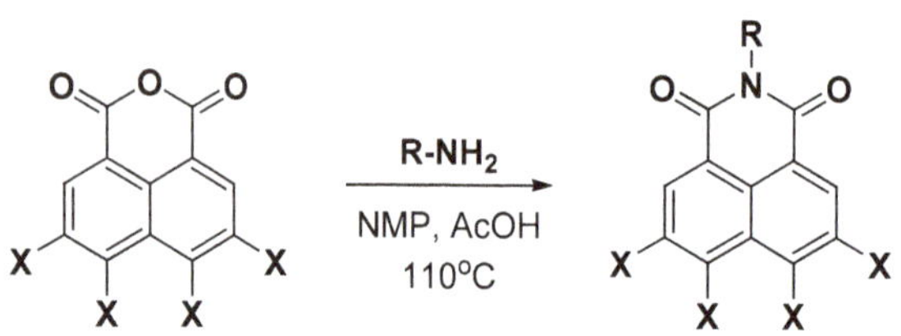

1-2 **3-6**

	R=	X=	Time, h	Yield, %
3	*n*-octyl	Cl	1	93
4	*n*-octyl	Br	1	90
5	*n*-butyl	Cl	2	91
6	*n*-butyl	Br	2	88

Scheme 1. Synthesis of imides **3–6**.

It is known that the chain length and branching are the two most important properties of the *N*-alkyl substituents in the naphthalimides, and they have a huge impact on their solubility. In general, they decrease dramatically when short and straight alkyl chains are used [31,32]. To test the solubility limits of the target disulfide compounds, we aimed to synthesize analogues of the *n*-octyl derivatives with shorter alkyl chain. For this purpose, we also reacted **1** and **2** with *n*-butylamine under similar conditions (Scheme 1) and successfully isolated the imides **5** and **6** in very high yields (91% and 88%, respectively), although in twice as long reaction time. After purification by column chromatography, imides **3–6** were characterized via ^{1}H and ^{13}C-NMR and elemental analysis. It is important to note that the crude imides **3–6** can also be easily purified by a simple reprecipitation procedure. The crude substances are dissolved in a minimal amount of dichloromethane, filtered, and reprecipitated with a large excess of methanol. Thus, they can be obtained in very high purity and used in the next step directly. Yields are reduced by only 3–5%, which makes this method very convenient.

The next step in our synthetic strategy was the formation of the new five-membered ring with two S-atoms bonded at positions 4 and 5 of the naphthalimides (Scheme 2). Based on our previous studies [25], we first attempted optimization of the synthetic procedure for this key step, both in terms of reaction conditions and scale. As a model reaction for dithiole heterocycle formation, we used imide **3**. The reaction yields were improved by slightly raising the temperature up to 175 °C and reducing the amount of sulfur to 2.5-fold compared to our previously reported similar reaction. This also resulted in significantly shorter reaction times, which were decreased by 40 to 50%. The reaction was further tested with respect to the use of an inert atmosphere, and the yield was found to practically not change when carried out under air or argon. To test the scale-up potential, after the optimization the reaction was carried out in 10, 20, and 40 mmol scales, yields of 93, 92, and 87% were achieved, respectively, which suggests promising scalability, with just a slight decrease of 1 to 6%, when increasing the scale by a factor of two and four, respectively.

	R=	X=	Yield, %
SCl8	*n*-octyl	Cl	93
SBr8	*n*-octyl	Br	90
SCl4	*n*-butyl	Cl	95
SBr4	*n*-butyl	Br	94

Scheme 2. Synthesis of dithiolo-annulated naphthalimides.

The optimized conditions were then applied to the rest of the tetrahalogen-substituted imides also. The reaction between **4** and sulfur afforded the final product, **SBr8**, in high yield (90%) on a gram scale. The *N*-(*n*-butyl)-substituted imides **5** and **6** also reacted with sulfur under the same conditions as their *n*-octyl analogues with even higher yields (Scheme 2). In this case, only imide **SCl4** had sufficient solubility to allow for purification via column chromatography. The other target product, **SBr4**, was purified via recrystallization from dioxane. All final products were obtained in very high yields and purity and have been characterized by means of ^{1}H, ^{13}C{1H}-NMR spectroscopic techniques, and elemental analysis (see Section 2 and Supplementary Materials).

3.3. Redox Properties of Peri-Substituted Dichalchogenides

Figure 4 compares the CV curves of *peri*-dithiolo-naphthalimides. The CV curves are recorded in a broad potential range (i.e., between 1.5 and 5.0 V) due to the stability of the electrolyte based on ionic liquids [33]. For all compositions, two redox bands can be distinguished in low- (i.e., below 2.5) and high-voltage (i.e., above 4.0 V) regions of the CV curves, respectively. The comparison of the CV curve profiles demonstrates that the low-voltage redox reaction proceeds more easily for compounds with shorter alkyl chain (**SCl4** and **SBr4**), while the longer alkyl chain gives rise to the high-voltage redox reaction. In comparison with the alkyl chain length, the presence of Cl and Br atoms at positions 3 and 6 predominantly affects the magnitude of the redox potential: going from Cl to Br substituents, the redox potential (expressed by $E_{1/2}$ measured at a scan rate of 20 mV/s) decreases from 4.26 V to 4.22 V and from 2.14 V to 2.10 V for **SCl4** and **SBr4**, while for **SCl8** and **SBr8**, the decrease is from 4.33 V to 4.20 V, respectively (the low-voltage band is not well resolved for these compounds).

Taking into account the previous data on the redox properties of organosulfur compounds [34], the low-voltage band can be associated with a redox reaction occurring between Li^+ ions and the disulfide moiety of the organic compounds. It is worth mentioning that for a simple organosulfur compound such as diphenyl disulfide, the potential of Li+ interaction is of 2.20 V [17], which approaches the experimentally determined one for **SCl4**. However, the possible participation of the attached –C=O groups in redox reaction with Li^+ cannot be excluded [35]: for example, the unsubstituted naphthalene diimides interact with 2 Li^+ at a potential of 2.55 V, which is much higher than that determined for **SCl4** (2.14 V). This signifies that the low potential reaction of **SCl4** with Li^+ takes place predominantly with the participation of disulfide bridge. According to the well-accepted classification of the organic redox compounds [36], the low-voltage reaction of *peri*-substituted compounds can be categorized as *n*-type ones.

Contrary to the low-voltage band, the high-voltage band could be attributed to the oxidation of the *peri*-substituted disulfides followed by bonding with anions from the electrolyte ($TFSI^-$). This reaction can be classified as a *p*-type. It is noticeable that for the *p*-type organic compounds, the highest potentials have so far been reported for phenazine-derived compounds, in which *N*-atoms are replaced by S and O: for example, 4.1 V for thianthrene and 4.2 V for dibenzodioxin, respectively [24,37]. The comparison shows that **SCl8** prepared by us displays even higher potential (i.e., $E_{1/2}$ = 4.3 V at a scan rate of 1 mV/s, Figure 4), which sets it apart from all previously reported *p*-type redox-active organic materials [36]. In general, taking into account the low- and high-potential redox reactions, the *peri*-substituted disulfides can be categorized as bipolar compounds exhibiting simultaneously n- and p-type interactions at around 2.1 and 4.3 V vs. Li/Li^+.

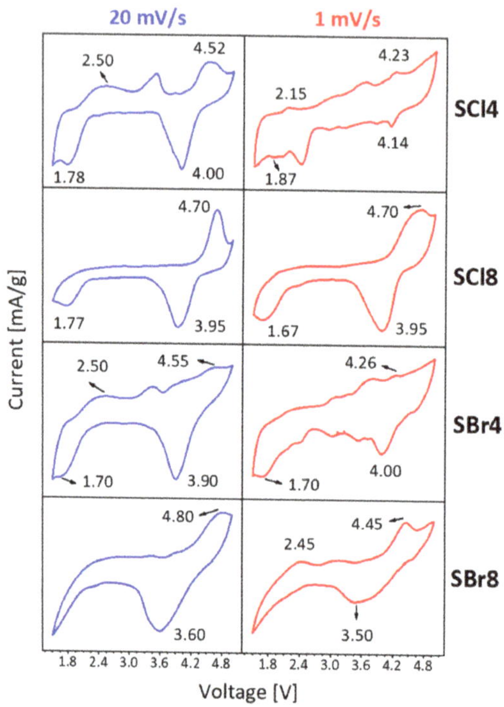

Figure 4. CV curves in a broad potential range (1.5–5.0 V) at a scanning rate of 20 mV/s (**left**) and 1 mV/s (**right**) for **SCl4**, **SCl8**, **SBr4**, and **SBr8**. As an electrolyte, ionic liquid comprising 1M LiTFSI-$Pyr_{1,3}$FSI solution is used.

To rationalize the different kinetics of low- and high-redox reactions, the CV curves are analyzed based on the dependence of the current (i) on the scan rate (v). By increasing the scan rate from 1 mV/s to 100 mV/s, the current of both low- and high-voltage bands increases obeying a $v^{1/2}$-dependence. This evidence the occurrence of diffusion-controlled reactions in low- and high-voltage regions. From the slope of the observed dependence ($di/dv^{1/2}$), the diffusion coefficient, D, can be estimated using the Randles–Ševčík equation [38]; at 25 °C, the D can be expressed by $D = [(di/dv^{1/2})/(2.69 \times 10^5\ n^{3/2}\ A\ C)]^2$, where n is the number of electrons, C reflects the charge concentration, and A is the electrode surface area. In order to avoid the uncertainties in defining the real values for n, C, and A parameters, we used the relative diffusion coefficient with an aim to compare more accurately the effect of the halogen substituents and the alkyl chain length on the redox reactions of *peri*-derived compounds: $D/D_{SCl4} = (di/dv^{1/2})^2/(di/dv^{1/2})^2_{SCL4}$, where D_{SCl4} is used as a reference value. The calculated values are given in Figure 4. Two features can be highlighted. First, the diffusion coefficient for high-voltage redox reaction exceeds more than two times that of the low-voltage reaction. This means that the high-voltage redox reaction (classified as p-type) proceeds faster than the low-voltage redox reaction classified as n-type. The established different kinetics for high- and low-voltage redox reactions are in good agreement with previous findings on p- and n-type organic electrode materials [36,39]. On the other hand, this supports once again the assignment of high-voltage redox reactions to oxidation of *peri*-substituted compounds followed by bonding with $TFSI^-$ electrolyte anions, while the low-voltage redox reaction comes from the reduction of the sulfur moiety, including reversible cleavage and formation of the S–S bond by Li^+ cations.

Furthermore, the comparison of data discloses that the highest diffusion coefficient is reached for the chloro-substituted compound with a short alkyl chain (Figure 5). In the high-voltage region, the diffusion coefficient decreases more than one order by extending the alkyl chain from 4 to 8. This trend is obeyed for the low-voltage redox reaction too (Figure 5). Both high- and low-voltage reactions exhibit slower kinetics when Cl atoms are replaced with Br atoms, this dependence is better expressed for the compounds with shorter alkyl chain. These interesting findings need further theoretical investigations.

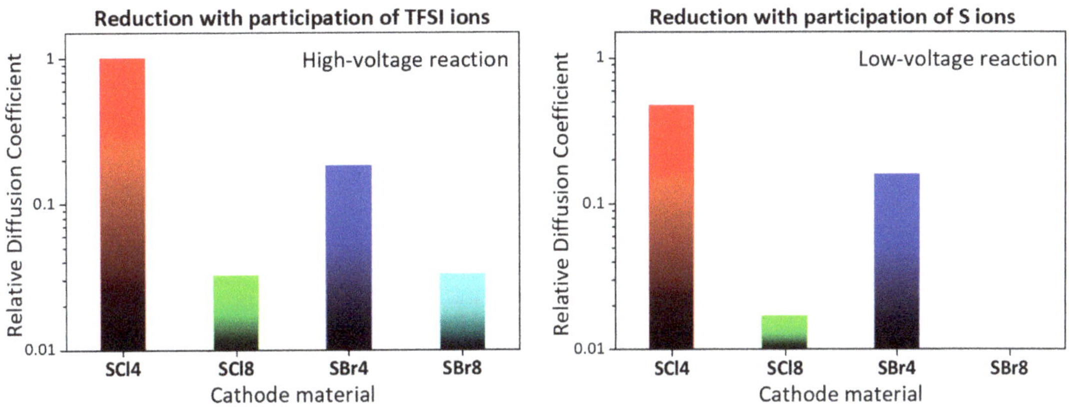

Figure 5. Relative diffusion coefficient for high- and low-voltage redox reactions ((**left**) and (**right**), respectively) determined from CV curves.

The next issue related to the redox properties of *peri*-substituted compounds is whether the CV curve profile depends on the potential limits where the redox reactions take place. Figure 6 compares the CV curve profiles of the **SCl4** recorded in two narrower potential regions: between 1.5 and 3.0 V and between 3.0 and 5.0 V. For the sake of better comparison, the CV curve profiles recorded in an extended potential range between 1.5 and 5.0 V are also given. As one can see, the high-voltage band (due to the *p*-type redox reaction) retains its shape and position in narrow and broad potential regions. It is worth mentioning that the high-voltage band becomes better resolved when the slower scan rate and narrower potential limits are used. In this case, the broad band is split into several peaks at 4.67/4.61 V, 4.30/4.12 V, and 3.87/3.87 V, thus suggesting a proceeding of multi-centre redox reaction. Contrary to the high-voltage redox reaction, the low-voltage reaction (due to the *n*-type interaction) is strongly dependent on the potential limits: in a narrow potential range between 1.5 and 3.0 V, one broad wave is more clearly resolved than the wave appearing after the recording a curve in an extended range of 1.5–5.0 V. This indicates that the *n*-type reaction undergoes some change (in terms of the activation of –C=O in addition to –S–S– redox interaction with Li^+) if the *p*-type redox reaction occurs. This change also reflects the reaction kinetics: the relative diffusion coefficient decreases by more than one order when the reaction proceeds in narrow potential range. However, the $E_{1/2}$ remains nearly the same: $E_{1/2}$ = 2.14 V, 2.17 V, 2.19 V, and 2.04 V for the potential limits of 1.5–5.0 V, 1.5–3.0 V, 1.5–3.75 V, and 1.0–3.75 V (Figure 6c).

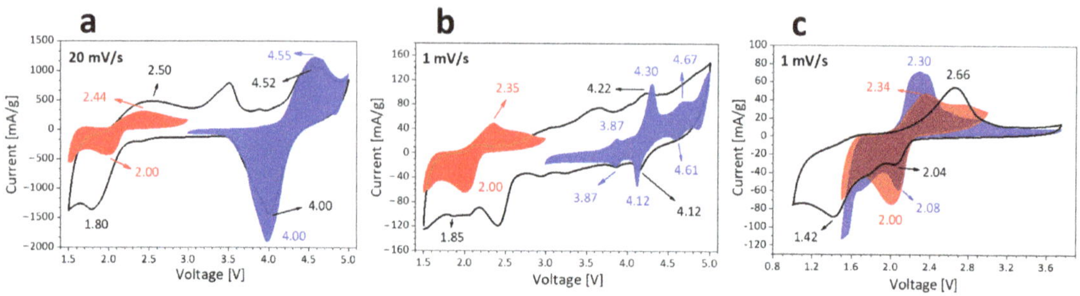

Figure 6. CV curves of **SCl4** recorded in low-voltage window between 1.5–3.0 V and high-voltage window between 3.0–5.0 V at a scan rate of 20 mV/s (**a**) and 1 mV/s (**b**). For the sake of comparison, the CV curve in an extended voltage window between 1.5 and5.0 V is also given. CV curves recorded in a potential range of 1.5–3.0 V (red), 1.5–3.75 V (blue), and 1.0–3.75 V (black lines) (**c**).

4. Conclusions

This study proposes a new concept for designing organosulfur compounds with regulated redox properties. As a proof-of-concept, we designed the *peri*-disulfo-substituted 1,8-naphthalimide derivatives with a fixed disulfide bridge, an added imide acceptor fragment, and an alkyl chain with a different length, as well as the presence of Cl or Br atoms at positions 3 and 6. The compounds are synthesized via a nucleophilic substitution of the halogen atoms at both *peri*-positions with elemental sulfur in a very-high yield and purity.

The redox properties of the new compounds were studied in model Li-ion cells with liquid ionic electrolyte. All compounds simultaneously display *n*- and *p*-type redox reactions between 1.5 and 5.0 V: at about 2.0 V vs. Li/Li^+, there is a reversible reduction of the sulfur bridge with Li^+; while above 4.0 V vs. Li/Li^+, the oxidation of the organic compound followed by a binding with $TFSI^-$ anions occurs. The high-voltage reaction is faster than the low-voltage one. The redox potentials of the *p*- and *n*-type redox reactions are mainly sensitive to whether Cl or Br substituents are available in the molecule architecture, while the alkyl chain length determines the kinetics of the redox reactions. As a result, the chloro-substituted compound with the shorter alkyl chain displays the best kinetics for both low- and high-voltage redox reactions.

In general, this study is the first report of peri-dithiolo-1,8-naphthalimides as bipolar organic redox compounds. Among them, the compound with the shorter alkyl chain and Cl-substitution (i.e., **SCl4**) appears to be the most appropriate candidate for electrochemical applications. Because of the big difference in the potentials of low- and high-voltage redox reactions, the reported organosulfur compounds present an interesting option for application as electrodes in symmetrical ion cells, which presents a new perspective in the future development of full organic ion batteries. In addition, the established potential of about 4.3 V vs. Li/Li^+ for SCl4 makes it an interesting option as an electrode in dual-ion batteries, which presents a new direction in the utilization of redox organic compounds.

Supplementary Materials: The following supporting information can be downloaded at https://www.mdpi.com/article/10.3390/ma16237471/s1: Figure S1: ^{1}H NMR and ^{13}C NMR spectra of **3–6** in $CDCl_3$; Figure S2: ^{1}H NMR and ^{13}C NMR spectra of **SCl8** in $CDCl_3$; Figure S3: ^{1}H NMR and ^{13}C NMR spectra of **SBr8** in $CDCl_3$; Figure S4: ^{1}H NMR and ^{13}C NMR spectra of **SCl4** in $C_2D_2Cl_4$; Figure S5: ^{1}H NMR and ^{13}C NMR spectra of **SBr4** in $C_2D_2Cl_4$.

Author Contributions: Conceptualization—D.M. and YZ.; Methodology—Y.Z. and M.M.; Validation—S.S. (Silva Stanchovska) and S.S. (Stanimir Stoyanov); Formal analysis—S.S. (Silva Stanchovska), M.M. and N.S.; Investigation—S.S. (Silva Stanchovska) and N.S.; Resources—D.M. and Y.Z.; Writing—Original Draft—D.M., Y.Z. and S.S. (Stanimir Stoyanov); Writing—Review and Editing—D.M., Y.Z. and S.S. (Stanimir Stoyanov); Visualization—S.S. (Silva Stanchovska), D.M. and S.S. (Stanimir Stoyanov); Supervision—D.M. and Y.Z.; Project administration—D.M. and Y.Z.; Funding acquisition—D.M. and Y.Z. All authors have read and agreed to the published version of the manuscript.

Funding: This research was funded by the Bulgarian National Science Fund, grant number КП-06-H69/1.

Institutional Review Board Statement: Not applicable.

Informed Consent Statement: Not applicable.

Data Availability Statement: Data are contained within the article.

Acknowledgments: The authors acknowledge the TwinTeam project D01-272 "European Network on Materials for Clean Technologies" for providing the opportunity to present the results at the SizeMat4 conference, as well as for the financial publication support. The authors are sincerely grateful to Radostina Stoyanova for knowledgeable remarks and discussions during the preparation of this article.

Conflicts of Interest: The authors declare no conflict of interest.

References

1. Kim, T.; Song, W.; Son, D.-Y.; Ono, L.K.; Qi, Y. Lithium-ion batteries: Outlook on present, future, and hybridized technologies. *J. Mater. Chem. A* **2019**, *7*, 2942–2964. [CrossRef]
2. Kim, H.-J.; Krishna, T.; Zeb, K.; Rajangam, V.; Gopi, C.V.V.M.; Sambasivam, S.; Raghavendra, K.V.G.; Obaidat, I.M. A Comprehensive Review of Li-Ion Battery Materials and Their Recycling Techniques. *Electronics* **2020**, *9*, 1161. [CrossRef]
3. Deng, C.; Li, X.; Chen, R.; Ye, K.; Lipton, J.; Maclean, S.A.; Wang, H.; Taylor, A.D.; Weng, G.M. Recent advances in rocking chair batteries and beyond. *Energy Storage Mater.* **2023**, *60*, 102820. [CrossRef]
4. Sironval, V.; Scagliarini, V.; Murugadoss, S.; Tomatis, M.; Yakoub, Y.; Turci, F.; Hoet, P.; Lison, D.; van den Brule, S. $LiCoO_2$ particles used in Li-ion batteries induce primary mutagenicity in lung cells via their capacity to generate hydroxyl radicals. *Part. Fibre Toxicol.* **2020**, *17*, 6. [CrossRef] [PubMed]
5. Arvidsson, R.; Chordia, M.; Nordelöf, A. Quantifying the life-cycle health impacts of a cobalt-containing lithium-ion battery. *Int. J. Life Cycle Assess.* **2022**, *27*, 1106–1118. [CrossRef]
6. Wentker, M.; Greenwood, M.; Leker, J. A Bottom-Up Approach to Lithium-Ion Battery Cost Modeling with a Focus on Cathode Active Materials. *Energies* **2019**, *12*, 504. [CrossRef]
7. Mrozik, W.; Rajaeifar, M.A.; Heidrich, O.; Christensen, P. Environmental impacts, pollution sources and pathways of spent lithium-ion batteries. *Energy Environ. Sci.* **2021**, *14*, 6099–6121. [CrossRef]
8. Li, L.; Yin, Y.; Hei, J.; Wan, X.; Li, M.; Cui, Y. Molecular Engineering of Aromatic Imides for Organic Secondary Batteries. *Small* **2021**, *17*, 2005752. [CrossRef]
9. Chen, Y.; Dai, H.; Fan, K.; Zhang, G.; Tang, M.; Gao, Y.; Zhang, C.; Guan, L.; Mao, M.; Liu, H.; et al. A Recyclable and Scalable High-Capacity Organic Battery. *Angew. Chem. Int. Ed.* **2023**, *62*, e202302539. [CrossRef]
10. Guo, W.; Fu, Y. Electrochemistry of Electrode Materials Containing S−Se Bonds for Rechargeable Batteries. *Chem.A Eur. J.* **2020**, *26*, 13322–13331. [CrossRef]
11. Lu, Y.; Zhang, Q.; Li, L.; Niu, Z.; Chen, J. Design Strategies toward Enhancing the Performance of Organic Electrode Materials in Metal-Ion Batteries. *Chem* **2018**, *4*, 2786–2813. [CrossRef]
12. Wang, D.-Y.; Guo, W.; Fu, Y. Organosulfides: An Emerging Class of Cathode Materials for Rechargeable Lithium Batteries. *Acc. Chem. Res.* **2019**, *52*, 2290–2300. [CrossRef] [PubMed]
13. Shadike, Z.; Tan, S.; Wang, Q.-C.; Lin, R.; Hu, E.; Qu, D.; Yang, X.-Q. Review on organosulfur materials for rechargeable lithium batteries. *Mater. Horizons* **2021**, *8*, 471–500. [CrossRef]
14. Visco, S.J.; Liu, M.; De Jonghe, L.C. Ambient Temperature High-Rate Lithium/Organosulfur Batteries. *J. Electrochem. Soc.* **1990**, *137*, 1191–1192. [CrossRef]
15. Li, F.; Si, Y.; Liu, B.; Li, Z.; Fu, Y. Lithium Benzenedithiolate Catholytes for Rechargeable Lithium Batteries. *Adv. Funct. Mater.* **2019**, *29*, 1902223. [CrossRef]
16. Bhargav, A.; Patil, S.V.; Fu, Y. A phenyl disulfide@CNT composite cathode for rechargeable lithium batteries. *Sustain. Energy Fuels* **2017**, *1*, 1007–1012. [CrossRef]
17. Wang, D.-Y.Y.; Si, Y.; Li, J.; Fu, Y. Tuning the electrochemical behavior of organodisulfides in rechargeable lithium batteries using N-containing heterocycles. *J. Mater. Chem. A* **2019**, *7*, 7423–7429. [CrossRef]
18. Sakai, N.; Mareda, J.; Vauthey, E.; Matile, S. Core-substituted naphthalenediimides. *Chem. Commun.* **2010**, *46*, 4225. [CrossRef]
19. Chen, S.; Jia, T.; Zhou, G.; Zhang, C.; Hou, Q.; Wang, Y.; Luo, S.; Shi, G.; Zeng, Y. A Cross-Linked Triphenylamine-Based Polymer Cathode Material with Dual Anion-Cation Reversible Insertion for Lithium Ion Battery. *J. Electrochem. Soc.* **2019**, *166*, A2543–A2548. [CrossRef]
20. Al Kobaisi, M.; Bhosale, S.V.; Latham, K.; Raynor, A.M.; Bhosale, S.V. Functional Naphthalene Diimides: Synthesis, Properties, and Applications. *Chem. Rev.* **2016**, *116*, 11685–11796. [CrossRef]

21. Zagranyarski, Y.; Mutovska, M.; Petrova, P.; Tomova, R.; Ivanov, P.; Stoyanov, S. Dioxin-annulated 1,8-naphthalimides—Synthesis, spectral and electrochemical properties, and application in OLED. *Dye. Pigment.* **2021**, *184*, 108585. [CrossRef]
22. Said, A.I.; Staneva, D.; Angelova, S.; Grabchev, I. Self-Associated 1,8-Naphthalimide as a Selective Fluorescent Chemosensor for Detection of High pH in Aqueous Solutions and Their Hg^{2+} Contamination. *Sensors* **2023**, *23*, 399. [CrossRef]
23. Banerjee, S.; Veale, E.B.; Phelan, C.M.; Murphy, S.A.; Tocci, G.M.; Gillespie, L.J.; Frimannsson, D.O.; Kelly, J.M.; Gunnlaugsson, T. Recent advances in the development of 1,8-naphthalimide based DNA targeting binders, anticancer and fluorescent cellular imaging agents. *Chem. Soc. Rev.* **2013**, *42*, 1601–1618. [CrossRef]
24. Lee, S.; Hong, J.; Jung, S.K.; Ku, K.; Kwon, G.; Seong, W.M.; Kim, H.; Yoon, G.; Kang, I.; Hong, K.; et al. Charge-transfer complexes for high-power organic rechargeable batteries. *Energy Storage Mater.* **2019**, *20*, 462–469. [CrossRef]
25. Abul-Futouh, H.; Zagranyarski, Y.; Müller, C.; Schulz, M.; Kupfer, S.; Görls, H.; El-khateeb, M.; Gräfe, S.; Dietzek, B.; Peneva, K.; et al. [FeFe]-Hydrogenase H-cluster mimics mediated by naphthalene monoimide derivatives of peri-substituted dichalcogenides. *Dalt. Trans.* **2017**, *46*, 11180–11191. [CrossRef] [PubMed]
26. Inamasu, T.; Yoshitoku, D.; Sumi-otorii, Y.; Tani, H.; Ono, N. Electrochemical Behaviors of Naphtho[1,8-cd][1,2]dithiol, Dibenzo[c,e][1,2]dithiin, and Naphtho[1,8-cd:4,5-c′d′]bis[1,2]dithiol. *J. Electrochem. Soc.* **2003**, *150*, A128. [CrossRef]
27. Liang, Y.; Tao, Z.; Chen, J. Organic electrode materials for rechargeable lithium batteries. *Adv. Energy Mater.* **2012**, *2*, 742–769. [CrossRef]
28. Mauger, A.; Julien, C.; Paolella, A.; Armand, M.; Zaghib, K. Recent Progress on Organic Electrodes Materials for Rechargeable Batteries and Supercapacitors. *Materials* **2019**, *12*, 1770. [CrossRef]
29. Yin, X.; Sarkar, S.; Shi, S.; Huang, Q.; Zhao, H.; Yan, L.; Zhao, Y.; Zhang, J. Recent Progress in Advanced Organic Electrode Materials for Sodium-Ion Batteries: Synthesis, Mechanisms, Challenges and Perspectives. *Adv. Funct. Mater.* **2020**, *30*, 1908445. [CrossRef]
30. Mutovska, M.; Skabeev, A.; Konstantinov, K.; Cabanetos, C.; Stoyanov, S.; Zagranyarski, Y. One-pot synthesis of fused-rings heterocyclic systems based on symmetrically benzofuran annulated 1,8-naphthalimides. *Dye. Pigment.* **2023**, *220*, 4–10. [CrossRef]
31. Seybold, G. New perylene and violanthrone dyestuffs for fluorescent collectors. *Dye. Pigment.* **1989**, *11*, 303–317. [CrossRef]
32. Langhals, H. Cyclic Carboxylic Imide Structures as Structure Elements of High Stability. Novel Developments in Perylene Dye Chemistry. *Heterocycles* **1995**, *40*, 477. [CrossRef]
33. Karuppasamy, K.; Theerthagiri, J.; Vikraman, D.; Yim, C.-J.; Hussain, S.; Sharma, R.; Maiyalagan, T.; Qin, J.; Kim, H.-S. Ionic Liquid-Based Electrolytes for Energy Storage Devices: A Brief Review on Their Limits and Applications. *Polymers* **2020**, *12*, 918. [CrossRef]
34. Guo, W.; Wang, D.; Chen, Q.; Fu, Y. Advances of Organosulfur Materials for Rechargeable Metal Batteries. *Adv. Sci.* **2022**, *9*, 2103989. [CrossRef] [PubMed]
35. Vadehra, G.S.; Maloney, R.P.; Garcia-Garibay, M.A.; Dunn, B. Naphthalene Diimide Based Materials with Adjustable Redox Potentials: Evaluation for Organic Lithium-Ion Batteries. *Chem. Mater.* **2014**, *26*, 7151–7157. [CrossRef]
36. Kye, H.; Kang, Y.; Jang, D.; Kwon, J.E.; Kim, B.-G. p-Type Redox-Active Organic Electrode Materials for Next-Generation Rechargeable Batteries. *Adv. Energy Sustain. Res.* **2022**, *3*, 2200030. [CrossRef]
37. Speer, M.E.; Kolek, M.; Jassoy, J.J.; Heine, J.; Winter, M.; Bieker, P.M.; Esser, B. Thianthrene-functionalized polynorbornenes as high-voltage materials for organic cathode-based dual-ion batteries. *Chem. Commun.* **2015**, *51*, 15261–15264. [CrossRef] [PubMed]
38. Bard, A.J.; Faulkner, L.R. *Electrochemical Methods: Fundamentals and Applications*, 2nd ed.; Wiley: New York, NY, USA, 2000; ISBN 978-0-471-04372-0.
39. Acker, P.; Rzesny, L.; Marchiori, C.F.N.; Araujo, C.M.; Esser, B. π-Conjugation Enables Ultra-High Rate Capabilities and Cycling Stabilities in Phenothiazine Copolymers as Cathode-Active Battery Materials. *Adv. Funct. Mater.* **2019**, *29*, 1906436. [CrossRef]

materials

Article

Graphene-Based Composites for Thermoelectric Applications at Room Temperature

Sonya Harizanova [1,*], Vassil Vulchev [2] and Radostina Stoyanova [1,*]

1 Institute of General and Inorganic Chemistry, Bulgarian Academy of Sciences, 1113 Sofia, Bulgaria
2 Faculty of Physics, University of Sofia, 1164 Sofia, Bulgaria; vdv@phys.uni-sofia.bg
* Correspondence: sonya@svr.igic.bas.bg (S.H.); radstoy@svr.igic.bas.bg (R.S.)

Abstract: The thermoelectric materials that operate at room temperature represent a scientific challenge in finding chemical compositions with three optimized, independent parameters, namely electrical and thermal conductivity and the Seebeck coefficient. Here, we explore the concept of the formation of hybrid composites between carbon-based materials and oxides, with the aim of modifying their thermoelectric performance at room temperature. Two types of commercially available graphene-based materials are selected: N-containing reduced graphene oxide (NrGO) and expanded graphite (ExGr). Although the NrGO displays the lowest thermal conductivity at room temperature, the ExGr is characterized by the lowest electrical resistivity and a negative Seebeck coefficient. As oxides, we choose two perspective thermoelectric materials: p-type $Ca_3Co_4O_9$ and n-type $Zn_{0.995}Al_{0.005}O$. The hybrid composites were prepared by mechanical milling, followed by a pelleting. The thermoelectric efficiency was evaluated on the basis of its measured electrical resistivity, Seebeck coefficient and thermal conductivity at room temperature. It was found that that 2 wt.% of ExGr or NrGO leads to an enhancement of the thermoelectric activity of $Ca_3Co_4O_9$, while, for $Zn_{0.995}Al_{0.005}O$, the amount of ExGr varies between 5 and 20 wt.%. The effect of the composites' morphology on the thermoelectric properties is discussed on the basis of SEM/EDS experiments.

Keywords: thermoelectric oxides; expanded graphite; graphene oxide; layered oxides; zinc oxide; multiphase composites

Citation: Harizanova, S.; Vulchev, V.; Stoyanova, R. Graphene-Based Composites for Thermoelectric Applications at Room Temperature. *Materials* **2023**, *16*, 7262. https://doi.org/10.3390/ma16237262

Academic Editor: Andres Sotelo

Received: 22 October 2023
Revised: 13 November 2023
Accepted: 16 November 2023
Published: 21 November 2023

1. Introduction

Oxide materials have been attracting attention due to their possible thermoelectric applications, since they exhibit a high Seebeck coefficient and thermal stability, together with low toxicity [1]. However, their relatively low electrical conductivity and high thermal conductivity cause obstacles for oxide commercialization as a thermoelectric material [1,2]. In general, the performance of thermoelectric materials is quantified by the dimensionless figure of merit $ZT = S^2T/(\rho\lambda)$, where S is the Seebeck coefficient, T is the absolute temperature, ρ is the electrical resistivity, and λ is the thermal conductivity [3,4]. Based on these relations, high thermoelectric activity could be reached when oxides simultaneously have high electrical conductivity and low thermal conductivity [3,4]. In this respect, the most intriguing thermoelectric oxides comprise Co-based, strongly correlated systems (perovskites, $LaCoO_3$, layered oxides, $NaCo_2O_4$ and $RBaCo_2O_{5+x}$, misfit layered oxides, $Ca_4Co_3O_9$), a Ti-based Ruddlesden–Popper phase ($SrTiO_3$), ZnO, etc. [5,6]. Given the diversity and flexibility of the crystal structures, the state-of-the-art approach for improving the thermoelectric performance of oxides is through selective metal substitution [5]. The metal substituents affect both the electronic structure and vibrational properties of oxides, thus contributing to augmented thermoelectric activity [5]. As each substituent only has an effect on one parameter (i.e., electrical or thermal conductivities), the use of more than one substituent enables the optimization of the several independent parameters and can help to achieve the precise control of thermoelectric properties [7,8]. Despite the effectiveness of this approach, the preparation of multiple substituted oxides is usually a complex and expensive process.

As an alternative to single and multiple substitution, the approach including the formation of composites between several distinct phases is becoming more attractive [9–12]. In multiphase composites, interfacial and texturized effects (the grain boundaries, confined particle sizes, oriented growth, etc.) give rise to the enhanced phonon-scattering and charge carrier mobility, which, in turn, lead to a better thermoelectric activity [13–16]. Two groups of multiphase composites need to be highlighted: multicomponent inorganic or organic composites and hybrid organic–inorganic composites [17]. The main feature of all organic or inorganic composites is their enhanced thermoelectric performance due to the combination of components with high electrical and low thermal conductivities [10,11,13,16]. However, the main disadvantage of the organic composites is their low Seebeck coefficients, while, for inorganic composites, their electrical conductivity is still far from the desired values. These disadvantages can be overcome by the formation of hybrid composites between inorganic and organic components, where the thermoelectric performance is dictated by the low thermal conductivity of the organic component and high Seebeck coefficient of the inorganic component [18]. In this respect, graphene has emerged as a competitive component thanks to its unique electrical conductivity and flexibility [19,20]. Unfortunately, the high thermal conductivity and low Seebeck coefficient restrict the graphene's merit in the range of 10^{-4} [15,16]. Given the difficulties in the synthesis of pure graphene, reduced graphene oxide (rGO) has recently become the subject of intensive studies due to its cheaper and easier means of synthesis [9,21]. Based on DFT calculations, it has been shown that the thermal conductivity can be effectively reduced by the attachment of oxygen atoms to the graphene layers (i.e., the formation of graphene oxide) [22,23]. Contrary to graphene, graphene oxide exhibits lower electrical conductivity, which leads to a worsening of its thermoelectric activity [19,24]. The enhancement of electrical conductivity is achieved through a mild reduction in graphene oxide (i.e., rGO) [20,25]. Thus, rGO has recently become a main subject of study due to its thermoelectrical applications [9,26]. It has been reported that rGO can also be used as an additive to oxides in order to control their thermoelectric performance [27,28]. In addition to graphene and rGO, expanded graphite (ExGr) has also recently attracted research interest [29,30]. ExGr has been utilized as a filler to multifunctional cement composites [31,32].

Despite the variety of studies of hybrid composites, there are still few investigations on thermoelectric materials that are able to operate at room temperature [33,34]. At present, the main room-temperature thermoelectric materials rely on BiTe-, S-, and Se-based alloys, which are highly toxic [29,30]. Contrary to the alloy-based materials, environmentally benign oxides performed worse at room temperature [35,36]. The question is whether the room-temperature thermoelectric efficiency could be improved by the formation of hybrid composites.

Here, we explore the concept of the formation of hybrid composites between carbon-based materials and oxides, with the aim of modifying their thermoelectric performance at room temperature. Three types of commercially available graphene-based materials are selected: reduced graphene oxide (rGO), N-containing reduced graphene oxide (NrGO) and expanded graphite (ExGr). As oxides, we chose *p*-type $Ca_3Co_4O_9$ and *n*-type $Zn_{0.995}Al_{0.005}O$. These oxides were selected due to their ability to modify the electrical and thermal properties through the regulation of the interface and boundary grains [37,38]. In addition, the thermal conductivity of oxides displays size-dependent effects, which allows for the suppression of thermal conductivity through effective phonon scattering [39]. Thus, we intend to modify the thermoelectric performance of these oxides at room temperature through the formation of composites with graphene-based materials. The hybrid composites were prepared by mechanical milling. The thermoelectric efficiency was evaluated on the basis of the measured electrical resistivity, Seebeck coefficient and thermal conductivity at room temperature. Based on SEM/EDS analysis, the effect of the composites' morphology on the thermoelectric activity is discussed. The study of thermoelectric parameters will provide the necessary information on the applicability of the studied materials to the direct conversion of thermal energy into electrical energy and the temperature range in which the materials are most promising.

2. Materials and Methods

Pristine oxides, carbon materials and synthetic procedures. We used $Ca(NO_3)_2.4H_2O$ (Sigma Aldrich, St. Louis, MO, USA), $Co(NO_3)_2.6H_2O$, $Zn(NO_3)_2.6H_2O$ (Sigma Aldrich, St. Louis, MO, USA), $Al(NO_3)_3.9H_2O$ (Sigma Aldrich, St. Louis, MO, USA), citric acid (p.a. Chem-Solutions, GmbH, Dimitrovgrad, Bulgaria) and NH_4HCO_3 (p.a. Chem-Solutions, GmbH, Dimitrovgrad, Bulgaria) as reagents. Layered $Ca_3Co_4O_9$ was obtained by a Pechini-type reaction, as described elsewhere [10]. The solution containing $Ca(NO_3)_2.4H_2O$, $Co(NO_3)_2.6H_2O$, citric acid and ethylene glycol (at a ratio of Ca:Co:CA:EG = 1:1:10:40) was heated at 90 °C until drying and polyesterification; then, the solid residue was heated at 400 °C for 3 h. Aluminum-doped zinc oxide ($Zn_{0.995}Al_{0.005}O$ with wurtzite-type structure) was prepared by co-precipitation from an aqueous solution of zinc and aluminum nitrates with NH_4HCO_3; the details are provided elsewhere [40]. The precursors were tableted and annealed at a specific temperature depending on the oxide composition: $Ca_3Co_4O_9$ was annealed at 800 °C for 20 h in an oxygen atmosphere, while $Zn_{0.995}Al_{0.005}O$ was obtained at 750 °C for 10 h in air. The reduced graphene oxide (rGO), N-doped reduced graphene oxide (NrGO) and expanded graphite (EXG 98 20/20 μm) are commercial products provided by Graphit Kropfmühl GmbH (Hauzenberg, Germany).

Preparation of composites. The composites between oxides ($Ca_3Co_4O_9$ or $Zn_{0.995}Al_{0.005}O$) and carbon additives (rGO, NrGO or ExCr) were fabricated through the mechanical mixing of the given oxide with a carbon material at a weight ratio of 98:2%, 95:5%, and 80:20%. The mixtures were pelleted and thermally treated at 200 °C for 5 h in argon atmosphere. For the sake of convenience, the composites will be denoted as follows: $Ca_{98}NrGO_2$, $Ca_{95}NrGO_5$ and $Ca_{80}NrGO_{20}$; $Ca_{98}ExGr_2$, $Ca_{95}ExGr_5$ and $Ca_{80}ExGr_{20}$; $Zn_{98}ExGr_2$, $Zn_{95}ExGr_5$ and $Zn_{80}ExGr_{20}$; $Zn_{98}NrGO_2$, $Zn_{95}NrGO_5$ and $Zn_{80}NrGO_{20}$.

Characterization. A structural analysis of samples was carried out using a powder X-ray diffractometer (Bruker Advance D8, Karlsruhe, Germany) equipped with LynxEye detector (CuKα). The morphology was analysed through scanning electron microscopy (SEM). SEM images of pellets along the surface and cross-section were monitored by a JEOL JSM 6390 microscope equipped with an EDS analyzer (Oxford INCA Energy 350) in a regime of secondary electron images (SEIs).

For the thermoelectrical characterization, square pellets with a dimension of 9 mm and thickness of about 1–2 mm of were fabricated. The pellet porosity was evaluated by a comparison of the pellet density (determined by the Archimedes method) with the theoretical density of $Ca_3Co_4O_9$ and $Zn_{0.995}Al_{0.005}O$. The results indicate that the pellet porosity varied at around 25%, irrespective of its phase composition. This allows for us to correctly compare the thermal properties of composites.

The electrical resistivity was determined by MMR's Variable Temperature Hall System (K2500-5SLP-SP) within the framework of the Van der Pauw method. Thermal conductivity was calculated using a C-Therm TCi Thermal Conductivity Analyzer (MTPS). The Seebeck coefficient of the samples was measured at room temperature using the in-house setup. The holder comprises two solid plates with thermocouples. One of the plates contains a gradient heater. The measured sample was inserted between the solid plates (Figure S1). Then, the Seebeck voltage and temperature difference were recorded upon reaching the stationary condition. All parameters (i.e., electrical resistivity, thermal conductivity and Seebeck coefficient) were measured at 298 K.

3. Results

3.1. Thermoelectric Properties of rGO, NrGO and ExGr

The thermoelectrical properties of expanded graphite, rGO, and NrGO are compared in Table 1. The comparison shows that the expanded graphite exhibits the lowest electrical resisitivity (i.e., around 9.6×10^{-4} Ω.cm) and Seebeck coefficient (i.e., −26.1 μ/K), resulting in the highest power factor (PF) even at room temperature (i.e., 71 μW/(m.K^2) at 20 °C). The magnitude of the PF is close to that determined for commercial-graphite-produced composite films (i.e., 87 μW/(m.K^2) at 25 °C and 94 μW/(m.K^2) at 150 °C) [29]. It is worth mentioning that the sign of the Seebeck coefficient is negative. Despite the relatively good magnitude of the PF, the high thermal conductivity of ExGr gives rise to a lower figure of merit. In comparison with the expanded graphite, both rGO and NrGO are characterized by their extremely low thermal conductivities (i.e., more than two orders lower than that of ExGr), but their figures of merit remain small, making them unsuitable for practical application. The close inspection of the data for rGO and NrGO indicates that both the power factor and figure of merit reach higher magnitudes for NrGO. It is of importance that the figure of merit for NrGO varies in the range of 10^{-3}, which is slightly higher from the previously reported data (i.e., about 10^{-4}) [19,20]. As far as we known, these are the first data on the thermoelectric activity of reduced graphene oxide where the O atoms are replaced by N. This motivated us to use only NrGO as a component in multiphase composites.

Table 1. Electrical resistivity, Seebeck coefficient, power factor, thermal conductivity, and figure of merit for ExGr, rGO, NrGO, $Ca_3Co_4O_9$, and $Zn_{0.995}Al_{0.005}O$, measured at 298 K.

Samples	ρ Ω.cm	S μV/K	PF = S^2/ρ μW/(m·K^2)	λ W/m.K	Figure of Merit $S^2T/(\rho \cdot \lambda)$
ExGr	9.6×10^{-4}	−26.1	71	0.838	0.0255
NrGO	4.45×10^{-2}	−2.8	1.7×10^{-2}	0.003	0.0017
rGO	7.98×10^{-2}	−0.6	3.9×10^{-4}	0.005	2×10^{-5}
$Ca_3Co_4O_9$	3×10^{-2}	94	29	0.620	0.0130
$Zn_{0.995}Al_{0.005}O$	4.6×10^{3}	−590	8×10^{-3}	0.300	8×10^{-6}

3.2. Multiphase Composites between Oxides and Carbon Additives

The ball-milling of oxides with carbon additives yields composites in which every individual component (oxide, NrGO, or ExGr) retains its structure. Figure 1 compares the XRD patterns of oxide–carbon composites. The indexation of XRD patterns shows the appearance of diffraction peaks due to $Ca_3Co_4O_9$ and expanded graphite phases. In the case of NrGO, the diffraction peak (at around 24.6°) is too broad, thus preventing its observation in the XRD patterns of composites. The lattice parameters for $Ca_3Co_4O_9$ and $Zn_{0.995}Al_{0.005}O$ components are not changed during the composites' formation and correspond to the previously reported parameters [10,37]: a = 4.8249 Å; b_1 = 4.5687 Å; c = 10.862 Å; b_2 = 2.8096 Å; β = 98.4° for $Ca_3Co_4O_9$ and a = 3.2504 Å; c = 5.2062 Å for $Zn_{0.995}Al_{0.005}O$.

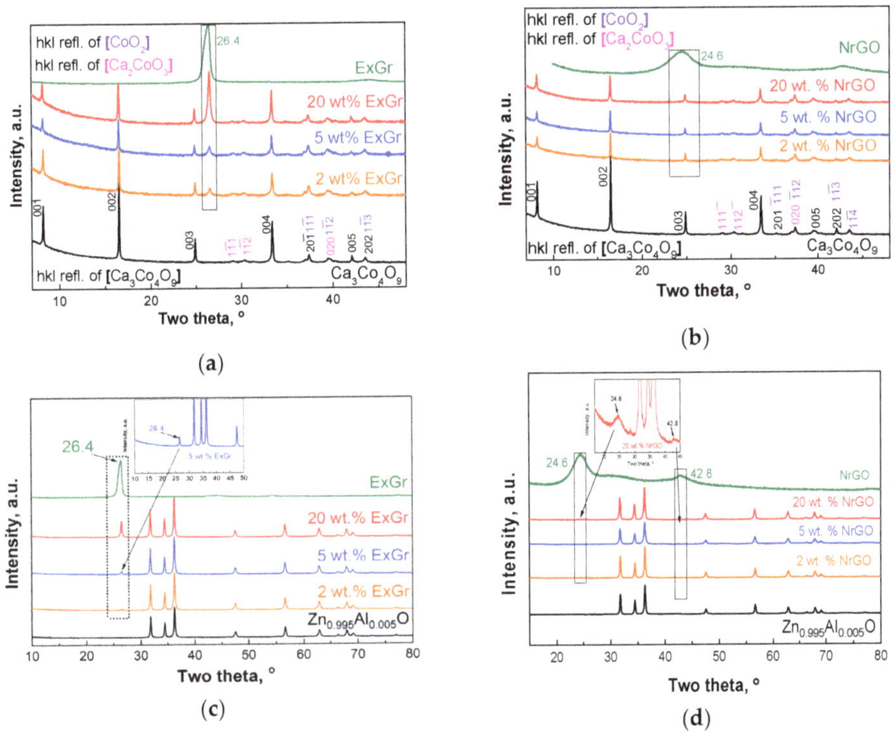

Figure 1. XRD patterns of composites of $Ca_3Co_4O_9$ (**a**,**b**) and $Zn_{0.995}Al_{0.05}O$ (**c**,**d**) with ExGr (**a**,**c**) and NrGO (**b**,**d**).

3.3. *Effect of NrGO and ExGr on the Thermoelectric Properties of* $Ca_3Co_4O_9$

The formation of multiphase composites enables to compare the effect of carbon components on the thermoelectric properties of $Ca_3Co_4O_9$. The ExGr has excellent electrical conductivity, which is about two orders higher than that of the oxide. In contrast, the electrical conductivity of the oxide slightly exceeds that of the NrGO (Table 1). Although the Seebeck coefficient has a positive sign for the oxide, the negative Seebeck coefficient is observed for ExGr and NrGO (Table 1). This reveals that different types of charge carriers are responsible for the electrical properties: for the oxide, the p-type is responsible, while for the ExGr and NrGO, the n-type is responsible. In terms of magnitude, the oxide Seebeck coefficient is the highest. Because of the excellent electrical conductivity, the power factor of the ExGr reaches the highest magnitude irrespective of its low Seebeck coefficient (Table 1). The thermal conductivity increases following the order NrGO < $Ca_3Co_4O_9$ < ExGr. Combining all parameters into the figure of merit, it appears that ExGr and $Ca_3Co_4O_9$ have comparable thermoelectric activities (0.025 and 0.013 at 20 °C, respectively), which outperform those of NrGO.

The addition of ExGr in amounts of up to 5 wt.% to $Ca_3Co_4O_9$ only slightly affects the electrical resistivity, while the next amount of ExGr significantly reduces the electrical resistivity (Figure 2a). Along with the observed trend in changes in the electrical resistivity, the Seebeck coefficient progressively decreases, but its sign remains positive up to 5 wt.% of ExGr (Figure 2b). At 20 wt.% of ExGr, the sign of Seebeck is converted from positive to negative, reaching the value of the ExGr component. The simultaneous changes in the electrical resistivity and Seebeck coefficient could be explained in terms of the opposite charge carrier types observed for the individual components $Ca_3Co_4O_9$ and ExGr (p- and n-type, respectively), as well as the average values of the composites' electrical resistivity, measured between $Ca_3Co_4O_9$ and ExGr components. Considering the electrical resistivity and Seebeck coefficient, the power factor of the $Ca_3Co_4O_9$-ExGr composites is lower than that of the individual $Ca_3Co_4O_9$ and ExGr components (Figure 2c).

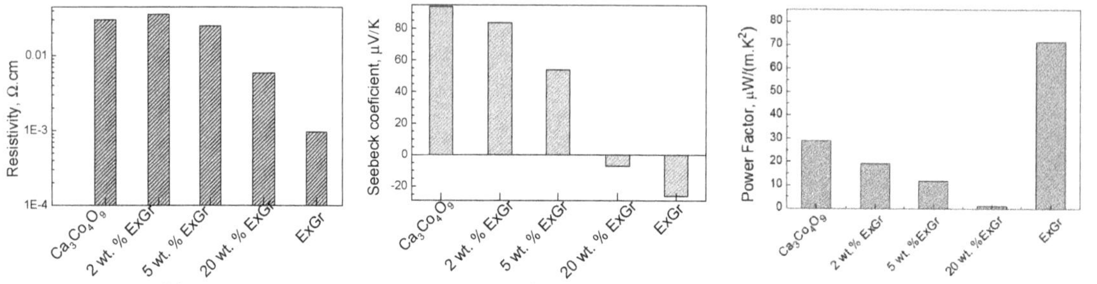

Figure 2. Electrical resistivity (**a**), Seebeck coefficient (**b**), and power factor (**c**) of the composites between $Ca_3Co_4O_9$ and ExGr.

In comparison with the ExCr additives, the addition of NrGO leads to an increase in the electrical resistivity of $Ca_3Co_4O_9$ (Figure 3a). It is interesting that the electrical resistivity of the $Ca_3Co_4O_9$-NrGO composites is slightly higher than that of the individual $Ca_3Co_4O_9$ and NrGO components. Up to 5 wt.% of NrGO, the Seebeck coefficient varies around the magnitude of the oxide component, followed by a drastic decrease after further increases in the NrGO content (Figure 3b). However, the sign of the Seebeck coefficient remains positive between 2 and 20 wt.% of NrGO, which is different to that of the ExGr-containing composites. The different trends in variation of the electrical resistivity and Seebeck coefficients of ExGr and NrGO-containing composites are, most probably, related to the specific manner of packing of the oxide particles using carbon additives: it appears that ExGr more strongly modifies the oxide in comparison with NrGO. (This will be discussed in the next section.) Irrespective of these different behaviors, the power factor of $Ca_3Co_4O_9$ is almost unchanged for up to 5 wt.% of NrGO addition, as in the case of ExGr-containing composites (Figure 3c). All these data disclose that small amounts of ExGr and NrGO (up to 5 wt.%) slightly affect the power factor of the oxide.

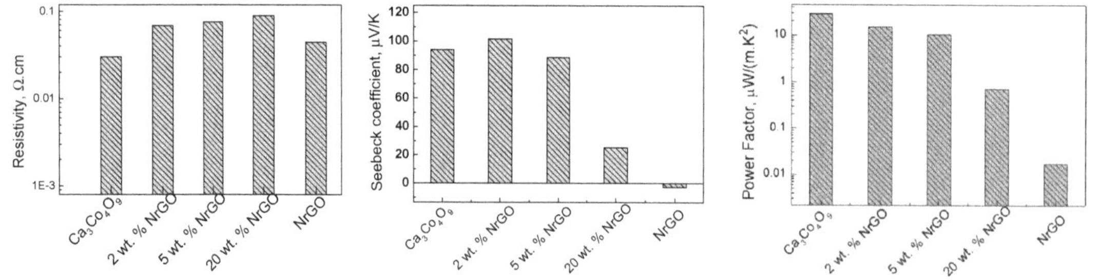

Figure 3. Electrical resistivity (**a**), Seebeck coefficient (**b**), and power factor (**c**) of the composites between $Ca_3Co_4O_9$ and NrGO.

The parameter that undergoes a strong change after the addition of carbon is the thermal conductivity (Figure 4). For the $Ca_3Co_4O_9$-ExGr composites, the thermal conductivity strongly decreases, so they become smaller than the individual $Ca_3Co_4O_9$ and ExGr components (Figure 4a). To rationalize this dependence, the contribution of the conductive carriers and phonon scattering to the overall thermal conductivity (λ) was taken into account. In the case when only one type of charge carrier contributes to the electron conductivity, the electrical thermal conductivity (λ_e) is inversely proportional to the electrical resistivity (ρ) according to the Wiedemann–Franz law: $\lambda_e = LT/\rho$, where L is the Lorentz number (2.45×10^{-8} V^2/K^2) [7]. Taking into account the experimental data on the electrical resistivity of composites, the calculated conductive-carrier-induced thermal conductivity of $Ca_3Co_4O_9$ increases with the ExGr content, as follows: 0.024, 0.021, 0.030, 0.125, and 0.764 W/(mK) for $Ca_3Co_4O_9$; $Ca_{98}ExGr_2$, $Ca_{95}ExGr_5$, $Ca_{80}ExGr_{20}$, and ExGr, respectively. The comparison shows that the calculated electrical thermal conductivity is significantly lower than the experimentally measured values for the $Ca_3Co_4O_9$ component (i.e., 0.62 W/(m.K)), signifying the leading role of phonon scattering in the overall thermal conductivity (about 96% of overall λ). Contrary to $Ca_3Co_4O_9$, the overall thermal conductivity of ExGr (i.e., 0.838 W/(m.K)) is governed by conductive carriers. For the $Ca_3Co_4O_9$-ExGr composites, phonon scattering remains a leading term in the overall thermal conductivity for up to 5 wt.% of ExGr (about 93% of the overall λ), while, at around 20 wt.% of ExGr, the role of conductive carriers dramatically increases, reaching 43% of the overall λ. Since the thermal conductivities of the composites are lower than those of the individual $Ca_3Co_4O_9$ and ExGr components, it is possible to assume that phonon-boundary scattering is an important factor. It is now recognized that, through increasing the phonon scattering at grain boundaries, it is possibly effectively to reduce the thermal conductivity [13,41]. In other words, this is a future aim for the modification of the oxide interfaces through the addition of ExGr. The smaller thermal conductivity suggests an improvement in the figure of merit, with this improvement being best for $Ca_{98}ExGr_2$ (Figure 4b).

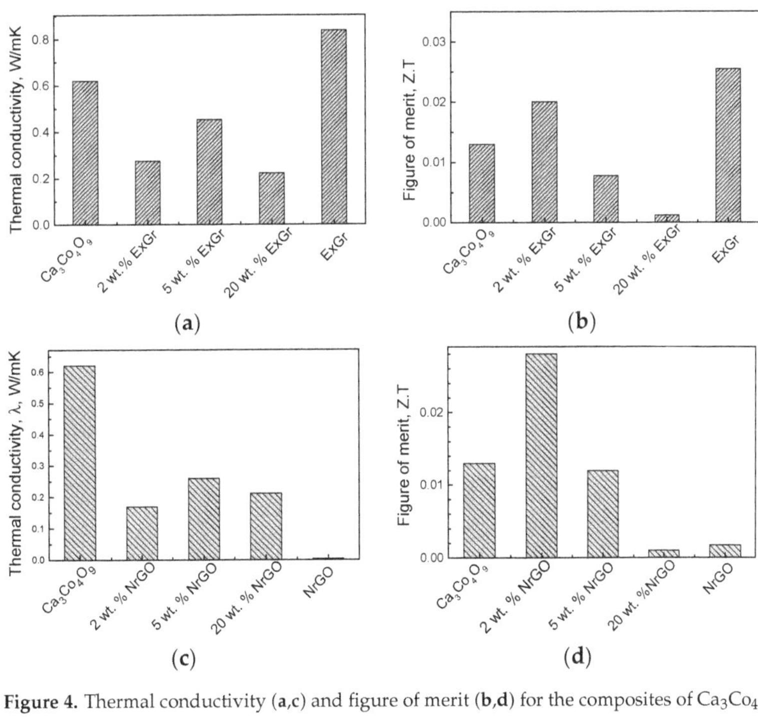

Figure 4. Thermal conductivity (**a**,**c**) and figure of merit (**b**,**d**) for the composites of $Ca_3Co_4O_9$ with ExGr (**a**,**b**) and with NrGO (**c**,**d**).

The NrGO has the same effect as the ExGr: the thermal conductivity of the $Ca_3Co_4O_9$-NrGO composites decreases after the addition of small amounts of NrGO (i.e., 2 wt.%) (Figure 4c), but remains higher than that of the NrGO component. It is worth mentioning that the overall thermal conductivity of NrGO is also governed by the conductive carriers as in the case of ExGr: thus, the higher electrical resistivity of NrGO determines its smaller thermal conductivity compared to that of ExGr (Table 1). However, for the $Ca_3Co_4O_9$-NrGO composites, the conductive-carrier-induced thermal conductivity makes a small contribution to the overall thermal conductivity (i.e., around 3–7% of λ) in the concentration range of 2–20 wt.%, a phenomenon that has already been observed for $Ca_3Co_4O_9$-ExGr composites. Because of the lowest thermal conductivity of NrGO, the thermal conductivities of NrGO-containing composites are lower in magnitude in comparison with those for $Ca_3Co_4O_9$-ExGr composites. This means that, irrespective of the stronger modification effect of ExGr on the oxide interface, the figure of merit of $Ca_3Co_4O_9$-NrGO composites will be higher than that of $Ca_3Co_4O_9$-ExGr composites. Thus, the best figure of merit is observed for the composite with 2 wt.% NrGO, which is higher than that of the composite containing 2 wt.% of ExGr (Figure 4d). It is of importance that even 2 wt.% of ExGr or NrGO is sufficient to improve the thermoelectric performance of the $Ca_3Co_4O_9$ oxide.

3.4. Effect of NrGO and ExGr on the Thermoelectric Properties of $Zn_{0.995}Al_{0.005}O$

Although $Ca_3Co_4O_9$ belongs to the p-type thermoelectric materials, the Al-doped ZnO is classified as n-type [42,43]. $Zn_{0.995}Al_{0.005}O$ displays a large Seebeck coefficient with a negative sign (-590 μV/K), but the electrical resistivity is extremely high (Table 1). It is worth mentioning that Al doping is needed to improve the electrical properties of ZnO [37–39]. In this case, the electrical resistivity remains higher in comparison with other thermoelectric oxides (i.e., $Ca_3Co_4O_9$). Thus, irrespective of the high magnitude of the Seebeck coefficient, the high electrical resistivity determines the small power factor for $Zn_{0.995}Al_{0.005}O$ (i.e., about 10^{-2} μW/(m.K^2), Table 1).

The addition of both ExGr and NrGO acts in the same manner. ExGr and NrGO yield a drastic decrease in the electrical resistivity of $Zn_{0.995}Al_{0.005}O$ (Figure 5). The optimal content of carbon additives is between 5 and 20 at %: in this case, the electrical conductivities of the composites approach those of the individual ExGr component or NrGO. Because of the low electrical resistivity of ExGr, the $Zn_{80}ExGr_{20}$ composite outperforms the analogue, with 20 wt.% NrGO (i.e., an electrical resistivity of 2.9×10^{-3} Ω.cm for $Zn_{95}ExGr_5$ versus 2.3×10^{-1} Ω.cm for $Zn_{80}NrGO_{20}$). Along with the electrical resistivity, the Seebeck coefficient is also decreased, but the sign always remains negative. It is noticeable that even small amounts of carbon additives provoke a dramatic decrease in the Seebeck coefficient, a phenomenon that is not observed for the $Ca_3Co_4O_9$-containing composites. Notwithstanding, due to the low electrical resistivity, the highest power factor is observed for the composites between $Zn_{0.995}Al_{0.005}O$ and ExGr in amounts between 5 to 20 wt.%. It is of importance that the magnitude of the power factor of $Zn_{95}ExGr_5$ and $Zn_{80}ExGr_{20}$ ((i.e., around 2 μW/(m.K^2)) is three orders higher than that of the individual $Zn_{0.995}Al_{0.005}O$ component.

In comparison with $Ca_3Co_4O_9$, $Zn_{0.995}Al_{0.005}O$ exhibits a lower thermal conductivity. However, as in the case of $Ca_3Co_4O_9$, the thermal conductivity of $Zn_{0.995}Al_{0.005}O$ is governed by phonon scattering (i.e., about 100% of overall λ). The addition of ExGr and NrGO to $Zn_{0.995}Al_{0.005}O$ displays different effects on the thermal conductivity (Figure 6). After the addition of ExGr, the thermal conductivity of composites slightly increases in comparison with that of $Zn_{0.995}Al_{0.005}O$, but always remains lower than that of ExGr (Figure 6a). Irrespective of the higher thermal conductivity, and thanks to the better electrical conductivities, the $Zn_{0.995}Al_{0.005}O$-ExGr composites still exhibit a better figure of merit that that of $Zn_{0.995}Al_{0.005}O$. The NrGO causes a decrease in the thermal conductivity of $Zn_{0.995}Al_{0.005}O$, and at 20 wt.% of NrGO, the thermal conductivity of the composite approaches that of NrGO (Figure 6c). Thus, the composite with 20 wt.% of NrGO is characterized by a slight improvement in the figure of merit. The important result suggests that the thermoelectric activity of $Zn_{0.995}Al_{0.005}O$ is improved thanks to the enhancement of the electrical conductivity of $Zn_{0.995}Al_{0.005}O$, especially after the addition of ExGr. Thus, the comparison evidences that ExGr plays a different role in the improvement in the thermoelectric properties of $Zn_{0.995}Al_{0.005}O$ and $Ca_3Co_4O_9$.

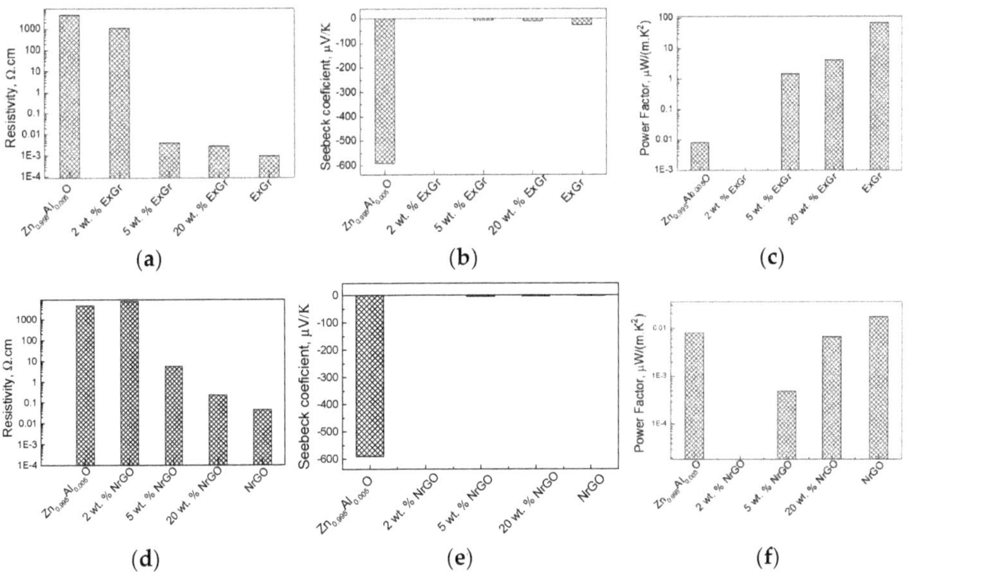

Figure 5. Electrical resistivity (**a**,**d**), Seebeck coefficient (**b**,**e**), and power factor (**c**,**f**) of the composites $Zn_{0.995}Al_{0.005}O$-ExGr (**top**) and $Zn_{0.995}Al_{0.005}O$-NrGO (**bottom**).

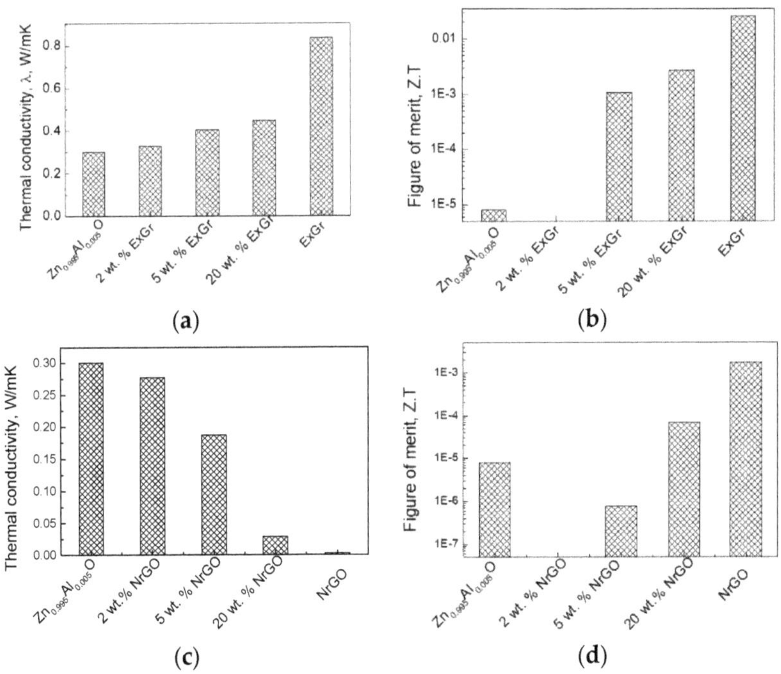

Figure 6. Thermal conductivity (**a**,**c**) and figure of merit (**b**,**d**) for the composites of $Zn_{0.995}Al_{0.005}O$ with ExGr (**a**,**b**) and with NrGO (**c**,**d**).

3.5. Composites Morphology

The dissimilar effects of ExGr and NrGO on the thermoelectric properties of oxides are closely related to the morphology of the composites. The morphologies of the individual oxide and carbon components are quite different (Figure 7). For $Ca_3Co_4O_9$, well-shaped micrometric particles can be observed, while for $Zn_{0.995}Al_{0.005}O$, small, irregular nanometric particles occur. The different morphologies of $Ca_3Co_4O_9$ and $Zn_{0.995}Al_{0.005}O$ can be used to explain their thermal conductivities (Table 1): the smaller particles of $Zn_{0.995}Al_{0.005}O$ ensure more effective routes for phonon scattering (such as grain boundaries), thus reducing the thermal conductivity. It has been reported that, for oxides with particle sizes below 100 nm, the thermal conductivity is reduced, which is of importance for their thermoelectric applications [44]. In comparison with oxides, the morphology of ExGr and NrGO consists of flake-like micrometric particles.

Figure 7. SEM images of the pellets containing $Ca_3Co_4O_9$ (**a**), ExGr (**b**), $Zn_{0.995}Al_{0.005}O$ (**c**) and NrGO (**d**). The top view of the pellet is shown for all compositions.

In comparison with individual components, the morphology of the composites is changed. Since the thermoelectrical properties are measured using a square pellet, Figure 8 compares the SEM images of composites taken on the top and cross-section of the pellet. For the composites of $Ca_3Co_4O_9$ with ExGr, the morphology on the pellet top consists of well-contacted micrometric and flake-like particles, with the relative part of the flake-like particles being increased according to the amount of ExGr. At the cross-sectional area of the pellet, the micrometric particles seem to dominate the flake-like ones. This picture is observed when ExGr is replaced with NrGO: on the pellet's top, the flake-like particles are clearly seen, while at the pellet cross-section, micrometric particles mainly appear (Figure 8).

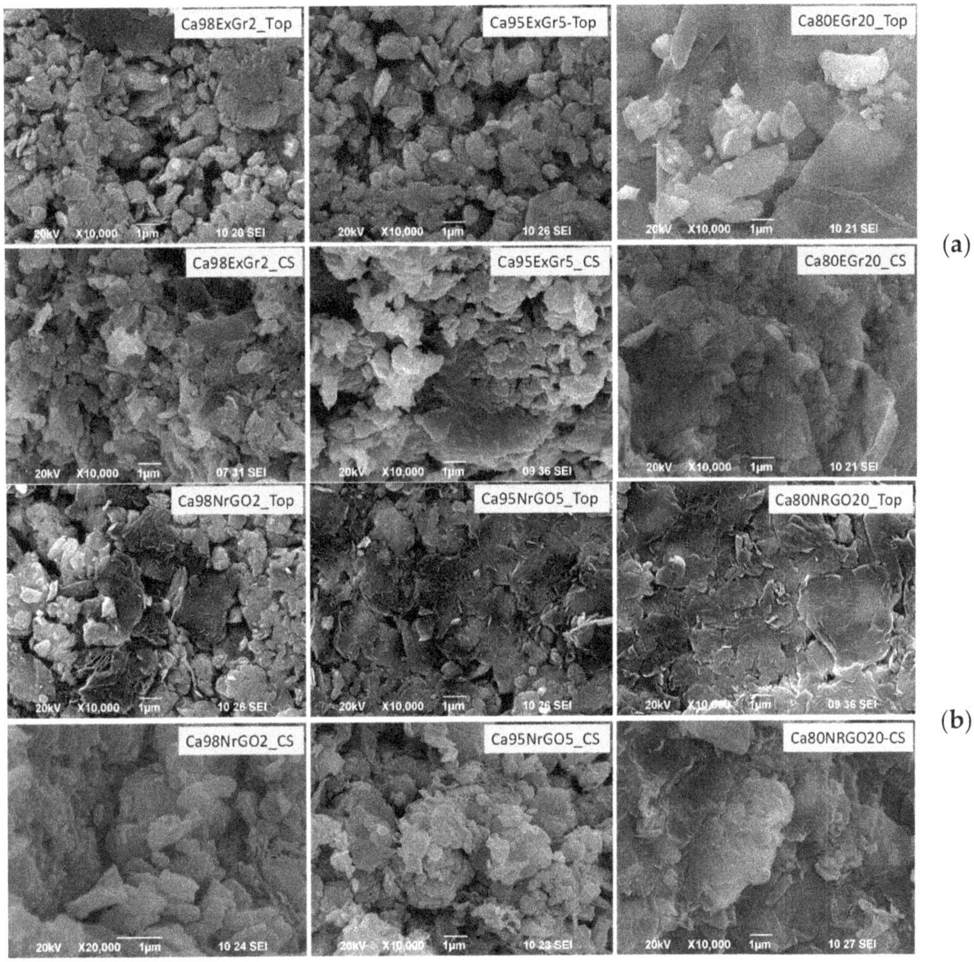

Figure 8. SEM images of the pellets (top and cross-sectional view) for the composites $Ca_3Co_4O_9$-ExGr (**a**) and $Ca_3Co_4O_9$-NrGO (**b**). The carbon additives increase from 2, 5 to 20 wt.% (from **left** to **right**).

To check the SEM observations, an EDS analysis is undertaken (Figure S2). Figure 9 compares the C-to-Co ratio determined on the top and inside of the pellet. For accurate comparison, the element content for all samples was estimated from one and same area: x = 50.7 μm and y = 38.2 μm. In general, EDS accesses different depth profiles depending on the element's nature, but these usually vary between 1 and 2 μm. (It should be noted that the pellet thickness is about 1–2 mm, which is about three orders higher than the EDS element's profile depth). Irrespective of the lower accuracy in the EDS determination of the light C in comparison with the heavier Co, the C-to-Co ratio allows for the element distribution on the top of and inside the pellet to be monitored. The comparison clearly shows that the C-to-Co ratio is higher on the pellet's top. This rule is obeyed for the oxide composites with ExGr and NrGO. The depth distribution of the carbon additives suggests the formation of a heterostructure along the pellet's thickness. This will be the next source of phonon scattering, contributing to a reduction in the thermal conductivity of composites.

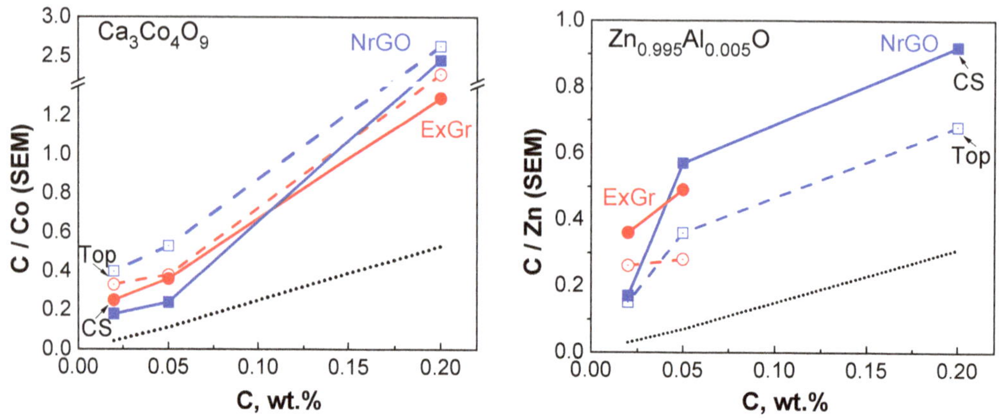

Figure 9. C-to-Co (**a**) and C-to-Zn ratio (**b**), determined from EDS-SEM of composites, as a function of the nominal carbon content in the given pellet. The open symbols correspond to the ratio determined on the top of the pellet, while the solid symbol reflects the ratio determined at the pellet's cross-section. The calculated C-to-Co and C-to-Zn values from the nominal carbon content are indicated as dotted lines.

The common feature of composites of $Ca_3Co_4O_9$ with ExGr and NrGO is the formation of a network comprising well-wrapped micrometric particles with flake-like particles. This is better manifested by the SEM images taken at a higher magnification (Figure 10): the oxide particles are well embedded into the NrGO's thinner flakes, while for ExGr, the thicker flakes serve as a binder between oxide particles. The manner of particle packing ensures good contact between oxide and carbon particles, which is of importance when regulating the thermoelectric properties of composites. In addition, the intrinsic properties of ExGr and NrGO also play a significant role. The best thermoelectric activity of Ca_{98}NrGO is mainly deu to the simultaneous decrease in the thermal conductivity and preservation of the Seebeck coefficient. The figure of merit of Ca_{98}NrGO reaches a magnitude of 0.03 at room temperature, which is lower than the highest reported one for the $Ba_{0.27}CoO_2$ thin film (i.e., figure of merit of 0.11 along in-plane) [45]. In comparison with Ca_{98}NrGO, Ca_{98}ExGr adopts a slightly lower figure of merit (i.e., about 0.025) due to its bigger thermal conductivity. This is related to the different role of ExGr and NrGO regarding the the interface modification. The lower electrical resistivity of ExGr determines its bigger thermal conductivity compared to that of NrGO (Table 1).

The next peculiarity is observed when $Ca_3Co_4O_9$ is replaced with $Zn_{0.995}Al_{0.005}O$ (Figure 11). In this case, it appears that ExGr wraps better the nanometric oxide particles when the carbon content is 5 wt.%. In addition, the top of the pellet appears to be richer in oxide particles in comparison with the pellet's cross-section. The addition of NrGO produces the same picture as is observed for ExGr additives: the oxide particles are well wrapped by NrGO at a content of 5 wt.% and they are more concentrated on the pellet top. It is noticeable that the morphological peculiarities of $Zn_{0.995}Al_{0.005}O$-based composites are opposite to those observed for $Ca_3Co_4O_9$-based composites. This is confirmed by an EDS analysis of the C-to-Zn ratio on the top of and inside in the pellet (Figure 9): the top layers of the pellet become richer in oxide particles (expressed by amount of Zn). Irrespective of the different distribution of carbon additives along the pellets, it is important that, for both type of oxide composites, the carbon particles ensure a good contact between oxide particles, which is of importance for the modification of their thermoelectric properties (Figure 11).

Figure 10. SEM images at higher magnifications for Ca_{98}ExGr (**a**), Ca_{98}NrGO (**b**), Zn_{80}ExGr (**c**), and Zn_{80}NrGO (**d**). These composites display the best figure of merit.

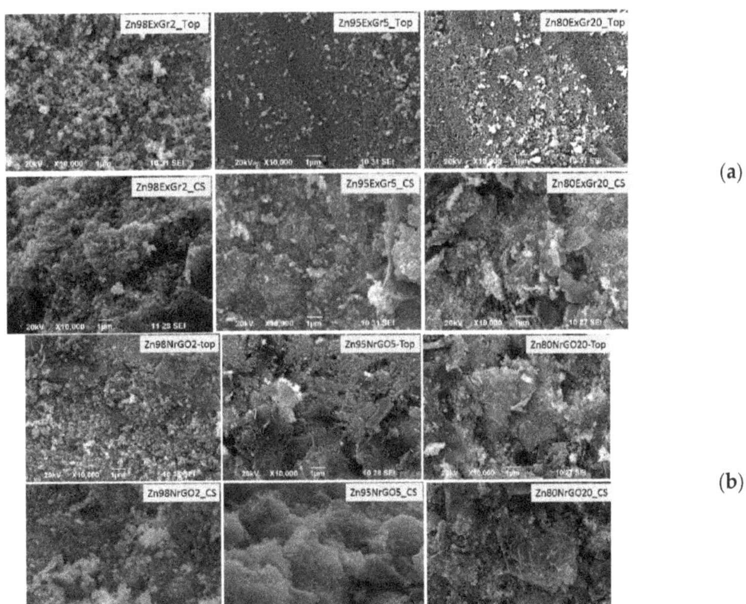

Figure 11. SEM images of the pellets (top and cross-sectional view) of $Zn_{0.995}Al_{0.005}O$-ExGr (**a**) and $Zn_{0.995}Al_{0.005}O$-NrGO composites (**b**). The carbon additives increase from 2 to 20 wt.% (from **left** to **right**).

4. Conclusions

Among the carbon-based materials, the expanded graphite (ExGr) exhibits the lowest electrical resistivity and negative Seebeck coefficient, thus converting it to an n-type thermoelectric material with an amazing power factor (PF), even at room temperature (71 μW/(m.K^2) at 20 °C). Despite the relatively good magnitude of the power factor, the thermal conductivity of ExGr is high, which gives rise to its low figure of merit. Contrary to ExGr, both rGO and NrGO are characterized by extremely low thermal conductivities (more than two orders lower than that of ExGr), but their figures of merit remain small due to their high electrical resistivity and low Seebeck coefficient. A comparison of the thermoelectric properties shows that NrGO outperforms the rGO analogue, thus motivating the use of only NrGO as a component in multiphase composites.

Through ball-milling, composites of oxides and carbon additives are formed, where every individual component retains its structure. The common feature of composites is the formation of a network of well-wrapped oxide particles with flake-like carbon-based particles, thus ensuring good contact between oxide and carbon particles. After pelleting the $Ca_3Co_4O_9$–carbon composites, the pellet top is enriched on carbon additives in comparison with the cross-sectional area. In the case of $Zn_{0.995}Al_{0.005}O$-carbon composites, the opposite trend is observed—the pellet top appears to be richer in oxide particles in comparison with the pellet's cross-sectional area. The different manner of oxide packing is related to their morphology: for $Ca_3Co_4O_9$, well-shaped micrometric particles dominate, while nanoparticles account for $Zn_{0.995}Al_{0.005}O$.

The manner of oxides' particle packing and the intrinsic properties of ExGr and NrGO regulate the thermoelectric properties of composites. The thermoelectric activity of p-type $Ca_3Co_4O_9$ is improved when 2 wt.% of ExGr or NrGO is added. The improved thermoelectric activity is a result of the simultaneous decrease in the thermal conductivity and preservation of the Seebeck coefficient. It is of importance that the thermal conductivity of the $Ca_3Co_4O_9$-based composites decreases even in the case when individual components $Ca_3Co_4O_9$ and ExGr exhibit high thermal conductivity. This is related to the enhanced phonon scattering in composites due to the heterostructural distribution of carbon–oxide phases. For the n-type $Zn_{0.995}Al_{0.005}O$, it is necessary to add more than 5 wt.% of carbon additives to enhance thermoelectric activity. Among ExGr and NrGO, only ExGr leads to a drastic augmentation of the figure of merit of $Zn_{0.995}Al_{0.005}O$ (i.e., more than three orders of magnitude) due to a strong decrease in the electrical resistivity.

The established correlations between the thermoelectric properties of composites and their morphology and the amount of individual components could be used for further optimization of the thermoelectric performance of oxide materials at room temperature.

Supplementary Materials: The following supporting information can be downloaded at: https://www.mdpi.com/article/10.3390/ma16237262/s1, Figure S1: In-house setup for the measurement of the Seebeck coefficient; Figure S2: SEM-EDS analysis of the element content of $Ca_{95}ExGr_5$ composite.

Author Contributions: Conceptualization, S.H. and R.S.; methodology, S.H. and V.V.; software, S.H. and V.V.; validation, S.H. and V.V.; investigation, S.H.; resources, R.S.; data curation, S.H. and V.V.; writing—original draft preparation, S.H. and R.S.; writing—review and editing, R.S.; visualization, S.H. and V.V.; funding acquisition, R.S. All authors have read and agreed to the published version of the manuscript.

Funding: This research was funded by Project Д01-272/02.10.2020—"European Network on Materials for Clean Technologies", funded by the Ministry of Education and Science under the National Program "European Scientific Networks". Project BG05M2OP001-1.001-0008, "National center of mechatronics and clean technologies" funded by the Operational Programme Science and Education for Smart Growth, co-financed by the European Union through the European Regional Development Fund.

Institutional Review Board Statement: Not applicable.

Informed Consent Statement: Not applicable.

Data Availability Statement: Data are contained within the article and Supplementary Materials.

Acknowledgments: Authors are thankful to Dr. Robert Feher from Graphit Kropfmühl GmbH for providing NrGO (EXG U 98300 FNH) and ExGr materials (EXG 98 20/20 μm).

Conflicts of Interest: The authors declare no conflict of interest.

References

1. Feng, Y.; Jiang, X.; Ghafari, E.; Kucukgok, B.; Zhang, C.; Ferguson, I.; Lu, N. Metal oxides for thermoelectric power generation and beyond. *Adv. Compos. Hybrid Mater.* **2018**, *1*, 114–126. [CrossRef]
2. Fergus, J.W. Oxide materials for high temperature thermoelectric energy conversion. *J. Eur. Ceram. Soc.* **2012**, *32*, 525–540. [CrossRef]
3. Wolf, M.; Hinterding, R.; Feldhoff, A. High Power Factor vs. High *zT*—A Review of Thermoelectric Materials for High-Temperature Application. *Entropy* **2019**, *21*, 1058. [CrossRef]
4. Zebarjadi, M.; Esfarjani, K.; Dresselhaus, M.S.; Ren, Z.F.; Chen, G. Perspectives on Thermoelectrics: From Fundamentals to Device Applications. *Energy Environ. Sci.* **2012**, *5*, 5147–5162. [CrossRef]
5. Kieslich, G.; Cerretti, G.; Veremchuk, I.; Hermann, R.P.; Panthofer, M.; Grin, J.; Tremel, W.A. Metal oxides with adaptive structures for thermoelectric applications. *Phys. Status Solidi A* **2016**, *213*, 808–823. [CrossRef]
6. Liu, Y.; Zhi, J.; Li, W.; Yang, Q.; Zhang, L.; Zhang, Y. Oxide Materials for Thermoelectric Conversion. *Molecules* **2023**, *28*, 5894. [CrossRef]
7. Vulchev, V.; Vassilev, L.; Harizanova, S.; Khristov, M.; Zhecheva, E.; Stoyanova, R. Improving of the Thermoelectric Efficiency of $LaCoO_3$ by Double Substitution with Nickel and Iron. *J. Phys. Chem. C* **2012**, *116*, 13507–13515. [CrossRef]
8. Harizanova, S.; Zhecheva, E.; Valchev, V.; Markov, P.; Khristov, M.; Stoyanova, R. Effect of Multiple Metal Substitutions for A- and B-perovskite Sites on the Thermoelectric Properties of $LaCoO_3$. *IJSRST* **2016**, *6*, 428–439.
9. Okhay, O.; Tkach, A. Impact of Graphene or Reduced Graphene Oxide on Performance of Thermoelectric Composites. *C J. Carbon Res.* **2021**, *7*, 37. [CrossRef]
10. Harizanova, S.; Faulques, E.; Corraze, B.; Payen, C.; Zając, M.; Wilgocka-Ślęzak, D.; Korecki, J.; Atanasova, G.; Stoyanova, R. Composites between Perovskite and Layered Co-Based Oxides for Modification of the Thermoelectric Efficiency. *Materials* **2021**, *14*, 7019. [CrossRef]
11. Butt, S.; Xu, W.; Farooq, M.U.; Ren, G.-K.; Mohmed, F.; Lin, Y.; Nan, C.W. Enhancement of Thermoelectric Performance in Hierarchical Mesoscopic Oxide Composites of $Ca_3Co_4O_9$ and $La_{0.8}Sr_{0.2}CoO_3$. *J. Am. Ceram. Soc.* **2015**, *98*, 1230–1235. [CrossRef]
12. Fortulan, R.; Aminorroaya Yamini, S. Recent Progress in Multiphase Thermoelectric Materials. *Materials* **2021**, *14*, 6059. [CrossRef]
13. Prasad, R.; Bhame, S.D. Review on texturization efects in thermoelectric oxides. *Mater. Renew. Sustain. Energy* **2020**, *9*, 1–22. [CrossRef]
14. Zhang, Y.; Zhang, Q.; Chen, G. Carbon and carbon composites for thermoelectric applications. *Carbon Energy* **2020**, *2*, 408–436. [CrossRef]
15. He, X.; Zhang, T.; Chen, F.; Jiang, J. Applications of Graphene in Composite Thermoelectric Materials. *Prog. Chem.* **2018**, *30*, 439–447. [CrossRef]
16. Yamini, S.A.; Santos, R.; Fortulan, R.; Gazder, A.A.; Malhotra, A.; Vashaee, D.; Serhiienko, I.; Mori, T. Room-Temperature Thermoelectric Performance of n-Type Multiphase Pseudobinary Bi_2Te_3–Bi_2S_3 Compounds: Synergic Effects of Phonon Scattering and Energy Filtering. *ACS Appl. Mater. Interfaces* **2023**, *15*, 19220–19229. [CrossRef]
17. Bao, Y.; Sun, Y.; Jiao, F.; Hu, W. Recent Advances in Multicomponent Organic Composite Thermoelectric Materials. *Adv. Electron. Mater.* **2023**, *9*, 2201310. [CrossRef]
18. Jin, H.; Li, J.; Locozzia, J.; Zeng, X.; Wei, P.-C.; Yang, C.; Li, N.; Liu, Z.; He, J.H.; Zhu, T.; et al. Hybrid Organic–Inorganic Thermoelectric Materials and Devices. *Angw. Chemie Int. Ed.* **2019**, *58*, 15206–15226. [CrossRef]
19. Zong, P.; Liang, J.; Zhang, P.; Wan, C.; Wang, Y.; Koumoto, K. Graphene-Based Thermoelectrics. *ACS Appl. Energy Mater.* **2020**, *3*, 2224–2239. [CrossRef]
20. Mulla, R.; White, A.O.; Dunnill, C.W.; Barron, R. The role of graphene in new thermoelectric materials. *Energy Adv.* **2023**, *2*, 606–614. [CrossRef]
21. Zong, P.; Chen, X.; Zhu, Y.; Liu, Z.; Zeng, Y.; Chen, L. Construction of a 3D-rGO network-wrapping architecture in a $Yb_yCo_4Sb_{12}$/rGO composite for enhancing the thermoelectric performance. *J. Mater. Chem. A* **2015**, *3*, 8643–8649. [CrossRef]
22. Chen, J.; Li, L. Thermal Conductivity of Graphene Oxide: A Molecular Dynamics Study. *JETP Lett.* **2020**, *112*, 117–121. [CrossRef]
23. Mu, X.; Wu, X.; Zhang, T.; Go, D.B.; Luo, T. Thermal Transport in Graphene Oxide-From Ballistic Extreme to Amorphous Limit. *Sci. Rep.* **2014**, *4*, 3909–3918. [CrossRef]
24. Wang, W.; Zhang, Q.; Li, J.; Liu, X.; Wang, L.; Zhu, Z.; Luo, W.; Jiang, W. An efficient thermoelectric material: Preparation of reduced graphene oxide/polyaniline hybrid composites by cryogenic grinding. *RSC Adv.* **2015**, *5*, 8988–8995. [CrossRef]

25. Okhay, O.; Gonçalves, G.; Dias, C.; Ventura, J.; Vieira, E.M.F.; Gonçlves, L.M.V.; Tkachb, A.J. Tuning electrical and thermoelectric properties of freestanding graphene oxide papers by carbon nanotubes and heat treatment. *J. Alloys Compd.* **2019**, *781*, 196–200. [CrossRef]
26. Mehmood, T.; Kim, J.-H.; Lee, D.-J.; Dizhur, S.; Odessey, R.; Hirst, E.S.; Osgood, R.M., III; Sayyad, M.H.; Munawar, M.A.; Xu, J. A microstructuring route to enhanced thermoelectric efficiency of reduced graphene oxide films. *Mater. Res. Express* **2019**, *6*, 075614–075627. [CrossRef]
27. Lin, Y.; Norman, C.; Srivastava, D.; Azough, F.; Wang, L.; Robbins, M.; Simpson, K.; Freer, R.; Kinloch, I.A. Thermoelectric power generation from lanthanum strontium titanium oxide at room temperature through the addition of graphene. *ACS Appl. Mater. Interfaces* **2015**, *7*, 15898–15908. [CrossRef]
28. Feng, X.P.; Fan, Y.C.; Nomura, N.; Kikuchi, K.; Wang, L.J.; Jiang, W.; Kawasaki, A. Graphene promoted oxygen vacancies in perovskite for enhanced thermoelectric properties. *Carbon* **2017**, *112*, 169–176. [CrossRef]
29. Sun, S.; Shi, X.-L.; Liu, W.-D.; Wu, T.; Wang, D.; Wu, H.; Zhang, X.; Wang, Y.; Liu, Q.; Chen, Z.-G. Cheap, Large-Scale, and High-Performance Graphite-Based Flexible Thermoelectric Materials and Devices with Supernormal Industry Feasibility. *ACS Appl. Mater. Interfaces* **2022**, *14*, 8066–8075. [CrossRef]
30. Wei, Q.J.; Zhang, L.L.; Zhao, L.; Yang Hao, C.L. Enhanced thermoelectric properties of carbon fiber reinforced cement composites. *Ceram Int.* **2016**, *42*, 11568–11573. [CrossRef]
31. Wei, J.; Zhao, L.; Zhang, Q.; Nie, Z.; Hao, L. Enhanced thermoelectric properties of cement-based composites with expanded graphite for climate adaptation and large-scale energy harvesting. *Energy Build.* **2018**, *159*, 66–74. [CrossRef]
32. Frąc, M.; Pichór, W.; Szołdra, P. Cement composites with expanded graphite as resistance heating elements. *J. Compos. Mater.* **2020**, *54*, 3821–3831. [CrossRef]
33. Han, Z.; Li, J.-W.; Jiang, F.; Xia, J.; Zhang, B.-P.; Li, J.-F.; Liu, W. Room-temperature thermoelectric materials: Challenges and a new paradigm. *J. Mater.* **2022**, *8*, 427–436. [CrossRef]
34. Jiang, J.; Zhu, H.; Niu, Y.; Zhu, Q.; Song, S.; Zhou, T.; Wang, C.; Ren, Z. Achieving high room-temperature thermoelectric performance in cubic AgCuTe. *J. Mater. Chem. A* **2020**, *8*, 4790–4799. [CrossRef]
35. Ohta, H.; Sugiura, K.; Koumoto, K. Recent progress in oxide thermoelectric materials: P-type $Ca_3Co_4O_9$ and n-type $SrTiO_3$. *Inorg. Chem.* **2008**, *47*, 8429–8436. [CrossRef]
36. Sales, B.C.; Jin, R.; Affholter, K.A.; Khalifah, P.; Veith, G.M.; Mandrus, D. Magnetic, thermodynamic, and transport characterization of Na 0.75 Co O 2 single crystals. *Phys. Rev. B* **2004**, *70*, 174419-1–174419-9. [CrossRef]
37. Shi, Z.; Su, T.; Zhang, P.; Lou, Z.; Qin, M.; Gao, T.; Xu, J.; Zhu, J.; Gao, F. Enhanced Thermoelectric Performance of $Ca_3Co_4O_9$ Ceramics through Grain Orientation and Interface Modulation. *J. Mater. Chem. A* **2020**, *8*, 19561–19572. [CrossRef]
38. Baghdadi, N.; Salah, N.; Alshahrie, A.; Koumoto, K. Microwave Irradiation to Produce High Performance Thermoelectric Material Based on Al Doped ZnO Nanostructures. *Crystals* **2020**, *10*, 610. [CrossRef]
39. Wu, X.; Lee, J.; Varshney, V.; Wohlwend, J.L.; Roy, A.K.; Luo, T. Thermal Conductivity of Wurtzite Zinc-Oxide from First-Principles Lattice Dynamics—A Comparative Study with Gallium Nitride. *Sci. Rep.* **2016**, *6*, 22504. [CrossRef]
40. Harizanova, S.; Zhecheva, E.; Valchev, V.; Khristov, M.; Markov, P.V.; Stoyanova, R. Effect of the particle sizes on the thermoelectric efficiency of metal substituted $LaCo_1$-xNixFexO$_3$ perovskites and $Zn_{1-x}Al_xO$ wurtzite. *Bulg. Chem. Commun.* **2017**, *49*, 100–106.
41. Sulaiman, S.; Izman, S.; Udayac, M.B.; Omar, M.F. Review on grain size effects on thermal conductivity in ZnO thermoelectric material. *RSC Adv.* **2022**, *12*, 5428–5438. [CrossRef]
42. Jood, P.; Mehta, R.J.; Zhang, Y.; Peleckis, G.; Wang, X.; Siegel, R.W.; Borca-tasciuc, T.; Dou, S.X.; Ramanath, G. Al-Doped Zinc Oxide Nanocomposites with Enhanced Thermoelectric Properties. *Nano. Lett.* **2011**, *11*, 4337–4342. [CrossRef]
43. Sulaiman, S.; Sudin, I.; Al-Naib, U.M.B.; Omar, M.F. Review of the Nanostructuring and Doping Strategies for High-Performance ZnO Thermoelectric Materials. *Crystals* **2022**, *12*, 1076. [CrossRef]
44. Zhang, D.B.; Li, H.Z.; Zhang, B.P.; Liang, D.D.; Xia, M. Hybrid-structured ZnO thermoelectric materials with high carrier mobility and reduced thermal conductivity. *RSC Adv.* **2017**, *7*, 10855–10864. [CrossRef]
45. Takashima, Y.; Zhang, Y.; Wei, J.; Feng, B.; Ikuhara, Y.; Jun Cho, H.; Ohta, H.J. Layered cobalt oxide epitaxial films exhibiting thermoelectric ZT = 0.11 at room temperature. *Mater. Chem. A* **2021**, *9*, 274–280. [CrossRef]

materials

MDPI

Article

Synthesis and Luminescent Properties of Barium Molybdate Nanoparticles

Maria Gancheva [1,*], Reni Iordanova [1], Iovka Koseva [1], Georgi Avdeev [2], Gergana Burdina [1] and Petar Ivanov [3]

[1] Institute of General and Inorganic Chemistry, Bulgarian Academy of Sciences, "Acad. G. Bonchev" Str., bl. 11, 1113 Sofia, Bulgaria; reni@svr.igic.bas.bg (R.I.); ikosseva@svr.igic.bas.bg (I.K.); g.naburdina@abv.bg (G.B.)

[2] Institute of Physical Chemistry "Acad. Rostislaw Kaischew", Bulgarian Academy of Sciences, "Acad. G. Bonchev" Str., bl. 11, 1113 Sofia, Bulgaria; g_avdeev@ipc.bas.bg

[3] Institute of Optical Materials and Technologies "Acad. Jordan Malinowski", Bulgarian Academy of Sciences, "Acad. G. Bonchev" Str., bl. 109, 1113 Sofia, Bulgaria; petar@iomt.bas.bg

* Correspondence: m.gancheva@svr.igic.bas.bg; Tel.: +359-2-979-3588

Abstract: $BaMoO_4$ was obtained via facile mechanochemical synthesis at room temperature and a solid-state reaction. An evaluation of the phase composition and structural and optical properties of $BaMoO_4$ was conducted. The influence of different milling speeds on the preparation of $BaMoO_4$ was explored. A shorter reaction time for the phase formation of $BaMoO_4$ was achieved using a milling speed of 850 rpm. A milling speed of 500 rpm led to partial amorphization of the initial reagents and to prolongation of the synthesis time of up to 3 h of milling time. Solid-state synthesis was performed via heat treatment at 900 °C for 15 h. X-ray diffraction analysis (XRD), infrared (IR) and UV diffuse reflectance (UV-Vis) and photoluminescence (PL) spectroscopy were carried out to characterize the samples. Independently of the method of preparation, the obtained samples had tetragonal symmetry. The average crystallite sizes of all samples, calculated using Scherrer's formula, were in the range of 240 to 1540 Å. IR spectroscopy showed that more distorted structural MoO_4 units were formed when the compound was synthesized via a solid-state reaction. The optical band gap energy of the obtained materials was found to decrease from 4.50 to 4.30 eV with increasing crystallite sizes. Green- and blue-light emissions were observed for $BaMoO_4$ phases under excitation wavelengths of 330 and 488 nm. It was established that the intensity of the PL peaks depends on two factors: the symmetry of MoO_4 units and the crystallite sizes.

Keywords: $BaMoO_4$; high-energy ball milling; solid-state reaction; nanoparticles; infrared absorption bands; optical band gaps; blue and green photoluminescence; color coordinates

Citation: Gancheva, M.; Iordanova, R.; Koseva, I.; Avdeev, G.; Burdina, G.; Ivanov, P. Synthesis and Luminescent Properties of Barium Molybdate Nanoparticles. *Materials* **2023**, *16*, 7025. https://doi.org/10.3390/ma16217025

Academic Editors: Alexander N. Obraztsov and Antonio Polimeni

Received: 20 September 2023
Revised: 31 October 2023
Accepted: 1 November 2023
Published: 3 November 2023

1. Introduction

$BaMoO_4$ is a compound that belongs to transition-metal oxides with a scheelite-type structure. This phase is an important material and has useful applications in different scientific and technical areas as photocatalysts [1], anode materials for lithium ion batteries (LIB) and sodium ion batteries (SIB) [2], solid-state lasers [3], phosphors [4,5] and host matrices for doping with Re^{3+} ions [5–7]. $BaMoO_4$ possesses good thermal and chemical stability. It is a semiconductor material with wide optical band gap and high luminescence at room temperature [1,3–9]. In the scheelite structure, Ba ions are coordinated with eight oxygen atoms, and Mo ions are connected to four oxygen atoms, forming MoO_4 groups [10]. The MoO_4 structural unit plays an important role in the electronic structural order and luminescence properties. J.C. Sczancoski et al. reported that the photoluminescence (PL) emission is adjudicated to the $MoO_4{}^{2-}$ complex, where cations like Ca^{2+}, Sr^{2+} and Ba^{2+} acts as lattice-modifying agents, affecting the photoluminescence property directly [11]. There are data that provide evidence that $BaMoO_4$ exhibits green and/or blue emissions at room temperature depending on its morphology, defects, crystallite size, different excitation sources and method of preparation [3–26]. The attention of researchers is focused on fast chemical or physical synthesis of $BaMoO_4$ under mild conditions and establishing the relationship between its structure, morphology and luminescence properties [10,20–25]. $BaMoO_4$ particles with shuttle-like microcrystals and an octahedral form exhibit blue emission at room temperature [15]. According to L. Ma et al., $BaMoO_4$ with various morphologies, such as ellipsoid-like, peanut-like, cube-like and flower-like, with sizes above 0.25 μm, exhibited the highest, high, middle and lowest photoluminescent emissions at 560 nm [19]. On the other hand, $BaMoO_4$ with nest-like and decahedron particles show strong green emission and weak red emission [16]. Micron-octahedron and micron-flower $BaMoO_4$ obtained via the sonochemical method exhibited a strong and broad green emission peak with a maximum at around 508 nm [17]. Y. Wang et al. investigated the influence of defects and Jahn–Teller distorted MoO_4 symmetry on the shape and position of the blue emission of $BaMoO_4$ [20]. The blue emission of $BaMoO_4$ powder or film was registered under excitation wavelengths of 250, 280, 350, 360 and 370 nm [9,14,23,25]. Photoluminescence in the green range was visible at the same wavelengths and at higher ones [16,17,21]. According to many authors, the appearance of blue or green emissions strongly depend on the types of defects that form during the experimental conditions and the morphology of the final products [4,12,13,20,22]. Several studies have reported that morphological control can be used to adjust the photoluminescence properties in the crystallite phases with the scheelite-type structure [15–17]. In the literature, there are data that show that the PL intensity and position of band emissions depend on the thermal treatment conditions of powders [24,25].

Different physicochemical routes have been applied for the preparation of $BaWO_4$, such as solid-state reactions [5], hydrothermal synthesis [2,4,9,18], the aqueous mineralization process [15], coprecipitation [13,21], the sonochemical method [1,17], microwave-assisted synthesis [14,23], sol-gel [6], the complex polymerization method [3,24], laser-induced synthesis [25] and mechanochemical synthesis [27–30]. For example, the $BaMoO_4$ phase obtained through the ball milling of initial mixtures of $BaCO_3$ + MoO_3 and $BaCO_3$ + Na_2MoO_4 can be used in the NIR pigment applications [27]. X.B. Chen et al. applied the high-temperature ball milling (HTBM) method for the preparation of $BaMoO_4$:Eu^{3+} red phosphors [28]. On the other hand, $AMoO_4$ (A = Ba, Sr) films were fabricated via a ball-rotation-assisted solid/solution reaction at room temperature [29]. R.C. Lima et al. [31] reported that additional milling led to an increase in the intensity of the emission spectra. Mechanochemical treatment has several advantages, such as shorter reaction time, cost-effectiveness, simplicity and reliability. The mechanochemical activation of solids leads to the appearance of fresh areas and active defects, which improves their catalytic, electrical and optical characteristics. The focus of our group is on the effects of different ball milling conditions for faster synthesis of a variety of inorganic compounds at ambient temperature [32–34]. This motivated us to use mechanochemical treatment for the preparation of $BaMoO_4$. The obtained structural and luminescent properties will be compared with those of $BaMoO_4$ obtained via a solid-state reaction.

2. Materials and Methods

2.1. Direct Mechanochemical Synthesis

The reagents used in mechanochemical treatment are $BaCO_3$ (Merck, purity 99.9%) and MoO_3 (Merck, purity 99.9%). The stoichiometric ratio of the starting materials was 1:1 and corresponded to crystal-phase $BaMoO_4$. High-energy ball milling of the initial mixture was carried out in a planetary ball mill (Fritsch, Premium line, Pulverisette No 7) with two different milling speeds: 500 and 850 rpm. The grinding ball was 2.5 mm in diameter and the ball-to-powder mass ratio was 10:1. Activation was performed in an air atmosphere. To minimize the temperature during milling, the process was carried out for periods of 15 min, with rest periods of 5 min [32–34]. The labels of the milled samples were as follows: $BaMoO_4$-I using a milling speed of 500 rpm, $BaMoO_4$-II using milling speed of 850 rpm.

2.2. Solid-State Reaction

The starting materials for the solid-state reaction were the same as those used for the mechanochemical activation, i.e., $BaCO_3$ (Merck, 99.99%, Rahway, NJ, USA) and MoO_3 (Merck, 99.99%). The stoichiometric proportions of both components were mixed and homogenized in an agate mortar at room temperature. Subsequently, the mixture was transferred to an alumina crucible and thermally treated at 900 °C for 15 h in an electrical furnace. The as-prepared sample was marked as $BaMoO_4$-III.

2.3. Characterizations

Phase identification of $BaMoO_4$ was confirmed via X-ray powder diffraction analysis (XRD). The X-ray powder diffraction study was performed using a Bruker D8 Advance instrument (Bruker, Billerica, MA, USA) equipped with a copper tube (CuKα) and a position-sensitive LynxEye detector. The crystallite size and cell parameters were calculated using HighScore plus 4.5 and ReX software (ReX v. 0.9.3 build ID 202308221535 (2023-08-22)). The crystallite size (D) was measured via diffraction line analysis using the Scherrer equation while taking into account the 2θ (angular positions), Int (integral intensity value) and FWHM (full width at half-maximum). LaB_6 (NIST standard 660c-https://tsapps.nist.gov/srmext/certificates/660c.pdf) was used as a standard [35]. The cell parameters were obtained after refinement of the diffraction lines using the pseudo-Voigt profile function.

Infrared spectra were registered in the range 1200–400 cm^{-1} on a Nicolet-320 FTIR spectrometer using the KBr pellet technique with a spectral resolution of 2 nm. The diffuse reflectance UV-Vis spectra were recorded using a Thermo Evolution 300 UV-Vis Spectrophotometer (Thermo Fisher Scientific, Waltham, MA, USA) equipped with a Praying Mantis device (Harrick Scientific, Pleasantville, NY, USA). For taking background measurements, we used Spectralon. The PL emission spectra were measured using a Horiba Fluorolog 3-22 TCS spectrophotometer (Horiba, Kyoto, Japan) equipped with a 450 W Xenon Lamp (Edinburgh Instruments, Livingston, UK) as the excitation source. This automated modular system has the highest sensitivity among those available on the market, allowing for the measurement of light emission for practically any type of sample. We used double-grating monochromators with emissions in the range of 200–950 nm; Ex. and Em. Bandpasses of 0–15 nm, continuously adjustable from a computer; wavelength accuracy of +/−0.5 nm; and scan speed of 150 nm/s. All spectra were measured at room temperature.

3. Results and Discussion

3.1. XRD Analysis

Comparisons of the structure, crystallite size and symmetry of the MoO_4 units and the optical properties of the samples produced via mechanochemical and solid-state methods were performed. The milling speed is an important parameter during mechanochemical activation, and it was studied. The effects of both milling speeds (500 and 850 rpm) on the reaction time and phase formation of $BaMoO_4$ were established via X-ray diffraction analysis (Figure 1A,B). The XRD patterns of the initial mixture before high-energy milling treatment show the principal peaks of orthorhombic MoO_3 (PDF-98-035-0609) and orthorhombic $BaCO_3$ (PDF-98-001-5196). A low milling speed at 500 rpm for 1h led to a decrease in intensity and broadening of the diffraction lines on the initial reagents. This is a result of a decrease in particle sizes, destruction of the long-range order and partial amorphization. In the same X-ray pattern, a new diffraction line at 26.35°, typical for tetragonal $BaMoO_4$ (PDF-01-089-4570), was observed. Increasing the milling time up to 3 h caused the appearance of additional reflections characteristic of $BaMoO_4$. The small intensity peak at 13.00° and amorphous halo indicate that the full reaction did not occur. The complete reaction between the activated reagents was finished after 5 h of milling time (Figure 1A). In order to verify the reaction time and phase formation of $BaMoO_4$, we used a milling speed of 850 rpm. The higher milling speed led to the appearance of the principal peaks of $BaMoO_4$ after a shorter time of activation (15 min). But diffraction lines typical of unreacted MoO_3 were observed at 2θ = 24° (Figure 1B). A single phase of tetragonal $BaMoO_4$ was synthesized after 30 min of milling time, which is shorter compared to the reaction time activation was applied at 500 rpm. The phase formation of $BaMoO_4$ at the higher milling speed is due to more effective solid-state diffusion between regents during the ball–material collisions. Previously, we pointed out that $BaMoO_4$ was obtained via different methods of preparation, which involved additives, solvents and prolonged heat treatment [4,9,13–18]. The obtained results demonstrated that activation via high-energy ball milling offers enough energy to generate a chemical reaction at room temperature. It was noted that the ball milling of the mixture of $BaCO_3$ and MoO_3 using the milling speed of 850 rpm produces tetragonal $BaMoO_4$ with a faster reaction time. These results demonstrate that mechanochemical treatment is a more substantial approach to the rapid preparation of powder materials. Figure 1C exhibits the XRD pattern of $BaMoO_4$ after heat treatment at 900 °C for 15 h. Remarkable narrowing of the diffraction lines was observed, which occurred due to the high crystallinity of $BaMoO_4$ compared to the mechanochemically synthesized $BaMoO_4$ powders. No additional diffraction lines were found, indicating that the obtained samples were a pure single phase when using both preparation techniques.

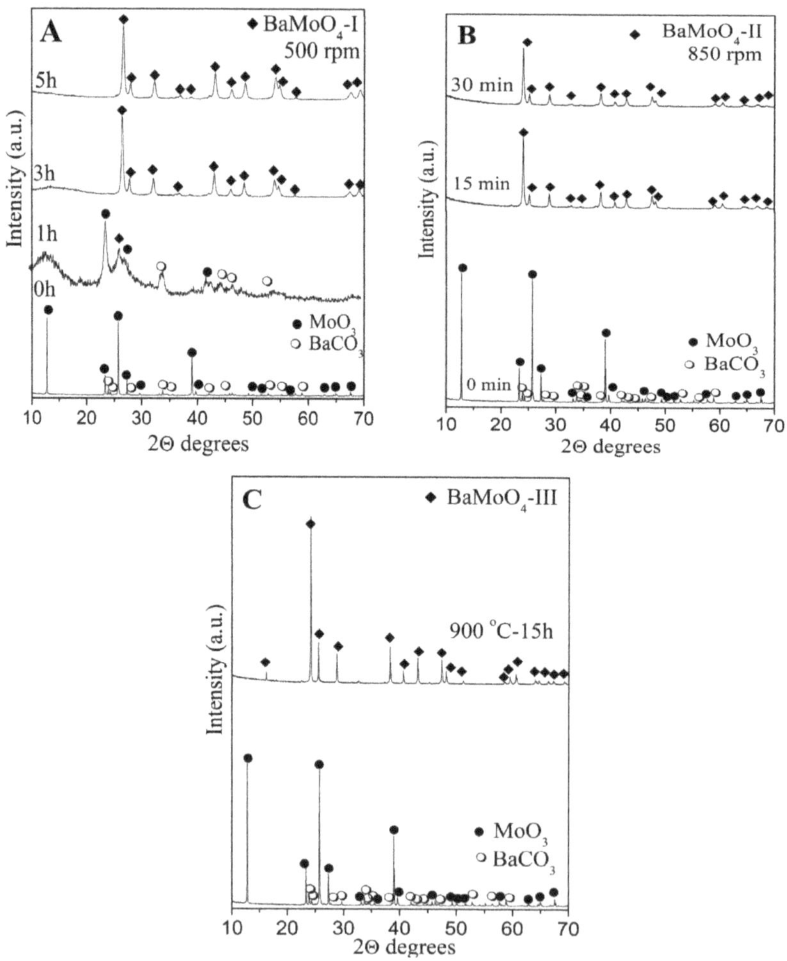

Figure 1. XRD patterns of the initial mixture and mechanochemically activated mixture at 500 rpm (**A**), mechanochemically activated mixture at 850 rpm (**B**) and $BaMoO_4$ obtained after solid-state reaction (**C**).

Table 1 presents lattice parameters a and c, the volume of the unit cells and the average crystallite sizes for the $BaMoO_4$ obtained via different methods of synthesis. A higher milling speed for a short time led to a slight increase in lattice parameters a and c, the volume of the unit cells and the average crystal size. This fact can be attributed to the increase in lattice defects due to higher energy caused by a higher milling speed. A rapid mechanochemical reaction induced a smaller crystallite size due to fast crystallite formation. In solid-state synthesis, the total number of defects significantly decreased, which is demonstrated by an increase in the crystallite size up to 1540 Å. Lattice parameter c decreased compared with those of $BaMoO_4$-I and $BaMoO_4$-II obtained after a shorter time of synthesis. This can be attributed to the clustering of point defects in the *a* plane, which was also indirectly confirmed via infrared analysis.

Table 1. Lattice parameter values, unit cell volumes and average crystallite sizes of $BaMoO_4$ prepared using both methods.

Samples	*a* = *b* (Å)	*c* (Å)	Unit Cell Volume (Å)	Average Crystallite Size (Å)
$BaMoO_4$-I 5 h/500 rpm	5.5718 (9)	12.800 (3)	397.3864	240 (5)
$BaMoO_4$-II 30 min/850 rpm	5.5811 (4)	12.819 (1)	399.2755	270 (5)
$BaMoO_4$-III 900 °C-15 h	5.6083 (2)	12.7019 (5)	399.5213	1540 (4)
PDF-$BaMoO_4$	5.58	12.82	399.17	-

3.2. *Infrared Analysis*

The phase formation of $BaMoO_4$ using different methods of synthesis was confirmed via IR spectroscopy (Figure 2). On the other hand, this analysis is suitable to investigate the local structure of metal ions in the crystallite phases. The infrared spectra display one absorption band in the range of 820 to 835 cm^{-1} due to the υ_3 vibration of the MoO_4 structural units forming the crystalline structure of $BaMoO_4$ [1,6,18,24]. The IR spectra of all $BaMoO_4$ samples are similar. Small differences in the frequency of the main absorption mode were detected. The band position at 825 is shifted up to 835 cm^{-1} and becomes more symmetric using the higher milling speed of 850 rpm. This fact can be attributed to the formation of more symmetric MoO_4 units. The partial amorphization and longer milling time at the speed of 500 rpm favors the formation of a distorted MoO_4 entity. The IR spectrum of $BaMoO_4$ prepared via the solid-state reaction exhibits a more narrow and intense absorption band (820 cm^{-1}), an indication of the higher crystallinity of sample, which is confirmed via XRD analysis (Figure 1C and Table 1). In this case, the appearance of a shoulder at 860 cm^{-1} is attributed to elimination of the ν_3 vibration degeneracy of MoO_4 tetrahedra with different local symmetry [36]. This result can be attributed to the formation of more distorted Mo-tetrahedral groups due to the long sintering time. We can conclude that the mechanochemical activation at higher milling speed leads to the formation of more symmetrical MoO_4 structural units. Bearing in mind the IR features, we expect that the obtained $BaMoO_4$ powders will possess different luminescence behavior according to J.C. Sczancoski et al. [11].

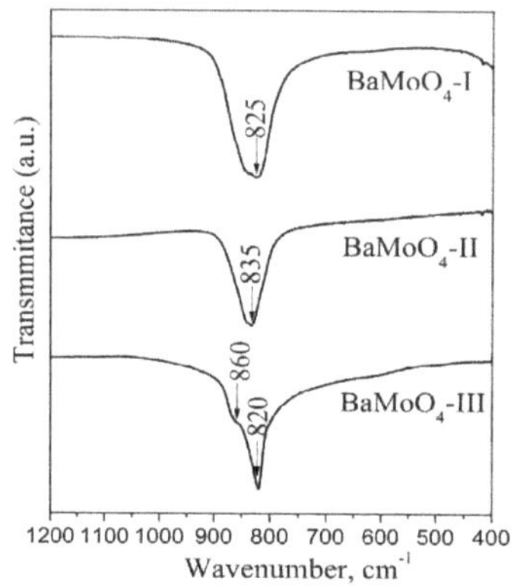

Figure 2. Infrared spectra of $BaMoO_4$ synthesized using different methods of preparation.

3.3. *Ultraviolet–Visible Spectroscopy*

Regarding the effects of the band gaps of inorganic materials on their luminescent and electrical properties, according to the literature data, the value of the optical band gap will increase with a decrease in crystallite size [12,37,38]. The UV-vis absorbance spectra of the obtained $BaMoO_4$ are shown in Figure 3. All samples exhibit one absorption peak in range 215–235 nm, which is attributed to the charge-transfer transitions within the $MoO_4{}^{2-}$ complex [1,3–8]. The observed UV-Vis spectra of our samples are closer to those of $BaMoO_4$ obtained via the hydrothermal method, the molten salt synthesis and the sonochemical route [4,8,17]. A slight shift in this peak is observed upon applying the different methods of preparation (Figure 3A). In the UV-Vis absorption spectrum of $BaMoO_4$-III prepared via the solid-state reaction, a band at 300 nm is also observed. Bearing in mind the reported data, this band corresponds to the creation of the excitonic state in A^{2+} ions (A^{2+} = Ba, Sr, Ca) [3,39]. The optical band gap energy (Eg) of the samples prepared via mechanochemical and solid-state methods was calculated using the Tauc equation [40].

$$\alpha h\nu = A\,(h\nu - Eg)^n,$$

where h is Planck's constant, α is the absorption coefficient, ν is the photon frequency, A is a constant, Eg is the optical band gap and $n = 1/2$ is used for $BaMoO_4$. Based on this equation, the band gaps of the investigated crystal phases are estimated to be above 4.30 eV, as illustrated in Figure 3B. The examined optical band gaps with respect to milling speed and time show a tendency to decrease in value with increasing the crystallite size ($BaMoO_4$-I with D = 240 Å and Eg = 4.50 eV; $BaMoO_4$-II with D = 270 Å and Eg = 4.47 eV). This tendency was also observed for $BaMoO_4$-III (Eg = 4.30 eV with D 1540 Å) prepared via the solid-state reaction. The smaller optical band gap suggests that the crystal structure of $BaMoO_4$-III has more defects, which is established via X-ray analysis and IR spectroscopy. The formation of defects led to the induction of additional electronic states between the valence band and conduction band in the materials, resulting in a reduction in the Eg value. The values of the calculated band gaps for milled samples subjected to different speeds are slightly higher than those of samples synthesized via heat treatment at 900 °C. The obtained values of the optical band gaps are higher than those of $BaMoO_4$ obtained via hydrothermal synthesis, the coprecipitation method, laser-induced synthesis and the microwave plasma method [4,19,24,41]. According to W. Wang et al., materials with lower crystallite size and with larger values of the optical band gap (Eg) possess higher luminescence intensity [37]. To better understand the correlation between the optical properties and crystal structure, we measured the PL spectra of the obtained $BaMoO_4$ samples.

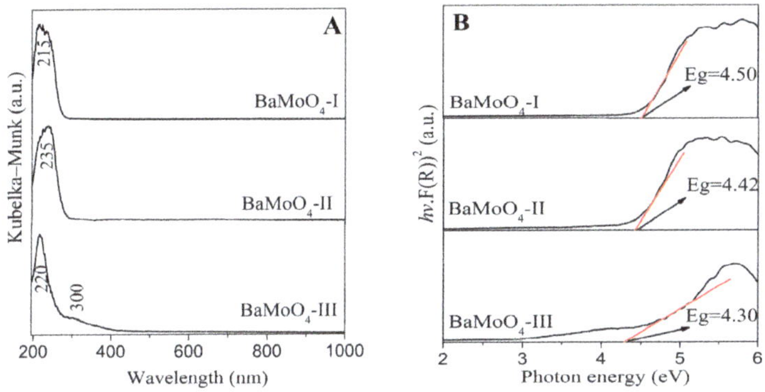

Figure 3. (**A**) UV-Vis spectra of $BaMoO_4$ obtained using different methods of synthesis; (**B**) Tauc's plot of $BaMoO_4$ obtained using different methods of synthesis.

3.4. Photoluminescence (PL) Analysis

The photoluminescence emission behavior of the three $BaMoO_4$ powders were investigated at room temperature and are presented in Figure 4. An asymmetric and intense blue peak with a maximum at 400 nm was recorded under excitation at 330 nm (Figure 4A). The profile of blue emission lines is similar to those of $BaMoO_4$ obtained via the hydrothermal and microwave-induced plasma methods [1,3]. The effect of ball milling speed/time on light emission was investigated and is discussed below. Significant differences were found for the investigated samples. The position of the maximum was changed from 400 to 405 nm for $BaMoO_4$-I obtained using the milling speed of 500 rpm. The PL in the blue region was broader after longer milling time at the same rotation speed ($BaMoO_4$-I). The lowest intensity was observed for $BaMoO_4$-III obtained via the traditional solid-state reaction. A broad green symmetric emission centered at about 560 nm was observed under excitation at 488 nm (Figure 4B). The shape of the PL line is typical of inorganic metal oxide with a scheelite-type structure [20,42]. The PL strength in the green region is almost the same for both samples obtained via mechanochemical activation using milling speeds of 500 and 850 rpm, respectively (Figure 4B). In this case, the broadest peak and lowest intensity were observed for $BaMoO_4$-III obtained via the solid-state route. S. Raghunath et al. reported that $BaMoO_4$ obtained via the precipitation method possesses lower emission intensity in the range of 510 to 590 nm [43]. According to the literature data, blue emission is attributed to charge-transfer transitions within the [MoO_4] complex, while the green emission is due to intrinsic distortion in the [MoO_4] tetrahedron group and can arise due to various factors, like crystallinity, morphology, surface defects, etc. [3,16,20,21,44]. In this instance, the variation in the PL intensity of our samples was caused by the different preparation methods, crystallite sizes and main structural units, i.e., MoO_4. The above results show that $BaMoO_4$-I prepared after a brief milling period using a milling speed of 850 rpm exhibited higher blue luminescence intensity than samples obtained at a lower milling speed of 500 rpm and solid-state reaction, respectively. The broader peak and lower intensity in the blue and green emissions of $BaMoO_4$-III prepared via the solid-state reaction are probably due to the presence of more distorted MoO_4 units according to the obtained IR results. This result is in good agreement with those reported by R. Künzel et al. [4]. The photoluminescence results indicated that the $BaMoO_4$ prepared via mechanochemical activation had a small number of defects. The other factor that affected the PL intensity was the crystallite size, which was clearly observed in both emissions. Our investigations show that the $BaMoO_4$-I and $BaMoO_4$-II obtained via direct mechanochemical synthesis with lower crystallite sizes (240 and 270 Å) possess higher luminescence strength. Similar results have been reported by other authors [37,45]. On the other hand, the lower values of the optical band gap (Eg) and PL emissions of $BaMoO_4$ produced via solid-state synthesis were impacted by the formation of distorted MoO_4 and higher crystallite sizes. From the obtained PL data, we can conclude that symmetric MoO_4 units and lower crystallite size are important factors for improving luminescence efficiency. Our future investigations will be focused on the mechanochemical synthesis $BaMoO_4$ doped with different rare-earth ions (Eu^{3+}, Dy^{3+} and Tb^{3+}) for obtaining materials with multiple colors. As $BaMoO_4$ matrix luminescence is blue or green, the first step will be the preparation of Dy^{3+}-doped $BaMoO_4$ to achieve the white light emission.

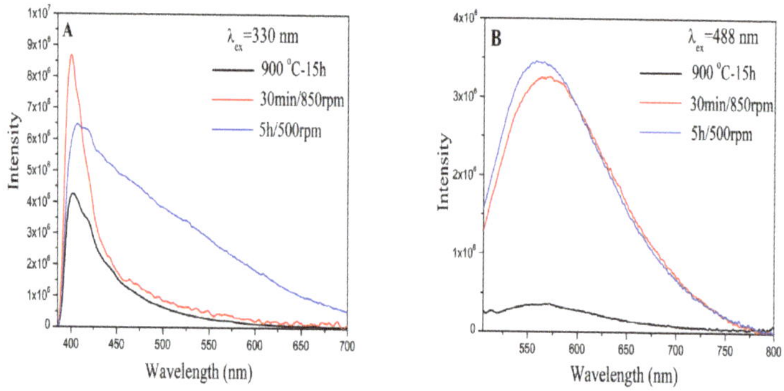

Figure 4. Photoluminescence emission spectra of $BaMoO_4$ obtained using different methods of synthesis under excitation at 330 nm (**A**) and 488 nm (**B**).

Color chromaticity coordinates are another feature that impacts optical properties. The values of the x and y coordinates of $BaMoO_4$ powders were calculated using a standard procedure from their emission spectra and are presented in Figure 5A,B. It can be seen from the figures that they fall within the blue and green areas, respectively. The values of the CIE parameters of the obtained $BaMoO_4$ samples are summarized in Table 2.

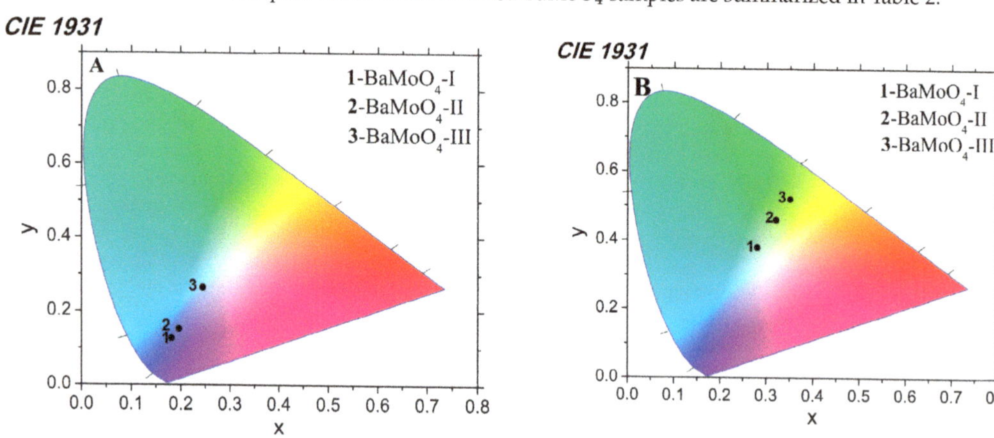

Figure 5. CIE color coordinates of obtained $BaMoO_4$ samples under excitation at 330 (**A**) and 488 nm (**B**).

Table 2. CIE color coordinates of obtained $BaMoO_4$ samples under excitation at 330 and 488 nm.

Samples	x, y (exc. 330 nm)	x, y (exc. 488 nm)
$BaMoO_4$-I	0.18, 0.12	0.28, 0.38
$BaMoO_4$-II	0.20, 0.15	0.32, 0.46
$BaMoO_4$-III	0.25, 0.26	0.35, 0.52

4. Conclusions

In this work we investigate the correlation between the method of synthesis and the structural and optical properties of $BaMoO_4$. Nanoparticles of the $BaMO_4$ with a scheelite-type structure were prepared via a mechanochemical approach and a solid-state reaction. It was established that the milling speed is a crucial parameter for the rapid synthesis of $BaMoO_4$ at room temperature. The crystallite size of both materials obtained at different speeds of mechanochemical activation (500 and 850 rpm) were in the nanoscale range, up to 300 Å. The calculated optical band gaps were wider (above 4.47 eV). The deformation of the structural units that formed the $BaMoO_4$ compound was established via IR spectroscopy. It was found that the optical properties depend on the applied method of synthesis, and therefore, on the structural entity distortion and the crystallite size. Mechanochemically prepared $BaMoO_4$ had stronger and more symmetric photoluminescence spectra in the blue and green regions compared to the sample prepared via solid-state synthesis. The symmetry of MoO_4 structural units and crystallite size are both factors that affected the emission intensity. The obtained results suggest the potential use of $BaMoO_4$ in the production of optoelectronic devices.

Author Contributions: Conceptualization, M.G., I.K. and R.I.; methodology, M.G.; software, G.A. and XRD; investigation, G.B.; writing—original draft preparation, M.G., writing—review and editing, I.K. and R.I.; PL measurements and visualization, P.I. All authors have read and agreed to the published version of the manuscript.

Funding: The authors acknowledge the project Д01-272-"European Network on Materials for Clean Technologies" for providing the opportunity to present the results at the SizeMat4 conference as well as for the financial publication support.

Institutional Review Board Statement: Not applicable.

Informed Consent Statement: Not applicable.

Data Availability Statement: Not applicable.

Acknowledgments: The authors acknowledge the project Д01-272-"European Network on Materials for Clean Technologies" for providing the opportunity to present the results at the SizeMat4 conference as well as for the financial publication support. Research equipment from the Distributed Research Infrastructure INFRAMAT, part of Bulgarian National Roadmap for Research Infrastructures, supported by the Bulgarian Ministry of Education and Science, was used in this investigation.

Conflicts of Interest: The authors declare no conflict of interest.

References

1. Bazarganipou, M. Synthesis and characterization of $BaMoO_4$ nanostructures prepared via a simple sonochemical method and their degradation ability of methylene blue. *Ceram. Int.* **2016**, *42*, 12617–12622. [CrossRef]
2. Ma, X.; Zhao, W.; Wu, J.; Jia, X. Preparation of flower-like $BaMoO_4$ and application in rechargeable lithium and sodium ion batteries. *Mater. Lett.* **2017**, *188*, 248–251. [CrossRef]
3. Marquesa, A.P.A.J.; Meloa, D.M.A.; Longo, E.; Paskocimasa, C.A.; Pizanic, P.S.; Leite, E.R. Photoluminescence properties of $BaMoO_4$ amorphous thin films. *J. Solid State Chem.* **2005**, *178*, 2346–2353. [CrossRef]
4. Künzel, R.; Latini, R.M.; Umisedo, N.K.; Yoshimura, E.M.; Okuno, E.; Courrol, L.C.; Marques, A.P.A. Effects of beta particles irradiation and thermal treatment on the traps levels structure and luminescent properties of $BaMoO_4$ phosphor. *Ceram. Int.* **2019**, *45*, 7811–7820. [CrossRef]
5. Wu, B.; Yang, W.; Liu, H.; Huang, L.; Zhao, B.; Wang, C.; Xu, G.; Lin, Y. Fluorescence spectra and crystal field analysis of $BaMoO_4$: Eu3+ phosphors for white light-emitting diodes. *Spectrochim. Acta A Mol. Biomol. Spectrosc.* **2014**, *123*, 12–17. [CrossRef]
6. Jena, P.; Gupta, S.K.; Natarajan, V.; Sahu, M.; Satyanarayana, N.; Venkateswarlu, M. Structural characterization and photoluminescence properties of sol–gel derived nanocrystalline $BaMoO_4$:Dy^{3+}. *J. Lum.* **2015**, *158*, 203–210. [CrossRef]
7. Sharath, R.A.; Rahulan, K.M.; Flower, N.A.L.; Sujatha, R.A.; Vinitha, G.; Rejeev, L.R.; Saha, T.; Prakashbabu, D. Third-order nonlinear optical characteristics of Er^{3+}-doped $BaMoO_4$ nanostructures. *J. Mater. Sci. Mater. Electron.* **2022**, *33*, 8542–8550. [CrossRef]
8. Afanasiev, P. Molten salt synthesis of Barium molybdate and tungstate microcrystals. *Mater. Lett.* **2017**, *61*, 4622–4626. [CrossRef]

9. Alencar, L.D.S.; Mesquita, A.; Feitosa, C.A.C.; Balzer, R.; Probst, L.F.D.; Marcelo, D.C.B.; Rosmaninho, G.; Fajardo, H.V.; Bernardi, M.I.B. Preparation, characterization and catalytic application of Barium molybdate ($BaMoO_4$) and Barium tungstate ($BaWO_4$) in the gas-phase oxidation of toluene. *Ceram. Int.* **2017**, *43*, 4462–4469. [CrossRef]
10. Oliveira, M.C.; Gracia, L.; Nogueira, I.C.; Gurgel, M.F.C.; Mercury, J.M.R.; Longo, E.; Andres, J. On the morphology of $BaMoO_4$ crystals: A theoretical and experimental approach. *Cryst. Res. Technol.* **2016**, *51*, 634–644. [CrossRef]
11. Sczancoski, J.C.; Bomio, M.D.R.; Cavalcante, L.S.; Joya, M.R.; Pizani, P.S.; Varela, J.A.; Longo, E.; Siu, L.M.; Andres, J.A. Morphology and Blue Photoluminescence Emission of $PbMoO_4$ Processed in Conventional Hydrothermal. *J. Phys. Chem.* **2009**, *C113*, 5812–5822. [CrossRef]
12. Marques, A.P.A.; Picon, F.C.; Melo, D.M.A.; Pizani, P.S.; Leite, E.R.; Varela, J.A.; Longo, E. Effect of the order and disorder of $BaMoO_4$ powders in photoluminescent properties. *J. Fluoresc.* **2008**, *18*, 51–59. [CrossRef] [PubMed]
13. Phuruangat, A.; Thongtem, T.; Thongtem, S. Precipitate synthesis of $BaMoO_4$ and $BaWO_4$ nanoparticles at room temperature and their photolumincnescence properties. *Superlattices Microstruct.* **2012**, *52*, 78–83. [CrossRef]
14. Lim, C.S. Microwave-assisted synthesis and photolumincescence of $MMoO_4$ (M Ca, Ba) particles via a metathetic reaction. *J. Lum.* **2012**, *132*, 1774–1780. [CrossRef]
15. Wu, X.; Du, J.; Li, H.; Zhang, M.; Xi, B.; Fan, H.; Zhu, Y.; Qian, Y. Aqueous mineralization process to synthesize uniform shuttle-like BaMoO4 microcrystals at room temperature. *J. Solid State Chem.* **2017**, *180*, 3288–3295. [CrossRef]
16. Luo, Z.; Li, H.; Shu, H.; Wang, K.; Xia, J.; Yan, Y. Synthesis of $BaMoO_4$ nestlike nanostructures under a new growth mechanism. *Cryst Growth Des.* **2008**, *8*, 2275–2281. [CrossRef]
17. Zhang, J.; Li, L.; Zi, W.; Zou, L.; Gana, S.; Ji, G. Size-tailored synthesis and luminescent properties of three-dimensional $BaMoO_4$, $BaMoO_4$:Eu^{3+} micron-octahedrons and micron-flowers via sonochemical route. *Luminescence* **2015**, *30*, 280–289. [CrossRef]
18. Luo, Y.S.; Zhang, W.D.; Dai, X.J.; Yang, Y.; Fu, S.Y. Facile Synthesis and luminescent properties of novel flowerlike $BaMoO_4$ nanostructures by a simple hydrothermal route. *J. Phys. Chem. C* **2009**, *113*, 4856–4861. [CrossRef]
19. Ma, L.; Sun, Y.; Gao, P.; Yin, Y.; Qin, Z.; Zhou, B. Controlled synthesis and photoluminescence properties of hierarchical $BaXO_4$ (X = Mo, W) nanostructures at room temperature. *Matter. Lett.* **2010**, *64*, 1235–1237. [CrossRef]
20. Wang, Y.; Gao, H.; Wang, S.; Fang, L.; Chen, X.; Yu, C.; Tang, S.; Liu, H.; Yi, Z.; Yang, H. Facile synthesis of $BaMoO_4$ and $BaMoO_4/BaWO_4$ heterostructures with type -I band arrangement and enhanced photoluminescence properties. *Adv. Powder Technol.* **2021**, *32*, 4186–4197. [CrossRef]
21. Cavalcante, L.S.; Sczancoski, J.C.; Tranquilin, R.L.; Joya, M.R.; Pizani, P.S.; Varela, J.A.; Longo, E. $BaMoO_4$ powders processed in domestic microwave-hydrothermal: Synthesis, characterization and photoluminescence at room temperature. *J. Phys. Chem. Solids* **2008**, *69*, 2674–2680. [CrossRef]
22. Sczancoski, J.C.; Cavalcante, L.S.; Marana, N.L.; Silva, R.O.; Tranquilin, R.L.; Joya, M.R.; Pizani, P.S.; Varela, J.A.; Sambrano, J.R.; Li, M.S.; et al. Electronic structure and optical properties of $BaMoO_4$ powders. *Curr. Appl. Phys.* **2010**, *10*, 614–624. [CrossRef]
23. Ryu, J.H.; Yoon, J.W.; Lim, C.S.; Shim, K.B. Microwave-assisted synthesis of barium molybdate by a citrate complex method and oriented aggregation. *Mater. Res. Bull.* **2005**, *40*, 1468–1476. [CrossRef]
24. Marques, A.P.A.; Melo, D.M.A.; Longo, E.; Paskocimas, C.A.; Pizani, P.S.; Leite, E.R. Photoluminescent $BaMoO_4$ nanopowders prepared by complex polymerization method (CPM). *J. Solid State Chem.* **2006**, *179*, 671–678. [CrossRef]
25. Ryu, J.H.; Kim, K.M.; Mhim, S.W.; Park, G.S.; Eun, J.W.; Shim, K.B.; Lim, C.S. Laser-induced synthesis of $BaMoO_4$ nanocolloidal suspension and its optical properties. *Appl. Phys.* **2008**, *A92*, 407–412. [CrossRef]
26. Pereira, W.S.; Sczancoski, J.C.; Lango, E. Tailoring the photoluminescence of $BaMoO_4$ and $BaMoO_4$ heirarachical architectures via precipitation induced by a fast precursor injection. *Mater. Lett.* **2021**, *293*, 129681. [CrossRef]
27. Baby, O.M.; Balamurugan, S.; Ashika, S.A.; Fathima, T.K.S. Synthesis and characterization of high NIR refecting ecofriendly $BaMoO_4$ pigments in scheelite family. *Emergent Mater.* **2022**, *5*, 1213–1225. [CrossRef]
28. Chen, X.B.; Shao, Z.B.; Tian, Y.W. New method for preparation of luminescent lanthanide materials $BaMoO_4$:Eu^{3+}. *Mater. Technol.* **2011**, *26*, 67–70. [CrossRef]
29. Rangappa, D.; Fujiwara, T.; Watanabe, T.; Yoshimura, M. Fabrication of $AMoO_4$ (A = Ba, Sr) film on Mo substrate by solution reaction assisted ball-rotation. *Mater. Res. Bull.* **2008**, *43*, 3155–3163. [CrossRef]
30. Xiao, E.C.; Li, J.; Wang, J.; Xing, C.; Guo, M.; Qiao, H.; Wang, Q.; Qi, Z.M.; Dou, G.; Shi, F. Phonon characteristics and dielectric properties of $BaMoO_4$ ceramic. *J. Mater.* **2018**, *4*, 383–389. [CrossRef]
31. Lima, R.C.; Anicete-Santos, M.; Orhan, E.; Maurera, M.A.M.A.; Souza, A.G.; Pizani, P.S.; Leite, E.R.; Varela, J.A.; Lango, E. Photoluminescent property of mechanically milled $BaWO_4$ powder. *J. Lim.* **2007**, *126*, 741–746. [CrossRef]
32. Dimitriev, Y.; Gancheva, M.; Iordanova, R. Effects of the mechanical activation of zinc carbonate hydroxide on the formation and properties of zinc oxides. *J. Alloys Compd.* **2012**, *519*, 161–166. [CrossRef]
33. Gancheva, M.; Szafraniak-Wiza, I.; Iordanova, R.; Piroeva, I. Comparative analysis of nanocrystalline $CuWO_4$ obtained by high-energy ball milling. *J. Chem. Techn. Metall.* **2019**, *54*, 1233–1239. Available online: https://journal.uctm.edu/node/j2019-6/14_18-220_p_1233-1239.pdf (accessed on 18 September 2023).
34. Gancheva, M.; Iordanova, R.; Mihailov, L. Direct mechanochemical synthesis and characterization of $SrWO_4$ nanoparticle. *Bulg. Chem. Commun.* **2022**, *54*, 337–342. Available online: http://bcc.bas.bg/BCC_Volumes/Volume_54_Number_4_2022/bcc-54-4-2022-337-342-gancheva-nc02.pdf (accessed on 18 September 2023).

35. Black, D.; Mendenhall, M.; Henins, A.; Filliben, J.; Cline, J. The Certification of Standard Reference Material 660C for Powder Diffraction, Powder Diffraction. 2020. Available online: https://tsapps.nist.gov/srmext/certificates/660c.pdf (accessed on 2 November 2023).
36. Iordanova, R.S.; Milanova, M.K.; Kostov, K.L. Glass formation in the MoO_3–CuO system. *Phys. Chem. Glas. Eur. J. Glass Sci. Technol. B.* **2006**, *47*, 631–637. Available online: https://www.ingentaconnect.com/content/sgt/ejgst/2006/00000047/00000006/art00003 (accessed on 18 September 2023).
37. Wang, W.; Pan, Y.; Zhang, W.; Liu, X.; Li, L. The size effect to O^{2-}—Ce^{4+} charge transfer emission and band gap structure of Sr_2CeO_4. *Luminescence* **2018**, *33*, 907–912. [CrossRef]
38. Bandi, S.; Vidyasagar, D.; Adil, S.; Singhd, M.K.; Basud, J.; Srivastav, A.K. Crystallite size induced bandgap tuning in WO_3 derived from nanocrystalline tungsten. *Scr. Mater.* **2020**, *176*, 47–52. [CrossRef]
39. Spassky, D.A.; Ivanov, S.N.; Kolobanov, V.N.; Mikhailin, V.V.; Zemskov, V.N.; Zadneprovski, B.I.; Potkin, L.I. Optical and luminescent properties of the lead and barium molybdates. *Radiat. Meas.* **2004**, *38*, 607–610. [CrossRef]
40. Tauc, J. Absorption edge and internal electric fields in amorphous semiconductors. *Mater. Res. Bull.* **1970**, *5*, 721–729. [CrossRef]
41. Klinbumrung, A.; Phuruangrat, A.; Thongtem, T.; Thongtem, S. Synthesis, Characterization and optical properties of $BaMoO_4$ synthesized by microwave plasma method. *Russ. J. Inorg. Chem.* **2018**, *63*, 725–731. [CrossRef]
42. Cavalcante, L.S.; Sczancoski, J.C.; Espinosa, J.W.M.; Varela, J.A.; Pizani, P.S.; Longo, E. Photoluminescent behavior of $BaWO_4$ powders processed in microwave-hydrothermal. *J. Alloys Compd.* **2009**, *474*, 195–200. [CrossRef]
43. Raghunath, S.; Balan, R. Solvent assisted synthesis and characterization of $AMoO_4$ (A = Ca, Sr & Ba) nanomaterials. *Mater. Today Proc.* **2021**, *46*, 2930–2933.
44. Benchikhi, M.; Ouatib, R.E.; Guillement-Fritsch, S.; Er-Rakho, L.; Durand, B. Characterization and photoluminescence properties of ultrafine copper molybdate (α-$CuMoO_4$) powders via a combustion—Like process. *Int. J. Mener. Metall. Mater.* **2016**, *23*, 13401345. [CrossRef]
45. Sabri, N.S.; Yahya, A.K.; Talari, M.K. Emission properties of Mn doped ZnO nanoparticles prepared by mechanochemical processing. *J. Lim.* **2012**, *132*, 1735–1739. [CrossRef]

materials

MDPI

Article

Preparation, Characterization, and Antibacterial Properties of Cu-Fibreboards

Lyubomir Aleksandrov [1,*], Nadezhda Rangelova [2], Nevena Lazarova-Zdravkova [3], Nelly Georgieva [3], Mirela Dragnevska [4] and Sanchi Nenkova [4]

1 Institute of General and Inorganic Chemistry, Bulgarian Academy of Sciences, 1113 Sofia, Bulgaria
2 Department of Industrial Safety, University of Chemical Technology and Metallurgy, 1756 Sofia, Bulgaria; rangelovang@gmail.com
3 Department of Biotechnology, University of Chemical Technology and Metallurgy, 1756 Sofia, Bulgaria; n.a.lazarova@gmail.com (N.L.-Z.); neli@uctm.edu (N.G.)
4 Department of Pulp, Paper and Printing Arts, University of Chemical Technology and Metallurgy, 1756 Sofia, Bulgaria; mireladr@mail.bg (M.D.); nenkova@uctm.edu (S.N.)
* Correspondence: lubomirivov@gmail.com; Tel.: +359-888-785-420

Abstract: In the present study, copper modified fibreboards were prepared and their existing phase, morphology, and antibacterial behaviour were investigated. The copper content and the physical and mechanical properties of fibreboards (thickness, bending strength, and swelling) were determined. X-ray diffraction analysis (XRD) showing diffraction peaks typical for cellulose, Cu_2S, and Na_2SO_4, depended on the preparation conditions. The average size of the Cu_2S crystals varied between 20 and 50 nm. The morphology of the obtained fibreboards, as well as the size and shape of copper particles, were observed by scanning electron microscopy (SEM) and transition electron microscopy (TEM). The antibacterial activity was tested against Gram-positive (*Bacillus subtilis* 3562) and Gram-negative (*Escherichia coli* K12 407) bacteria. The tests showed that the materials had higher antibacterial activity against *E. coli*, which depended on their preparation conditions. Based on these results, the obtained copper fibreboards can be used as antibacterial agents in the packaging and building industry.

Keywords: fibreboards; modification; copper nanoparticles; antibacterial activity

Citation: Aleksandrov, L.; Rangelova, N.; Lazarova-Zdravkova, N.; Georgieva, N.; Dragnevska, M.; Nenkova, S. Preparation, Characterization, and Antibacterial Properties of Cu-Fibreboards. *Materials* **2023**, *16*, 6936. https://doi.org/10.3390/ma16216936

Academic Editor: Aivaras Kareiva

Received: 16 September 2023
Revised: 12 October 2023
Accepted: 27 October 2023
Published: 28 October 2023

1. Introduction

Fibreboards are structural and decorative materials, fibrous homogeneous panels made from lignocellulosic materials that are combined most commonly with a synthetic resin and then bonded together under heat and pressure [1–3]. It is believed that, of all wood materials, the panel of fibreboards has the uniform structure because they are made of very fine and flexible particles, which determine the specific physical and mechanical parameters [4]. The manufactured fibreboards can find applications in various industries, such as the furniture industry, the automobile industry, the building industry, packaging, and electronic device applications [5–7]. An essential part of the woodworking sector is the production of wood composites, including wood fibreboards. The consumption of such composite materials in 2022 reached about 14.71 million m3, mainly determined in different industries [8]. The main component of the fibreboards is fibrous materials; as with lignocellulose materials, many sources can be used: corn biomass, wood fibres, lignin, and others [1–3,9,10]. Lignosulfonates are successfully used as binders used in the production of fibreboards. The new tendency is the manufacturing of fibreboards, where the waste technical hydrolysis lignin is used as an additive to the wood mass and its physical and mechanical properties are investigated [2,11,12]. In order to improve the thermal, physical, and mechanical properties of the fibreboards, various additives are often used (CuO, ZnO, Al_2O_3, nanoparticles, multi carbon nanotubes, and others). The improved properties are the result of the final composite becoming stronger due to the covering of unwanted cracks and voids by the used additive [10,13–17]. For instance, Alabduljabbar et al. [10] reported that when the concentration of alumina nanoparticles increased, the mechanical properties of the panels were significantly positively affected.

Due to contact with water, fibreboards used in humid environments have low durability. For increased durability and antimicrobial attack reduction, materials with metallic nanoparticles (such as ZnO) were developed [13]. Kandelbauer and Widsten described [18] the basic strategies to achieve antimicrobial properties on the surface of materials. As protection coatings against the growth of microorganisms, antimicrobial additives of organic and inorganic origin were used [18]. Inorganic additives include metals such as silver, copper, and zinc in different forms (salt; nano-sized), TiO_2, or others. It was reported that the addition of nanoparticles to the surface of materials leads to improved antimicrobial resistance [18,19]. In order to form particles and avoid clustering, it is necessary to stabilize them in a suitable matrix, which will ensure a homogeneous structure and their uniform distribution [20–25]. It was reported that copper nanoparticles are a new generation of wood preservative materials that work against decay fungi when they are used instead of conventional copper [17]. Many authors reported the antimicrobial effect of copper-based composite coatings against Gram-positive and Gram-negative bacteria [26,27]; others focused on films containing copper nanoparticles (CuNPs) [28]. At the same time there is scarce of information about the antibacterial effect of nanoparticles with the Cu_2S active form. Recently, the antibacterial activity of CuS/Cu_2S under near-infrared (NIR) irradiation was reported [29]. In our previous studies, the mechanism of the incorporation of Cu_2S into wood fibres by the coordinative bonding of copper ions with lignocelluloses materials and technology for wood fibre plates was reported [1,30]. The good prerequisite for the successful incorporation of copper particles can be explained by the fact that lignocellulosic materials contain a large number of functional groups from the building blocks of wood [30]. Considering the availability of few scientific works related to the antibacterial activity of wood-modified fibreboards, as well as our previous experience with the antimicrobial efficiency of various composite materials, motivated us to check the ability to modify fibreboards with copper and to clarify their antibacterial activity at different preparation conditions.

2. Materials and Methods

2.1. Material and Fibreboards Preparation

Five types of fibreboards were obtained using as start materials wood fibres (WF, Welde Bulgaria AD, Troyan, Bulgaria) and a two-component cupri reduction system as described previously [1,27]. $CuSO_4*5H_2O$ (Merck KGaA, Amsterdam, The Netherlands) was used as the metal precursor to produce CuNps. On the other hand, $Na_2S_2O_3*5H_2O$ (Valerus Ltd., Sofia, Bulgaria) was used as a reductor.

The main stages of obtaining of the fibreboards are summarized as follows:

The supplied wood fibres were diluted to a concentration of 4–6% with water. The next step was gluing, where 15% phenol formaldehyde resin as a binding agent (46.0–47.0% dry solids content, viscosity 300–450 mPa.s, Dynea, Bucharest, Romania) and 10% paraffin suspension as a water-repelling agent (Lukiol, Burgas, Bulgaria) in 1% of absolute dry fibres were used. After that, the wood fibre mass was further diluted by water to 1.3–1.8%. Using a dewatering process, the wood carpet was obtained. The moulding had a size of 30 × 30 cm, and the water solutions of the modifying mixture were applied to the surface of the obtained wood carpet [1]. The modification process was performed at 45% of the mixture to the absolute dry wood fibres and corresponding to the molar ratio between the components ($CuSO_4*5H_2O$:$Na_2S_2O_3*5H_2O$): 1:1, 1:1.5, and 1:2. The duration of residence of the applied modifying mixture on the carpet before the pressing process was 7, 12, and 17 min, respectively. The resulting modified wood carpets were pressed in a laboratory press at 170 °C for 10 min at high pressure (~4 MPa). The density of the obtained fibreboards was not less than 900 kg/m^3 [6].

2.2. Physical and Mechanical Characterizations of the Fibreboards

Physical and mechanical properties such as bending strength, thickness, and swelling were determined in accordance with national standards: BDS EN 310:1999, BDS EN 317:1998, and BDS EN 324-1:2001 [31–33]. The copper content in the samples was determined by inductively coupled plasma atomic emission spectroscopy ((ICP)—Prodigy High Dispersion ICP Spectrometer—Teledyne Leeman Labs, Mason, OH, USA), by thermal and acid decomposition methods. The samples were first immersed in concentrated HNO_3 solutions, ensued to dissolve all metal nanoparticles in the solutions. The resultant solutions were then sampled and quantified on ICP. All tests were performed three times and the average values are reported.

2.3. Structural and Morphological Characterizations of the Fibreboards

X-ray diffraction patterns were used to investigate the crystalline phases that appear in the materials. A Bruker D8 Advance diffractometer, Karlsruhe, Germany, was used at Cu Kα radiation in the $10 < 2\theta < 80$ range. The diffraction peaks of the samples were assigned using HighScore XRD analysis software version 3.0.4. The Debye Scherrer equation, $D = K\lambda/\beta\cos\theta$, was used to calculate the crystalline size of the crystals phase, where D is the crystalline size, K represents the Scherrer constant (0.98), λ is the X-ray wavelength (1.5418), and β is the full width at half maximum. The thermal stability of the obtained fibreboards was examined by parallel differential thermal analysis and thermogravimetry (DTA/TG) with the Seteram Labsysis Evo 1600 instrument, Lyon, France, in the temperature range of 25–650 °C at the heating rate of 10 K/min in the air atmosphere. Optical microscopy images were recorded using a Light Microscope BOEKO, Boeckel & Co. GmbH & Co. KG, Hamburg, Germany, at magnification of 5×. SEM images were recorded using the microscope JEOL JSM 6390, Oxford instrument, JEOL Ltd., Tokyo, Japan, at an accelerating voltage of 20 kV. Prior to analysis, the samples were coated with gold using a JEOL JFC-1200 coater in order to improve the topographic examination of the samples. The energy dispersive X-ray spectroscopy (EDX) images were obtained using the INCA Oxford instrument. TEM images were recorded using the transmission electron micrographs using a JEOL-2100, Oxford instrument, JEOL Ltd., Tokyo, Japan, at an accelerating voltage of 200 kV. The specimen was prepared by grinding and dispersing the powder in ethanol by ultrasonic treatment for 6 min. The suspension was dripped on a standard carbon/Cu grid.

2.4. Antibacterial Activity of the Fibreboards

The model bacterial strains used in this study included the Gram-positive bacteria *Bacillus subtilis* NBIMCC 3562 and the facultative anaerobic Gram-negative *Escherichia coli* K12 NBIMCC 407. The strains were obtained from the Bulgarian National Bank of Industrial Microorganisms and Cell Cultures. The cultures were grown, subcultured, and maintained in Luria–Bertani (LB) medium and stored at 4 °C. For the experiment, a single colony of each organism was inoculated into 50 mL of LB broth and incubated overnight (24 h) at 30 °C for *B. subtilis* 3562 and 37 °C for *E. coli* K 12 with shaking at 200 rpm. Then, 10 mg from each tested material were placed in a flask with the investigated strains. The control contained only a bacterial culture.

Bacterial growth in the presence of the tested materials was monitored by optical density measurements at 610 nm (OD_{610}) over 24 h at hourly intervals using a UV-Vis spectrophotometer (VWR UV-1600, VWR Corporate Headquarters, Radnor, PA, USA). The growth curves of the test organisms were analyzed graphically as a plot of OD_{610} versus contact time. The bacterial growth-inhibiting effect of the obtained materials was further confirmed by plating on LB agar plates, and the assay was based on the reduction in viable cells after exposure to the materials. Following 24 h of strain cultivation in the presence of the test materials, 100 μL of bacterial broth suspension was seeded on LB agar plates. Samples containing only bacteria were used as controls. The plates were incubated for 24 h at 30 °C for *B. subtilis* and 37 °C for *E. coli* K12, followed by counting the number of colonies on the plate. The measurement of the antibacterial activity of fibreboards was calculated according to Rangelova et al. [21]:

$$\text{Cell reduction} = \left(1 - \frac{\text{Test sample(CFU/mL)}}{\text{Control(CFU/mL)}}\right) \times 100\%.$$

The model strains were cultivated for 24 h with the fibreboards, and the release of copper ions in the suspensions was determined with the help of ICP analysis. All determinations were performed three times in duplicate sets and the average values are reported.

3. Results and Discussion

Modified fibreboards containing copper ions were developed using the two-component reduction system $CuSO_4*5H_2O:Na_2S_2O_3*5H_2O$ at the different ratios mentioned in Table 1. Visually, the colour on the surface of the obtained fibreboards changed from brown for the samples without copper to black depending on the preparation condition of the samples. The fibreboard preparation conditions and average results for copper content, physical, and some mechanical properties are given in Table 1. As can be seen, the molar ratio of the modifying mixture ($CuSO_4*5H_2O:Na_2S_2O_3*5H_2O$) has an effect on the copper content of the samples. Fibreboards F3 and F4 were characterized with lowest copper content and were obtained at the same molar ratio (1:2). This means that the duration of residence of the applied modifying mixture on the carpet before pressing did not significantly influence the copper content in the samples. The most suitable conditions for the highest copper content were obtained by applying the conditions for preparation of F1 fibreboard (see Table 1).

Table 1. Fibreboard production conditions (copper reduction system; duration of residence of the applied modifying mixture on the carpet before pressing) and characteristics.

Fibreboards	Conditions 45% Reagents to WF and 170 °C	Copper Content in Samples, %	Physical and Mechanical Properties		
			Thickness, mm	Bending Strength, MPa	Swelling, %
F0	-	0.01	2.8	49.78	34.48
F1	ratio 1:1; 17 min	1.07	3.0	33.30	19.35
F2	ratio 1:1.5; 12 min	0.76	2.9	41.62	25.92
F3	ratio 1:2; 7 min	0.36	2.8	44.01	32
F4	ratio 1:2; 17 min	0.40	2.9	40.84	34.61

In Table 1, some physical and mechanical properties of the obtained fibreboards can also be found. If it is compared, there are no significant differences in the values of the thickness (mm) between the samples. The other two parameters (bending strength and swelling) decreased compared to the unmodified fibreboards, but nevertheless, the values were within the accepted range of requirements according to European and national standards [34,35]. Similar results for the reduction in some physical properties for fibreboards modified by other metal nanoparticles (zinc and silver) were reported [36]. It was noticeable that sample F1 has the lowest value of bending strength and swelling due to the stable interaction between the matrix and the modifier agent. We consider that with the increase in copper content, free functional groups of lignocellulose materials are occupied with copper ions, and in this case, it is more difficult for phenol formaldehyde resin to affect a better connection with WF.

XRD patterns on the surface of the investigated fibreboards are presented in Figure 1. As can be seen, typical sharp peaks around 2θ = 14, 16, and 22, which are characteristic for cellulose, appeared for all samples except for sample F2 [37–39]. On the other hand, the modified fibreboards containing copper showed some diffraction lines that correspond to the following crystal phases: Cu_2S (ICSD 98-062-8818) and Na_2SO_4 (ICSD 98-010-0458). The obtained phases proved the proposed probable mechanism of the reduction process that was reported previously by us [30]. For the samples, the F3 and F4 diffraction peaks characteristic of the cellulose were dominating compared to the other characteristic diffraction lines. For the F2 fibreboard, it was difficult to identify the cellulose diffraction peaks, but several crystal phases appeared. Clearly, sample F1 was characterized by a predominating Cu_2S crystal phase compared to the other fibreboards. Obviously, for the copper content, the ratio of the modified mixture was a critical parameter. On the other hand, for successful preparation of Cu_2S, both parameters, the ratio of the modified mixture and the duration of residence of the applied modifying mixture on the carpet before pressing, were important. In addition, the crystal size of the formed Cu_2S was calculated by the Debye Scherrer equation. The average size of the crystals from copper sulfide varied between 20 and 50 nm. The smallest particles were calculated in sample F4, around 20 nm. For sample F1, the average size of Cu_2S was 31 nm, while in samples F2 and F3, the particles had sizes at around 35 and 50 nm, respectively.

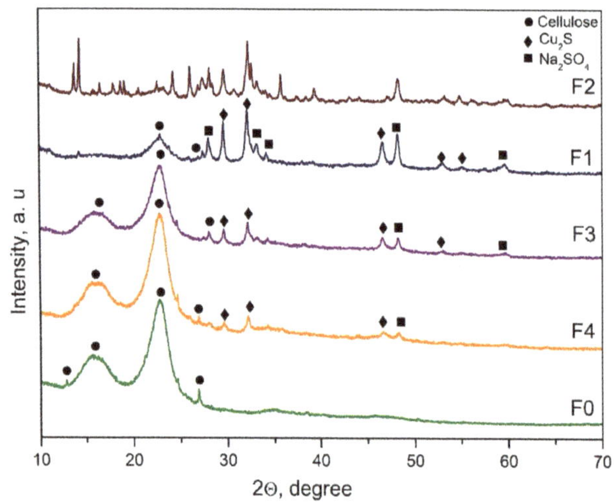

Figure 1. XRD patterns of unmodified and copper modified fibreboards.

In order to determine the thermal stability of modified fibreboards the DTA/TG analysis was performed (Figure 2). The small endo effect characteristic for all DTA curves (Figure 2a) at ~70 °C was due to the moisture and volatile matters evaporation. This effect corresponded to the initial reduction in weight in the samples that appears in TG curves (Figure 2b) and the mass loss varied from 2 to 5 wt.% depending on the copper content. The next broad and intense exothermic effects may correspond to the simultaneous degradation of the lignocellulose components [40–43]. Cellulose degradation occurred in the temperature range between 200 °C and 400 °C, while lignin degradation was achieved at higher temperatures [36–38]. All samples showed quite similar thermogravimetry curves and thermal degradation behaviour. The mass loss varied from 88 to 95 wt.%.

The microscope reflection images of the surface of the obtained fibreboards are presented in Figure 3. The unmodified sample (F0) was characterized by a typical fibrous structure. A non-homogeneous distribution of copper on the sample surface for the F4 fibreboard was observed (see white rings for sample F4). To a lesser extent, this problem is also seen in sample F3. A densification of the surface and a more even distribution of the layer without the presence of aggregates for F1 and F2 fibreboards were observed. According to the presented images, the F1 fibreboard was characterized by the most homogeneous, flat, and dense surface. Obviously, the ratio between the modified mixture played an important role for homogenous distribution of copper on the samples surface.

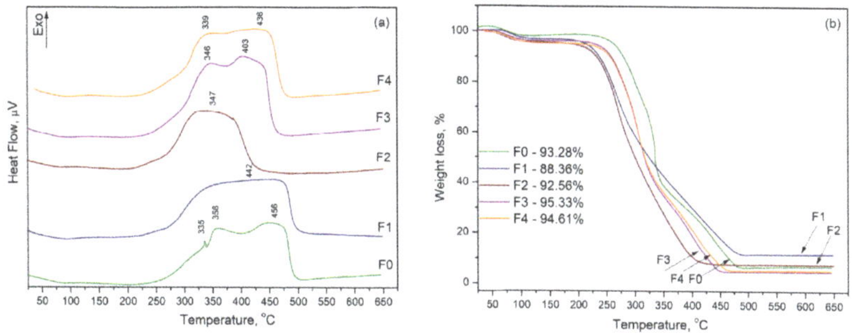

Figure 2. DTA curves (**a**) and TG curves (**b**) of unmodified and copper modified fibreboards.

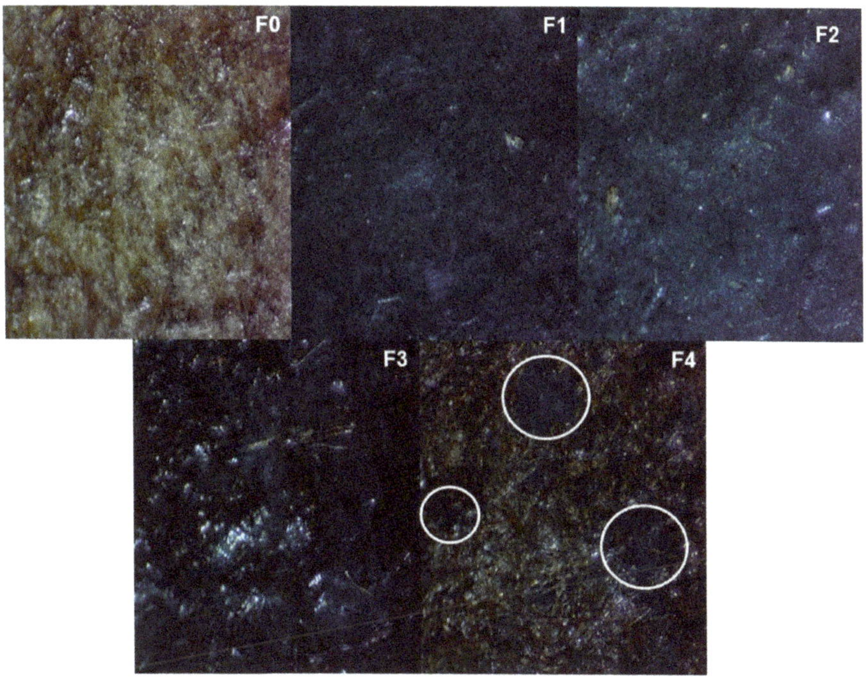

Figure 3. Optical microscopy images on the surface of unmodified and copper modified fibreboards at magnification 5×.

The morphology of the fibreboards and the size and shape of the copper crystallites were observed by SEM and TEM analyses. Due to some limitations, we decided to compare the unmodified fibreboard (F0) and the sample with the best homogeneity (F1). The SEM micrographs and EDX analysis are presented in Figure 4. The observed surfaces look like a pressed "carpet" with a fibrous structure. The difference between the investigated samples was in the presence of homogenously distributed glowing objects, which corresponded to the copper particles. It is also seen that copper nanoparticles were able to cover unwanted cracks and voids very well, as mentioned in the literature [10,13–17]. The provided elemental analysis showed the presence of carbon and oxygen on the surface of the F0 fibreboard while for sample F1, the presence of copper and sulphur was also found.

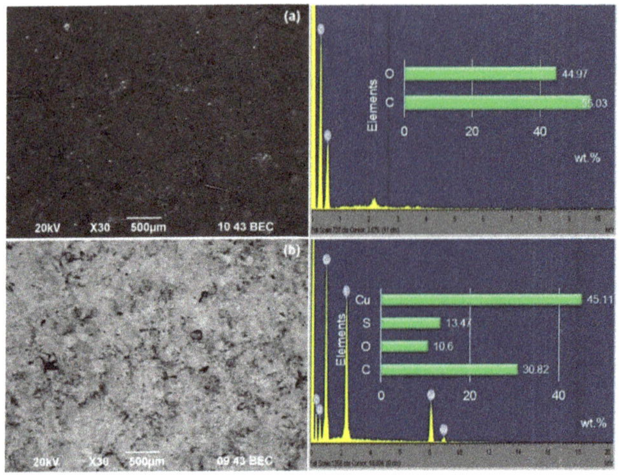

Figure 4. SEM and EDX images of unmodified—F0 (**a**) and copper-modified—F1 (**b**) fibreboards.

In addition to SEM analysis, in order to determine the size and shape of copper particles, TEM analysis was performed (Figure 5). The TEM image of powdered flakes from the surface of the F1 fibreboard showed spherical-shaped nanoparticles with an average size of 35 nm. As can be seen, the majority of nanoparticles have the same shape and the same size. The obtained results are in good agreement with the results of the X-ray analysis. Indeed, the fibreboard used and the modification methodology used were suitable for the formation and deposition of copper nanoparticles.

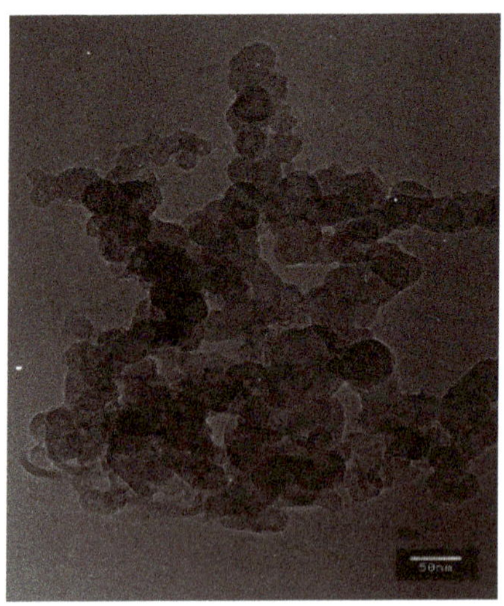

Figure 5. TEM image of copper-modified fibreboard F1 (bar is 50 nm).

The antibacterial activity of the obtained fibreboards was investigated against the model strains of the Gram-positive (*B. subtilis*) and Gram-negative (*E. coli* K12) bacteria. For this purpose, two methods were used: one based on building growth curves and the other on tracking the number of the colony-forming unit (CFU) of microorganisms after treatment with the analyzed materials.

The first method for determining the antimicrobial properties of the obtained materials was by tracking the growth of microorganisms in a batch culture and constructing growth curves. In Figure 6 are presented the batch growth curves of *E. coli* and *B. subtilis* in the presence of the fibreboards for 24 h. The growth of both strains in the presence of fibreboard F0 was in the typical growth phases for microorganisms; the lag and exponential phases were clearly expressed. The presence of fibreboards of different copper content (from 1 to 4) changed the form of the growth phases, in which a significant lengthening of the lag phase and shortening of the exponential phase were observed in both studied strains. The most notable change in the growth phases was observed in fibreboard F1, where individual phases were difficult to distinguish.

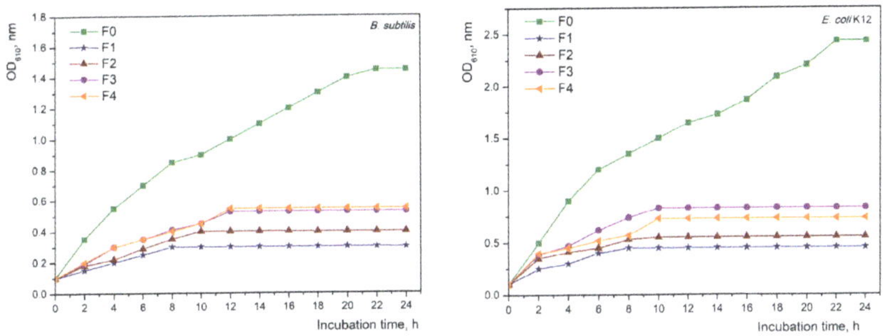

Figure 6. Bacterial growth curves of *B. subtilis* and *E. coli* K12 in the presence of fibreboards.

It is clear from the growth curves for both strains that the presence of the resultant boards (in all variants) had a greater impact on *E. coli* growth than *B. subtilis* growth. This is most likely caused by the varied accessibility of copper ions in the fibreboards as well as the different cell wall composition of Gram-positive and Gram-negative bacteria. The obtained results suggest that *E. coli* is more sensitive than *B. subtilis*. It was noted that the existence of copper nanoparticles has an impact on the development of microorganisms, as well as the contact time between them. The increased amount of the metal ions in the materials caused the growth phases to last longer and the growth to be delayed, which was proof of their inhibitory impact.

Another method of monitoring the antimicrobial activity of wood fibreboards was by monitoring the CFU counts (Figure 7). Petri dishes with single colonies of *E. coli* and *B. subtilis*, respectively, grown after exposure to boards with different copper contents are presented. The results show that the most cells grew in the control variant (F0) and fibreboard F4, which had the lowest copper content. In the remaining photos (fibreboards 1 to 3), a directly proportional decrease in the number of cells with increasing copper concentration was observed. These results correlate with the results from the growth curves, and it can be seen that E. coli exhibited a much higher sensitivity to the materials compared to *B. subtilis*. The highest inhibition was observed for fibreboard F1, where for *E. coli*, it was 84.6%, and for *B. subtilis*, it was 70.5%, respectively. In Table 2 are presented the results for all of the samples, together with the concentration of the copper nanoparticles in the medium. The similar results for copper nanoparticles highly sensitive to *E. coli* compared to the other microorganism were reported by Ramyadevi and co-workers [44].

The mechanism of the biocidal action of CuNps may be explained by the fact that CuNps release Cu(II) ions upon contact with moisture. These copper ions could bind with the -COOH and -SH groups of protein molecules of the bacterial cell wall. M. Crace et al., 2009 [45] obtained similar results by investigating copper alginate-cotton cellulose fibres. The surface of bacterial cells is negatively charged due to excess amounts of carboxyl groups in lipoproteins. At the same time, copper ions obtained from the nanoparticles in the liquid nutrient medium are positively charged. Therefore, adhesion and bioactivity are assumed to occur due to electrostatic forces. Peptidoglycans are negatively charged molecules that also bind copper ions released in the liquid growth medium. As a Gram-negative bacterium, *E. coli* may allow a greater amount of Cu^{2+} to reach the plasma membrane, but in most cases, it is less susceptible to antibacterial agents and antibiotics than Gram-positive bacteria [46].

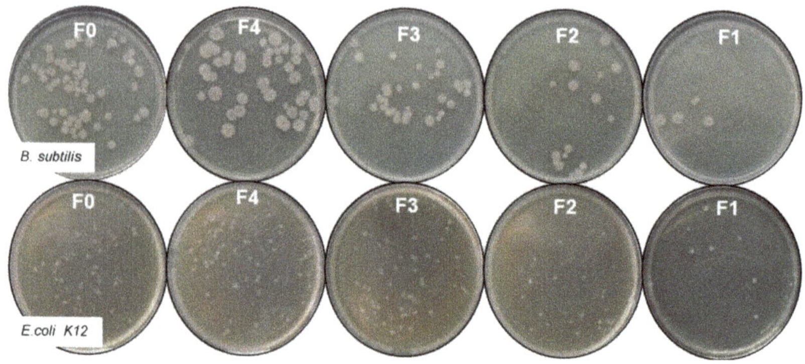

Figure 7. Antibacterial activity of tested fibreboards against model bacteria.

Table 2. Antibacterial test results and copper release after cultivation for 24 h.

Sample	*B. subtilis*			*E. coli*		
	CFU/mL	Inhibition, %	Copper Release, mg/L	CFU/mL	Inhibition, %	Copper Release, mg/L
F0	68×10^8	0	0.044	2.6×10^{10}	0	0.027
F1	20.1×10^8	70.5	12.67	0.4×10^{10}	84.6	11.9
F2	35×10^8	48.5	8.909	0.6×10^{10}	77	8.093
F3	47.2×10^8	31	2.050	1.6×10^{10}	38.5	1.769
F4	39.2×10^8	42.4	2.80	1.6×10^{10}	38.5	3.797

The highest inhibition was observed for fibreboard F1, where for *E. coli* it was 84.6% and for *B. subtilis* it was 70.5%, respectively. In Table 2 are presented the results for all of the samples, together with the concentration of the cooper nanoparticles in the medium.

As it is known, copper is an essential biocide. It is important to mention that copper alone fails to protect wood against fungi. For copper to be active, a second metal, such as chromium or arsenic, needs to be present in the system [47]. It was reported that the presence of copper nanoparticles instead of conventional copper improved the durability of wood against decay fungi, and it is not necessary for a second metal to be present in the system [17,47]. Spherical nanoparticles were reported to have a well-developed surface that results in good antibacterial properties [48,49]. Very recently, the excellent antibacterial activity of nano-derived CuS/Cu_2S against E. coli under near-IR light was reported [29]. The authors explained this excellent antibacterial activity with the presence of efficient interfacial charge separation and the generation of much more reactive oxygen species (ROS) [29]. Obviously, for efficient antibacterial ability, the copper content, the shape, and nano dimension were the key parameters. The presence of homogenously distributed nano Cu_2S was also important since this crystalline phase was characterized by excellent antibacterial properties [29]. The results show that the non-stoichiometric ratio of the modifying agents ($CuSO_4$*$5H_2O$:$Na_2S_2O_3$*$5H_2O$) led to the presence of several copper-containing crystalline phases that were not identified by us. This means that the stoichiometric amounts of the modifying mixture also affect the successful synthesis of Cu_2S. Apparently, F1 fibreboard possessed the best nano-crystallinity regarding Cu_2S crystal phases (Figure 1), as well as the best stoichiometric ratio. On the other hand, this sample had enhanced thermal stability and a flat, homogeneous distribution of Cu_2S, and a denser surface compared to the other investigated specimens. Moreover, the TEM analysis

(Figure 5) of fibreboard F1 showed the presence of spherical copper nanoparticles, with an average size of 35 nm (31 nm calculated by Debye Scherrer equation), formed by self-organizational processes. It can be summarized that the degree of inhibition of the tested microorganisms was directly dependent on the concentration of copper nano-particles presented in the medium as well as the contact time between them. Considering our results, the efficient antibacterial behaviour of the F1 fibreboard was obtained even without additional light exposure [29].

The results show that the obtained materials are suitable for manufacturing of packaging and furniture. Due to contact with water, fibreboards used in humid environments have low durability against fungal attacks. This fact gives us the idea to investigate the antifungal activity of the obtained fibreboards in the future.

4. Conclusions

The copper-modified fibreboards were prepared and their morphology, thermal stability, and antibacterial behaviour were investigated. The modification process was carried out by varying the copper reduction system. The copper content was determined and the physical and mechanical properties of fibreboards (thickness, bending strength, and swelling) were also presented. It was found that as the copper content increased, the bending strength and swelling decreased. Nevertheless, all mechanical parameters were in accordance with the European and national standards. Spherical nanoparticles, with an average size of 35 nm, were observed on the surface of the F1 fibreboard. The antibacterial activity was tested against model Gram-positive and Gram-negative bacteria. The results indicate that the antimicrobial activity of the materials depended on the copper content and preparation conditions. On the other hand, the copper content depended on the ratio of the modified mixture applied to the surface of the obtained wood carpet. In this regard, the fibreboards F1 showed the best antibacterial activity. The tests indicated that the materials had pronounced antibacterial performance against *E. coli*.

Author Contributions: Conceptualization, S.N.; methodology, L.A., N.R. and M.D.; software, N.R. and L.A.; validation, M.D. and N.L.-Z.; formal analysis, M.D.; investigation, L.A., N.R., N.L.-Z. and N.G.; resources, M.D. and N.R.; data curation, L.A., N.R. and N.L.-Z.; writing—L.A., N.R. and N.L.-Z.; visualization, L.A., N.R. and N.L.-Z.; supervision, S.N. All authors have read and agreed to the published version of the manuscript.

Funding: The authors acknowledge to the TwinTeam project Д01-272 "European Network on Materials for Clean Technologies" for providing the opportunity to present the results at SizeMat4 conference as well as for the financial publication support.

Institutional Review Board Statement: Not applicable.

Informed Consent Statement: Not applicable.

Data Availability Statement: Not applicable.

Conflicts of Interest: The authors declare no conflict of interest.

References

1. Dragnevska, M.; Velev, P.; Nenkova, S.; Garvanska, R. Development of a continuous method for production of copper-sulphide wood-polymer plate nanocomposite materials. *J. Univ. Chem. Technol. Metall.* **2011**, *46*, 349–356.
2. Yotov, N.; Valchev, I.; Petrin, S.; Savov, V. Lignosulphonate and waste technical hydrolysis lignin as adhesives for eco-friendly fiberboard. *Bul. Chem. Commun.* **2017**, *49*, 92–97.
3. Theng, D.; Arbat, G.; Delgado-Aguilar, M.; Vilaseca, F.; Ngo, B.; Mutjé, P. All-lignocellulosic fiberboard from corn biomass and cellulose nanofibers. *Ind. Crops Prod.* **2015**, *76*, 166–173. [CrossRef]
4. Akhtarq, M.; Kenealy, W.R.; Hom, E.G.; Swaney, R.S.; Winandy, J. Metod of Making Medium Density Faberboard. U.S. Patent 2008/0264588 A1, 30 October 2008.
5. Grzegorzewska, E.; Sedliačiková, M.; Drábek, J.; Behún, M. Evaluating the international competitiveness of polish furniture manufacturing industry in comparison to the selected EU countries. *Acta Fac. Xylologiae* **2020**, *62*, 149–164. [CrossRef]
6. Gul, W.; Alrobei, H. Effect of Graphene Oxide Nanoparticles on the Physical and Mechanical Properties of Medium Density Fiberboard. *Polymers* **2021**, *13*, 1818. [CrossRef]

7. Available online: http://welde.bg/index.php?m=593&lang=2 (accessed on 8 September 2023).
8. Available online: https://unece.org/forestrytimber/documents/2022/12/post-session-documents/market-forecast-tables-2022 (accessed on 24 October 2023).
9. Antov, P.; Jivkov, V.; Savov, V.; Simeonova, R.; Yavorov, N. Structural Application of Eco-Friendly Composites from Recycled Wood Fibres Bonded with Magnesium Lignosulfonate. *Appl. Sci.* **2020**, *10*, 7526. [CrossRef]
10. Alabduljabbar, H.; Alyousef, R.; Gul, W.; Akbar Shah, S.R.; Khan, A.; Khan, R.; Alaskar, A. Efect of Alumina Nano-Particles on Physical and Mechanical Properties of Medium Density Fiberboard. *Materials* **2020**, *13*, 4207. [CrossRef]
11. Valchev, I.; Yordanov, Y.; Savov, V.; Antov, P. Optimization of the hot-pressing regime in the production of eco-friendly fibreboards bonded with hydrolysis lignin. *Periodica Polytech. Chem. Eng.* **2022**, *66*, 125–134. [CrossRef]
12. Savov, V.; Valchev, I.; Antov, P.; Yordanov, Y.; Popski, Z. Effect of the Adhesive System on the Properties of Fiberboard Panels Bonded with Hydrolysis Lignin and Phenol-Formaldehyde Resin. *Polymers* **2022**, *14*, 1768. [CrossRef]
13. Silva, A.P.S.; Ferreira, B.S.; Favarim, H.R.; Silva, M.F.F.; Silva, J.V.F.; Azambuja, M.A.; Campos, C.I. Physical Properties of Medium Density Fiberboard Produced with the Addition of ZnO Nanoparticles. *BioResources* **2019**, *14*, 1618–1625. [CrossRef]
14. Gul, W.; Alrobei, H.; Akbar Shah, S.R.; Khan, A.; Hussain, A.; Asiri, A.M.; Kim, J. Effect of Embedment of MWCNTs for Enhancement of Physical and Mechanical Performance of Medium Density Fiberboard. *Nanomaterials* **2021**, *11*, 29. [CrossRef] [PubMed]
15. Gul, W.; Alrobei, H.; Akbar Shah, S.R.; Khan, A. Effect of Iron Oxide Nanoparticles on the Physical Properties of Medium Density Fiberboard. *Polymers* **2020**, *12*, 2911. [CrossRef]
16. Gul, W.; Akbar Shah, S.R.; Khan, A.; Ahmad, N.; Ahmed, S.; Ain, N.; Mehmood, A.; Salah, B.; Ullah, S.S.; Khan, R. Synthesis of graphene oxide (GO) and reduced graphene oxide (rGO) and their application as nano-fillers to improve the physical and mechanical properties of medium density fiberboard. *Front. Mater.* **2023**, *10*, 1206918. [CrossRef]
17. Jasmani, L.; Rusli, R.; Khadiran, T.; Jalil, R.; Adnan, S. Application of Nanotechnology in Wood-Based Products Industry: A Review. *Nanoscale Res. Lett.* **2020**, *15*, 207. [CrossRef]
18. Kandelbauer, A.; Widsten, P. Antibacterial melamine resin surfaces for wood-based furniture and flooring. *Prog. Org. Coat.* **2009**, *65*, 305–313. [CrossRef]
19. Nosál, E.; Reinprecht, L. Anti-bacterial and Anti-mold Efficiency of ZnO Nanoparticles Present in Melamine-laminated Surfaces of Particleboards. *BioResources* **2017**, *12*, 7255–7267. [CrossRef]
20. Rangelova, N.; Nenkova, S.; Lazarova, N.; Georgieva, N. Copper-based nanostructured lignocellulose materials with antibacterial activity. *Bul. Chem. Commun.* **2015**, *47*, 9–44.
21. Su, C.; Yuan, Q.; Gan, W.; Dai, D.; Huang, J.; Huang, Y. Study on a Composite Fiberboard with Multiple Electromagnetic Shielding Effectiveness. *Open Mater. Sci. J.* **2012**, *6*, 44–49. [CrossRef]
22. Lee, M.; Park, S.B.; Mun, S.P. One-step preparation of visible-light-responsive TiO2 with carbonized medium-density fiberboard for toluene degradation. *Wood Sci. Technol.* **2020**, *54*, 349–364. [CrossRef]
23. Rangelova, N.; Aleksandrov, L.; Angelova, T.; Georgieva, N.; Müller, R. Preparation and characterization of SiO2/CMC/Ag hybrids with antibacterial properties. *Carbohydr. Polym.* **2014**, *101*, 1166–1175. [CrossRef]
24. Paulose, P.I.; Jose, G.; Thomas, V.; Jose, G.; Unnikrishnan, N.V.; Warrier, M.K.R. Spectroscopic studies of Cu^{2+} ions in sol–gel derived silica matrix. *Bull. Mater. Sci.* **2002**, *25*, 69–74. [CrossRef]
25. Bejarano, J.; Caviedes, P.; Palza, H. Sol–gel synthesis and in vitro bioactivity of copper and zinc-doped silicate bioactive glasses and glass-ceramics. *Biomed. Mater.* **2015**, *10*, 025001. [CrossRef] [PubMed]
26. Montero, D.A.; Arellano, C.; Pardo, M.; Vera, R.; Gálvez, R.; Cifuentes, M.; Berasain, M.A.; Gómez, M.; Ramírez, C.; Vida, R.M. Antimicrobial properties of a novel copper-based composite coating with potential for use in healthcare facilities. *Antimicrob. Resist. Infect. Control* **2019**, *8*, 3. [CrossRef]
27. Sifri, C.D.; Burke, G.H.; Enfield, K.B. Reduced health care-associated infections in an acute care community hospital using a combination of self-disinfecting copper-impregnated composite hard surfaces and linens. *Am. J. Infect. Control* **2016**, *44*, 1565–1571. [CrossRef]
28. Thukkaram, M.; Vaidulych, M.; Kylian, O.; Rigole, P.; Aliakbarshirazi, S.; Asadian, M.; Nikiforov, A.; Biederman, H.; Coenye, T.; Du Laing, G.; et al. Biological activity and antimicrobial property of Cu/aC: H nanocomposites and nanolayered coatings on titanium substrates. *Mat. Sci. Eng. C* **2021**, *119*, 111513. [CrossRef]
29. Zhang, J.; Li, J.; Zhang, Q.; Guo, D. Constructing a novel CuS/Cu_2S Z-scheme heterojunction for highly-efficiency NIR light-driven antibacterial activity. *Appl. Surf. Sci.* **2023**, *624*, 156848. [CrossRef]
30. Nenkova, S.; Velev, P.; Dragnevska, M.; Nikolova, D.; Dimitrov, K. Lignocellulose nanocomposite containing copper sulfide. *BioResources* **2011**, *6*, 2356–2365. [CrossRef]
31. *BDS EN 310:1999*; Wood-Based Panels—Determination of Modulus of Elasticity in Bending and of Bending Strength. BDS: Sofia, Bulgaria, 1999.
32. *BDS EN 317:1998*; Particleboards and Fiberboards—Determination of Swelling in Thickness after Immersion in Water. BDS: Sofia, Bulgaria, 1998.
33. *BDS EN 324-1:2001*; Wood-Based Panels—Determination of Dimensions of Boards—Part 1: Determination of Thickness, Width and Length. BDS: Sofia, Bulgaria, 2001.
34. *BDS EN 622-1:2004*; Fibreboards—Specifications—Part 1: General Requirements. BDS: Sofia, Bulgaria, 2004.
35. *BDS EN 622-2:2004*; Fibreboards—Specifications—Part 2: Requirements for Hardboards. BDS: Sofia, Bulgaria, 2004.

36. Taghiyari, H.R.; Bayani, S.; Militz, H.; Papadopoulos, A.N. Heat treatment of pine wood: Possible effect of impregnation with silver nanosuspension. *Forests* **2020**, *11*, 466. [CrossRef]
37. Labidi, K.; Korhonen, O.; Zrida, M.; Hamzaoui, A.H.; Budtova, T. All-cellulose composites from alfa and wood fibers. *Ind. Crops Prod.* **2019**, *127*, 135–141. [CrossRef]
38. Chen, W.; Yu, H.; Liu, Y.; Chen, P.; Zhang, M.; Haib, Y. Individualization of cellulose nanofibers from wood using high-intensity ultrasonication combined with chemical pretreatments. *Carbohydr. Polym.* **2011**, *83*, 1804–1811. [CrossRef]
39. Das, D.; Hussain, S.; Ghosh, A.K.; Pal, A.K. Studies on cellulose nanocrystals extracted from musa sapientum: Structural and bonding aspects. *Cellul. Chem. Technol.* **2018**, *52*, 729–739.
40. Holy, S.; Temiz, A.; Demirel, G.K.; Aslan, M.; Amini, M.H.M. Physical properties, thermal and fungal resistance of Scots pine wood treated with nano-clay and several metal-oxides nanoparticles. *Wood Mater. Sci. Eng.* **2022**, *17*, 176–185. [CrossRef]
41. Sulaiman, N.S.; Hashim, R.; Hiziroglu, S.; Amini, M.H.M.; Sulaiman, O.; Selamat, M.E. Rubberwood particleboard manufactured using epichlorohydrin-modified rice starch as a binder. *Cellul. Chem. Technol.* **2016**, *50*, 329–338.
42. Boon, J.G.; Hashim, R.; Sulaiman, O.; Sugimoto, T.; Sato, M.; Salim, N.; Amini, M.H.M.; Ibrahım, N.I.; Ramle, S.F.M. Importance of lignin on the properties of binderless particleboard made from oil palm trunk. *ARPN J. Eng. Appl. Sci.* **2017**, *12*, 33–40.
43. Carrier, M.; Loppinet-Serani, A.; Denux, D.; Lasnier, J.-M.; Ham-Pichavant, F.; Cansell, F.; Aymonier, C. Thermogravimetric analysis as a new method to determine the lignocellulosic composition of biomass. *Biomass Bioenergy* **2011**, *35*, 298–307. [CrossRef]
44. Ramyadevi, J.; Jeyasubramanian, K.; Marikani, A.; Rajakumar, G.; Rahuman, A.A. Synthesis and antimicrobial activity of copper nanoparticles. *Mater. Lett.* **2012**, *71*, 114–116. [CrossRef]
45. Grace, M.; Chand, N.; Bajpai, S. Copper Alginate-Cotton Cellulose (CACC) Fibers with Excellent Antibacterial Properties. *J. Eng. Fibers Fabr.* **2009**, *4*, 24–35. [CrossRef]
46. Raffi, M.; Mehrwan, S.; Bhatti, T.M.; Akhter, J.I.; Hameed, A.; Yawar, W.; Masood ul Hasan, M. Investigations into the antibacterial behavior of copper nanoparticles against *Escherichia coli*. *Ann. Microbiol.* **2010**, *60*, 75–80. [CrossRef]
47. Kartal, S.N.; Green, I.F.; Clausen, C.A. Do the unique properties of nanometals affect leachability or efficacy against fungi and termites? *Int. Biodeterior. Biodegrad.* **2009**, *63*, 490–495. [CrossRef]
48. Bagchi, B.; Kar, S.; Dey, S.K.; Bhandary, S.; Roy, D.; Mukhopadhyay, T.K.; Dasa, S.; Nandy, P. In situ synthesis and antibacterial activity of copper nanoparticle loaded natural montmorillonite clay based on contact inhibition and ion release. *Colloids Surf. B.* **2013**, *108*, 358–365. [CrossRef]
49. Ma, X.; Zhou, S.; Xu, X.; Du, Q. Copper containing nanoparticles: Mechanism of antimicrobial effect and application in dentistry-a narrative review. *Front. Surg.* **2022**, *9*, 905892. [CrossRef] [PubMed]

materials

MDPI

Article

Hydrogen Storage Properties of Ball Milled MgH_2 with Additives- Ni, V and Activated Carbons Obtained from Different By-Products

Eli Grigorova [1,*], Pavel Markov [1], Boyko Tsyntsarski [2], Peter Tzvetkov [1] and Ivanka Stoycheva [2]

[1] Institute of General and Inorganic Chemistry, Bulgarian Academy of Sciences, Acad. G. Bonchev Str., Bld. 11, 1113 Sofia, Bulgaria; pvlmarkov@svr.igic.bas.bg (P.M.); ptzvetkov@svr.igic.bas.bg (P.T.)

[2] Institute of Organic Chemistry with Centre of Phytochemistry, Bulgarian Academy of Sciences, Acad. G. Bonchev Str., Bld. 9, 1113 Sofia, Bulgaria; boiko_sf@yahoo.com (B.T.); ivanka.stoycheva@orgchm.bas.bg (I.S.)

* Correspondence: egeorg@svr.igic.bas.bg

Abstract: The hydrogen sorption of materials based on 80 wt.% MgH_2 with the addition of 15 wt.% Ni or V and 5 wt.% activated carbons synthesized from polyolefin wax, a waste product from polyethylene production (POW), walnut shells (CAN), and peach stones (CPS) prepared by milling under an inert Ar atmosphere for a period of 1 h, is investigated. All precursors are submitted to pyrolysis followed by steam activation in order to obtain the activated carbons. The hydrogen sorption evaluations are carried out for absorption at 473 and 573 K with pressure of 1 MPa and for desorption at 623 and 573 K with pressure of 0.15 MPa. The composition of the samples after milling and hydrogenation is monitored by X-ray diffraction analyses. The 80 wt.% MgH_2–15 wt.%Ni–5 wt.% POW or CAN after absorption–desorption cycling and in a hydrogenated state at 573 K and 1 MPa are analyzed by TEM.

Keywords: sorption; hydrogen storage; metal hydrides; carbon materials

Citation: Grigorova, E.; Markov, P.; Tsyntsarski, B.; Tzvetkov, P.; Stoycheva, I. Hydrogen Storage Properties of Ball Milled MgH_2 with Additives- Ni, V and Activated Carbons Obtained from Different By-Products. *Materials* **2023**, *16*, 6823. https://doi.org/10.3390/ma16206823

Academic Editor: Ting Zhang

Received: 29 September 2023
Revised: 12 October 2023
Accepted: 20 October 2023
Published: 23 October 2023

1. Introduction

Materials based on solid-state hydrogen storage include adsorbents, liquid organics, complex and interstitial hydrides as well as chemical hydrogen. Each of them has advantages and disadvantages. Some of these materials have higher capacity but poor reversibility, and some have an elevated explosive risk such as, for example, in the case of complex hydrides. Other materials such as ammonia store hydrogen in an irreversible way, and their toxicity and recycling problems are very important. The highly porous adsorbents such as carbon nanotubes and MOFs adsorb hydrogen at a temperature below 100 K and high pressure [1]. Hydrogen storage in the form of metal hydrides provides important safety and high energy density advantages over the gas and liquid storage methods. Among a lot of metals and intermetallics that are capable to react with hydrogen in a reversible way magnesium–based materials as hydrogen storage media are more perspective. There has been intensive research regarding their hydrogen storage application due to the high theoretical hydrogen absorption capacity of magnesium (7.6 wt.%), very good reversibility, abundance, and low cost. These materials have also some fallibilities, such as slow kinetics, necessity of activation, and increased sorption temperatures, especially for the desorption process. Using a variety of dopants combined with milling in a planetary or vibratory mill leads to a decrease in hydrogen sorption temperatures and enhanced kinetics, and in general, to an improvement in the hydrogen absorption–desorption properties. Depending on the nature of the additives, their catalytic effect is different. Some of the additives can easily form hydrides; especially non-stoichiometric and others like iron, for example, act as active sites during the dissociative chemisorption of hydrogen. The amount of additives should be carefully balanced. A larger amount will reduce the hydrogen storage capacity of the material; however, to show a noticeable effect, they must be at least a few percent. When various carbon compounds are milled in mixtures based on Mg or MgH_2, the hydrogen sorption temperature is reduced, and better kinetics and the impeding of particles agglomeration during milling and hydrogen sorption cycling are observed [2–12]. Ball milling the MgH_2, which is brittle, is more effective than milling pure magnesium powder. Fuster et al. stated that the catalytic effect of graphite is more pronounced when it has increased content and is introduced at the start of milling. Another aspect of this study showed that the effect of graphite is not connected with its morphology [3]. An amelioration of dehydrogenation process occurs with the increasing carbon content, microwave power, and milling time when MgH_2 is catalyzed by different carbon materials under microwave irradiation [5]. Various carbon materials added to magnesium in reactive ball milling (milling under hydrogen) resulted in a significant reduction in hydrogenation time compared to pure magnesium [12]. During ball milling, a layer of carbon coating is formed between the particles, and this leads to the significant reduction in the particle size. This carbon layer prevents particles agglomeration and inhibits the restoration of the oxide layer on the surface during repeated hydrogen absorption and desorption. Among the additives used by Rud et al., amorphous carbon and fine-grained graphite powders promote a finer ball milling pulverization of magnesium than other carbon allotropes [12]. The Mg_2Ni has better sorption kinetics than Mg and absorbed–desorbed hydrogen at moderate temperatures [13–16]. The ternary hydride of Mg_2Ni-Mg_2NiH_4 crystalizes in the monoclinic and orthorhombic low-temperature phases and high temperature phase with cubic structure. These two low-temperature polymorph phases usually coexist, and their relative ratio depends on the mechanical and thermal history of the alloy and affects the dehydrogenation behavior, electric conductivity, and color of the obtained hydrides. The synthesis of Mg_2NiH_4 from MgH_2 and Ni is reported in [12,14,15], which showed that Mg_2NiH_4 has a noticeably higher catalytic efficiency than Ni and Mg_2Ni. Recently, some results regarding the synergetic effect on the MgH_2 hydrogen storage properties of two types of additives, with one being the carbon-containing material, have been reported [17,18]. Kajiwara et al. [17] proclaimed that adding carbon nanotubes expanded graphite and single-wall carbon nanotubes to MgH_2-Nb_2O_5 mixtures and resulted in better cycling stability, especially in the case of Mg-Nb_2O_5 carbon nanotubes compared to MgH_2-Nb_2O_5. Pawan Soni et al. [18] confirmed the better cycling stability of MgH_2 when two additives such as carbon spheres and Fe nanoparticles are used. Moreover, their results showed an improvement in the hydrogenation–dehydrogenation properties for MgH_2-Fe carbon spheres.

The MgH_2-based hydrogen storage tanks with two types of additives such as NbF_5 or TiF_4 and activated carbon or MWCNTs demonstrated very good reversibility by maintaining their hydrogen absorption capacity upon cycling [19–21]. Although due to deficient heat management and hydrogen diffusion the sorption performance inside the tank was not homogeneous, these obstacles could be solved by designing and fabricating an appropriate tank equipped with a heat exchanger and gas diffusion pathways.

Some results about the electrochemical activity of high-porosity adsorbents such as hydrogen storage media were published in [22–24]. The electrochemical hydrogen storage capacity for the nitrogen-doped graphene foam had an impressive value of 1916.5 mAh/g [22]. Definitely, these types of hydrogen adsorbents have perspectives such as hydrogen storage media.

In view of the above discussion, it seems that the synergetic effect of carbon-containing additives and some metals such as Ni or V could lead to an even greater improvement in the hydrogen absorption–desorption characteristics. We reported already some results about the hydrogen sorption properties of the composites based on MgH_2 or Mg and dopants as activated carbons from polyolefin wax or apricot stones and Ni, and in some parts, the quantity of the additives were varied [8,9,25]. The MgH_2-Ni activated carbon and MgH_2-V activated carbon composites are prepared by milling in a planetary mill under inert atmosphere of argon, and their hydrogen sorption characteristics are tested in this study. The dopants to MgH_2 are used in the following amounts: 5 wt.% carbon containing one synthesized from organic or agricultural waste precursors and 15 wt.% of Ni or V. Peach stones and walnut shells are widely affordable as precursors in Bulgaria and in the Balkan region and were chosen as plant precursors, and a polyolefin wax is used as organic waste for activated carbon preparation. The tasks of this work are the comparison between the effects of such dopants as Ni or V and the activated carbons from plants and organic wastes in the MgH_2-based materials, and the synthesis of Mg_2NiH_4 in less time, consuming less energy and using more delicate parameters of milling and hydrogenation than those reported to date.

2. Materials and Methods

Three different raw materials—polyolefin wax, walnut shells and peach stones—are subjected to pyrolysis first, and after that, they were activated by steam vapor to prepare the porous carbons. They will be denoted further in the text as POW, CAN and CPS. More detailed information about the synthesis procedure of these activated carbons can be found in [8,26]. Dried and crushed peach stones were subjected to pyrolysis and hydropyrolysis with water vapor in a tube furnace using a stainless-steel vertical reactor. The solid product was activated by 90 cm^3/min water vapor. The steam was introduced into the reactor at a temperature of 473 K. The final temperature was 1173 K (1 h, heating rate of 15 K/min). The same synthesis procedure was used for the walnut shells precursor [26]. The polyolefin wax sample is from a Bulgarian petroleum refinery in the city of Burgas, which produces polyethylene under low pressure, and it is situated on the Black Sea coast. Polyolefin wax melts at 388 K, its average molecular weight is 1100 and it decomposes at 633 K. Then, 200 g of polyolefin wax (POW) was heated to a melting point of 388 K. Drops of concentrated sulfuric acid were added while being stirred continuously up to solidification. The resulting solid was cleaned with distilled water before being dried at 423 K and carbonized at 873 K (in a covered silica crucible heated at a rate of 283 K/min filled with nitrogen gas). POW carbonizate was then activated by water vapor for one hour at 1073 K in a vertical stainless-steel reactor. The texture of the synthesized carbon material was characterized by N_2 adsorption at 77 K, which was carried out in an apparatus Quantachrome Autosorb iQ-C-XR/MP (Anton Paar GmbH, Graz, Austria). Before textural analyses, the sample was outgassed under vacuum at 623 K overnight. The isotherms were used to calculate the specific surface area (SBET), total pore (V_t) and micropore volumes (V_{mic}) by using the Dubinin–Radushkevich method. Analysis of the elements C, H, N and S was carried out with an Elementar Vario MACRO cube apparatus (Elementar Analysensysteme GmbH, Langenselbold, Germany). The oxygen amount in the samples was calculated by the difference.

The MgH_2 with purity of 98%, Ni powder < 50 μm (99.7%) and V powder 325 mesh (99.5%) are purchased by Sigma Aldrich (St. Louis, MO, USA). The used gases argon (99.999%) and hydrogen (99.999%) are purchased from Messer (Sofia, Bulgaria). Mixtures of MgH_2, Ni or V and the corresponding carbon additives with the composition of 80 wt.% MgH_2–15 wt.% Ni or V and 5 wt.% of the activated carbons are ball milled under Ar with applying the following parameters: 200 rpm for a period of 1 h, weight ratio of balls to sample 10:1, stainless steel vial with volume of 80 cm^3 and balls with 10 mm diameter. The materials were mixed in a glove box by hand, and after that, they were put in a ball milling vessel and closed under argon. For this purpose, a planetary mono mill Pulverisette 6 Fritsch (Fritsch, Idar, Oberstein, Germany) is used. The absorption/desorption characteristics of the composites are studied with a volumetric Sievert-type apparatus. The absorption measurements are accomplished at 473 and 573 K, using a pressure of 1 MPa, while desorption is performed at T = 623 and 573 K and P = 0.15 MPa. The composition after ball milling and hydrogenation is monitored by an XRD powder diffractometer D8 Advance with a LynxEye detector (Bruker, Karlsruhe, Germany) with Cu Kα radiation, vertical θ/θ goniometer, and a step size of 0.02 (2θ). The samples after 10 cycles of hydrogenation/dehydrogenation and in a hydrogenated state at 573 K and 1 MPa are analyzed by TEM HR STEM JEOL JEM 2100 with a CCD Camera GATAN Orius 832 SC1000 (JEOL, Tokyo, Japan) at an accelerating voltage of 200 kV. The specimens are prepared for TEM analyses by dispersing them in acetone. The suspensions are dripped on standard holey carbon/Cu grids purchased from Agar Scientific Ltd., Stansted, Essex, UK and are exposed for a short period of time to air until their transfer into a TEM holder. The samples are kept under argon in a glove box and were exposed to air only shortly for characterization by XRD and TEM.

3. Results and Discussion

3.1. *Synthesis and Characterization of Activated Carbons*

In the Table 1, data about the textural parameters of all three synthesized activated carbons are shown. The highest specific surface area has the activated carbon derived from peach stones (CPS), but the other two from polyolefin wax (POW) and walnut shells (CAN) have also a high specific surface area and similar total pore volume. The lowest micropores volume has the activated carbon derived from walnut shells (CAN). The activated carbons from plant precursors contain oxides and carbonates of magnesium and calcium and have increased ash percentage. The only one synthesized from organic waste is POW, and it has very low ash content and the highest oxygen and hydrogen percentage (Table 2). The ash and carbon in POW are 0.1% and 87.4%. The activated carbons differ not only by the pore volume and specific surface area but also by content of ash and carbon and the type of functional groups on the surface. Moreover, the chemical characterization of the activated carbon derived from POW confirms the high content of oxygen surface groups, which characterizes it as a potentially effective adsorbent for gases as well as heavy metals and organics in an aqueous solution [8]. Only the activated carbons derived from plant precursors contain nitrogen.

Table 1. Textural parameters of the activated carbons.

Activated Carbons	Specific Surface Area, m^2/g	V_{tot}, cm^3/g	V_{mic}, cm^3/g
Walnut shells (CAN)	743	0.67	0.21
Polyolefin wax (POW)	800	0.60	0.27
Peach stones (CPS) [26]	1257	0.63	0.44

Table 2. Analyses of the elements.

Type of Activated Carbon	Ash, wt.%	C, wt.%	H, wt.%	S, wt.%	N, wt.%	O, wt.%
Walnut shells (CAN)	2	89.5	2.4	0.8	0.9	6.4
Polyolefin wax (POW)	0.1	87.4	3.5	0.5	-	8.6
Peach stones (CPS) [26]	2.6	88.0	2.5	0.5	1.1	7.9

3.2. *Hydrogen Sorption Properties*

The maximal absorption capacity value is 5.3 wt.% H_2 after 1 h of hydrogenation at 573 K and 1 MPa, and it is the highest one for the sample with Ni and POW additives. For the 80 wt.% MgH_2–15 wt.% Ni–5wt.% CAN, the absorption capacity is 4.9 wt.% H_2, and for that with V and CPS, it is 3.8 wt.% H_2 under the same conditions. The three specimens show a very short activation period. For example, the 80 wt.% MgH_2–15 wt.% Ni–5wt.% CAN reaches in its first hydrogenation cycle at 573 K and 1 MPa a hydrogen absorption capacity of 0.7 wt.%, while in the second cycle, it reaches 2.4 wt.%, and already in the third one, the kinetics is improved significantly and 4.5 wt.% is obtained. The other two samples' behavior is similar, and they also need a few cycles of activation. This shows that the complex hydride Mg_2NiH_4 is relatively easy to be formed, and it is present in the samples only after several cycles of hydrogenation and dehydrogenation. Reducing the temperature of absorption to 473 K results in a substantial deterioration of the kinetics and decrease in absorption capacity (Figure 1a,b). The rate of absorption is very fast, and all three materials absorb for 5 min. from 2 wt.% to up to 2.73 wt.% of hydrogen. The composite containing vanadium is tested for hydrogen absorption for longer period of 2 h at 573 K and 1 MPa, and its hydrogen absorption capacity after that is 4.3 wt.%H_2. At 473 K and 1 MPa after 1 h of hydrogen absorption, the three composites reached absorption capacity as follows: 1.25 wt.% for the 80 wt.% MgH_2–15 wt.% Ni–5wt.% POW and 1.04 wt.% for the 80 wt.% MgH_2–15 wt.% Ni–5wt.% CAN and for the 80 wt.% MgH_2–15 wt.% V–5 wt.% CPS. The Mg_2NiH_4 is less stable than MgH_2, e.g., the equilibrium pressure plateau for this ternary hydride is higher. The transformation of the cubic high-temperature phase of Mg_2NiH_4 into the low-temperature forms occurs at 508 K. This hydride acts like a "hydrogen pump", accelerating the hydrogen absorption and desorption of Mg. The capacity of the desorption process at 623 K and 0.15 MPa has a similar value for both samples with Ni, but the reaction rate is higher for the one containing POW. As in the absorption process, the desorption process at a lower temperature of 573 K is faster, and a higher capacity is reached for the POW and nickel-containing sample. Up to 10 min from the start of the desorption reaction at 573 K, its rate is comparable with that at 623 K for the same sample with POW and Ni.

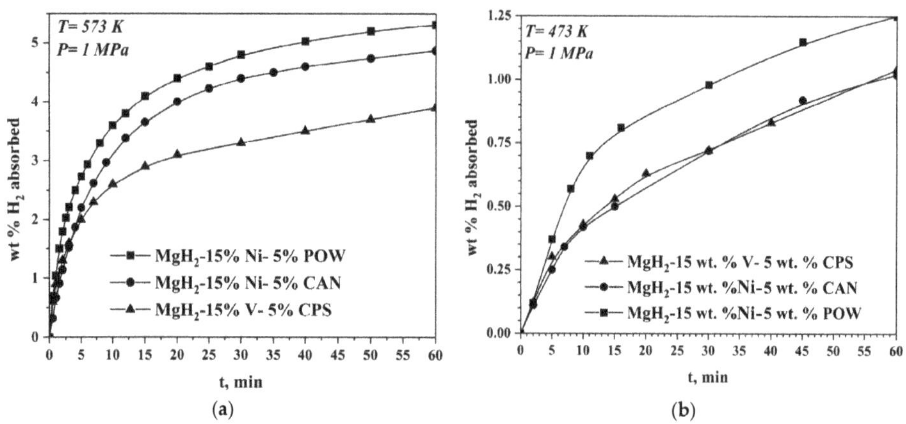

Figure 1. Hydrogen absorption curves of the samples at different temperatures: (**a**) 573 K and (**b**) 473 K and 1 MPa H_2. The data about the hydrogen absorption curves of the MgH_2–15 wt.% Ni—5 wt.% POW are published in [25].

The above-mentioned fact that Mg_2NiH_4 is less stable than MgH_2 and desorbs hydrogen at the beginning of the reaction is clearly visible in Figure 2. The desorption curves of both samples with Ni have rates up to 5 min faster than the one with V. Decreasing the temperature of desorption with only 323 K leads to a significant deterioration of the hydrogen desorption properties of studied materials, and after 1 h of desorption process, only about 20% of absorbed hydrogen is desorbed, and for the composites with activated carbons from plant precursors, this percentage is even lower. At the lower temperature of 533 K, the desorption was not detected for any of the three specimens.

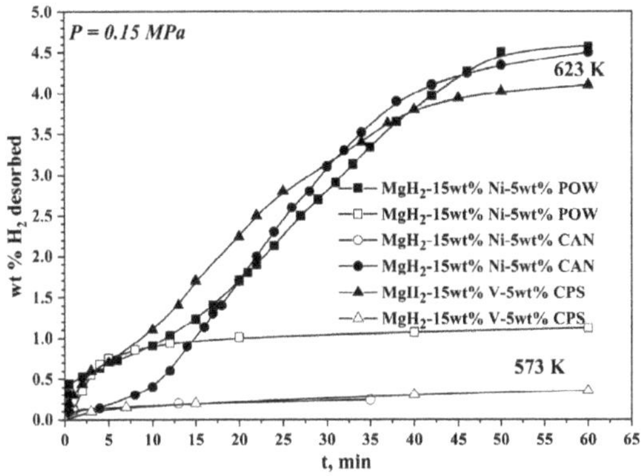

Figure 2. Hydrogen desorption curves of the samples at different temperatures and 0.15 MPa H_2. The data about the hydrogen desorption curves of the MgH_2–15 wt.% Ni–5 wt.% POW are published in [25].

The formation of the Mg_2NiH_4 hydride has a bidirectional effect. On the one hand, this hydride affects favorably the kinetics of hydrogenation and dehydrogenation, but on the other hand, Mg_2NiH_4 reduces the obtained capacity, because the theoretical hydrogen storage capacity of the Mg_2Ni is more than twofold lower than that of Mg, i.e., 3.6 wt.% and 7.6 wt.%.

The comparison of our results with some already published shows similar hydrogen sorption capacity of about 5.33 wt.% H_2 after cycling for the MgH_2–Nb_2O_5–carbon nanotube but a better rate of the reactions [17]. It should be pointed out that the absorption is tested at 1.9 MPa; the preparation methods, the composition and one of the additives are different. For the MgH_2 with 5 wt.% NiO/Al_2O_3 layered hybrid, the obtained absorption capacity is similar, e.g., around 5 wt.%, but at 523 K and 3 MPa. Another composite prepared and studied in this paper is MgH_2 with a 5 wt.% Ni/Al_2O_3 layered hybrid, which absorbed close to 6 wt.% H_2 at the same temperature and pressure [27]. Definitely, the higher pressure of absorption and the elevated percentage of MgH_2 are the reason for the better sorption properties of the MgH_2 with 5 wt.% Ni/Al_2O_3 layered hybrid than those reported in this study.

3.3. Characterization—XRD and TEM Analyses

X-ray diffraction patterns of the ball milled under argon for a period of 1 h with a composition of 80 wt.% MgH_2–15 wt.% Ni–5 wt.% POW, 80 wt.% MgH_2–15 wt.%–Ni-5 wt.%–CAN and 80 wt.% MgH_2–15 wt.% V–5 wt.% CPS mixtures do not show appearance of new phases, but regarding the relatively short milling time, it is not unexpected. The detected phases in the ball milled composites are mainly tetragonal MgH_2 and also Ni or V and some Mg (not presented). These grinding conditions were chosen because of the goal of the ball milling, which was mostly used for good homogenization of the studied materials.

The characterization of these types of materials by TEM is somehow challenging because of their oxygen and moisture sensitivity and the elevated energy of the electron beam, which can negatively affect them. The characterization by TEM of these sensitive materials, particularly high-resolution transmission electron microscopy (HRTEM) and selected area electron diffraction (SAED), is obtained by low beam current values and as short as possible exposition times. The XRD and the TEM analyses, e.g., SAED and HRTEM at different areas of the samples after hydrogenation and ten cycles of hydrogen absorption and desorption, confirmed the presence of MgH_2 as the main phase and Mg_2NiH_4 with a monoclinic and orthorhombic structure, which are the two polymorphic configurations of a low-temperature phase along with some graphite, Mg, Ni and MgO (Figures 3–5).

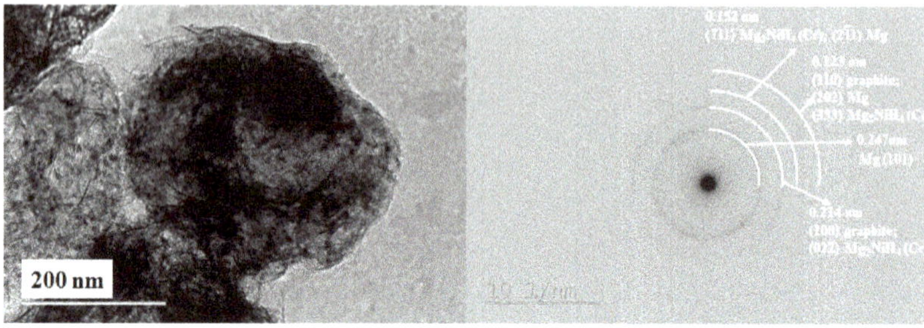

Figure 3. TEM image (**left**) and polycrystalline electron diffraction (**right**) of the 80 wt.% MgH_2–15 wt.% Ni–5 wt.% POW hydrogenation and cycling at 573 K and 1 MPa H_2 [25].

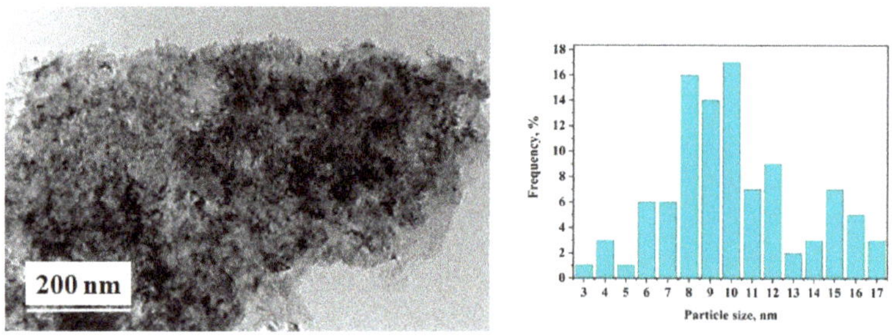

Figure 4. TEM image (**left**) and particles size distribution (**right**) of the 80 wt.% MgH_2–15 wt.% Ni–5 wt.% CAN after hydrogenation and cycling at 573 K and 1 MPa H_2.

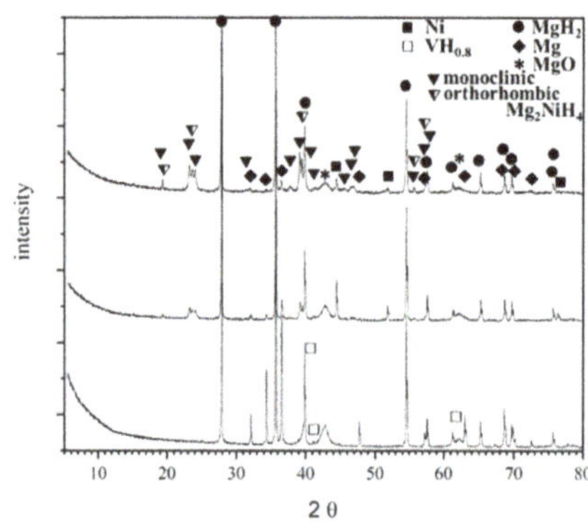

Figure 5. X-ray diffraction patterns obtained after hydrogenation and cycling at T = 573 K and P = 1 MPa of the: (a) MgH_2–15 wt.% Ni–5 wt.% POW; (b) MgH_2–15 wt.% Ni–5 wt.% CAN and (c) MgH_2–15 wt.% V–5 wt.% CPS.

For the composite 80 wt.% MgH_2–15 wt.% V–5 wt.% CPS, instead of ternary hydride Mg_2NiH_4, the non-stoichiometric hydride of vanadium VH_x is detected after hydrogenation and cycling (Figure 5). Using the Image J program for image analysis, the particle size distribution is obtained for the 80 wt.% MgH_2–15 wt.% Ni–5 wt.% POW or CAN, and the average particles size diameter is 8 nm for that containing POW and 9 nm for that with CAN (Figure 4). The particles size for both composites is between a few nm and less than 20 nm. After hydrogenation and cycling, the composites are more sensitive to oxidation, and the presence of MgO is very clearly visible on the XRD patterns. Another observation is that for the 80 wt.% MgH_2–15 wt.% V–5 wt.% CPS, there is much more nonhydrogenated magnesium compared to the other two composites after 1 h of hydrogenation at a temperature 573 K and pressure of 1MPa (Figure 5).

The comparison between the composite based on MgH_2 and only with activated carbons addition with these in the current study shows that the doping with nickel or vanadium as well as a carbon-containing additive leads to some desorption at a lower temperature of 573 K, as well as a comparable absorption rate, but it also leads to a lower capacity value. It should be noted that MgH_2 composites containing only activated carbons as an additive have a higher theoretical capacity, and the 95 wt.% MgH_2–5 wt.% AS (activated carbon derived from apricot stones) reached 5.8 wt.%, and for the 95 wt.% MgH_2–5 wt.% POW, the capacity value was 5.4 wt.% at 573 K and 1 MPa. Despite the higher absorption capacity of the composites only with the addition of activated carbons to MgH_2, their kinetics of absorption/desorption is slower and the desorption process was detectable only at 623 K [9]. Using plants and other waste precursors for obtaining activated carbons with high surface and developed pore structure is a very important topic in a view of ecology and more specifically in a waste recovery. The obtained materials are nanosized and showed relatively high absorption capacity values, especially this one with Ni and POW additives.

4. Conclusions

The maximal obtained absorption capacity value is 5.3 wt.% H_2 for the sample with the addition of POW and Ni, 4.9 wt.% H_2 for that with CAN and Ni, and 3.8 wt.% for the last one with V and CPS at a temperature of 573 K and a pressure of 1 MPa after 1 h of hydrogenation. Only after 3 min. of hydrogenation at 573 K, the MgH_2-based composite with Ni and POW absorbs 2.2 wt.% H_2. The hydrogen desorption capacities at a temperature of 623 K and a pressure of 0.15 MPa are with close values for the samples with Ni and with an increased rate of desorption for that containing POW. The presence of Ni in two of the studied composites leads to the formation of ternary hydride of Mg_2Ni and a low temperature form of Mg_2NiH_4 with monoclinic and orthorhombic structures, which is more favorable to the sorption kinetics and the absorption capacity than the vanadium addition and formation of its nonstoichiometric hydride. The combination between these two types of dopants with different natures leads to the conclusion that a more pronounced positive effect is activated carbon from organic waste with a high specific surface area, a more developed pore structure, lower ash content and a higher presence of oxygen containing surface groups and Ni because of the Mg_2NiH_4 formation. The activated carbons produced from biomass or plant wastes have some advantages such as eco-friendliness and low coast. Our results also showed that a more pronounced impact on the hydrogen storage properties of magnesium-based materials is that there are activated carbons derived from organic wastes. Moreover, the metal type of additive has a superior effect on the absorption–desorption characteristics of magnesium than the type of activated carbon.

Author Contributions: Conceptualization, E.G. and B.T.; methodology, E.G.; investigation, E.G., P.M., B.T., P.T. and I.S.; writing—original draft preparation, E.G. and B.T.; writing—review and editing, P.M., P.T. and I.S.; visualization, P.M. and P.T.; project administration, E.G. and B.T.; funding acquisition, B.T. All authors have read and agreed to the published version of the manuscript.

Funding: This research was funded by the National Science Fund of the Bulgarian Ministry of Education and Science (Project N° DFNI-KП-06-H27/9, 8 December 2018). Some of the experiments were carried out using the equipment from the National Infrastructure SI ESHER, which is supported by the Bulgarian Ministry of Education and Science under grant agreement N° DO1-161/28.07.22. The "European Network on Materials for Clean Technologies" TwinTeam project N° DO1-272 gave the authors the chance to discuss their results at the SizeMat4 conference and provided financial support for the publication.

Institutional Review Board Statement: Not applicable.

Informed Consent Statement: Not applicable.

Data Availability Statement: The data presented in this study are available on request from the corresponding author.

Conflicts of Interest: The authors declare no conflict of interest.

References

1. Balasubramanian, V. Chapter 10—Hydrogen Storage. In *Energy Sources: Fundamentals of Chemical Conversion Processes and Applications*; Elsevier: Amsterdam, The Netherlands, 2017; pp. 185–212.
2. Yartys, V.A.; Lototskyy, M.V.; Akiba, E.; Albert, R.; Antonov, V.E.; Ares, J.R.; Baricco, M.; Bourgeois, N.; Buckley, C.E.; Bellost von Colbe, J.M.; et al. Magnesium based materials for hydrogen based energy storage: Past, present and future. *Int. J. Hydrog. Energy* **2019**, *44*, 7809–7859. [CrossRef]
3. Wu, C.Z.; Wang, P.; Yao, X.; Liu, C.; Chen, D.M.; Lu, G.Q.; Cheng, H.M. Effect of carbon/noncarbon addition on hydrogen storage behaviors of magnesium hydride. *J. Alloys Compd.* **2006**, *414*, 259–264. [CrossRef]
4. Fuster, V.; Castro, F.J.; Troiani, H.; Urretavizcaya, G. Characterization of graphite catalytic effect in reactively ball milled MgH_2-C and Mg-C composites. *Int. J. Hydrog. Energy* **2011**, *36*, 9051–9061. [CrossRef]
5. Gattia, D.M.; Montone, A.; Pasquini, L. Microstructure and morphology changes in MgH_2/expanded natural graphite pellets upon hydrogen cycling. *Int. J. Hydrog. Energy* **2013**, *38*, 1918–1924. [CrossRef]
6. Awad, A.S.; Tayeh, T.; Nakhl, M.; Zakhour, M.; Ourane, B.; Le Troëdec, M.; Bobet, J.-L. Effect of carbon type (graphite, CFs and diamond) on the hydrogen desorption of Mg-C powder mixtures under microwave irradiation. *J. Alloys Compd.* **2014**, *607*, 223–229. [CrossRef]
7. Alsabawi, K.; Webb, T.A.; Gray, E.M.; Webb, C.J. The effect of C60 additive on magnesium hydride for hydrogen storage. *Int. J. Hydrog. Energy* **2015**, *40*, 10508–10515. [CrossRef]
8. Grigorova, E.; Khristov, M.; Stoycheva, I.; Tsyntsarski, B. Effect of activated carbon from polyolefin wax on the hydrogen sorption properties of magnesium. *Int. J. Hydrog. Energy* **2017**, *42*, 26872–26876. [CrossRef]
9. Grigorova, E.; Khristov, M.; Stoycheva, I.; Tsyntsarski, B.; Nihtianova, D.; Markov, P. Effect of activated carbons derived from apricot stones or polyolefin wax on hydrogen sorption properties of MgH_2. *Bulg. Chem. Commun.* **2017**, *49*, 109–114.
10. Pal, P.; Agarwal, S.; Tiwari, A.; Ichikawa, T.; Jain, A.; Dixit, A. Improved hydrogen desorption properties of exfoliated graphite and graphene nanoballs modified MgH_2. *Int. J. Hydrog. Energy* **2022**, *47*, 41891–41897. [CrossRef]
11. Verma, S.K.; Shaz, M.A.; Yadav, T.P. Enhanced hydrogen absorption and desorption properties of MgH_2 with graphene and vanadium disulfide. *Int. J. Hydrog. Energy* **2023**, *48*, 21383–21394. [CrossRef]
12. Rud, A.D.; Lakhnik, A.M. Effect of carbon allotropes on the structure and hydrogen sorption during reactive ball-milling of Mg-C powder mixtures. *Int. J. Hydrog. Energy* **2012**, *37*, 4179–4187. [CrossRef]
13. Cermak, J.; David, B. Catalytic effect of Ni, Mg_2Ni and Mg_2NiH_4 upon hydrogen desorption from MgH_2. *Int. J. Hydrog. Energy* **2011**, *36*, 13614–13620. [CrossRef]
14. Varin, R.A.; Czujko, T. Overview of processing of nanocrystalline hydrogen storage intermetallics by mechanical alloying/milling. *Mater. Manuf. Process.* **2002**, *17*, 129–156. [CrossRef]
15. Polanski, M.; Nielsen, T.K.; Kunce, I.; Norek, M.; Płociński, T.; Jaroszewicz, L.R.; Gundlach, C.; Jensen, T.R.; Bystrzycki, J. Mg_2NiH_4 synthesis and decomposition reactions. *Int. J. Hydrog. Energy* **2013**, *38*, 4003–4010. [CrossRef]
16. Martínez-Coronado, R.; Retuerto, M.; Torres, B.; Martínez-Lope, M.J.; Fernández-Díaz, M.T.; Alonso, J.A. High-pressure synthesis, crystal structure and cyclability of the Mg_2NiH_4 hydride. *Int. J. Hydrog. Energy* **2013**, *38*, 5738–5745. [CrossRef]
17. Kajiwara, K.; Sugime, H.; Noda, S.; Hanada, N. Fast and stable hydrogen storage in the porous composite of MgH_2 with Nb_2O_5 catalyst and carbon nanotube. *J. Alloys Compd.* **2022**, *893*, 162206. [CrossRef]
18. Soni, P.K.; Bhatnagar, A.; Shaz, M.A. Enhanced hydrogen properties of MgH_2 by Fe nanoparticles loaded hollow carbon spheres. *Int. J. Hydrog. Energy* **2023**, *48*, 17970–17982. [CrossRef]
19. Thongtan, P.; Dansirima, P.; Thiangviriya, S.; Thaweelap, N.; Suthummapiwat, A.; Plerdsranoy, P.; Utke, R. Reversible hydrogen sorption and kinetics of hydrogen storage tank based on MgH_2 modified by TiF_4 and activated carbon. *Int. J. Hydrog. Energy* **2018**, *43*, 12260–12270. [CrossRef]
20. Thiangviriya, S.; Plerdsranoy, P.; Sitthiwet, C.; Dansirima, P.; Thongtan, P.; Eiamlamai, P.; Utke, O.; Utke, R. MgH_2-TiF_4-MWCNTs based hydrogen storage tank with central tube heat exchanger. *Int. J. Hydrog. Energy* **2019**, *44*, 20173–20182. [CrossRef]
21. Thongtan, P.; Thiangviriya, S.; Utke, O.; Utke, R. MgH_2-based hydrogen storage tank: Kinetics, reversibility, and MWCNTs content. *J. Phys. Chem. Solids* **2022**, *163*, 11057. [CrossRef]
22. Aghajani, H.; Tabrizi, A.T.; Ghorbani, R.; Behrangi, S.; Stupavska, M.; Abdian, N. Evaluation of electrochemical hydrogen storage capability of three-dimensional nano-structured nitrogen-doped graphene. *J. Alloys Compd.* **2022**, *906*, 164284. [CrossRef]
23. Aghjehkohal, A.R.; Tabrizi, A.T.; Yildiz, M. Electrochemical hydrogen storage of synthesized heterostructure of hexagonal boron nitride-carbon nano tube. *J. Alloys Compd.* **2023**, *962*, 171159. [CrossRef]
24. Tabrizi, A.T.; Aghajani, H.; Mashtizadeh, A. Determining the effect of porosities on the hydrogen adsorption capacity of 3D printed PEEK. *Int. J. Hydrog. Energy* **2023**, *in press*. [CrossRef]
25. Grigorova, E.; Nihtianova, D.; Tsyntsarski, B.; Stoycheva, I. Investigation of Hydrogen Storage Characteristics of MgH_2 Based Materials with Addition of Ni and Activated Carbon. *Inorganics* **2020**, *8*, 12. [CrossRef]

26. Tsoncheva, T.; Mileva, A.; Tsyntsarski, B.; Paneva, D.; Spassova, I.; Kovacheva, D.; Velinov, N.; Karashanova, D.; Georgieva, B.; Petrov, N. Activated carbon from Bulgarian peach stones as a support of catalysts for methanol decomposition. *Biomass Bioenergy* **2018**, *109*, 135–146. [CrossRef]
27. Liu, Z.; Liu, J.; Wu, Z.; Tang, Q.; Zhu, Y.; Zhang, J.; Liu, Y.; Zhang, Y.; Ba, Z.; Hu, X.; et al. Enhanced hydrogen sorption kinetics of MgH_2 catalyzed by a novel layered Ni/Al_2O_3 hybrid. *J. Alloys Compd.* **2022**, *895*, 162682.

materials

MDPI

Article

Network Structure and Luminescent Properties of $ZnO–B_2O_3–Bi_2O_3–WO_3:Eu^{3+}$ Glasses

Aneliya Yordanova [1], Margarita Milanova [1,*], Reni Iordanova [1], Margit Fabian [2], Lyubomir Aleksandrov [1] and Petia Petrova [3]

1 Institute of General and Inorganic Chemistry, Bulgarian Academy of Sciences, Acad. G. Bonchev Str., bld. 11, 1113 Sofia, Bulgaria; a.yordanova@svr.igic.bas.bg (A.Y.); reni@svr.igic.bas.bg (R.I.); lubomir@svr.igic.bas.bg (L.A.)

2 Centre for Energy Research, 29-33 Konkoly Thege Street, 1121 Budapest, Hungary; fabian.margit@ek.hun-ren.hu

3 Institute of Optical Materials and Technologies "Acad. Jordan Malinowski", Bulgarian Academy of Sciences, Acad. G. Bonchev Str., bld. 109, 1113 Sofia, Bulgaria; petia@iomt.bas.bg

* Correspondence: margi71@abv.bg

Abstract: In this study, we investigated the influence of Bi_2O_3 and WO_3 on both structure and optical properties of $50ZnO:(49 - x)B_2O_3:1Bi_2O_3:xWO_3$; x = 1, 5, 10 glasses doped with 0.5 mol% Eu_2O_3. IR spectroscopy revealed the presence of trigonal $BØ_3$ units connecting superstructural groups, $[BØ_2O]^-$ metaborate groups, tetrahedral $BØ_4^-$ units in superstructural groupings (Ø = bridging oxygen atom), borate triangles with nonbridging oxygen atoms, $[WO_4]^{2-}$ tetrahedral, and octahedral WO_6 species. Neutron diffraction experimental data were simulated by reverse Monte Carlo modeling. The atomic distances and coordination numbers were established, confirming the short-range order found by IR spectra. The synthesized glasses were characterized by red emission at 612 nm. All findings suggest that Eu^{3+} doped zinc borate glasses containing both WO_3 and Bi_2O_3 have the potential to serve as a substitute for red phosphor with high color purity.

Keywords: glass structure; europium; IR; photoluminescence; density

Citation: Yordanova, A.; Milanova, M.; Iordanova, R.; Fabian, M.; Aleksandrov, L.; Petrova, P. Network Structure and Luminescent Properties of $ZnO–B_2O_3–Bi_2O_3–WO_3:Eu^{3+}$ Glasses. *Materials* **2023**, *16*, 6779. https://doi.org/10.3390/ma16206779

Academic Editor: Francesco Baino

Received: 21 September 2023
Revised: 13 October 2023
Accepted: 16 October 2023
Published: 20 October 2023

1. Introduction

Currently, white-light-emitting diodes are being investigated extensively as the next-generation solid-state light source owing to the advantages they bring to the table, including safety, an environmentally friendly nature, high stability, low power consumption, and long operational lifetime [1]. One of the ways in which white light can be obtained is by combining tri-color phosphors, such as $ZnS:Cu^+$, Al^{3+} (green) [2], $BaMgAl_{10}O_{17}:Eu^{2+}$ (blue) [3], and Y_2O_2S: Eu^{3+} (red) [4], coated on InGaN-based LED chip, emitting around 400 nm (near-UV). However, the red-emitting phosphor shows lower efficiency (eight times lower) compared to the blue and green phosphors, as well as exhibiting chemical instability under UV radiation which may cause environmental pollution due to the release of sulfide gas. The other commercially applied red phosphors, such as $Y_2O_3:Eu^{3+}$ and $YVO_4:Eu^{3+}$, also cannot achieve high emission efficiency [5]. Therefore, red-emitting phosphor with chemical and thermal stability and high efficiency upon near-UV excitation remains to be found.

Europium (III) ion is being considered as a suitable activator for red emission, resulting from its $^5D_0 \rightarrow {}^7F_j$ (j = 0–4) transitions in the visible range [6]. Unfortunately, Eu^{3+}-doped materials cannot be efficiently excited by the present LED chips, because its excitation peaks are weak in nature due to parity-forbidden f–f transitions. Searching for host materials that can overcome the weak Eu^{3+} absorptions is important for achieving high excitation and emission efficiency of the red luminescence. A possibility is introducing sensitizers into the host composition, such as Ce^{3+}, Bi^{3+}, Tb^{3+}, etc. [7,8]. It is well known that the luminescent properties of Eu^{3+}-doped materials can be modified by changing the host structure and composition. Glass materials are suitable matrixes for doping with lanthanide ions due to their chemical stability, high optical homogeneity, absence of absorbing particles, and low nonlinear refractive indices. Among many potential glass materials for luminescence applications, binary $ZnO–B_2O_3$ glasses have been attracting continuous scientific interest. Homogeneous binary zinc–borate glasses are formed in a very narrow range of compositions because of the existence of a very large region of immiscibility of two liquids in a $ZnO–B_2O_3$ system [9]. However, these glasses are characterized by good chemical and thermal stability, high mechanical strength, low dispersion, and low glass transition temperature. They possess high transparency (up to 90%) from the visible to mid-infrared region of the spectrum [10]. Eu^{3+}-doped zinc borate glasses yield very strong orange/red photoluminescence by UV excitation, especially for low europium concentrations ($<10^{19}$ cm^{-3}) [11]. It has been reported that the addition of WO_3 and/or Bi_2O_3 to $ZnO–B_2O_3$ glasses induces the expansion of the glass-forming region and also lowers the phonon energy [12,13]. In our recent works, we reported, for the first time, the preparation of tungsten-containing $ZnO–B_2O_3$ glasses doped with Eu^{3+} active ion and their luminescent properties [14,15]. The obtained results from glass structure, physical, thermal, and optical properties indicate the suitability of the $50ZnO:40B_2O_3:10WO_3$ glass network for the luminescence performance of Eu^{3+} ions. The positive effect of the addition of WO_3 on the luminescence intensity is proven by the stronger Eu^{3+} emission of the zinc–borate glass containing WO_3 compared to the WO_3-free zinc–borate glass, a phenomenon engendered mainly by the energy transfer from tungstate groups to the Eu^{3+} ions (sensitizing effect). The most intense luminescence peak observed at 612 nm and the high-integrated emission intensity ratio (R) of the $^5D_0 \rightarrow {}^7F_2/{}^5D_0 \rightarrow {}^7F_1$ transitions at 612 nm and 590 nm of 5.77 suggest that the glasses have the potential for red emission materials.

Another desirable component for luminescent glass hosts is Bi_2O_3 oxide, commonly used as an activator, emitting in the spectral region of 380–700 nm due to $^3P_1 \rightarrow {}^1S_0$ transition upon NUV excitation. Among many studies, the Bi^{3+} ion is also recognized as a favored sensitizer, which can greatly enhance the luminescence of the rare-earth ions (Eu^{3+}, Sm^{3+}, Tb^{3+}) through resonant energy transfer [16,17]. High bright red emission in Eu^{3+} containing zinc–borate glasses codoped with Bi^{3+} was observed, enhanced by 346 nm excitation ($^1S_0$-$^3P_1$ of Bi^{3+} ions) due to the sensitization effect of Bi^{3+} codopant [18]. Zinc bismuth borate glasses doped with different Eu^{3+} concentrations (1, 3, 5, 7, and 9 mol%) were prepared, and the systematic analysis of the results suggested that the glass doped with a Eu^{3+} concentration of 5 mol% is suitable for LED and display device applications [19].

More recently, we prepared zinc–borate glasses modified with Bi_2O_3. Bulk, transparent, dark brownish glasses with composition $50ZnO:(40 - x)B_2O_3:10Bi_2O_3:0.5Eu_2O_3:xWO_3$, x = 0 and 0.5, were synthesized. The obtained structural and optical data indicate that a zinc–borate glass network containing Bi_2O_3 provides highly asymmetric sites of Eu^{3+} ions, leading to high emission intensity. Moreover, the presence of WO_3 also leads to the increase in emission intensity of the rare-earth Eu^{3+} ion, as a result of the nonradiative energy transfer from the glass host to the active ion [20]. These data above show that the $ZnO–B_2O_3$ glass system containing both bismuth and tungstate oxides is a particularly interesting host for the europium ions in red phosphors applications.

Here, we continued our investigations by preparing such a glass composition with increasing WO_3 content and with the addition of low Bi_2O_3 concentrations (1 mol%) in order to meet the requirement of colorless glasses for optical application. The aim was to obtain bulk, colorless glasses with compositions $50ZnO:(49 - x)B_2O_3:1Bi_2O_3:xWO_3:0.5Eu_2O_3$, x = 1, 5, and 10 mol%, and to establish the influence of Bi_2O_3 and WO_3 on glass formation, structure, and optical properties.

2. Materials and Methods

2.1. Sample Preparation

Glasses of the compositions in mol% 50ZnO:(49 − x)B_2O_3:1Bi_2O_3:xWO_3; x = 1, 5, 10, doped with 0.5 mol% Eu_2O_3 were obtained by applying the melt quenching method, using reagent-grade ZnO (Merck KGaA, Amsterdam, The Netherlands), WO_3 (Merck KGaA, Darmstadt, Germany), B_2O_3 (SIGMA-ALDRICH, St. Louis, MO, USA), Bi_2O_3 (Alfa Aesar, Karlsruhe, Germany), and Eu_2O_3 (SIGMA-ALDRICH, St. Louis, MO, USA) as raw materials. B_2O_3 enriched with ^{11}B isotope (99.6%) was used in order to avoid the high neutron absorption cross-section of the ^{10}B isotope. Further in the text, samples' names are abbreviated to ZBBW1:Eu, ZBBW5:Eu, and ZBBW10:Eu, where the number refers to the WO_3 content in the compositions. The homogenized batches were melted at 1200 °C for 20 min in a platinum crucible in air. The melts were cast into graphite molds to obtain bulk glass samples. Glasses obtained by us earlier with the same compositions (50ZnO:(50 − x)B_2O_3:xWO_3; x = 1, 5, 10) doped with 0.5 mol% Eu_2O_3 without Bi_2O_3, and denoted as ZBW1:Eu, ZBW5:Eu, and ZBW10:Eu, were also included here in order to establish the influence of bismuth addition on the structure and luminescence properties of the new glass compositions. The glasses without Bi_2O_3 containing 1 and 5 mol% WO_3 were reported in the 6th International Conference on Optics, Photonics, and Lasers (OPAL' 2023), while the glass with the highest WO_3 amount of 10 mol% was represented in ref. [15].

2.2. Characterization Techniques

The phase formation of the samples was established by X-ray phase analysis with a Bruker D8 advance diffractometer, Karlsruhe, Germany, using Cu K_α radiation in the $10 < 2\theta < 60$ range. The thermal stability of the obtained glasses was examined by differential scanning calorimetry (DSC) with a Netzsch 404 F3 Pegasus instrument, Selb, Germany, in the temperature range 25–750 °C at a heating rate of 10 K/min in an argon atmosphere. The density of the obtained glasses at room temperature was measured by the Archimedes principle using toluene (ρ = 0.867 g/cm^3) as an immersion liquid on a Mettler Toledo electronic balance of sensitivity 10^{-4} g. The IR spectra of the glasses were measured using the KBr pellet technique on a Nicolet-320 FTIR spectrometer, Madison, WI, USA, with a resolution of ±4 cm^{-1}, by collecting 64 scans in the range 1600–400 cm^{-1}. A random error in the center of the IR bands was found to be ±3 cm^{-1}. The EPR analyses were carried out in the temperature range 120–295 K in X band at frequency 9.4 GHz on a spectrometer Bruker EMX Premium, Karlsruhe, Germany. Optical absorption spectra (UV–VIS–NIR) in the range 190–1500 nm were obtained with an error < 1% using a commercial double-beam spectrometer (UV-3102PC, Shimadzu, Kyoto, Japan). Photoluminescence (PL) excitation and emission spectra at room temperature for all glasses were measured with a Spectrofluorometer FluoroLog3-22, Horiba JobinYvon, Longjumeau, France. Neutron diffraction measurements were carried out in the momentum transfer range, Q = 0.45–9.8 Å^{-1}, for 24 h using neutrons of de Broglie wavelength, λ = 1.069 Å, at the 2-axis PSD diffractometer of Budapest Neutron Centre. The powder glass samples were mounted in a thin-walled cylindrical vanadium can with a diameter of 8 mm for the neutron diffraction experiments. The neutron diffraction data were corrected for detector efficiency, background scattering, and absorption effects, and normalized with vanadium [21]. The total structure factor, $S(Q)$, was calculated using local software packages.

2.3. The Reverse Monte Carlo Simulation

Reverse Monte Carlo (RMC) simulations were performed on neutron diffraction datasets of the experimental total structure factor, $S(Q)$, to determine the short-range structural properties of glasses by using RMC++ software (https://www.szfki.hu/~nphys/rmc++/downloads.html, accessed on 15 October 2023) [22]. The RMC technique minimizes the squared difference between the experimental $S(Q)$ and the simulated one from a 3-dimensional atomic configuration by using the following equations:

$$S(Q) = \sum_{i,j}^{k} w_{ij} S_{ij}(Q) \tag{1}$$

$$S_{ij}(Q) = 1 + \frac{4\pi\rho_0}{Q} \int_0^{r_{max}} r\left[g_{ij}(r) - 1\right] sinQr\, dr \tag{2}$$

$$w_{ij} = \frac{c_i c_j b_i b_j}{\left[\sum_{i,j}^{k} c_i b_j\right]^2} \tag{3}$$

where c_i and b_i are the molar fraction and coherent neutron scattering length for atoms of type i, the $S_{ij}(Q)$ denotes the partial structure factors, and w_{ij} are the neutron scattering weight factors for the 21 atomic pairs for the ZBBW:Eu series (explanation: $k = 5$, thus $k(k + 1)/2 = 15$ different atomic pairs are present). RMC simulations were used to generate partial atomic pair correlation function, $g_{ij}(r)$, and coordination number distributions. The simulation was started with an initial random configuration by building a box that contained 10.000 atoms of Zn, B, Bi, W, Eu, and O, with the atomic density, ρ_o, values of 0.0947 Å^{-3}, 0.0919 Å^{-3}, and 0.0888 Å^{-3} for the samples ZBBW1:Eu, ZBBW5:Eu, and ZBBW10:Eu, respectively. The RMC model box lengths for the three samples were 23.63 Å, 23.87 Å, and 24.14 Å for ZBBW1:Eu, ZBBW5:Eu, and ZBBW10:Eu samples, respectively.

In the RMC simulation procedure, constraints were used for the minimum interatomic distances between atom pairs (cut-off distances) to avoid unreasonable atom contacts. For each sample, about fifty RMC configurations were obtained with more than 2,600,000 accepted configurations of atoms.

3. Results

3.1. XRD Analysis and DSC Studies

Bulk, transparent, slightly colored glasses (insets, Figure 1a) of 50ZnO:(49 − x)B_2O_3:1Bi_2O_3:xWO_3:0.5Eu_2O_3, x = 1, 5, and 10 mol%, were obtained in this study. The measured X-ray diffraction patterns are shown in Figure 1a, and confirm the amorphous nature of the prepared materials. Glasses without Bi_2O_3 (50ZnO:(50 − x)B_2O_3:xWO_3:0.5Eu, x = 1, 5, and 10 mol%) were obtained earlier. The XRD patterns of the glass samples having 1 and 5 mol% WO_3 were present in the 6th International Conference on Optics, Photonics, and Lasers (OPAL' 2023). The photograph of the glass with the highest WO_3 amount of 10 mol% was represented in ref. [15].

DSC curves of the glass samples 50ZnO:(49 − x)B_2O_3:1Bi_2O_3:xWO_3:0.5Eu_2O_3, x = 1, 5, and 10 mol%, obtained are presented in Figure 2. Glasses are characterized with two humps corresponding to the two glass transition temperatures, T_{g1} and T_{g2}. The two glass transition effects observed are connected with the presence of two amorphous phases with different compositions in the investigated glasses. In the DSC curve of the glass having the highest WO_3 concentration of 10 mol%, an exothermic peak due to the glass crystallization at temperature Tc = 683 °C was observed. For the glasses containing lower WO_3 concentration of 5 and 1 mol%, glass crystallization effects did not appear, evidencing that these glasses possess higher thermal stability, which decreases with increasing WO_3 content. The DSC analysis shows that the thermal parameters of glasses present here do not differ significantly from those obtained for the glasses with the similar compositions without Bi_2O_3 reported in Ref. [15]. However, the thermal stability of glasses containing 1 mol% Bi_2O_3 was slightly lowered, most probably because of the increased structural heterogeneity and, hence, higher crystallization ability of the composition.

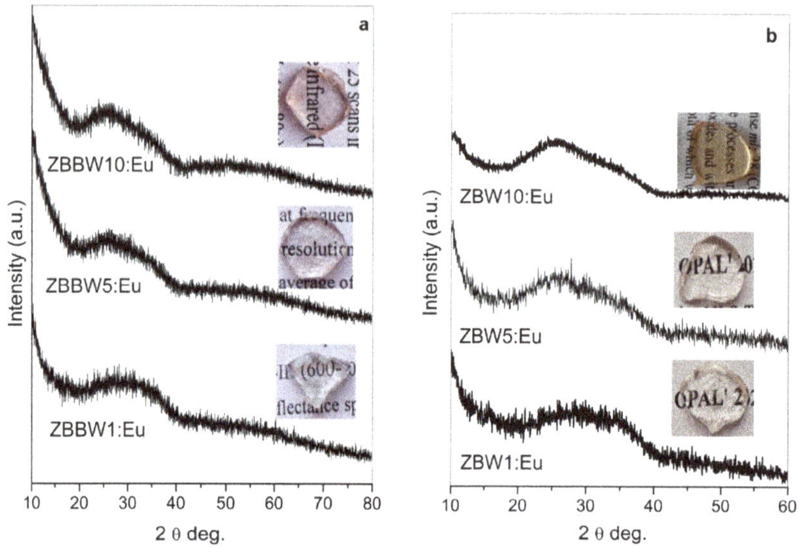

Figure 1. XRD patterns of (**a**) 50ZnO:(49 − x)B_2O_3:1Bi_2O_3:xWO_3:0.5Eu_2O_3, x = 1, 5, and 10 mol%; (**b**) 50ZnO:(50 − x)B_2O_3:xWO_3:0.5Eu_2O_3, x = 1, 5, and 10 mol%.

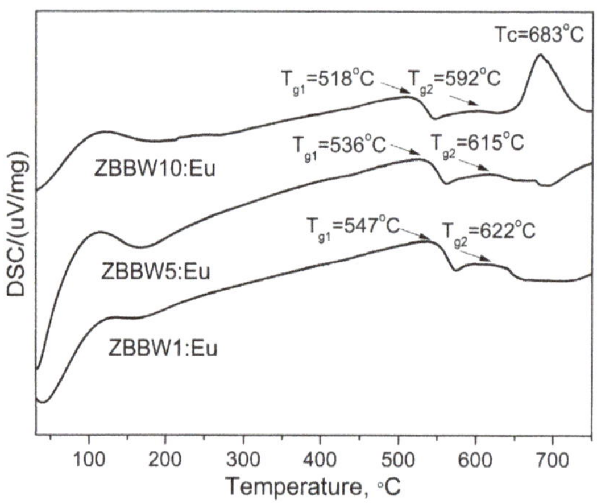

Figure 2. DSC curves of glasses 50ZnO:(49 − x)B_2O_3:1Bi_2O_3:xWO_3:0.5Eu_2O_3, x = 1, 5, and 10 mol%.

3.2. *IR Spectral Analysis*

Structural information of the studied glasses was obtained by comparative analysis of IR spectra of glasses with the same compositions without Bi_2O_3 and containing 1 mol% Bi_2O_3, which are shown in Figure 3.

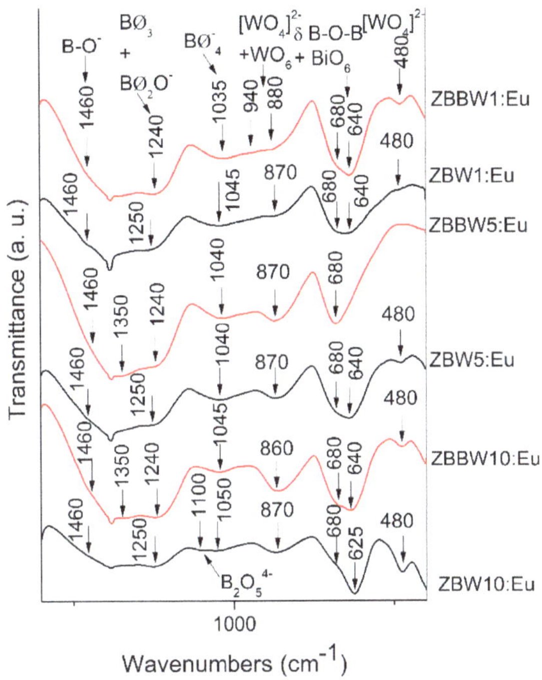

Figure 3. IR spectra of 50ZnO:(49 − x)B_2O_3:1Bi_2O_3:xWO_3:0.5Eu_2O_3, x = 1, 5, and 10 mol% (in red); and 50ZnO:(50 − x)B_2O_3:xWO_3:0.5Eu_2O_3, x = 1, 5, and 10 mol% (in black).

The IR spectra of Bi_2O_3-free glasses with the lower WO_3 concentration of 5 and 1 mol% were previously reported by us in the 6th International Conference on Optics, Photonics, and Lasers (OPAL' 2023), while the glass with the highest WO_3 amount of 10 mol% was discussed earlier in ref. [15]. The IR spectra of the glasses without Bi_2O_3 contain bands characteristic to the metaborate groups $BØ_2O^-$ (shoulder at 1460 cm^{-1}; band at 625–640 cm^{-1}), pyroborate dimmers, $B_2O_5{}^{4-}$ (weak band at 1110 cm^{-1}; band at 625–640 cm^{-1}), and superstructural groupings with $BØ_3$ and $BØ_4{}^-$ species (bands at 1250 cm^{-1} and at 1040–1050 cm^{-1}, band at 680 cm^{-1}) [14,23–25]. There are also WO_6 (bands at 870 cm^{-1} and at 680 cm^{-1}) and $[WO_4]^{2-}$ tetrahedra (band at 480 cm^{-1}) [23]. The addition of Bi_2O_3 to the glass compositions causes the $BO_3 \rightarrow BO_4$ transformation, resulting in the increase in the number of superstructural units (increased intensity of the bands at 1240 and 1045–1035 cm^{-1} and band at 680 cm^{-1}) [23]. For both the glass series, a partial $[WO_4]^{2-} \rightarrow WO_6$ transformation with increasing WO_3 concentration occurs, manifested by the disappearance of the bands at 940 cm^{-1} and 880 cm^{-1} ($\nu_3[WO_4]^{2-}$), and formation of an intense and well-formed band at 860–870 cm^{-1} (νWO_6). The tungstate octahedral species sharing common corners (W–O–W) and edges (W_2O_2) are supposed, having in mind the structural and IR data of crystalline Bi_2WO_6, $Bi_2W_2O_9$, and $ZnWO_4$ phases [26–31]. These compounds consist of corner- or edge-shared WO_6, whose IR spectra contain strong bands situated in the same spectral regions, from 870–800 cm^{-1} and 700–600 cm^{-1}. Tungstate and borate species are charge-balanced by Bi^{3+}, Zn^{2+}, and Eu^{3+} ions via Bi–O–W, B–O–Bi, Zn–O–W, Zn–O–B, Eu–O–W, and Eu–O–B bonding. Additionally, the new high frequency band at 1350 cm^{-1} observed in glasses having higher WO_3 of 5 and 10 mol% (Figure 2, ZBBW5:Eu and ZBBW10:Eu) is attributed to stretching of B–O$^-$ bonds in $BØ_2O^-$ triangles, and its presence suggests stronger interaction between Bi^{3+} ions and nonbridging oxygens [32]. Bi^{3+} ions are incorporated in the structure of investigated glasses as BiO_6 octahedra (bands at 640 and at 480 cm^{-1}). Thus, the addition of Bi_2O_3 to glasses

$50ZnO{:}(49 - x)B_2O_3{:}xWO_3{:}0.5Eu_2O_3$, x = 1, 5, and 10 mol%, leads to the formation of more stable and reticulated glass structure, compared with the glasses with the same composition without Bi_2O_3. The Zn^{2+} ions generally participate in the borate glasses as ZnO_4 tetrahedra with characteristic Zn^{2+} motion at 225 cm^{-1} [33]. The frequency of the Eu–O vibration, ν(Eu–O), has been measured at about 280 cm^{-1} for glasses $(1 - 2x)Eu_2O_3 - x(SrO - B_2O_3)$ [34].

The detailed assignments of the bands observed in the IR spectra of the present glasses are summarized in Table 1.

Table 1. Infrared bands (in cm^{-1}) and their assignments for glasses $50ZnO{:}(49 - x)B_2O_3{:}1Bi_2O_3{:}xWO_3{:}0.5Eu_2O_3$, x = 1, 5, and 10 mol%.

Infrared Bands Position (cm^{-1})	Assignment	Ref.
475	$\nu_4[WO_4]^{2-}$ + Bi–O vibrations in the BiO_6 groups	[32,35]
680	Bending vibrations of B–O–B bonds in superstructural	[32]
640–625	Bending vibrations of B–O–B bonds in meta- and pyroborates + Bi–O vibrations in the BiO_6 groups	[32]
860–870	νWO_6	[14,23]
940; 880	$\nu_3[WO_4]^{2-}$ in distorted tetrahedra	[23]
1050–1035	$\nu_{as}BØ_4^-$ involved in superstructural units	[14,23]
1100	ν_{as}(B–O–B); B–O–B bridge in pyroborate units, $B_2O_5^{4-}$	[24,35]
1245	ν_{as}(B–O–B); B–O–B bridges connect BO_3 units + BO_3 stretch in meta-, pyro-, orthoborate units	[15,36]
1350	$\nu(B–O^-)$ stretch in $BØ_2O^-$ units charge balanced by Bi^{3+}	[32]
1460	$\nu(B–O^-)$ stretch in $BØ_2O^-$ units	[23]

3.3. Density, Molar Volume, Oxygen Packing Density, and Oxygen Molar Volume

A structural information of two series of glasses was also gained by density (ρ_g) measurement, on which basis the values of several physical parameters listed in Table 2 (molar volume (V_m), oxygen molar volume (V_o), and oxygen packing density (OPD)) are evaluated, using the conventional formulae [37]. Bi_2O_3 containing glasses are characterized with the higher density as compared with the respective Bi_2O_3-free glasses because of the replacement of lighter B_2O_3 (molecular weight 69.62 g/mol) with heavier Bi_2O_3 (molecular weight 465.96 g/mol). The V_m and V_o values of glasses having 1 mol% Bi_2O_3 are lower, while their OPD values are higher as compared with the values of the same parameters established for the glasses without Bi_2O_3, evidencing better packing and bonding in the Bi_2O_3-containing glass network and lower number of nonbridging oxygens (NBOs) [38].

Table 2. Values of physical parameters of glasses $50ZnO{:}(49 - x)B_2O_3{:}1Bi_2O_3{:}xWO_3{:}0.5Eu_2O_3$, x = 1, 5, and 10 mol%: density (ρ_g), molar volume (V_m), oxygen molar volume (V_o), oxygen packing density (OPD).

Sample ID	ρ_g (g/cm³)	V_m (cm³/mol)	V_o (cm³/mol)	OPD (g atom/L)
ZBW1:Eu	3.475± 0.002	22.70	11.29	88.55
ZBW5:Eu	3.689± 0.002	23.14	11.51	86.86
ZBW10:Eu *	3.910 ± 0.002 *	23.91 *	11.73 *	84.27 *
ZBBW1:Eu	3.679 ± 0.001	22.57	11.20	89.28
ZBBW5:Eu	3.889 ± 0.005	23.01	11.42	87.57
ZBW10:Eu	4.175 ± 0.001	23.38	11.60	86.18

* The physical parameters of glass ZBW10:Eu were previously reported in Ref. [15].

3.4. RMC Modeling and Results

The RMC technique provided an excellent fit of the simulated structure factors (S(Q)-1) with the experimental one for $50ZnO{:}(49 - x)B_2O_3{:}1Bi_2O_3{:}xWO_3{:}0.5Eu_2O_3$, x = 1, 5, and 10 mol% (Figure 4).

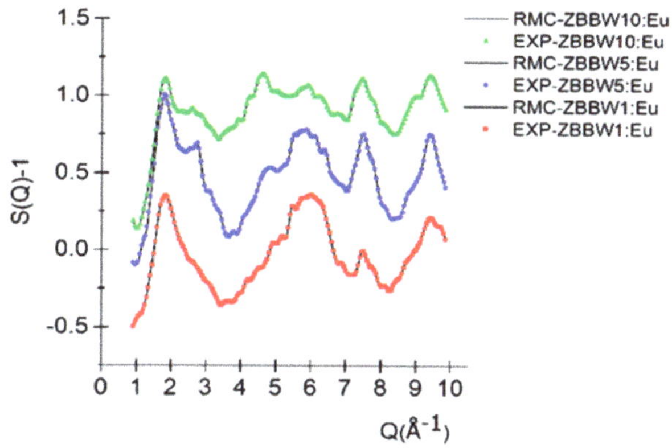

Figure 4. Experimental (color) and RMC (black line)-simulated neutron scattering structure factors for $50ZnO:(49 - x)B_2O_3:1Bi_2O_3:xWO_3:0.5Eu_2O_3$, x = 1, 5, and 10 mol% glasses.

From the RMC simulation, partial atomic pair-correlation functions, $g_{ij}(r)$, and average coordination number distributions, CN_{ij}, were revealed, with good stability and statistics. The Zn–O distribution functions show symmetrical peaks centered in the range of 1.95 ± 0.01 Å (Table 3).

Table 3. Interatomic X–O distances, $g_{ij}(r)$, obtained from RMC simulation. The errors are estimated from the reproducibility of various RMC runs.

Title 1	Zn–O $g_{ij}(r)$ (Å)	B–O $g_{ij}(r)$ (Å)	Bi–O $g_{ij}(r)$ (Å)	W–O $g_{ij}(r)$ (Å)	Eu–O $g_{ij}(r)$ (Å)	O–O $g_{ij}(r)$ (Å)
ZBBW1:Eu	1.95 ± 0.01	1.40/1.80 ± 0.05	2.00 ± 0.05	1.75 ± 0.05	2.20 ± 0.05	2.35 ± 0.03
ZBBW5:Eu	1.95 ± 0.01	1.40/1.80 ± 0.05	2.00 ± 0.05	1.75 ± 0.05	2.20 ± 0.05	2.35 ± 0.03
ZBB10:Eu	1.95 ± 0.01	1.40/1.80 ± 0.05	2.00 ± 0.05	1.75 ± 0.05	2.20 ± 0.05	2.35 ± 0.03

In function of $g_{ij}(r)$, we specify a range in r over which atoms are counted as neighbors. This can be understood as defining coordination shells. Introducing a *min* point (positions of minimum values on the lower) and *max* point (the upper side of the corresponding peak), these are presented in Table 4, where we present the average coordination numbers (summarized in Table 4).

Table 4. Average coordination numbers, CN_{ij}, calculated from RMC simulation. In brackets, the interval is indicated, where the actual coordination number was calculated.

Sample	Zn–O CN_{ij}	B–O CN_{ij}	W–O CN_{ij}	O–O CN_{ij}
ZBBW1:Eu	4.01 ± 0.05 (min: 1.80–max: 2.20)	3.90 ± 0.05 (min: 1.20–max: 1.65)	6.20 ± 0.1 (min: 1.65–max: 2.23)	5.63 ± 0.1 (min: 2.20–max: 2.60)
ZBBW5:Eu	3.99 ± 0.05 (min: 1.80–max: 2.20)	3.52 ± 0.05 (min: 1.20–max: 1.65)	6.42 ± 0.1 (min: 1.60–max: 2.25)	5.32 ± 0.1 (min: 2.20–max: 2.60)
ZBBW10:Eu	3.97 ± 0.05 (min: 1.80–max: 2.20)	3.48 ± 0.05 (min: 1.20–max: 1.65)	6.73 ± 0.1 (min: 1.60–max: 2.25)	5.54 ± 0.1 (min: 2.20–max: 2.60)

The average coordination number of the Zn–O was obtained from the RMC analysis, and it was found that Zn^{4+} was tetrahedrally coordinated with oxygens in the glassy network for all studied samples (Table 4). The B–O distribution function showed a relatively broad first neighbor distance at 1.40 ± 0.05 Å, and a slight shoulder at 1.80 ± 0.1 Å appeared in function of concentration in the $50ZnO:(49 - x)B_2O_3:1Bi_2O_3:xWO_3:0.5Eu_2O_3$, x = 1, 5, and 10 mol%, glass samples. The B–O coordination was in the range of 3.48 ± 0.05 to

4.00 ± 0.05, and obtained the changes in BO_3/BO_4 ratio. The boron atoms were coordinated mostly by three and four oxygen atoms, forming trigonal BO_3 and tetrahedral BO_4 units, in agreement with coordination numbers in SiO_2–Na_2O–B_2O_3 glasses [39], in MoO_3–ZnO–B_2O_3 glasses [40], and in ZnO–B_2O_3–Li_2O–Al_2O_3 glasses [41]. The nearest W–O distances showed characteristic peaks at 1.75 ± 0.05 (Table 3) for both series. The average coordination number of W–O was in a wide range, from 6.20 ± 0.1 to 6.73 ± 0.1 (see Table 3) within the limits of experimental uncertainty. Based on the coordination numbers, we can predict that the W–O network consists of WO_4 and WO_6 units. The WO_4/WO_6 ratio changed with the WO_3 concentration, and samples with highest WO_3 concentration (10 mol%) had mostly WO_6 units with fewer WO_4 units. In the case of Bi–O and Eu–O, thanks to the very low Bi_2O_3 and Eu_2O_3 concentration, it was not relevant to obtain reliable numbers for the coordination.

3.5. EPR Spectroscopy

The EPR analyses of ZBBW5:Eu were carried out at 295 K and 120, and in Figure 5 the obtained spectra are shown. As seen, the spectra contain multiple signals with different intensities and g-factors. The most prominent features are assigned to impurities of isolated Fe^{3+} ions (the signal with $g = 4.25$) and isolated Mn^{2+} ions (six hyperfine structure lines marked with *, inset).

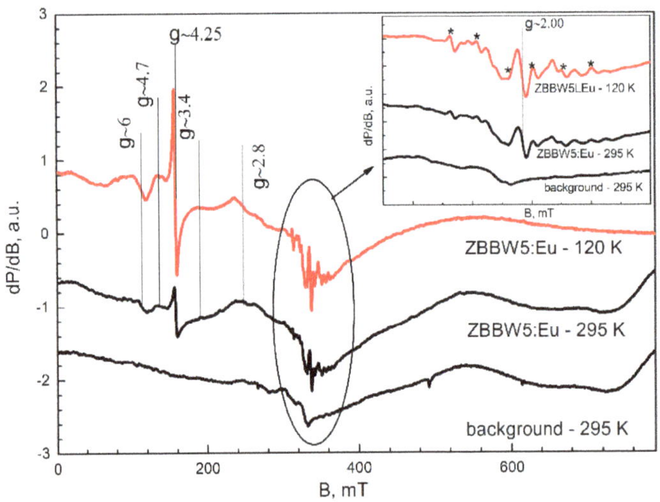

Figure 5. EPR spectra of ZBBW5:Eu, recorded at 120 and 295 K. The quartz tube background is represented at the bottom. EPR measurement conditions: Att = 16 (5 mW); MA = 20.

Eu^{2+} ions possess electron spin S = 7/2, and in their EPR spectrum seven signals were observed, corresponding to seven allowed transitions occurring between four Kramers' doublets (ms = ±1/2, ms = ±3/2, ms= ±5/2, and ms = ±7/2) according to the selection rule [42]. Depending on the zero-field splitting parameters (D, E) and crystal field symmetry, the EPR spectra of Eu^{2+} usually include a part of these signals. In the spectra of ZBBW5:Eu are discernable a set of not-well-resolved and low-intensive signals with g factors about 6.0, 4.7, 3.4, and 2.8 located in the range 0–300 mT. These signals could be assigned to Eu^{2+} ions [42,43] localized in a low-symmetry crystal field with large zero-field splitting (D > hν).

In addition, the central region of the spectrum shows a signal with $g = 2.00$. The assignment of this signal is somewhat difficult, as it could derive both from Fe^{3+} ion impurities existing in the sample and Gd^{3+} ions in a highly symmetric environment. That is why its attribution remains unclear.

To summarize, the EPR spectra recorded for sample ZBBW5:Eu confirm the presence of Eu^{2+} ions, with their concentration being assessed as extremely low based on the comparison between the background spectrum and the analyzed spectrum.

3.6. Luminescent Properties

Figure 6 shows the photoluminescent excitation spectra of Eu^{3+}-doped 50ZnO:(49 − x)B_2O_3:1Bi_2O_3:xWO_3, x = 1, 5, and 10 mol%, glasses monitoring the $^5D_0 \rightarrow {}^7F_2$ red emission of Eu^{3+} at 612 nm [6].

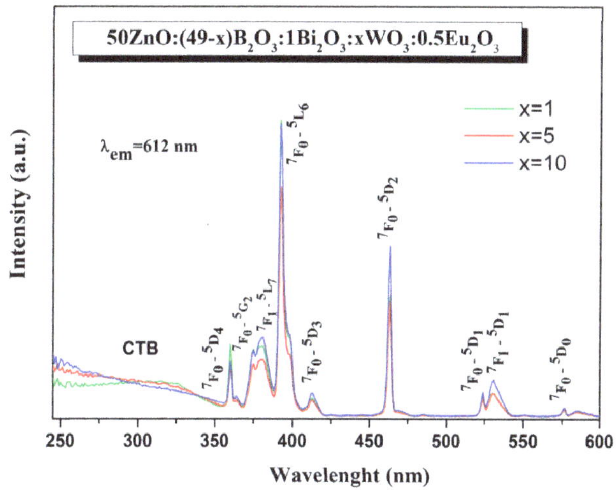

Figure 6. Excitation spectra of 50ZnO:(49 − x)B_2O_3:1Bi_2O_3:xWO_3:0.5Eu_2O_3 (x = 1, 5, and 10 mol%) glasses.

The low-intensive broad band below 350 nm is ascribed to the characteristic absorption of Bi^{3+} ($^1S_0 \rightarrow {}^3P_1$) [17], the charge transfer bands (CTB), resulting from energy transitions from O^{2-} to W^{6+} in WO_4 and WO_6 groups and O^{2-} to Eu^{3+}, as well as the ground 4f state of Eu^{3+} to W^{6+} [6,44–48]. The existence of the excitation band of host lattice absorption at Eu^{3+} emission (612 nm) implies the existence of nonradiative energy transfer from Bi^{3+} and WO_n groups to the active rare-earth ion [48,49]. The sharp lines in 350–600 nm range correspond to the f → f intraconfigurational forbidden transitions of Eu^{3+} from the ground state ($^7F_0$) and from the first excited state ($^7F_1$): $^7F_0 \rightarrow {}^5D_4$ (360 nm), $^7F_0 \rightarrow {}^5G_2$ (375 nm), $^7F_1 \rightarrow {}^5L_7$ (381 nm), $^7F_0 \rightarrow {}^5L_6$ (392 nm), $^7F_0 \rightarrow {}^5D_3$ (412 nm), $^7F_0 \rightarrow {}^5D_2$ (463 nm), $^7F_0 \rightarrow {}^5D_1$ (524 nm), $^7F_1 \rightarrow {}^5D_1$ (531 nm), and $^7F_0 \rightarrow {}^5D_0$ (576 nm) [6]. Among them, the electronic transition at 392 nm is the strongest one and was used as an excitation wavelength. Compared to the CTB, the intensity of the narrow f–f lines is stronger. This is favorable for Eu^{3+}-doped luminescent materials, since, in general, the intensity of these Eu^{3+} transitions is weak due to the parity-forbidden law. Thus, the obtained glasses can be effectively excited by near-UV and blue light, which is compatible with the present LED chips.

The emission spectra of Eu^{3+}-doped 50ZnO:(49 − x)B_2O_3: 1Bi_2O_3:xWO_3, x = 1, 5, and 10 mol%, glasses under the excitation of λ_{ex} = 392 nm consist of five emission peaks centered at 578, 592, 612, 651, and 700 nm, originating from $^5D_0 \rightarrow {}^7F_J$ (J = 0, 1, 2, 3, 4) intraconfigurational transitions of Eu^{3+} (Figure 7) [6].

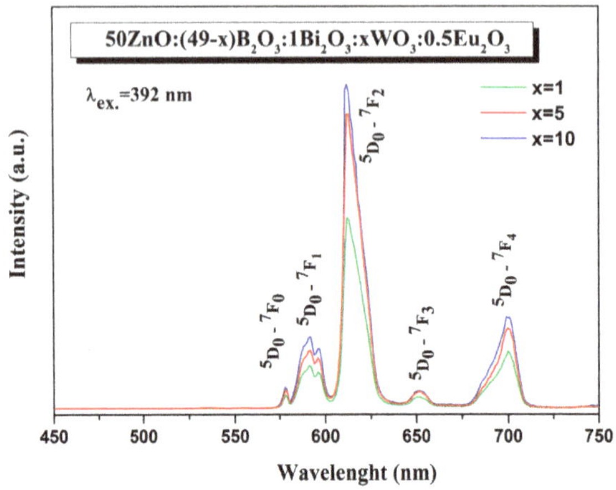

Figure 7. Emission spectra of 50ZnO:(49 − x)B_2O_3: 1Bi_2O_3:xWO_3:0.5Eu_2O_3 (x = 1, 5, and 10 mol%) glasses.

The characteristic single broad band emission of Bi^{3+}, originating from $^3P_1 \rightarrow {}^1S_0$ transition is located at 380–700 nm [50]. In the same spectral region, we also registered the broad emission band of the WO_n group [51]. The general requirement for energy transfer from both WO_3 and Bi_2O_3 to the rare-earth ion is satisfied, i.e., there exists a spectral overlap between the excitation peaks of Eu^{3+} (Figure 6) and the emission band of Bi^{3+} and WO_3 (Figure 7). As a result, both oxides can act as sensitizers, transferring the emission energy nonradiatively to the activator Eu^{3+} by quenching their luminescence. Moreover, an indication of energy transfer is the absence of characteristics of WO_3 and Bi_2O_3 emission bands [15,20,48]. As can be seen, with the increase of WO_3 up to 10 mol% (Figure 7) and with the introduction of Bi_2O_3 in the 50ZnO:40B_2O_3:10WO_3:0.5Eu_2O_3 glass composition (Figure 8), a significant enhancement of the emission intensities was achieved.

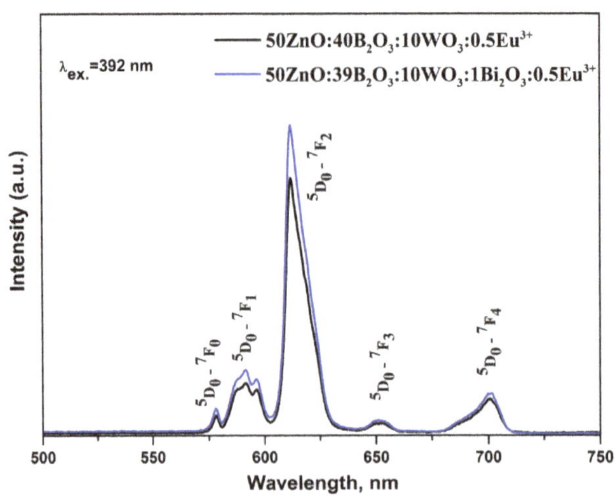

Figure 8. Emission spectra of 50ZnO:(40 − x)B_2O_3: xBi_2O_3:10WO_3:0.5Eu_2O_3 (x = 0 and 1 mol%).

The most intensive emission peak, observed at 612 nm, corresponds to the hypersensitive to the site symmetry electric dipole (ED) transition $^5D_0 \rightarrow {}^7F_2$, while the second-most intensive magnetic dipole (MD) $^5D_0 \rightarrow {}^7F_1$ one is insensitive to the site symmetry and is considered almost constant [6,44,52,53]. The integrated emission intensity ratio (R) of these two transitions $^5D_0 \rightarrow {}^7F_2/{}^5D_0 \rightarrow {}^7F_1$ is used to estimate the degree of symmetry around Eu^{3+} ions and the strength of covalence of the europium–oxygen bond. The higher R values indicate more site asymmetry of the rare-earth ion, a high covalency between Eu^{3+} and O^{2-} ions, and an enhanced emission intensity [6,54,55]. The intensity ratios, R, of the present glasses (R = 4.7–5.7) (Table 5) are higher than most of the other reported Eu^{3+}-doped glasses and have close values to Eu^{3+}:Y_2O_3 and Eu^{3+}:Y_2O_2S, indicating that the synthesized glasses are characterized by a more distorted environment of the Eu^{3+} ion and a high covalent bonding between Eu^{3+} and the surrounding ligands, thus achieving an enhanced Eu^{3+} emission intensity [15,18,20,36,56–64].

Table 5. Comparison of the luminescence intensity ratio (R) of $^5D_0 \rightarrow {}^7F_2$ to $^5D_0 \rightarrow {}^7F_1$ transition of Eu^{3+}-doped oxide glasses.

Glass Composition	R Values	Ref.
$50ZnO{:}48B_2O_3{:}1Bi_2O_3{:}1WO_3{:}0.5Eu_2O_3$	4.61	Present work
$50ZnO{:}44B_2O_3{:}1Bi_2O_3{:}5WO_3{:}0.5Eu_2O_3$	5.04	Present work
$50ZnO{:}\ 40B_2O_3{:}10WO_3{:}0.5Eu_2O_3$	5.57	[12]
$50ZnO{:}39B_2O_3{:}1Bi_2O_3{:}10WO_3{:}0.5Eu_2O_3$	5.73	Present work
$50ZnO{:}40B_2O_3{:}10WO_3{:}xEu_2O_3$ ($0 \leq x \leq 10$)	4.54–5.77	[15]
$50ZnO{:}40B_2O_3{:}5WO_3{:}5Nb_2O_5{:}xEu_2O_3$ (x = 0, 0·1, 0·5, 1, 2, 5 and 10)	5.09–5.76	[36]
$50ZnO{:}(40-x)B_2O_3{:}10Bi_2O_3{:}0.5Eu_2O_3{:}xWO_3$, x = 0 and 0.5	3.58; 3.79	[20]
$20ZnO{:}8Al_2O_3{:}(12-x)Bi_2O_3{:}60B_2O_3{:}xEu_2O_3$	1.951–2.78	[56]
$39.5Li_2O{:}59.5SiO_2{:}1Eu_2O_3$	3.20	[57]
$4ZnO{:}3B_2O_3$ 0.5 ÷ 2.5 mol% Eu^{3+}	3.94–2.74	[58]
Eu^{3+}: $45B_2O_3$-5ZnO-49PbO	3.03	[59]
$15PbF_2{:}25WO_3{:}(60-x)TeO_2{:}xEu_2O_3$ x = 0.1, 0.5, 1.0 and 2.0 mol%	2.37–2.78	[60]
$40ZnO{:}(30-x)\ B_2O_3{:}30P_2O_5{:}xEu_2O_3$ ($0.1 \leq x \leq 0.9$)	2.96–3.65	[61]
$60ZnO{:}(40x)B_2O_3{:}0.2Eu_2O_3{:}xBi_2O_3$ (x = 0, 0.1, 0.2, 0.5, 1.0)	2.98	[18]
$(100-x){:}(0.2Bi_2O_3{-}0.8GeO_2){:}xEu_2O_3$ (x = 0.5, 1, 1.5, 2 mol%)	3.94–4.21	[62]
Eu^{3+}:Y_2O_3	3.8–5.2	[63]
Eu^{3+} doped Y_2O_2S	6.45–6.62	[64]

The R values were found to increase from 4.61 to 5.73 (Table 5) as the WO_3 concentration raised from 1 to 10 mol%. The incorporation of small amounts of Bi_2O_3 (1 mol%) into the glass structure also led to an increase in the asymmetric ratio from 5.57 for the glass $50ZnO{:}40B_2O_3{:}10WO_3{:}0.5Eu_2O_3$ to 5.73 for the glass $50ZnO{:}39B_2O_3{:}1Bi_2O_3{:}10WO_3{:}0.5Eu_2O_3$. These R values were much higher as compared to the R value for glass with high Bi_2O_3 content (10 mol%) ($50ZnO{:}(40 - x)B_2O_3{:}10Bi_2O_3{:}0.5Eu_2O_3{:}xWO_3$, x = 0 and 0.5) (R = 3.79) [20]. This result shows that we have found an appropriate glass composition for hosting an active rare-earth ion that provide a high Eu^{3+} emission intensity.

4. Discussion

In this study, by analyzing the IR spectra of zinc–borate glasses containing 1, 5, and 10 mol% WO_3, with and without Bi_2O_3, it was found that 1 mol% Bi_2O_3 leads to an increase in the number of the BO_4^- involved in superstructural groupings and formation of Bi–O–B, Bi–O–W, Zn–O–B, Zn–O–W, Eu–O–W, and Eu–O–B cross-links in the glass structure. With WO_3 loading in the Bi^{3+}-containing glasses, a partial $[WO_4]^{2-} \rightarrow WO_6$ transformation took place. The tungstate octahedra shared common corners and/or edges, forming W–O–W and W_2O_2 bonds. There were also B–O–B bonds with different numbers with the WO_3 content. The RMC modeling also revealed the presence of both WO_6 and WO_4 units and trigonal BO_3 and tetrahedral BO_4 units varying in amount with the composition. Thus, the bismuth-containing glasses were characterized by a more reticulated and rigid network that ensures low symmetry sites of Eu^{3+} ions, which is favorable for the luminescence emission of the active Eu^{3+} ion. The structural features revealed by IR analysis agreed well with the measured density and calculated physical parameters. Bi_2O_3-containing glasses were characterized with the lower molar and oxygen molar volume and higher oxygen packing density, as compared with Bi_2O_3-free glasses with the same compositions, as a result of the formation of highly cross-linked structure and the presence of new mixed bonding with participation of Bi^{3+} ions. A significant enhancement of the Eu^{3+} emission was established in the glasses $50ZnO{:}(49 - x)B_2O_3{:}1Bi_2O_3{:}xWO_3{:}0.5Eu_2O_3$, (a) x = 1 mol%, (b) x = 5 mol%, (c) x = 10 mol%, in the presence of low Bi_2O_3 content (1 mol%) and with the increase of WO_3 content (up to 10 mol%). The photoluminescent spectra of the new glasses showed an intensive red luminescence at 612 nm as well as a very large value of the luminesce ratio R (over 5), both evidencing that Eu^{3+} ions occupied distorted sites in the created glass network. Particularly, the glass with 10 mol% WO_3 showed the strongest emission of the active ions as a result of the structural features established and also because of an energy transfer from tungstate and bismuthate groups to the active ion.

5. Conclusions

The results of this investigation show that the zinc borate glass matrix with the simultaneous presence of both Bi_2O_3 and WO_3 is very suitable for implementing the active Eu^{3+} as it possesses a reticulated and rigid glass structure, ensuring a more asymmetrical local structure around Eu^{3+} sites, accordingly yielding a higher luminescence of the incorporated Eu^{3+} ions. On the other hand, both bismuth and tungsten oxides have a synthesizer effect by transferring the emission energy nonradiatively to the activator Eu^{3+}, which additionally improves its luminesce properties. This suggests that the obtained glasses are potential candidates for red-light-emitting phosphors.

Author Contributions: Conceptualization, A.Y., M.M. and R.I.; methodology, A.Y., M.M. and R.I.; software, A.Y., M.M. and M.F.; validation, A.Y., M.M. and M.F.; formal analysis, A.Y., M.M. and M.F.; investigation, A.Y., M.M., M.F., L.A. and P.P.; resources, R.I. and M.F.; data curation, A.Y., M.M., R.I. and M.F.; writing—original draft preparation, A.Y., M.M. and M.F.; writing—review and editing, R.I.; visualization, A.Y., M.M. and R.I.; supervision, R.I.; project administration, R.I. and M.F.; funding acquisition, R.I. and M.F. All authors have read and agreed to the published version of the manuscript.

Funding: This work is supported by the National Science Programme "European research network" under project TwinTeam Д01-272/02.10.2020 г. and by joint research project within the framework of an international scientific cooperation between BAS and MTA IC-HU/01/2022-2023.

Institutional Review Board Statement: Not applicable.

Informed Consent Statement: Not applicable.

Data Availability Statement: Not applicable.

Acknowledgments: Special thanks are due to R. Kukeva for EPR measurements.

Conflicts of Interest: The authors declare no conflict of interest.

References

1. Cho, J.; Park, J.H.; Kim, J.K.; Schubert, E.F. White light-emitting diodes: History, progress, and future. *Laser Photonics Rev.* **2017**, *11*, 1600147. [CrossRef]
2. Chen, Y.Y.; Duh, J.G.; Chiou, B.S.; Peng, C.G. Luminescent mechanisms of ZnS: Cu: Cl and ZnS: Cu: Al phosphors. *Thin Solid Films* **2001**, *392*, 50–55. [CrossRef]
3. Kim, K.B.; Kim, Y.I.; Chun, H.G.; Cho, T.Y.; Jung, J.S.; Kang, J.G. Structural and optical properties of $BaMgAl_{10}O_{17}$: Eu^{2+} phosphor. *Chem. Mater.* **2002**, *14*, 5045–5052. [CrossRef]
4. Trond, S.S.; Martin, J.S.; Stanavage, J.P.; Smith, A.L. Properties of Some Selected Europium-Activated Red Phosphors. *J. Electrochem. Soc.* **1969**, *116*, 1047–1050. [CrossRef]
5. Feldmann, C.; Jüstel, T.; Ronda, C.R.; Schmidt, P.J. Inorganic luminescent materials: 100 years of research and application. *Adv. Funct. Mater.* **2003**, *13*, 511–516. [CrossRef]
6. Binnemans, K. Interpretation of europium (III) spectra. *Coord. Chem. Rev.* **2015**, *295*, 1–45. [CrossRef]

7. Setlur, A.A. Sensitizing Eu^{3+} with Ce^{3+} and Tb^{3+} to make narrow-line red phosphors for light emitting diodes. *Electrochem. Solid-State Lett.* **2012**, *15*, J25. [CrossRef]
8. Hong, J.H.; Zhang, Z.G.; Cong, C.J.; Zhang, K.L. Energy transfer from Bi^{3+} sensitizing the luminescence of Eu^{3+} in clusters embedded into sol–gel silica glasses. *J. Non-Cryst. Solids* **2007**, *353*, 2431–2435. [CrossRef]
9. Spadaro, F.; Rossi, A.; Laine, E.; Hartley, J.; Spencer, N.D. Mechanical and tribological properties of boron oxide and zinc borate glasses. *Phys. Chem. Glas. Eur. J. Glass Sci. Technol. B* **2016**, *57*, 233–244. [CrossRef]
10. Ehrt, D. Zinc and manganese borate glasses-phase separation, crystallization, photoluminescence and structure. *Phys. Chem. Glas. Eur. J. Glass Sci. Technol. B* **2013**, *54*, 65–75.
11. Ivankov, A.; Seekamp, J.; Bauhofer, W. Optical properties of Eu^{3+}-doped zinc borate glasses. *J. Lumin.* **2006**, *121*, 123–131. [CrossRef]
12. Taki, Y.; Shinozaki, K.; Honma, T.; Dimitrov, V.; Komatsu, T. Electronic polarizability and interaction parameter of gadolinium tungsten borate glasses with high WO_3 content. *J. Solid. State Chem.* **2014**, *220*, 191–197. [CrossRef]
13. Ali, A.A.; Fathi, A.M.; Ibrahim, S. Material characteristics of WO_3/Bi_2O_3 substitution on the thermal, structural, and electrical properties of lithium calcium borate glasses. *Appl. Phys. A* **2023**, *129*, 299. [CrossRef]
14. Milanova, M.; Aleksandrov, L.; Iordanova, R. Structure and luminescence properties of tungsten modified zinc borate glasses doped with Eu^{3+} ions. *Mater. Today Proc.* **2022**, *61*, 1206–1211. [CrossRef]
15. Milanova, M.; Aleksandrov, L.; Yordanova, A.; Iordanova, R.; Tagiara, N.S.; Herrmann, A.; Gao, G.; Wondraczek, L.; Kamitsos, E.I. Structural and luminescence behavior of Eu^{3+} ions in ZnO-B_2O_3-WO_3 glasses. *J. Non-Cryst. Solids* **2023**, *600*, 122006. [CrossRef]
16. Yu, B.; Zhou, X.; Xia, H.; Chen, B.; Song, H. Novel Bi^{3+}/Eu^{3+} co-doped oxyfluoride transparent KY_3F_{10} glass ceramics with wide tunable emission and high optical temperature sensitivity. *J. Lumin.* **2021**, *239*, 118366. [CrossRef]
17. Giraldo, O.G.; Fei, M.; Wei, R.; Teng, L.; Zheng, Z.; Guo, H. Energy transfer and white luminescence in Bi^{3+}/Eu^{3+} co-doped oxide glasses. *J. Lumin.* **2020**, *219*, 116918. [CrossRef]
18. Sontakke, A.D.; Tarafder, A.; Biswas, K.; Annapurna, K. Sensitized red luminescence from Bi^{3+} co-doped Eu^{3+}: ZnO–B_2O_3 glasses. *Phys. B Condens. Matter* **2009**, *404*, 3525–3529. [CrossRef]
19. Kaewkhao, J.; Boonin, K.; Yasaka, P.; Kim, H.J. Optical and luminescence characteristics of Eu^{3+} doped zinc bismuth borate (ZBB) glasses for red emitting device. *Mater. Res. Bull.* **2015**, *71*, 37–41. [CrossRef]
20. Aleksandrov, L.; Yordanova, A.; Milanova, M.; Iordanova, R.; Fabian, M. Doping effect of WO_3 on the structure and luminescent properties of Zno-B_2O_3-Bi_2O_3:Eu^{3+} glass. *J. Chem. Technol. Metall.* **2023**, *58*, 707–715.
21. Svab, E.; Meszaros, G.; Deak, F. Neutron powder diffractometer at the Budapest research reactor. *Mat. Sci. Forum* **1996**, *228*, 247–252. [CrossRef]
22. Gereben, O.; Jovari, P.; Temleitner, L.; Pusztai, L. A new version of the RMC++ Reverse Monte Carlo programme, aimed at investigating the structure of covalent glasses. *J. Optoelec. Adv. Mater.* **2007**, *9*, 3021–3027.
23. Iordanova, R.; Milanova, M.; Aleksandrov, L.; Khanna, A. Structural study of glasses in the system B_2O_3-Bi_2O_3-La_2O_3-WO_3. *J. Non-Cryst. Solids* **2018**, *481*, 254–259. [CrossRef]
24. Kamitsos, E.I.; Patsis, A.P.; Karakassides, M.A.; Chryssikos, G.D. Infrared reflectance spectra of lithium borate glasses. *J. Non-Cryst. Solids* **1990**, *126*, 52–67. [CrossRef]
25. Ouis, M.A.; El Batal, F.H.; Azooz, M.A. FTIR, optical and thermal studies of gadmium borate glass doped with Bi_2O_3 and effect of gamma irradiation. *J. Aust. Ceram. Soc.* **2020**, *56*, 283–290. [CrossRef]
26. Islam, M.; Lazure, S.; Vannier, R.; Nowgorodski, G.; Mairesse, G. Structural and computational studies of Bi_2WO_6 based oxygen ion conductors. *J. Mater. Chem.* **1998**, *8*, 655–660. [CrossRef]
27. Buttrey, D.; Vogt, T.; White, B. High-temperature incommensurate-to-commensurate phase transition in the Bi_2MoO_6 catalyst. *J. Solid State Chem.* **2000**, *155*, 206–208. [CrossRef]
28. Sleight, A.W. Accurate cell dimensions for ABO_4 molybdates and tungstates. *Acta Cryst.* **1972**, *28*, 2899–2991. [CrossRef]
29. Mączka, M.; Macalik, L.; Hermanovicz, K.; Kepinski, L.; Tomaszewski, P. Phonon properties of nanosized bismuth layered ferroelectric material—Bi_2WO_6. *J. Raman Spectrosc.* **2010**, *41*, 1059–1066. [CrossRef]
30. Mączka, M.; Macalik, L.; Hanuza, J. Raman and IR spectra of the cation-deficient Aurivillius layered crystal $Bi_2W_2O_9$. *J. Raman Spectrosc.* **2009**, *40*, 2099–2103.
31. Zhao, X.; Yao, W.; Wu, Y.; Zhang, S.; Yan, H.; Zhi, Y. Fabrication and photoelectrochemical properties of porous $ZnWO_4$ film. *J. Solid State Chem.* **2006**, *179*, 2562–2570. [CrossRef]
32. Varsamis, C.P.; Makris, E.N.; Valvi, C.; Kamitsos, E.I. Short-range structure, the role of bismuth and property-structure correlations in bismuth borate glasses. *Phys. Chem. Chem. Phys.* **2021**, *23*, 10006–10020. [CrossRef] [PubMed]
33. Yao, Z.Y.; Möncke, D.; Kamitsos, E.I.; Houizot, P.; Celarie, F.; Rouxel, T.; Wondraczek, L. Structure and mechanical properties of copper-lead and copper-zinc borate glasses. *J. Non-Cryst. Solids* **2016**, *435*, 55–68. [CrossRef]
34. Winterstein-Beckmann, A.; Moncke, D.; Palles, D.; Kamitsos, E.I.; Wondraczek, L. Structure and Properties of Orthoborate Glasses in the Eu_2O_3–(Sr,Eu)O–B_2O_3 Quaternary. *J. Phys. Chem. B* **2015**, *119*, 3259–3272. [CrossRef] [PubMed]
35. Iordanova, R.; Milanova, M.; Aleksandrov, L.; Shinozaki, K.; Komatsu, T. Structural study of WO_3-La_2O_3-B_2O_3-Nb_2O_5 glasses. *J. Non-Cryst. Solids* **2020**, *543*, 120132. [CrossRef]

36. Aleksandrov, L.; Milanova, M.; Yordanova, A.; Iordanova, R.; Nedyalkov, N.; Petrova, P.; Tagiara, N.S.; Palles, D.; Kamitsos, E.I. Synthesis, structure and luminescence properties of Eu^{3+}-doped $50ZnO.40B_2O_3.5WO_3.5Nb_2O_5$ glass. *Phys. Chem. Glas. Eur. J. Glass Sci. Technol. B* **2023**, *64*, 101–109.
37. El-Fayoumi, M.A.K.; Farouk, M. Structural properties of Li-borate glasses doped with Sm^{3+} and Eu^{3+} ions. *J. Alloys Compd.* **2009**, *482*, 356–360. [CrossRef]
38. Villegas, M.A.; Fernández Navarro, J.M. Physical and structural properties of glasses in the TeO_2–TiO_2–Nb_2O_5 system. *J. Eur. Ceram. Soc.* **2007**, *27*, 2715–2723. [CrossRef]
39. Fabian, M.; Araczki, C. Basic network structure of SiO_2-Na_2O-B_2O_3 glasses from diffraction and reverse Monte Carlo simulation. *Phys. Scrip.* **2016**, *91*, 054004. [CrossRef]
40. Fabian, M.; Svab, E.; Krezhov, K. Network structure with mixed bond-angle linkages in MoO_3-ZnO-B_2O_3 glasses: Neutron diffraction and reverse Monte Carlo modelling. *J. Non-Cryst. Solids* **2016**, *433*, 6–13. [CrossRef]
41. Othman, H.; Valiev, D.; Polisadova, E. Structural and mechanical properties of zinc aluminoborate glasses with different content of aluminium oxide. *J. Mater. Sci Mater. Electron.* **2017**, *28*, 4647–4653. [CrossRef]
42. Brodbeck, M.; Iton, L.E. The EPR spectra of Gd^{3+} and Eu^{3+} in glassy systems. *J. Chem. Phys.* **1985**, *83*, 4285–4299. [CrossRef]
43. Nandyala, S.; Hungerford, G.; Babu, S.; Rao, J.L.; Leonor, I.B.; Pires, R.; Reis, R.L. Time resolved emission and electron paramagnetic resonance studies of Gd^{3+} doped calcium phosphate glasses. *Adv. Mater. Lett.* **2016**, *7*, 277–281. [CrossRef]
44. Blasse, G.; Grabmaier, B.C. *Luminescent Materials*, 1st ed.; Springer: Berlin/Heidelber, Germany, 1994; p. 18.
45. Hoefdraad, H.E. The charge-transfer absorption band of Eu^{3+} in oxides. *J. Solid State Chem.* **1975**, *15*, 175–177. [CrossRef]
46. Parchur, A.K.; Ningthoujam, R.S. Behaviour of electric and magnetic dipole transitions of Eu^{3+}, $^5D_0 \rightarrow ^7F_0$ and Eu–O charge transfer band in Li^+ co-doped YPO_4:Eu^{3+}. *RSC Adv.* **2012**, *2*, 10859–10868. [CrossRef]
47. Mariselvam, K.; Juncheng, L. Synthesis and luminescence properties of Eu^{3+} doped potassium titano telluroborate (KTTB) glasses for red laser applications. *J. Lumin.* **2021**, *230*, 117735. [CrossRef]
48. Dutta, P.S.; Khanna, A. Eu^{3+} activated molybdate and tungstate based red phosphors with charge transfer band in blue region. *ECS J. Solid State Sci. Technol.* **2013**, *2*, R3153–R3167. [CrossRef]
49. Thieme, C.; Herrmann, A.; Kracker, M.; Patzig, C.; Höche, T.; Rüssel, C. Microstructure investigation and fluorescence properties of Europium-doped scheelite crystals in glass-ceramics made under different synthesis conditions. *J. Lumin.* **2021**, *238*, 118244. [CrossRef]
50. Minquan, W.; Xianping, F.; Guohong, X. Luminescence of Bi^{3+} ions and energy transfer from Bi^{3+} ions to Eu^{3+} ions in silica glasses prepared by the sol-gel process. *J. Phys. Chem. Solids* **1995**, *56*, 859–862. [CrossRef]
51. Sungpanich, J.; Thongtem, T.; Thongtem, S. Large-scale synthesis of WO_3 nanoplates by a microwave-hydrothermal method. *Ceram. Int.* **2012**, *38*, 1051–1055. [CrossRef]
52. Shionoya, S.; Yen, W.M.; Yamamoto, H. *Phosphor Handbook*, 2nd ed.; CRC Press: London, UK; New York, NY, USA, 2018; p. 177.
53. Gandhi, Y.; Kityk, I.V.; Brik, M.G.; Rao, P.R.; Veeraiah, N. Influence of tungsten on the emission features of Nd^{3+}, Sm^{3+} and Eu^{3+} ions in ZnF_2–WO_3–TeO_2 glasses. *J. Alloys Compd.* **2010**, *508*, 278–291. [CrossRef]
54. Devi, C.H.B.; Mahamuda, S.; Swapna, K.; Venkateswarlu, M.; Rao, A.S.; Prakash, G.V. Compositional dependence of red luminescence from Eu^{3+} ions doped single and mixed alkali fluoro tungsten tellurite glasses. *Opt. Mater.* **2017**, *73*, 260–267. [CrossRef]
55. Nogami, M.; Umehara, N.; Hayakawa, T. Effect of hydroxyl bonds on persistent spectral hole burning in Eu^{3+} doped BaO-P_2O_5 glasses. *Phys. Rev. B* **1998**, *58*, 6166–6171. [CrossRef]
56. Swapna, K.; Mahamuda, S.; Rao, A.S.; Sasikala, T.; Packiyaraj, P.; Moorthy, L.R.; Vijayaprakash, G. Luminescence characterization of Eu^{3+} doped Zinc Alumino Bismuth Borate glasses for visible red emission applications. *J. Lumin.* **2014**, *156*, 80–86. [CrossRef]
57. Raju, B.D.; Reddy, C.M. Structural and optical investigations of Eu^{3+} ions in lead containing alkali fluoroborate glasses. *Opt. Mater.* **2012**, *34*, 1251–1260. [CrossRef]
58. Bettinelli, M.; Speghini, A.; Ferrari, M.; Montagna, M. Spectroscopic investigation of zinc borate glasses doped with trivalent europium ions. *J. Non-Cryst. Solids* **1996**, *201*, 211–221. [CrossRef]
59. Lakshminarayana, G.; Buddhudu, S. Spectral analysis of Eu^{3+} and Tb^{3+}:B_2O_3-ZnO-PbO glasses. *Mater. Chem. Phys.* **2007**, *102*, 181–186. [CrossRef]
60. Babu, A.M.; Jamalaiah, B.C.; Suhasini, T.; Rao, T.S.; Moorthy, L.R. Optical properties of Eu^{3+} ions in lead tungstate tellurite glasses. *Solid State Sci.* **2011**, *13*, 574–578. [CrossRef]
61. Bindu, S.H.; Raju, D.S.; Krishna, V.V.; Rao, T.R.; Veerabrahmam, K.; Raju, C.L. UV light induced red emission in Eu^{3+}-doped zincborophosphate glasses. *Opt. Mater.* **2016**, *62*, 655–665. [CrossRef]
62. Gökçe, M. Development of Eu^{3+} doped bismuth germanate glasses for red laser applications. *J. Non-Cryst. Solids* **2019**, *505*, 272–278. [CrossRef]

63. Kabir, M.; Ghahari, M.; Afarani, M.S. Co-precipitation synthesis of nano Y_2O_3: Eu^{3+} with different morphologies and its photoluminescence properties. *Ceram. Int.* **2014**, *40*, 10877–10885. [CrossRef]
64. Som, S.; Mitra, P.; Kumar, V.; Kumar, V.; Terblans, J.J.; Swart, H.C.; Sharma, S.K. The energy transfer phenomena and colour tunability in Y_2O_2S: Eu^{3+}/Dy^{3+} micro-fibers for white emission in solid state lighting applications. *Dalton Trans.* **2014**, *43*, 9860–9871. [CrossRef] [PubMed]

materials

MDPI

Article

Effect of TiO_2 on Pd/La_2O_3-CeO_2-Al_2O_3 Systems during Catalytic Oxidation of Methane in the Presence of H_2O and SO_2

Ralitsa Velinova [1], Silviya Todorova [2], Daniela Kovacheva [1], Hristo Kolev [2], Yordanka Karakirova [2], Pavel Markov [1], Katerina Tumbalova [1], Georgi Ivanov [1] and Anton Naydenov [1,*]

[1] Institute of General and Inorganic Chemistry, Bulgarian Academy of Sciences, Acad. G. Bonchev Str., bl. 11, 1113 Sofia, Bulgaria; raligeorgieva@svr.igic.bas.bg (R.V.); didika@svr.igic.bas.bg (D.K.); pvlmarkov@svr.igic.bas.bg (P.M.); katerinatumbalova@mail.bg (K.T.); geoivanov@yahoo.com (G.I.)

[2] Institute of Catalysis, Bulgarian Academy of Sciences, Acad. G. Bonchev Str., bl. 11, 1113 Sofia, Bulgaria; todorova@ic.bas.bg (S.T.); hgkolev@ic.bas.bg (H.K.); daniepr@ic.bas.bg (Y.K.)

* Correspondence: naydenov@svr.igic.bas.bg

Abstract: New results on the effect of TiO_2 on Pd/La_2O_3-CeO_2-Al_2O_3 systems for catalytic oxidation of methane in the presence of H_2O and SO_2 have been received. Low-temperature N_2-adsorption, XRD, SEM, HRTEM, XPS, EPR and FTIR techniques were used to characterize the catalyst. The presence of Ce^{3+} on the catalytic surface and in the volume near the lantana was revealed by EPR and XPS. After aging, the following changes are observed: (i) agglomeration of the Pd-clusters (from 8 nm to 12 nm); (ii) transformation of part of the TiO_2 from anatase to larger particles of rutile; and (iii)—the increase in PdO/Pd—ratio above its optimum. The modification by Ti of the La_2O_3-CeO_2-Al_2O_3 system leads to higher resistance towards the presence of SO_2 most likely due to the prevailing formation of unstable surface sulfites instead of thermally stable sulfates. Based on kinetic model calculations, the reaction pathway over the Pd/La_2O_3-CeO_2-TiO_2-Al_2O_3 catalyst follows the Mars–van Krevelen mechanism. For evaluation of the possible practical application of the obtained material, a sample of Pd/La_2O_3-CeO_2-TiO_2-Al_2O_3, supported on rolled aluminum-containing stainless steel (Aluchrom VDM®), was prepared and tested. Methane oxidation in an industrial-scale monolithic reactor was simulated using a two-dimensional heterogeneous reactor model.

Keywords: methane oxidation; Pd/La_2O_3-CeO_2-TiO_2-Al_2O_3; deactivation; sulfur poisoning; catalyst regeneration; TiO_2

Citation: Velinova, R.; Todorova, S.; Kovacheva, D.; Kolev, H.; Karakirova, Y.; Markov, P.; Tumbalova, K.; Ivanov, G.; Naydenov, A. Effect of TiO_2 on Pd/La_2O_3-CeO_2-Al_2O_3 Systems during Catalytic Oxidation of Methane in the Presence of H_2O and SO_2. *Materials* **2023**, *16*, 6784. https://doi.org/10.3390/ma16206784

Academic Editor: Lubomira Tosheva

Received: 19 September 2023
Revised: 10 October 2023
Accepted: 18 October 2023
Published: 20 October 2023

1. Introduction

Methane is the main component of natural gas and recent investigations indicate that it has an even more significant impact on global warming than previously thought [1,2]. The sources of pollution by methane are agriculture landfills and the combustion of coal and natural gas [3]. Recently, the need for methane incineration is also due to problems arising from the production of electricity and the great concern regarding the protection of the environment [4–8].

A very effective technology for reducing methane emissions is catalytic combustion; however, its major drawback is connected with the deactivation of the catalysts (mainly based on palladium). Among the catalysts used for complete oxidation, palladium supported on γ-Al_2O_3 stands out due to its large surface area and cost-effectiveness [9–11]. It is known that PdO/Al_2O_3 catalysts are unstable at high temperatures, resulting in reduced alumina surface area and the transformation of palladium oxide into Pd^0 [12,13]. Deactivation is also accelerated by palladium sintering at elevated temperatures and the presence in the gases of sulfur compounds and water vapor at high concentrations. Therefore, there is a pressing need to develop catalysts that not only exhibit high activity but also demonstrate excellent thermal stability and resistance to water and sulfur compounds. To enhance the properties of palladium-based catalysts, various additives have been explored.

Lanthanum, for instance, has been commonly used to prevent the deactivation of catalysts by retarding the conversion of palladium oxide into palladium and improving stability against alumina surface area loss [5,9].

As reported by Ozawa et al. [14], adding La_2O_3 stabilizes the surface area of alumina and further modification by CeO_2 prevents the transformation of palladium oxide to palladium. It was suggested that the addition of La into CeO_2 decreases the particle size and, as a result, inhibits the sintering of CeO_2. In general, Ce–La-based compounds exhibit excellent catalytic performance due to the remarkable oxygen storage capacity; however at elevated temperatures (above 1000 °C), CeO_2 sinters result in catalyst deactivation [11]. In our previous study [15], we demonstrated the high activity and thermal stability of the Pd/ La_2O_3-CeO_2-Al_2O_3 catalyst; however, its resistance to sulfur dioxide can be considered unsatisfactory.

The modification by TiO_2 is based on the fact that TiO_2 is only weakly and reversibly sulfated in the presence of SO_2 and oxygen [16]. To enhance the sulfur resistance of catalysts, TiO_2 as a material has been introduced into exhaust gas catalysts [17,18]. The use of TiO_2 aims to promote sulfur desorption and significant progress in sulfur tolerance is observed [18]. Its employment as a catalytic support is associated with increased activity due to phase–support interactions [19].

TiO_2 can increase the mobility of oxygen by creating oxygen vacancies through a more efficient reduction process (from Ti^{4+} to Ti^{3+}), supplying the Lewis acid sites for the adsorption and dissociation of molecules [20].

The three crystalline forms of TiO_2 (anatase, rutile, and brookite) possess different properties, thus providing the possibility for the TiO_2-supported catalysts to perform different catalytic behavior [21,22]. It should be pointed out that for environmental applications, anatase is the most frequently used [23]. The combination of TiO_2, which acts as a scavenger for SO_2 and H_2O, along with inert SiO_2, which facilitates the removal of poisoning compounds after exposure to SO_2 and H_2O, has proven effective in improving resistance to poisoning and the catalyst's regenerative potential. This behavior reveals the Pd—TiO_2-based catalyst could be an attractive material for further practical implementation [24].

Based on existing research, an effective catalyst for methane combustion should primarily consist of (i) γ—Al_2O_3 as a carrier matrix; (ii) Pd as a catalytically active component; (iii) La_2O_3 for thermal stabilization of γ—Al_2O_3; (iv) CeO_2 to improve oxygen exchange to keep the optimal ratio between Pd and PdO; and (v) TiO_2—for improving poisoning tolerance and catalyst regeneration. Regarding the choice of synthesis methods, it can be pointed out that the sol–gel method offers an advanced way to create solids with a high specific area, well-defined porosity, and high resistance to deactivation in a single step [25]. This technique permits the physical and chemical properties of the final solid to be controlled throughout the synthesis steps; thus, the method allows the incorporation of the catalytic active component during the gelation step and ensures effective metal–support interaction [26].

The current study aims to investigate a Pd/La_2O_3-CeO_2-TiO_2-Al_2O_3 model system prepared using the sol–gel method evaluation on catalytic activity, thermal stability, water vapor effect, sulfur resistance, and the possibility of use of the synthesized material as an active component for creating monolithic catalysts for application in reduction of methane emissions.

2. Materials and Methods

2.1. Catalysts Synthesis

The La_2O_3-CeO_2-TiO_2-Al_2O_3 support was synthesized via the sol–gel method by the procedure reported elsewhere [27]. The powdered $Al[OCH(CH_3)_2]_3$ (≥98%, Sigma-Aldrich, St. Louis, MO, USA) was dissolved in water, and after that, amounts of 1M HNO_3, cerium (III) nitrate hexahydrate (99%, Sigma-Aldrich, St. Louis, MO, USA), lanthanum(III) nitrate hexahydrate (≥99.9%, Sigma-Aldrich) and titanium(IV) isopropoxide (97%, Sigma-Aldrich, St. Louis, MO, USA) were included to produce a mixture containing 80 wt.% aluminum oxide, 8 wt.% cerium (IV) oxide, 4 wt.% lanthnum(III) oxide and 8% titanium dioxide. The prepared mixture was homogenized at 373 K, then refluxed for 72 h and the obtained gel was dried at ambient temperature before heat treatment in air for 4 h at 500 °C. The synthesized catalytic support was impregnated with an aqueous solution of palladium(II) nitrate hydrate (99.8%, Thermo Scientific Chemicals, Waltham, MA, USA) and treated for 2 h in 0.1 vol.% CH_4 in N_2 gas mixture at 450 °C (in the absence of O_2) to produce a catalyst with a nominal palladium content of 2.0% (w/w).

In order to obtain data approaching the operation of the catalyst in the form of a monolithic catalytic element, experiments with Aluchrome VDM® (VDM Metals International GmbH, Werdohl, Germany) were performed. The preparation of a single monolithic channel (D = 3.5 mm, L = 118 mm, thickness of 0.2 mm) involves the following stages: (i) thermal treatment of the alloy at 920 °C during 25 h, aiming at the formation of α-Al_2O_3 whiskers on the steel surface; (ii) application of a primary bonding layer of γ-Al_2O_3; (iii) coating with a sol containig La_2O_3-CeO_2-TiO_2-Al_2O_3; and (iv) impregnation with palladium (II) nitrate dihydrate.

In the present study, the catalyst was denoted as Pd/La_2O_3-CeO_2-TiO_2-Al_2O_3.

2.2. Characterization Techniques

The physical nitrogen adsorption/desorption isotherms were examined at 77 K using a NOVA 1200e Surface & Pore Analyzer (Quantachrome, Boynton Beach, FL, USA). The Brunauer–Emmett–Teller (BET) equation was applied for a specific surface area estimation [28], the total pore volume being determined at a relative pressure of 0.99. The pore size distributions were determined using the desorption branch of the isotherms, employing the Barrett–Joyner–Halenda (BJH) method [29]. The volume of the micropores was derived by the V-t-method [30].

The X-ray diffraction (XRD) diffractograms of pure support as well as fresh, used, after sulfur poisoning and thermally aged catalysts were determined within the 2Θ range 10–80° on a Bruker D8 Advance diffractometer (Cu Kα radiation, LynxEye detector, Karlsruhe, Germany). The determination of phase composition was carried out using the EVA software package, which utilized the ICDD-PDF2(2014) database. To quantify and determine the average crystallite size, the Topas-4.2 program (Karlsruhe, Germany) was employed.

The surface morphology and elemental composition of the catalysts were examined on a scanning electron microscope SEM/FIB LYRA I XMU, TESCAN (Brno—Kohoutovice, Czech Republic) connected with an energy dispersive spectroscope (EDS) (QUANTAX 200, Bruker, Germany).

Transmission electron micrographs (TEM) were performed on a JEOL JEM 2100 microscope (JEOL Ltd., Tokyo, Japan) operating at 200 kV. The catalyst specimens were suspended in ethanol via an ultrasonic bath and subsequently placed onto holey C/Cu grids.

The catalysts' composition and electronic structure were examined using X-ray photoelectron spectroscopy (XPS). The chemical composition of the samples was measured by monitoring the areas and binding energies of C1s, O1s, La3d, Ce3d, Al2p, Pd3d, Ti2p, and S2p photoelectron peaks. The photoelectron peaks measurements have been carried out on the ESCALAB MkII (VG Scientific, now Thermo Scientific, Waltham, MA, USA) electron spectrometer. More about the setup of the spectrometer and data processing can be found in [31].

The Electron paramagnetic resonance (EPR) spectra were recorded by the JEOL JES-FA 100 EPR spectrometer. The spectrometer is equipped with a cylindrical resonator (TE011) and operates at X–band frequency (9.5 GHz). The catalysts were put in special quartz tubes and positioned in the cavity center. The EPR measurements were performed at temperatures from 123 to 323 K using a Varied Temperature Controller ES-DVT4 (JEOL Ltd., Tokyo, Japan). The low temperatures can be easily obtained by sending liquid nitrogen to the sample area. The following conditions were used: modulation frequency—100 kHz, microwave power 1 mW, modulation amplitude 0.2 mT, sweep 500 mT, time constant 0.3 s, and sweep time 2 min.

Fourier transform infrared spectroscopy (FTIR) was performed using a Nicolet 6700 FTIR spectrometer (Thermo Electron Corporation, Madison, WI, USA). The measurements were carried out in transmission mode, and the spectral resolution was set at 4 cm^{-1}.

2.3. Catalytic Tests

The catalytic activity study was performed by using a flow reactor under the following specified conditions: catalyst volume of 0.7 cm^3 (0.5 cm^3 catalyst sample and 0.2 cm^3 quartz–glass particles with the same size as the catalyst), irregular shaped particles with a diameter of 0.45 $\pm$ 0.15 mm. The inner reactor diameter is 6.0 mm. The gaseous hourly space velocity ($GHSV_{STP}$) was 60,000 h^{-1}. The inlet concentrations of reagents were varied as follows: CH_4 concentrations: 5 $\times$ 10^{-2}, 1.0 $\times$ 10^{-1} and 2.7 $\times$ 10^{-1} vol.%, O_2 on levels of 0.9, 5.0 and 20.0 vol.%, additional H_2O on levels of 0, 1.2 and 2.2 vol.%, and balance to 100 vol.% by N_2 (4.6). The standard deviation of the experimental data was estimated based on six repeated measurements. Gas analysis was carried out by using of an on-line gas-analyzers for $CO/CO_2/O_2$ (Maihak-Sick Mod. S 710, V.1.31, Hamburg, Germany), THC-FID (analyzer for total organic content in gas phase, Thermo FID-TG, SK Elektronik GmbH, Leverkusen, Germany) and for SO_2 measurement (MultiGas FTIR Gas Analyzer 2030G, MKS Instruments Inc., Andover, MA, USA).

3. Results and Discussion

3.1. Catalytic Experiments

The data from the tests on total methane oxidation in the absence and presence of water vapor are shown in Figure 1. As can be seen, the light-off temperature (T_{50}) in dry gas feed is about 328 °C, while in humid gas feed, the effect of the water leads to an increase of about 40 °C.

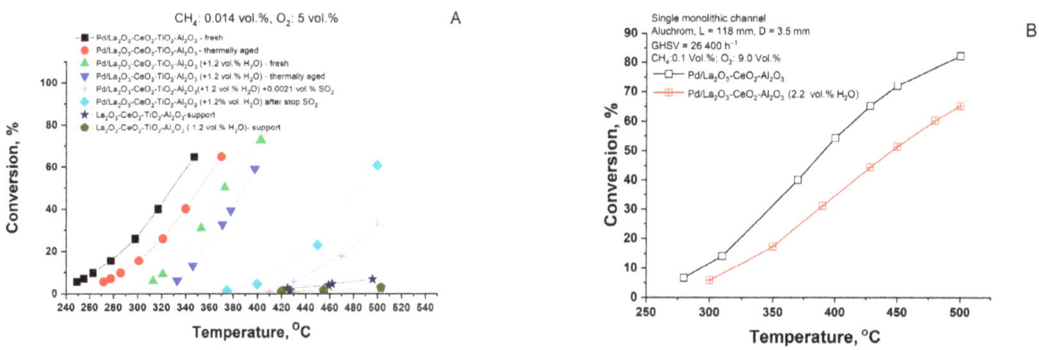

Figure 1. Catalytic activity of the Pd/La_2O_3-CeO_2-TiO_2-Al_2O_3 catalyst during the reaction of total methane oxidation in dry and humid gas feed, sulfur dioxide and after thermal aging (**A**) and tests on the Pd/La_2O_3-CeO_2-TiO_2-Al_2O_3 catalyst, prepared as single monolithic channel (**B**).

For possible practical application, two identical samples of the obtained catalyst were subjected to treatment in air as follows: (i) 170 h in the air at 500 °C in the absence and presence of 1.2 vol.% H_2O (denoted thermally aged) and (ii) in the presence of sulfur dioxide (0.0021 vol.%) performed in the catalytic reactor for 48 h at 450 °C. It can be observed that after the thermal aging, the activity measured in dry gas feed shows a shift in T_{50} to higher temperatures of about 20 °C, while in the presence of 1.2 vol.% H_2O it leads to further increase by another 40 °C. Data on the catalytic activity of La_2O_3-CeO_2-TiO_2-Al_2O_3 support in dry and humid gas feed are also represented for comparative analysis. It is evident that the pure support exhibits notably low catalytic activity. However, it is important to emphasize that its contribution should not be neglegted.

The difference between the measured T_{50} for the fresh and for the corresponding stability tested Pd/La_2O_3-CeO_2-Al_2O_3 and Pd/La_2O_3-CeO_2-TiO_2-Al_2O_3 samples are shown in Figure 2.

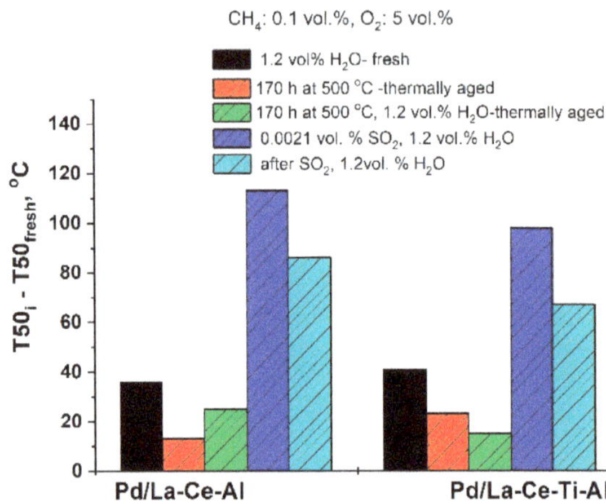

Figure 2. Comparative data on T_{50}, measured with fresh samples and after tests for thermal aging and sulfur dioxide resistance.

It is seen (Figure 2) that in humid gas feed, the effect of the water vapor in the Pd/La_2O_3-CeO_2-TiO_2-Al_2O_3 fresh sample is negligible when compared with the effect of the water on the Pd/La_2O_3-CeO_2-Al_2O_3 sample. However, after the thermal aging, the activity of the Ti-containing sample in the humid gas has been improved, while in the dry gas its the activity is slightly lower as compared with the sample without Ti. Regarding sulfur deactivation, an increase in the temperatures (after the reaction in the presence of SO_2) in T_{50} (apprx. 100–110 °C) was observed. Further testing without sulfur dioxide in the gas shows that the samples restore part of their initial activity, i.e., the Pd/La_2O_3-CeO_2-TiO_2-Al_2O_3 poses higher resistance to sulfur poisoning as compared with the Pd/La_2O_3-CeO_2-Al_2O_3 sample.

Furthermore, the characterization of the catalyst was performed using a variety of methods.

3.2. Low-Temperature Nitrogen Adsorption

The adsorption/desorption isotherm and pore size distribution (PSD) of the synthesized La_2O_3-CeO_2-TiO_2-Al_2O_3 support and the fresh, used, after sulfur poisoning and thermally aged catalysts prepared by the sol–gel method are presented in Figure 3A,B.

The obtained data indicate that all of the samples displayed similar characteristics, corresponding to a typical isotherm of type IV according to the IUPAC classification [32]. The prepared material is mesoporous of type H1 hysteresis, featuring clearly defined cylindrical-like pore channels or clusters of compact, nearly uniform spheres.

According to the BJH method from the desorption branch of the isotherms, the PSD was calculated (Figure 3B). All samples showed a bimodal structure. It was reported that support with a distinct bimodal pore structure shows significant benefits in catalytic reactions. This is attributed to the fact that the presence of large pores facilitates the molecular transport pathway, while the small pores offer a substantial surface area for supporting the active phase, as indicated in reference [33]. The samples exhibit a mesopore volume of approximately 0.2 cm^3/g, with an average pore diameter ranging from 4.9 nm to 8.1 nm, depending on the applied test conditions (Table 1). The specific surface area is significantly decreased and the average pore diameter is increased as a result of the Pd deposition.

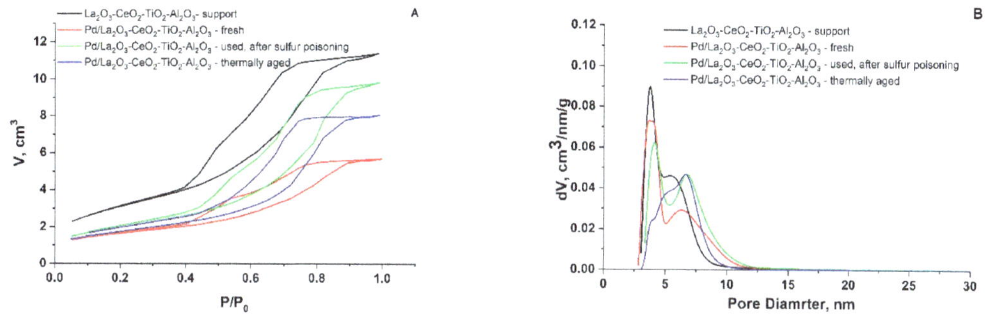

Figure 3. Adsorption/desorption isotherms (**A**) and PSD (**B**) of support and fresh, used, after sulfur poisoning and thermally aged catalysts.

Table 1. Specific surface areas and pore properties of La_2O_3-CeO_2-TiO_2-Al_2O_3 support and Pd/La_2O_3-CeO_2-TiO_2-Al_2O_3 catalysts.

Sample	S_{BET} m^2/g	S_{ext} m^2/g	S_{micro} m^2/g	V_{micro} cm^3/g	Total Pore Volume cm^3/g	Average Pore Diameter nm
La_2O_3-CeO_2-TiO_2-Al_2O_3-support	210	134	76	0.04	0.26	4.9
Pd/La_2O_3-CeO_2-TiO_2-Al_2O_3-fresh	106	103	3	0.02	0.19	7.4
Pd/La_2O_3-CeO_2-TiO_2-Al_2O_3-used, after sulfur poisoning	100	-	-	-	0.22	8.6
Pd/La_2O_3-CeO_2-TiO_2-Al_2O_3-thermally aged	82	-	-	-	0.16	8.1

For evaluation of the possible extent of micropores, the V-t method has been applied. The external surface area, denoted as a S_{ext} of the microporous samples, was determined through the slope of the t-plot. Consequently, the micropore surface area (S_{micro}) was calculated using the formula $S_{micro} = S_{BET} - S_{ext}$. The obtained results show no significant presence of micropores for the used, after sulfur poisoning and thermally aged catalysts. Therefore, the reported data are based on the assumption that the total (BET) surface area is practically equal to the external surface (V-t method).

3.3. Powder X-ray Diffraction

The data from the XRD patterns of the obtained support, fresh, used, after sulfur poisoning and thermally aged catalysts are presented in Figure 4. The diffraction pattern of the support is broad, indicating a relatively low degree of the crystallinity of the phases. In this material, the two phases AlO(OH) (Bohemite)(ICDD PDF 83-2384) together with γ-Al_2O_3 (ICDD-PDF- 70-9085) are identified.

The pattern of fresh catalyst demonstrates the presence of the AlO(OH), γ-Al_2O_3, and PdO (ICDD-PDF 41-1107). The basic diffraction peak of palladium oxide is very small and broad, suggesting a significant degree of dispersion of the palladium across the catalytic surface. After catalytic tests, the phase composition of the support is changed. AlO(OH) is transformed to γ-Al_2O_3, which remains the primary component of the support along with the introduction of a CeO_2-type phase (ICDD-PDF-81-792).

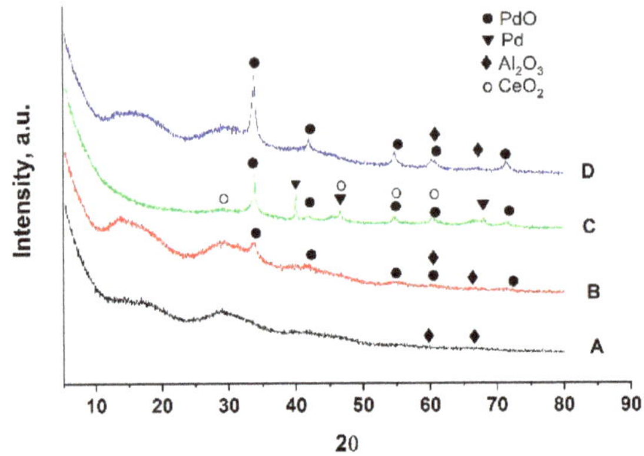

Figure 4. XRD patterns of La_2O_3-CeO_2-TiO_2-Al_2O_3 support (A) Pd/La_2O_3-CeO_2-TiO_2-Al_2O_3-fresh (B); Pd/La_2O_3-CeO_2-TiO_2-Al_2O_3-used, after sulfur poisoning (C) and Pd/La_2O_3-CeO_2-TiO_2-Al_2O_3-thermally aged (D).

The PdO phase remains in the used, largely unchanged sample. Nevertheless, a novel, highly crystalline metal Pd phase (ICDD-PDF-46-1043) appears after the catalytic experiments. After the thermal aging at 500 °C for 170 h, the cubic metal palladium phase is not detectable by XRD. The calculated crystallite size of the deposited palladium oxide phase and palladium are provided in Table 2.

Table 2. Phase composition and average crystallite size.

Sample	Phase Composition According to XRD	Phase Composition According to HRTEM	PdO (nm) [a]	Pd (nm) [a]	Mean Crystalline Size (nm) [b]
La_2O_3-CeO_2-TiO_2-Al_2O_3-support	AlO(OH), γ-Al_2O_3,	-	-	-	
Pd/La_2O_3-CeO_2-TiO_2-Al_2O_3-fresh	AlO(OH), γ-Al_2O_3, PdO	AlO(OH), γ-Al_2O_3, PdO, Pd, CeO_2, TiO_2-anatase	12	-	8
Pd/La_2O_3-CeO_2-TiO_2-Al_2O_3-used, after sulfur poisoning	γ-Al_2O_3, CeO_2,PdO, Pd	AlO(OH), γ-Al_2O_3, PdO, Pd, CeO_2, TiO_2-rutile	11	40	8
Pd/La_2O_3-CeO_2-TiO_2-Al_2O_3-thermally aged	AlO(OH), γ-Al_2O_3, PdO	AlO(OH), γ-Al_2O_3, PdO, Pd, CeO_2, TiO_2-rutile	19	-	12

[a] Crystallite size calculated by the size-strain analysis tool implemented in the Topas 4.2 program. [b] Determinate from TEM.

Observation reveals that in the fresh and used samples after the sulfur poisoning, the crystalline size of PdO is around 11–12 nm, while after thermal aging, it becomes 19 nm.

3.4. Scanning Electron Microscopy

Figure S1 represents the SEM photographs of the studied catalysts. SEM analysis of fresh Pd/La_2O_3-CeO_2-TiO_2-Al_2O_3 reveals an inhomogeneous grainy structure, which after the catalytic test becomes homogeneous. The chemical composition of fresh and used catalytic samples after sulfur poisoning was examined by EDX. The data are presented in Table S1.

The obtained results are in agreement with the applied nominal ratio for the used elements. In the used, after sulfur poisoning of the catalyst, sulfur was detected, which is evidence of the formation of some sulfate or sulfite compounds on the surface of the catalyst after prolonged exposure to SO_2 and H_2O.

3.5. Transmission Electron Microscopy

The morphology and phase composition of Pd/La_2O_3-CeO_2-TiO_2-Al_2O_3 were examined by high resolution transmission electron microscopy (HRTEM). The catalysts exhibit uniform dispersion of palladium, appearing as dark spots across all samples (as depicted in Figure 5). The mean particle size, determined by analyzing 200 randomly chosen nanoparticles in a fresh prepared catalyst, was found to be 8 nm.

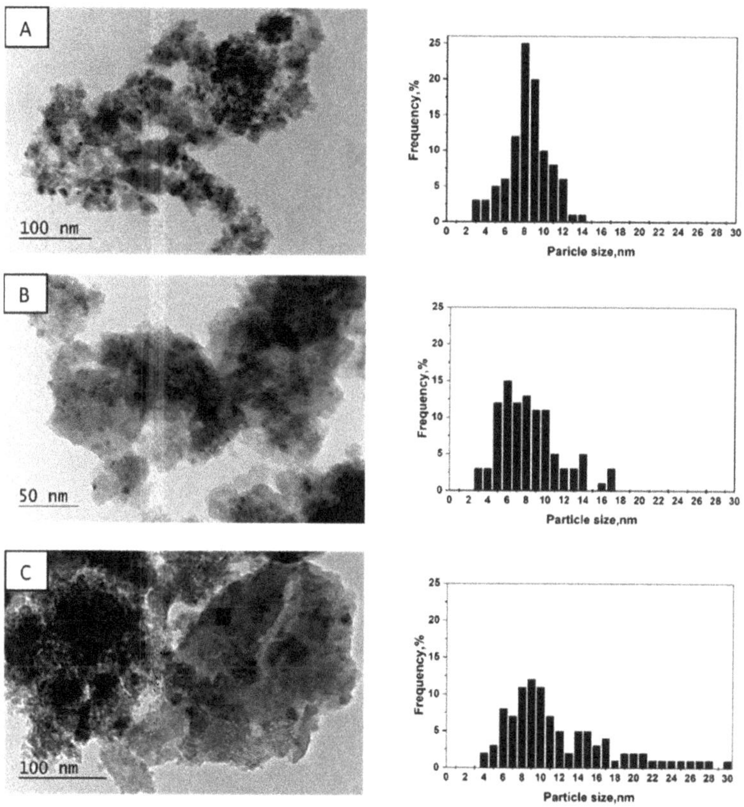

Figure 5. TEM images and corresponding particle size distribution histograms of Pd/La_2O_3-CeO_2-TiO_2-Al_2O_3-fresh (**A**); Pd/La_2O_3-CeO_2-TiO_2-Al_2O_3-used, after sulfur poisoning (**B**) and Pd/La_2O_3-CeO_2-TiO_2-Al_2O_3-thermally aged (**C**).

After catalytic tests, the average size in Pd/La_2O_3-CeO_2-TiO_2-Al_2O_3 remains the same, which can be evidence of high dispersion, and, therefore, of the thermal stability of Pd particles. After thermal aging, the enlargement of Pd particle size and some agglomerates is observed (Figure 5C). The main crystallite size has increased to 12 nm. Certain research findings [34] have indicated that alteration in the Pd catalyst morphology during the reaction can be impacted by changes in the support materials, potentially leading to an influence on the overall performance of the catalyst.

Within all samples (fresh, used, after sulfur poisoning, and thermally aged) the selected area electron diffraction (SAED) patterns show the presence of PdO (PDF-41-1107), Pd (PDF-46-1043), γ-Al_2O_3 (PDF-70-9085), AlO(OH) (PDF-83-2384) and CeO_2 (PDF-81-0792), (Figures 6B, 7B and 8B). Additionally, in the Pd/La_2O_3-CeO_2-TiO_2-Al_2O_3-fresh sample the formation of TiO_2—anatase (PDF-83-2243) was confirmed by the HRTEM and SAED analyses (Figure 6B,D), while in used, after sulfur poisoning and thermally aged samples, the TiO_2—rutile (PDF-88-1175) and (PDF-87-0920) was detected (Figures 7B,D and 8B).

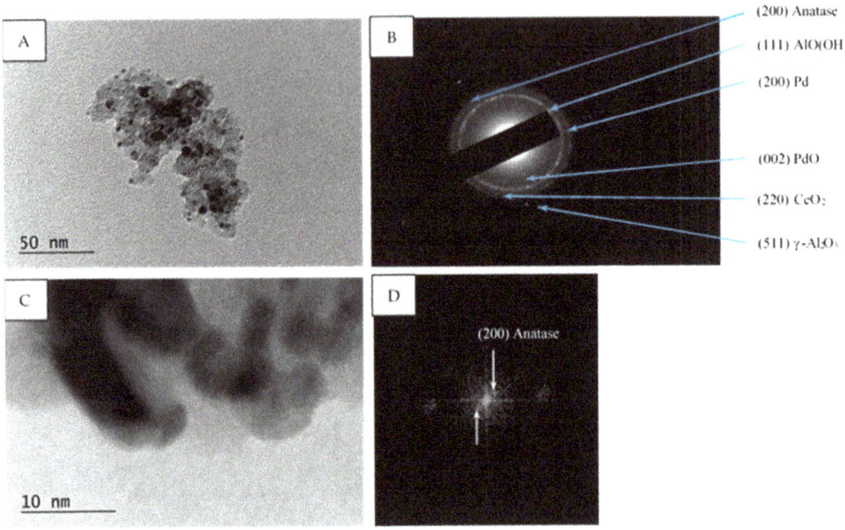

Figure 6. TEM micrograph (**A**), corresponding SAED pattern (**B**), HRTEM micrograph (**C**), and corresponding FFT (Fast Fourier Transform) (**D**) of Pd/La_2O_3-CeO_2-TiO_2-Al_2O_3 fresh catalyst. The identified (200) reflexes correspond to d = 1.8900 Å crystal planes of anatase (PDF 83-2243).

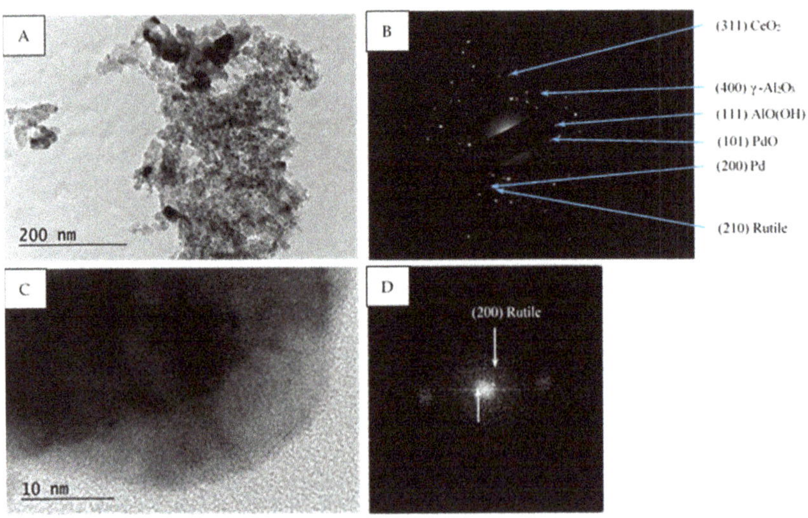

Figure 7. TEM micrograph (**A**), corresponding SAED pattern (**B**), HRTEM micrograph (**C**), and corresponding FFT (**D**) of Pd/La_2O_3-CeO_2-TiO_2-Al_2O_3 used after, sulfur poisoning catalyst. The identified (200) reflexes correspond to d = 2.2968E (Rutile PDF 87-0920).

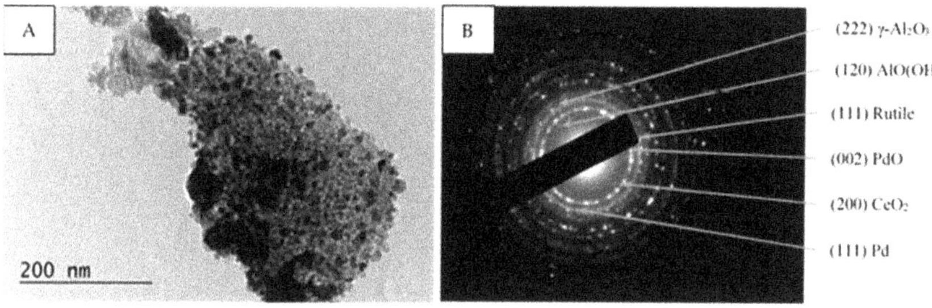

Figure 8. TEM micrograph (**A**) and corresponding SAED pattern (**B**) of sample Pd/La_2O_3-CeO_2-TiO_2-Al_2O_3-thermally aged.

3.6. X-ray Photoelectron Spectroscopy

The XPS results are shown in Table 3, Figures 9 and 10. The binding energies for all investigated catalysts are distinguished as follows: in the interval of 335.7–335.1, BE is attributed to Pd^0 species from metal palladium particles, while in the interval of 337.1–336.3 eV, they are assigned to palladium oxide [35].

Table 3. Surface atomic concentration, at.%.

Sample	O1s	Al2p	La3d	Ce3d	Pd3d	Ti2p	S2p
S2Pd/La_2O_3-CeO_2-TiO_2-Al_2O_3-fresh	56.48%	36.14%	0.53%	1.45%	4.15%	1.24%	-
Pd/La_2O_3-CeO_2-TiO_2-Al_2O_3-used, after sulfur poisoning	58.40%	31.63%	0.53%	1.38%	6.43%	1.33%	0.30%
Pd/La_2O_3-CeO_2-TiO_2-Al_2O_3-thermally aged	60.00%	30.02%	0.49%	1.41%	6.68%	1.41%	-

As can be seen from Figure 9B, the concentration of Pd^{2+} slightly increases in the used and thermally treated samples. A similar phenomenon was observed in our previous studies [36,37], and this effect is explained by the oxidation of Pd to PdO. In the case of the studied system Pd/La_2O_3-CeO_2-TiO_2-Al_2O_3, the variation of the surface concentration of Pd^{2+} is insignificant, which supports the stability of the catalysis.

According to published data, the 3.5 eV splitting between the primary peak and the satellite of La3$d_{5/2}$ is typical for $La(OH)_3$, and the splitting of 4.5 eV is characteristic for La_2O_3 (Figure 9A). These results imply that both La_2O_3 and $La(OH)_3$ are present on the surfaces of both the fresh and used catalysts, after sulfur poisoning catalysts. It is likely that the La_2O_3 is covered with $La(OH)_3$, as previous research [38] has suggested that lanthanum oxide tends to spontaneously react with water vapor at ambient temperature, resulting in the formation of $La(OH)_3$ [39]. In the case of the aged sample, $La(OH)_3$ is the only species on the surface.

The XPS spectra in the Ce3d region are presented in Figure 9C. A well-known fact is that some Ce-containing samples are sensitive to X-rays in a vacuum, which leads to a change of the oxidation state of Ce. This complicates quantitative XPS analysis and defining the oxidation state of ceria and eventually the ratio Ce^{3+}/Ce^{4+}. In our experimental data, we have provided a careful measurement and each scan was recorded separately. After a comparative analysis of the shape of the recorded curves and binding energies, we did not observe any changes between scans. To increase the signal-to-noise ratio, we have performed summation of separate scans. The curve fitting procedure was applied to the obtained spectra to determine their composition and to estimate the quantitative ratio between Ce^{3+} and Ce^{4+} [40]. The curve-fitted XPS spectra of Ce3d are shown in Figure 9C. It can be seen that the concentration of Ce^{3+} increases after aging and after testing with SO_2. The presence of Ce^{3+} has been proven through EPR spectroscopy (see discussion below). The obtained binding energies for Ti 2p peak for the fresh sample and after the catalytic test in the presence of SO_2 is at 460 eV. This binding energy is ascribed to Ti^{4+} in TiO_2 particles.

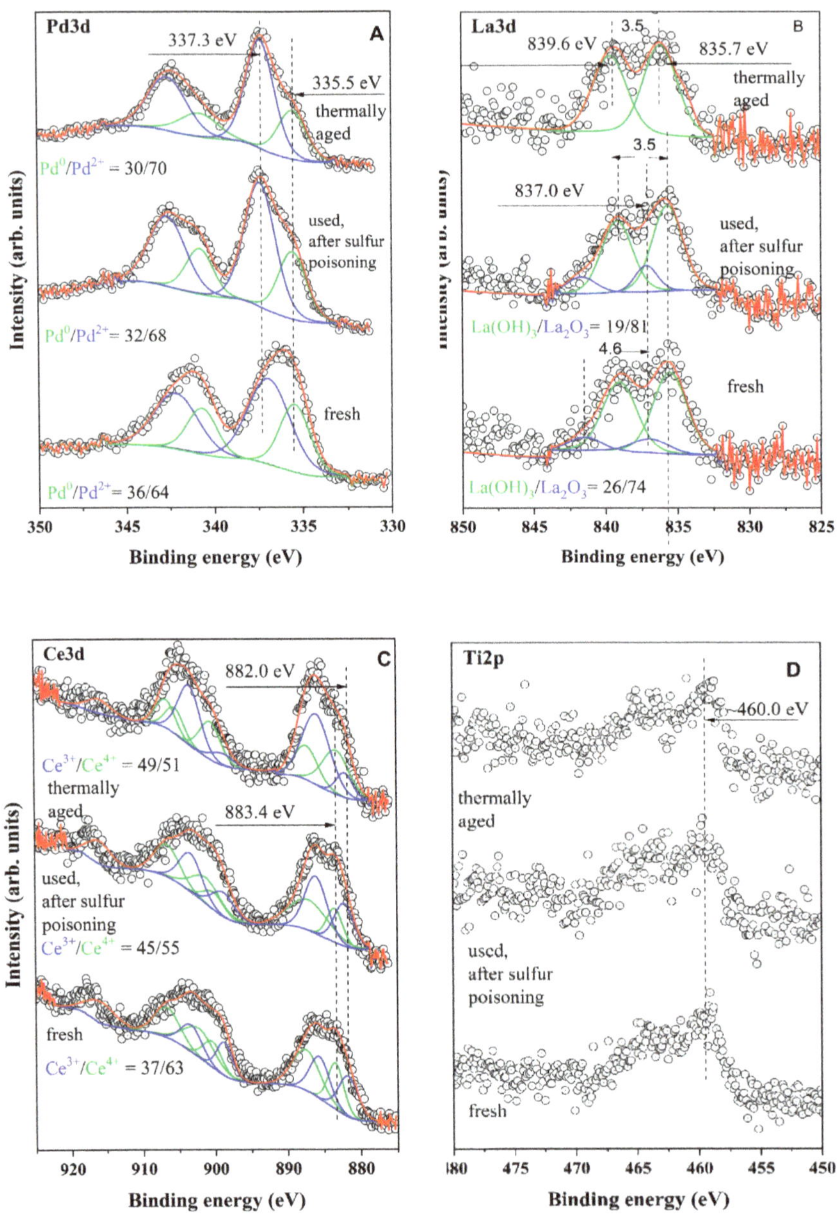

Figure 9. X-ray photoelectron spectra of La3d (**A**), Pd3d (**B**), Ce3d (**C**) and Ti2p (**D**) for Pd/La_2O_3-CeO_2-TiO_2-Al_2O_3-fresh, Pd/La_2O_3-CeO_2-TiO_2-Al_2O_3-used, after sulfur poisoning and Pd/La_2O_3-CeO_2-TiO_2-Al_2O_3-thermally aged. (**A**) The open circles represents experimental data both oxidation states are in green (Pd^0) and blue (Pd^{2+}). The red line is used for resulting curve of curve fitting procedure to be compared with experimental data. (**B**) The open circles represents experimental data. $La(OH)_3$ is shown in green, whereas, the blue line is used for La_2O_3. The red line is used for resulting curve of curve fitting procedure to be compared with experimental data. (**C**) The open circles represents experimental data both oxidation states are in green (Ce^{4+}) and blue (Ce^{3+}). The red line is used for resulting curve of curve fitting procedure to be compared with experimental data.

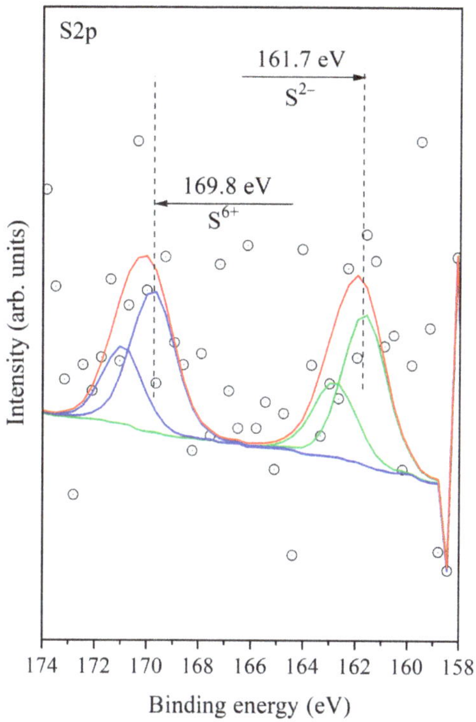

Figure 10. X-ray photoelectron spectra of S2p for Pd/La_2O_3-CeO_2-TiO_2-Al_2O_3-used, after sulfur poisoning catalyst. The open circles represents experimental data both oxidation states are in green (S^{2-}) and blue (S^{6+}). Because we are measuring 2p core level of sulfur it is doublet peak representing the standard spectra of each oxidation state. The red line is used for resulting curve of curve fitting procedure to be compared with experimental data.

S2p spectra (Figure 10) show two peaks at ~169.8 eV and ~161.7 eV. The bands with the same positions have been observed in publication [41] when Pt/Al_2O_3 reacts with the sulfur dioxide + oxygen mixture and have been attributed to the simultaneous formation of the sulfate species and the sulfide species. In our case, aluminum oxide predominates in the support, and we can attribute the band at ~169.8 eV and ~161.7 eV to the formation of sulfates and the sulfide species as well. The S $2p_{3/2}$ binding energy typically ranges from 160 eV to 164 eV in metal sulfides [42]. The presence of sulfates is also confirmed by FTIR spectroscopy. Taking the references mentioned above, we can assume that sulfides are formed on the surface of palladium in our case as well.

3.7. Electron Paramagnetic Resonance

The EPR spectra of Pd/La_2O_3-CeO_2-TiO_2-Al_2O_3 catalysts at temperature 123 K are shown in Figure 11.

The EPR spectra are complex and are composed of superposition of several overlapping EPR lines. In all spectra, a line with a g value of 4.23 is recorded. This line is related to Fe^{3+}, which very often is contained as an impurity in the starting substances, and it is recorded because of the very high sensitivity of EPR spectroscopy. In the fresh sample (Figure 11a), the EPR lines located at g = 2.4294, 2.1757 and 2.051 are due to the presence of paramagnetic palladium particles, which can be Pd^+ or Pd^{3+}. The g values are slightly different from those reported in the literature probably because of different conditions and the environment of the palladium in the present work [43]. It should be pointed out that the palladium species in the higher than +2 oxidation state were detected by XPS, as reported in our previous investigation [15]. The fact that in this study they were observed only with EPR spectroscopy gives us reason to suggest that they are localized in volume.

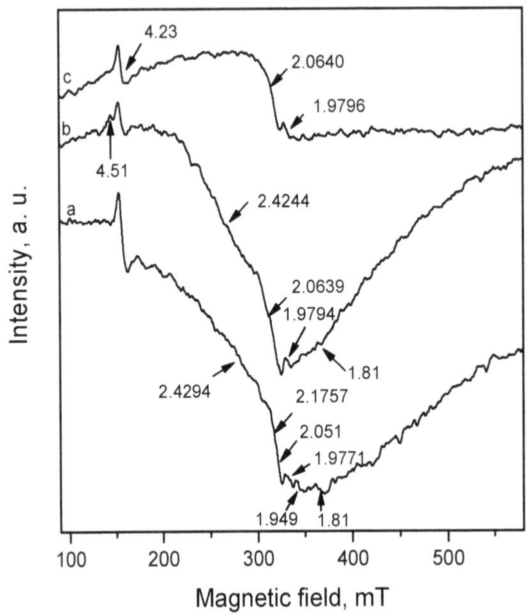

Figure 11. EPR spectra of: (a) Pd/La_2O_3-CeO_2-TiO_2-Al_2O_3-fresh (a); Pd/La_2O_3-CeO_2-TiO_2-Al_2O_3-used, after sulfur poisoning (b) and Pd/La_2O_3-CeO_2-TiO_2-Al_2O_3-thermally aged (c).

The EPR lines with g values 1.9771 and 1.949 are assigned to Ce^{3+} or Ti^{3+}. Both ions have similar EPR parameters and the simultaneous presence in the system makes their separation difficult. Two narrow signals with practically the same g values ($g_\perp$ = 1.967 and g_{II} = 1.944) are attributed to Ce^{3+} ions associated with an anion vacancy or electrons trapped at anion vacancies partially delocalized onto orbitals of cerium ion [44].

It should be pointed out that similar EPR parameters were reported for Ti^{3+} in anatase [45]. The XPS data show the presence of Ce^{3+}. From the XPS data, it can be argued unambiguously that the presence of Ti^{3+} is due to the very low intensity signal in the Ti2p region. Taking into account that the anatase is established by SAED analysis in the fresh sample and EPR is a highly sensitive technique that allows investigation of paramagnetic species [46]; thus, we cannot exclude the presence of Ti^{3+}.

The EPR lines due to Ce^{3+} or Ti^{3+} in the fresh sample, in the EPR spectra after thermal aging are maintained and positioned at g factor 1.9796 (Figure 11c). This shows that the paramagnetic ions, which are responsible for it are located in the volume. In addition, a weak signal with g = 1.81 is detected, which, according to the literature data, is connected with Ce^{3+} in close range to La [47]. Ce^{3+}, as a $4f^1$ ion, is characterized by strong spin-orbital coupling leading to large deviations from the g factor of free electron (2.0023). Moreover, due to the short relaxation times, it is detectable usually at low temperatures. The EPR line with g = 1.81 disappears after thermal aging, which proves that these Ce ions are on the surface of La_2O_3.

After catalytic tests, the EPR spectra do not change significantly. Again, signals for palladium paramagnetic ions are observed, but this time the temperature dependence of the signal with g 2.4244 shows the presence of the superparamagnetic palladium particles.

This is because, with the decrease of the recording temperature, the EPR line is broadening and is moved to the lower magnetic field (Figure 12). This behavior is typical for superparamagnetic particles. Nonlinear behavior shows that the particles have various sizes but the superparamagnetic state remains. In addition, a line with a g value of 4.51 is recorded, which is due to Ce^{3+}. This fact, together with the increase in the intensity of the EPR signal with g = 1.9794, shows that during the reaction, a reduction of the cerium ions is taking place.

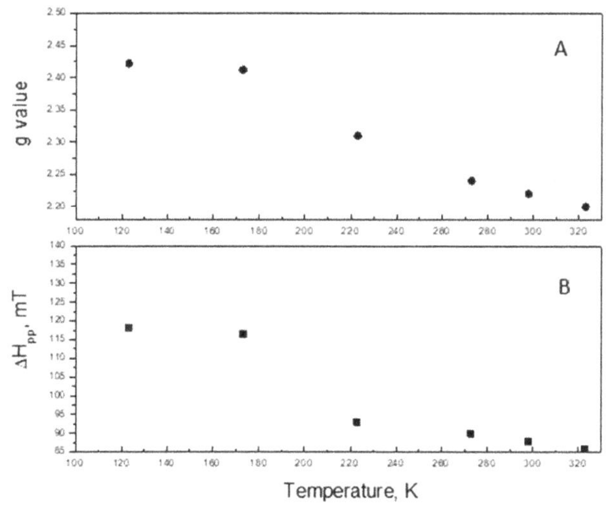

Figure 12. The variation of the g factor (**A**) and line width (**B**) of the EPR spectra changing in the recording temperature.

3.8. FTIR Spectroscopy

FTIR spectra of a fresh Pd/La_2O_3-CeO_2-TiO_2-Al_2O_3 sample and after the catalytic test with sulfur dioxide are presented in Figure 13. Low-resolved bands centred at 1150 cm^{-1} and 1070 cm^{-1} are visible in the IR spectra. The band centred at 1150 cm^{-1} is ascribed to the sulphates in bulk according to data in the literature [48]. There is no evidence of the presence of aluminium sulfate, either X-ray or TEM, which can be considered as evidence that the formed sulphates are amorphous.

The band at 1070 cm^{-1} is very weak and strongly overlaps with the band of the support. According to Schoonheydt [49], the vibration at 1070 cm^{-1} is assigned to a SO_3^{2-} species coordinated through its sulfur.

In this investigation, as in the previous publication [15], no band was observed at 1435 cm^{-1} for sulphate groups formed on the palladium particles. The adsorption bands at 595 cm^{-1} and 669 cm^{-1} most likely are the result of PdO [50,51].

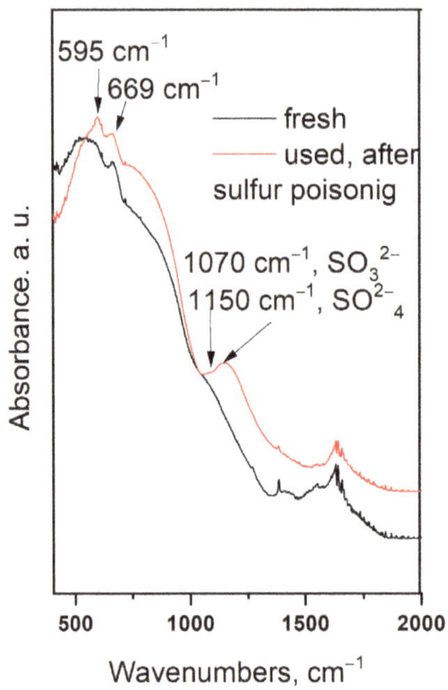

Figure 13. Infrared spectra of Pd/La_2O_3-CeO_2-TiO_2-Al_2O_3-fresh and Pd/La_2O_3-CeO_2-TiO_2-Al_2O_3-used, after sulfur poisoning catalysts.

3.9. Reaction Kinetics

To extend further the analysis of the studied catalysts, an investigation on the kinetics and mechanism of the reaction has been carried out. The kinetics parameters calculations were performed by multivariate analysis. For these calculations, data from the temperature-conversion curves were used for direct integration of the reaction rates. To fit the experimentally measured rates with kinetics parameters, a special computation program for a numerical (nonlinear) optimization was used. The minimized residual sum of squares between the measured experimental points and the corresponding predictions of the model (RSS) and the squared correlation coefficient (R^2) were selected as optimization criteria for the model's consistency. Details on the calculation procedure were published earlier [52,53].

The values for the calculated reaction order towards the oxygen lead to the suggestion of a significant role of the oxygen chemisorption (Table 4, Power law kinetics model). The reaction order towards the water vapor (−0.33) reveals a very significant inhibition effect.

Table 4. Kinetics parameters based on power law model.

	PWL $r = kC_{voc}^{m}C_{ox}^{n}C_{water}^{p}$						
	E_a	k_o	m (CH_4)	n (O_2)	p (H_2O)	RSS	R^2
Pd/La_2O_3-CeO_2-TiO_2-Al_2O_3	108.0	3.65×10^9	0.94	0.02	−0.33	10.6	1.00

E_{ai}, kJ/mol; k_{oi}, mol.s^{-1}.m^{-3}; $k_{oi,pwl}$, mol.s$^{-1-[1-(m+n+p)]}$; E_{ai}, kJ/mol.

The relevance of the mechanistic models used for the kinetic calculations towards the experimental data set is presented in Figure 14 and the calculated results are given in Tables 5 and 6. One could see that the lowest values for RSS criteria and the highest correlation between the model and experiment are obtained for the Mars–van Kerevelen model with the assumption that the water adsorbs on oxidized and reduced sites and slow desorption of products occur (MVK-SDP). Therefore, this mechanism should be considered as more consistent with the experimental results than the alternative Langmuir–Hinshelwood mechanism, where the water competes with oxygen and methane.

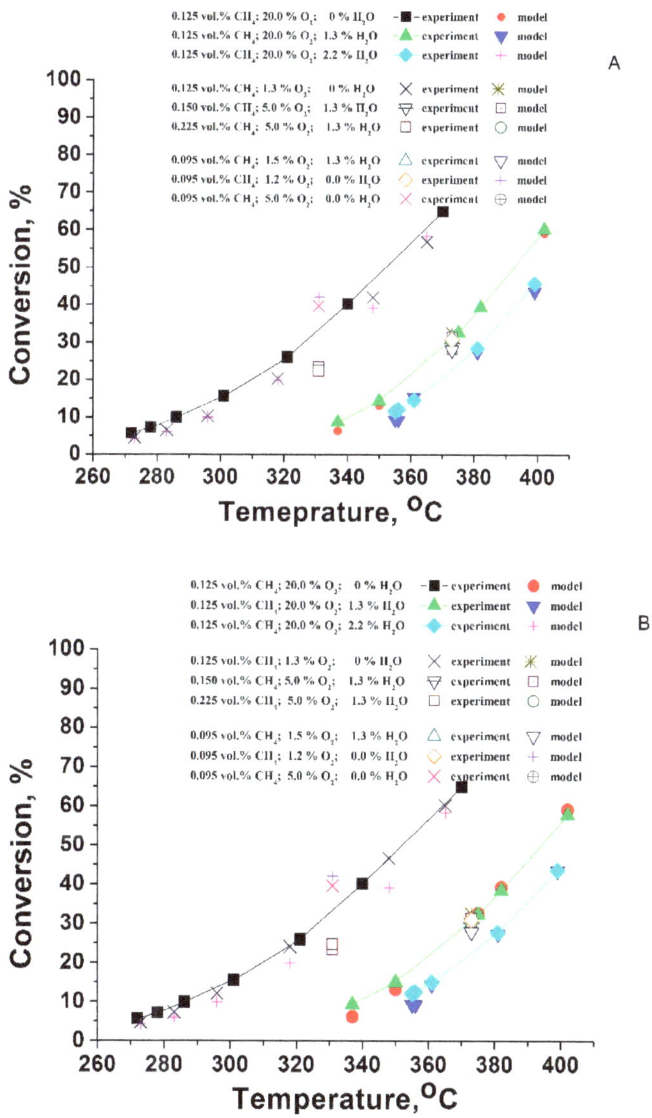

Figure 14. Comparison between the experimentally measured conversions at different conditions and the model prediction by the Mars–van Krevelen mechanism (**A**) and Langmuir–Hinshelwood mechanism (**B**).

Table 5. Reaction rate expressions and kinetics parameters for applied MVK-SDP model.

Model: MVK-SDP, (Water Adsorbs on Oxidized and Reduced Sites, Slow Desorption of Products) $r = \frac{k_{red}k_{ox}C_{voc}C_{ox}}{\gamma k_{red}C_{voc}(1+K_{water-voc}.C_{water-voc})+k_{ox}C_{ox}(1+K_{water-ox}.C_{water-ox})+(k_{red}k_{ox}/k_{des})C_{voc}C_{ox}}\ \gamma = 2$												
	$E_{a\cdot ox}$	$k_{o\cdot ox}$	$E_{a\cdot red}$	$k_{o\cdot red}$	$-\Delta H_{w\cdot ox}$	$k_{o\cdot water\cdot ox}$	$\Delta H_{w\cdot red}$	$k_{o\cdot w,\ red}$	$E_{a\cdot des}$	$k_{o\cdot des}$	RSS	R^2
Pd/La_2O_3-CeO_2-TiO_2-Al_2O_3	124.1	4.49×10^{11}	51.3	1.02×10^6	94.8	3.01×10^{-7}	74.6	1.89×10^{-7}	95.8	1.56×10^8	3.8	1.00

E_{ai}, kJ/mol; ΔH_i. kJ/mol; k_{oi}, m^3/mol; $k = k_o.\ \exp(-E_a/RT)$; $K_{i(voc,ox,\ w)} = k_{o(voc,ox,\ w)}.\ \exp(-\Delta H_{i,voc,ox,\ w}/RT)$; $-\Delta H_i = E_{des} - E_{ads}$.

Table 6. Reaction rate expressions and kinetics parameters for applied Langmuir–Hinshelwood model.

LH-DS-D: water compete with oxygen and methane $r = \frac{kK_{voc}C_{voc}K_{ox}^{1/2}C_{ox}^{1/2}}{(1+K_{voc}C_{voc}+K_{water-voc}C_{water})(1+K_{ox}^{1/2}C_{ox}^{1/2}+K_{water-ox}C_{water})}$												
	E_a	k_o	$-\Delta H_{voc}$	$k_{o\cdot voc}$	$-\Delta H_{ox}$	$k_{o\cdot ox}$	ΔH_{water}	$k_{o\cdot water}$	$\Delta H_{water\cdot red}$	$k_{o\cdot water\cdot red}$	RSS	R^2
Pd/La_2O_3-CeO_2-TiO_2-Al_2O_3	136.6	6.60×10^{11}	169.2	1.22×10^3	77.4	5.23×10^{-6}	79.5	7.96×10^{-1}	75.8	8.08×10^{-6}	4.8	0.98

E_{ai}, kJ/mol; ΔH_i. kJ/mol; k_{oi}, m^3/mol; $k = k_o.\ \exp(-E_a/RT)$; $K_{i(voc,ox,\ w)} = k_{o(voc,ox,\ w)}.\ \exp(-\Delta H_{i,voc,ox,\ w}/RT)$; $-\Delta H_i = E_{des} - E_{ads}$, RSS—squared sum of residuals. R^2—squared correlation coefficient.

Summarizing, the deposition of Pd to the La_2O_3- CeO_2- TiO_2-Al_2O_3 system leads to a decrease in the specific surface area, accompanied by an increase in the average pore diameter from 4.9 nm to about 8 nm, the pore-size distribution being transformed from a mono- to bimodal structure. Based on the literature data [37], this morphological structure offers significant benefits when these materials are used as catalytic support. More specifically, the small pores ensure a large surface area for better dispersion of the supported active phase (noble metal, for instance). The larger pores provide conditions for improved internal mass transfer within the catalyst.

As reported [54], catalyst thermal deactivation can occur due to several factors: (i) reduction in the active surface area due to the enlargement of palladium particles, (ii) pore collapse of the active phase; (iii) decrease in the catalytic support area; and (iv) alterations in the chemical composition of active catalytic phases into less active phases.

Typically, the sintering occurs at high temperatures (>500 °C) and is accelerated in the presence of water vapor [55], the driving force being the minimization of the surface energy, reduced by the transport and increase in the particles [56]. Within our study, the thermal deactivation behavior of the catalyst may be explained by phase changes of TiO_2. It is known that anatase irreversibly transforms to rutile at elevated temperatures. This transformation does not have a fixed temperature. Pure bulk anatase is transformed irreversibly to rutile in air at 600 °C; however, the reported transition temperatures vary in the range 400–1200 °C [57–59], owing to the use of different methods of determining the transition temperatures, raw materials, and processing methods.

During the heat treatment part of TiO_2, anatase may be transformed to rutile and the rutile grains coarsen at the expense of neighbouring anatase until the large rutile grains begin to impinge on each other [60,61]. This increase in grain size leads to a decrease in surface area and a further decrease in activity [62–64]. Additionally, it is reported that calcination above 465 °C has always resulted in the phase rutile [65]. The phase transition is associated with increased crystal size, resulting in a significant decrease in specific surface area [66].

In our case, the concentration of TiO_2 (8 wt.%) is not sufficient for reliable XRD analysis, and the obtained XPS data show low intensity broad peaks, the only possible determination of the changes was made by HRTEM. Within the present study, the results from HRTEM analysis reveal that the decrease in the specific surface is connected with the transformation of part of the anatase to larger particles of rutile (whose process is reported to proceed at temperatures above 465 °C [65].

In parallel, the deactivation could be related to the growth of the palladium particles from 8 nm to 12 nm and the increased PdO/Pd ratio, more specifically, this ratio is higher than its optimal value for the applied reaction conditions. As reported by Su at all [67], small palladium particles enhance the activity of PdO in methane combustion by facilitating the dissociation of CH_4 without being oxidized under the given reaction conditions. It should be pointed out that the presence of metallic Pd in contact with PdO facilitates the reduction of PdO by CH_4, i.e., methane activation proceeds more on metallic palladium than palladium oxide [15]. Pd plays a crucial role in dissociating CH_4 more effectively when compared with PdO, with the resulting reaction products diffusing towards the Pd—PdO interface, where PdO is converted into metallic Pd.

At the same time, the modification with TiO_2 of the La_2O_3-CeO_2-Al_2O_3 system results in higher resistance towards the presence of SO_2 due to prevailed formation of unstable surface sulfites instead of the thermally stable sulfates, as observed with the system without TiO_2.

3.10. Monolithic Reactor Tests and Modeling

The methane combustion processes were described using a two-dimensional heterogeneous model of a monolithic channel. The conversion degree at the outlet of the monolithic channel was calculated by using the method of mixing-cup average concentration. It consists of multiplying the concentrations of the laminar streamlines by the corresponding volumetric flows and summing up over all the streamlines and dividing this sum by the total volumetric flow. The reported results are for calculated conversion degree and temperature profiles inside the monolithic channel using the obtained data for reactions in cases of isothermal (experiment) conditions and then the behavior of the reactor at adiabatic reactor operation is based on simulation by using the reactor model. The heat transfer within the entire monolithic element ocurrs by conduction trough the channels in the radial direction and by fluid convection in the axial direction. The catalytic element is modeled by assuming that the heat is transferred trough a thin thermal boundary layer with a driving force, proportional to the difference between the temperature in the first to the wall channel and the temperature of the reactor wall. Of course, in the theoretical case, one could suppose complete thermal insulation and the behavior of the monolithic reactor is fully adiabatic. However, in most of the applications, one should expect some extent of heat exchange with the ambient environment and therefore, the effect of the wall temperature has been taken into account. For convenience, the temperature and concentration profiles are colored according to the calculated values, i.e., from blue color for low temperatures or conversions towards the red color for their high values (passing through mixed colors within the intermedia values). A second-order approximation is used for the numerical solution; further details are provided in the literature [68–73].

The simulation results (Figure 15) demonstrate the possibilities for the abatement of 2400 Nm^3/h methane-containing gas (CH_4: 0.25 vol.%, H_2O: 2.2 vol.%, 9 vol.% O_2). The model prediction shows that the required dimensions of the monolith for achieving 99% conversion are the following: D = 1.0 m and L = 0.6 m. Therefore, for effective neutralization of methane in presence of water, the reactor should operate adiabatically at GHSV of 5100 h^{-1}.

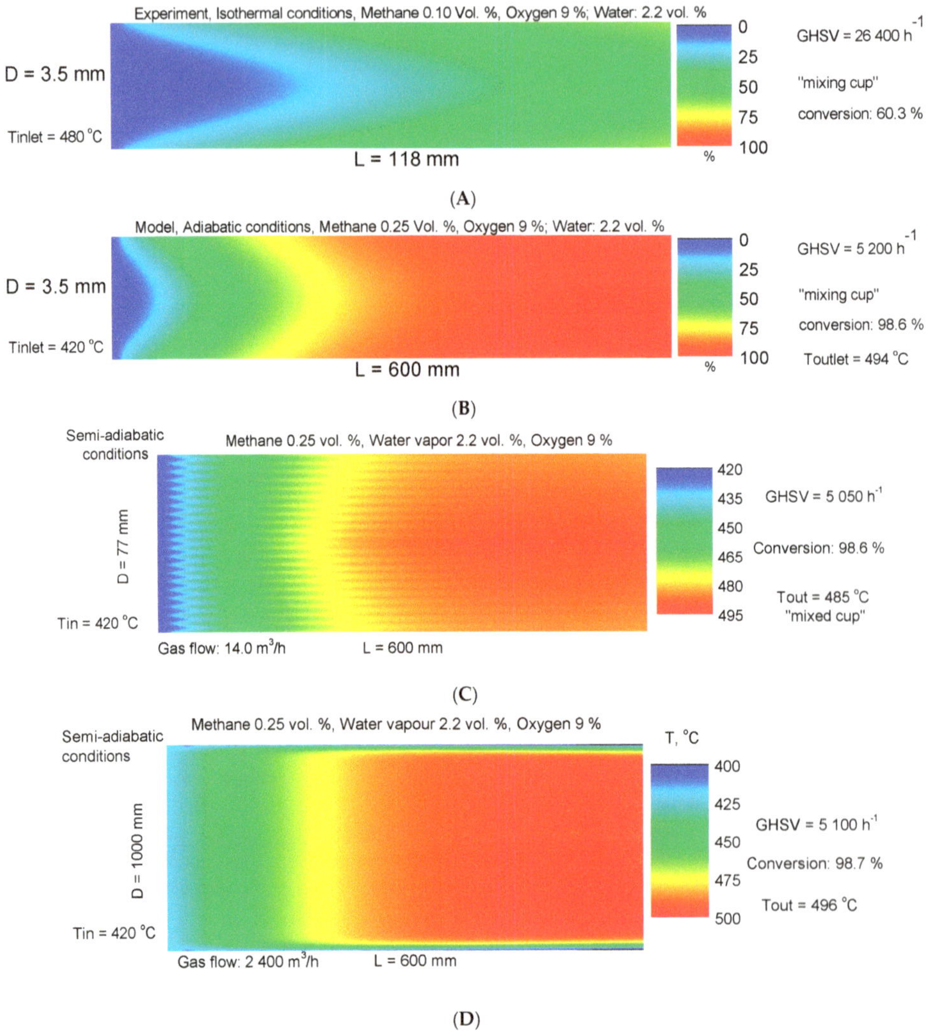

Figure 15. Experimentally measured conversion degrees and temperature profiles in a single monolithic channel at isothermal conditions (**A**); simulated conversion degrees and temperature profiles within a single adiabatic channel for ensuring of 99% methane conversion (**B**); pilot-scale simulation of methane combustion using the kinetic data, obtained at isothermal conditions (**C**); and full-scale reactor model for methane combustion accouning for the heat loss at the reactor wall (**D**).

The practical significance of these results is closely tied to challenges arising from mobile sources of pollution, specifically stemming from the release of unburned hydrocarbons in the exhaust emissions of internal combustion engines utilizing natural gas (primarily composed of approximately 95% methane) as their fuel source. Consequently, there exists a notable interest in the development of novel catalytic converter materials that can ensure highly efficient reduction of methane emissions.

4. Conclusions

The deposition of palladium to the La_2O_3-CeO_2-TiO_2-Al_2O_3 system leads to a decrease in the specific surface area, accompanied by an increase in the average pore diameter, with the pore-size distribution transforming from a mono- to bimodal structure. The methane complete oxidation reaction occurs at temperatures exceeding 220 °C. T_{50} in the absence of water vapor is 328 °C. However, in the presence of 1.2% water vapor (with 5% oxygen content and $GHSV_{STP}$ of 60,000 h^{-1}), T_{50} increases to 370 °C.

Most likely, the reaction of complete oxidation of methane follows the Mars−van Krevelen mechanism, where the water molecules adsorb on both oxidized and reduced sites. A slow desorption of the products (CO_2, H_2O) is also suggested and implemented into the rate equation.

The thermal aging at 500 °C leads to lowering the catalytic activity, which is due to the (i) agglomeration of the Pd-clusters (from 8 nm to 12 nm); (ii) transformation of part of the TiO_2 from anatase to larger particles of rutile, resulting in a decreased specific surface area; and (iii) increased ratio PdO/Pd above its optimal value, which is specific for the applied reaction conditions. The presence of Ce^{3+} on the catalytic surface and in the volume was revealed by EPR and XPS. Most likely, the Ce^{3+} ions on the surface are near lantana.

The modification with Ti leads to improved activity in the presence of water vapor after thermal aging and a slightly decrease in the dry gas mixture. The effect of water vapor in terms of calculated reaction order is −0.33.

The simultaneous formation of sulfats, sulfites, and sulfides in the studied system is suggested. The benefit of the modification with TiO_2 of the La_2O_3-CeO_2-Al_2O_3 system is the higher resistance towards the presence of SO_2, most likely due to the prevailing formation of unstable surface sulfites instead of the thermally stable sulfates, as observed with the system without TiO_2.

The results from the kinetic model calculation show that the reaction pathway over the Pd/La_2O_3-CeO_2-TiO_2-Al_2O_3 catalyst follows the Mars–van Krevelen mechanism. A sample of Pd/La_2O_3-CeO_2-TiO_2-Al_2O_3, supported on rolled stainless steel with aluminum content (Aluchrom VDM®), was produced and tested to assess the potential practical applications of the obtained material. A two-dimensional heterogeneous model of a monolithic channel was employed to simulate methane oxidation within an industrial scale monolithic reactor.

Supplementary Materials: The following supporting information can be downloaded at: https://www.mdpi.com/article/10.3390/ma16206784/s1, Figure S1. SEM images of (a) Pd/La_2O_3-CeO_2-TiO_2-Al_2O_3-fresh and (b) Pd/La_2O_3-CeO_2-TiO_2-Al_2O_3-used, after sulfur poisoning catalysts and Table S1. Surface composition from the area of Pd/La_2O_3-CeO_2-TiO_2-Al_2O_3 catalysts obtained by SEM/EDX analysis.

Author Contributions: Conceptualization, A.N., S.T. and R.V.; methodology, A.N., S.T. and R.V.; formal analysis, R.V., S.T., D.K., H.K., Y.K. and P.M.; investigation, R.V., S.T., D.K., H.K., Y.K., G.I. and P.M.; data curation, R.V., H.K., Y.K., K.T. and P.M., writing—original draft preparation, A.N., S.T. and R.V.; writing—review and editing, A.N., S.T. and R.V.; visualization, R.V., A.N., K.T. and S.T. All authors have read and agreed to the published version of the manuscript.

Funding: The authors acknowledge the project D01-272-"European Network on Materials for Clean Technologies" for providing the opportunity to present the results at the SizeMat4 conference as well as for the financial publication support.

Institutional Review Board Statement: Not applicable.

Informed Consent Statement: Not applicable.

Data Availability Statement: Data sharing is not applicable to this article.

Acknowledgments: The authors acknowledge the project D01-272 "European Network on Materials for Clean Technologies" for providing the opportunity to present the results at SizeMat4 conference as well as for the financial publication support. In this investigation, research equipment of the distributed research infrastructure INFRAMAT supported by the Bulgarian Ministry of Education and Science (contract D01-172/28.07.2022) was used. The Thermo FID-TG analyzer and MultiGas FTIR Gas Analyzer 2030G were used within the framework of the National Center of Mechatronics and Clean Technologies (BG05M2OP001-1.001-0008).

Conflicts of Interest: The authors declare no conflict of interest.

References

1. Amoatey, P.; Izady, A.; Al-Maktoumi, A.; Chen, M.; Al-Harthy, I.; Al-Jabri, K.; Msagati, T.A.; Nkambule, T.T.; Baawain, M.S. A critical review of environmental and public health impacts from the activities of evaporation ponds. *Sci. Total Environ.* **2021**, *796*, 149065. [CrossRef]
2. Moumen, A.; Azizi, G.; Chekroun, K.; Baghour, M. The effects of livestock methane emission on the global warming: A review. *Int. J. Glob. Warm.* **2016**, *9*, 229–253. [CrossRef]
3. Getabalew, M.; Alemneh, T.; Akeberegn, D. Methane Production in Ruminant Animals: Implication for Their Impact on Climate Change. *Concepts Dairy Vet. Sci.* **2019**, *2*, 204–211. [CrossRef]
4. Arendt, E.; Maione, A.; Klisinska, A.; Sanz, O.; Montes, M.; Suarez, S.; Blanco, J.; Ruiz, P. Structuration of $LaMnO_3$ perovskite catalysts on ceramic and metallic monoliths: Physico-chemical characterisation and catalytic activity in methane combustion. *Appl. Catal. A Gen.* **2008**, *339*, 1–14. [CrossRef]
5. Barrera, A.; Fuentes, S.; Dıaz, G.; Gymez-Cortes, A.; Tzompantzi, F.; Molina, J.C. Methane oxidation over Pd catalysts supported on binary Al_2O_3–La_2O_3 oxides prepared by the sol–gel method. *Fuel* **2012**, *93*, 136–141. [CrossRef]
6. Zhang, B.; Li, D.; Wang, X. Catalytic performance of La–Ce–O mixed oxide for combustion of methane. *Catal. Today* **2010**, *158*, 348–353. [CrossRef]
7. Persson, K.; Jansson, K.; Järås, S. Characterisation and microstructure of Pd and bimetallic Pd–Pt catalysts during methane oxidation. *J. Catal.* **2007**, *245*, 401–414. [CrossRef]
8. Guo, Y.; Lu, G.; Zhang, Z.; Jiang, L.; Wang, X.; Li, S.; Zhang, B.; Niu, J. Effects of ZrO_2/Al_2O_3 properties on the catalytic activity of Pd catalysts for methane combustion and CO oxidation. *Catal. Today* **2007**, *126*, 441–448. [CrossRef]
9. Xiao, L.; Sun, K.; Xu, X. Catalytic combustion of methane over CeO_2–MO_x(M = La^{3+}, Ca^{2+}) Solid Solution Promoted Pd/c-Al_2O_3 Catalysts. *Acta Phys.-Chim. Sin.* **2008**, *24*, 2108–2113. [CrossRef]
10. Tao, J.; Liu, Y.; Deng, J.; Jing, L.; Hou, Z.; Wei, L.; Wang, Z.; Dai, H. Methane combustion over zeolite-supported palladium-based catalysts. *Catalysts* **2023**, *13*, 1251. [CrossRef]
11. Baohua, Y.; Renxian, Z.; Yuejuan, W.; Xiaoming, Z. Effect of rare earths (La, Pr, Nd, Sm, and Y) on the methane combustion over Pd/Ce–Zr/Al_2O_3 catalysts. *Appl. Catal. A Gen.* **2005**, *295*, 31–39. [CrossRef]
12. Arai, H.; Fukuzawa, H. Research and development on high temperature catalytic combustion. *Catal. Today* **1995**, *26*, 217–221. [CrossRef]
13. Furuya, T.; Sasaki, K.; Hanakata, Y.; Ohhashi, T.; Yamada, M.; Tsuchiya, T.; Furuse, Y. Development of a hybrid catalytic combustor for a 1300 °C class gas turbine. *Catal. Today* **1995**, *26*, 345–350. [CrossRef]
14. Ozawa, Y.; Tochihara, Y.; Watanabe, A.; Nagai, M.; Omi, S. Stabilizing effect of Nd_2O_3, La_2O_3 and ZrO_2 on Pt•PdO/Al_2O_3 during catalytic combustion of methane. *Appl. Catal. A Gen.* **2004**, *258*, 261–267. [CrossRef]
15. Velinova, R.; Todorova, S.; Drenchev, B.; Ivanov, G.; Shipochka, M.; Markov, P.; Nihtianova, D.; Kovacheva, D.; Larine, A.V.; Naydenov, A. Complex study of the activity, stability and sulphur resistance of Pd/La_2O_3- CeO_2-Al_2O_3 system as a monolithic catalyst for abatement of methane. *Chem. Eng. J.* **2019**, *368*, 865–876. [CrossRef]
16. Forzatti, P. Present status and perspectives in de-NOx SCR catalysis. *Appl. Catal. A Gen.* **2001**, *222*, 221–236. [CrossRef]
17. Kanno, Y.; Hihara, T.; Watanabe, T.; Katoh, K. *Low Sulfate Generation Diesel Oxidation Catalyst*; SAE Technical Paper; 2004-01-1427; SAE International: Warrendale, PA, USA, 2004. [CrossRef]
18. Hirata, H.; Hachisuka, I.; Ikeda, Y.; Tsuji, S.; Matsumoto, S.I. NO_x storage-reduction three-way catalyst with improved sulfur tolerance. *Top. Catal.* **2001**, *16*, 145–149. [CrossRef]
19. Satterfield, C.N. *Heterogeneous Catalysis in Industrial Practice*, 2nd ed.; McGraw-Hill: New York, NY, USA, 1991.
20. Giraudon, J.M.; Nguyen, T.B.; Leclercq, G.; Siffer, S.; Lamonier, J.F.; Aboukais, A.; Vantomme, A.; Su, B.L. Chlorobenzene total oxidation over palladium supported on ZrO_2, TiO_2 nanostructured supports. *Catal. Today* **2008**, *137*, 379–384. [CrossRef]
21. Kawahara, T.; Konishi, Y.; Tada, H.; Tohge, N.; Nishii, J.; Ito, S. A patterned TiO_2(anatase)/TiO_2(rutile) bilayer-type photocatalyst: Effect of the anatase/rutile junction on the photocatalytic activity. *Angew. Chem.* **2002**, *114*, 2935–2937. [CrossRef]
22. Diebold, U. The surface science of titanium dioxide. *Surf. Sci. Rep.* **2003**, *48*, 53–229. [CrossRef]
23. Rui, Z.; Wu, S.; Peng, C.; Ji, H. Comparison of TiO_2 Degussa P25 with anatase and rutile crystalline phases for methane combustion. *Chem. Eng. J.* **2014**, *243*, 254–264. [CrossRef]
24. Di Carlo, G.; Melaet, G.; Kruse, N.; Liotta, F.L.; Pantaleoc, G.; Veneziac, M.A. Combined sulfating and non-sulfating support to prevent water and sulfur poisoning of Pd catalysts for methane combustion. *Chem. Commun.* **2010**, *46*, 6317–6319. [CrossRef] [PubMed]
25. Pecchi, G.; Morales, M.; Reyes, P. Pt/SiO_2 catalysts obtained by the sol-gel method. Influence of the pH of gelation on the surface and catalytic properties. *React. Kinet. Catal. Lett.* **1997**, *61*, 237–244. [CrossRef]
26. Pecchi, G.; Reyes, P.; López, T.; Gómez, R.; Moreno, A.; Fierro, J.L.G.; Martínez-Arias, A. Catalytic Combustion of methane on fe-tio_2 catalysts prepared by sol-gel method. *J. Solgel. Sci. Technol.* **2003**, *27*, 205–214. [CrossRef]
27. Yoldas, B.E. Alumina gels that form porous transparent Al_2O_3. *J. Mater. Sci.* **1975**, *10*, 1856–1860. [CrossRef]
28. Gomez-Serrano, V.; Gonzalez-Garcia, C.; Gonzalez-Martın, M. Nitrogen adsorption isotherms on carbonaceous materials, Comparison of BET and Langmuir surface areas. *Powder Technol.* **2001**, *116*, 103–108. [CrossRef]
29. Barrett, E.P.; Joyner, L.G.; Halenda, P.P. The determination of pore volume and area distributions in porous substances. I. Computations from nitrogen isotherms. *J. Am. Chem. Soc.* **1951**, *73*, 373–380. [CrossRef]

30. Galarneau, A.; Villemot, F.; Rodriguez, J.; Fajula, F.; Coasne, B. Validity of the t-plot method to assess microporosity in hierarchical micro/mesoporous materials. *Langmuir* **2014**, *30*, 13266–13274. [CrossRef]
31. Rosmini, C.; Tsoncheva, T.; Kovatcheva, D.; Velinov, N.; Kolev, H.; Karashanova, D.; Dimitrov, M.; Tsyntsarski, B.; Sebastián, D.; Jesús Lázaro, M. Mesoporous Ce–Fe–Ni nanocomposites encapsulated in carbon-nanofibers: Synthesis, characterization and catalytic behavior in oxygen evolution reaction. *Carbon* **2022**, *196*, 186–202. [CrossRef]
32. Gregg, S.J.; Sing, K.S.W. *Adsorption, Surface Area, and Porosity*; Academia Press: London, UK, 1991; pp. 111–194.
33. Levenspiel, J. *Chemical Reaction Engineering*, 2nd ed.; Wiley: New York, NY, USA, 1972; p. 496.
34. Lyubovsky, M.; Pfefferle, L.; Datye, A.; Bravo, J.; Nelson, T. TEM study of the microstructural modifications of an alumina-supported palladium combustion catalyst. *J. Catal.* **1999**, *187*, 275–284. [CrossRef]
35. Wenge, L.; Deyong, G.; Xin, X. Research progress of palladium catalysts for methane combustion. *China Petrol. Proc. Petrochem. Technol. Rev.* **2012**, *14*, 1–9.
36. Todorova, S.; Naydenov, A.; Velinova, R.; Kolev, H.; Larin, A.V.; Stoyanova, D.; Shopska, M.; Tenchev, K.; Stefanov, P. Pd–MeOx/Al_2O_3 (Me=Co, La, Ce) catalysts for methane combustion. *Reac. Kineti. Mech. Catal.* **2019**, *126*, 663–678. [CrossRef]
37. Stefanov, P.; Todorova, S.; Naydenov, A.; Tzaneva, B.; Kolev, H.; Atanasova, G.; Stoyanova, D.; Karakirova, Y.; Aleksieva, K. On the development of active and stable Pd–Co/γ-Al_2O_3 catalyst for complete oxidation of methane. *Chem. Eng. J.* **2015**, *266*, 329–338. [CrossRef]
38. Tyuliev, G.; Panayotov, D.; Avramova, I.; Stoichev, D.; Marinova, T. Thin-film coating of Cu-Co oxide catalyst on lanthana/zirconia films electrodeposited on stainless steel. *Mat. Sci. Eng. C* **2003**, *23*, 117–121. [CrossRef]
39. Burroughs, P.; Hamnett, A.; Orchard, A.F.; Thornton, G. Satellite structure in the X-ray photoelectron spectra of some binary and mixed oxides of lanthanum and cerium. *J. Chem. Soc. Dalton Trans.* **1976**, *17*, 1686–1698. [CrossRef]
40. Matolín, V.; Cabala, M.; Cháb, V.; Matolínová, I.; Prince, K.C.; Škoda, M.; Šutara, F.; Skála, T.; Veltruská, K. A resonant photoelectron spectroscopy study of Sn(Ox) doped CeO_2 catalysts. *Surf. Interf. Anal.* **2008**, *40*, 225–230. [CrossRef]
41. Smirnov, M.; Kalinkin, A.V.; Pashis, A.V.; Prosvirin, I.P.; Bukhtiyarov, V.I. Interaction of SO_2 with Pt model supported catalysts studied by XPS. *J. Phys. Chem. C* **2014**, *118*, 22120–22135. [CrossRef]
42. Moulder, J.F.; Stickle, W.F.; Sobol, P.E.; Bomben, K.D.M. *Handbook of X-ray Photoelectron Spectroscopy*; Perkin-Elmer Corporation: Eden Prairie, MN, USA, 1992.
43. Michalik, J.; Narayana, M.; Kevan, L. Studies of the interaction of Pd^{3+} and Pd^{+} with organic adsorbates, water, and molecular oxygen in Pd-Ca-X zeolites by ESR and Electron-spin echo modulation spectroscopy. *J. Phys. Chem.* **1985**, *89*, 4553–45601. [CrossRef]
44. Gamarra, D.; López Cámara, A.; Monte, M.; Rasmussen, S.B.; Chinchilla, L.E.; Hungría, A.B.; Munuera, G.; Gyorffy, N.; Schay, Z.; Cortés Corberán, V.; et al. Preferential oxidation of CO in excess H_2 over CuO/CeO_2 catalysts: Characterization and performance as a function of the exposed face present in the CeO_2 support. *Appl. Catal. B Environ.* **2013**, *130–131*, 224–238. [CrossRef]
45. Cybula, A.; Priebe, J.B.; Pohl, M.-M.; Sobczak, J.W.; Schneider, M.; Zielinska-Jurek, A.; Brückner, A.; Zaleska, A. The effect of calcination temperature on structure and photocatalytic properties of Au/Pd nanoparticles supported on TiO_2. *Appl. Catal. B Environ.* **2014**, *152–153*, 202–211. [CrossRef]
46. Xiong, L.-B.; Li, J.-L.; Yang, B.; Yu, Y. Ti^{3+} in the surface of titanium dioxide: Generation, properties and photocatalytic application. *J. Nanomater.* **2012**, *2012*, 831524. [CrossRef]
47. Abragam, A.; Bleaney, B. *Book: Electron Paramagnetic Resonance of Transition Metals*; Clarendon Press: Oxford, UK, 1970; p. 222.
48. Luo, T.; Gorte, R.J. Characterization of SO_2-poisoned ceria-zirconia mixed oxides. *Appl. Catal. B* **2004**, *53*, 77–85. [CrossRef]
49. Schoonheydt, R.A.; Lunsford, J.H. Infrared spectroscopic investigation of the adsorption and reactions of SO_2 on MgO. *J. Catal.* **1972**, *26*, 261–271. [CrossRef]
50. Trura, M.J.; Regull, P.; Victorit, L.; Dolors de Castellar, M. XPS and IR (ATR) analysis of Pd oxide films obtained by electrochemical methods. *Surf Interface Anal.* **1988**, *11*, 4477449. [CrossRef]
51. Kliche, G. Far-infrared reflection spectra of PdO, PdS, PdSe and PtS. *Infrared Phys.* **1985**, *25*, 381–383. [CrossRef]
52. Todorova, S.; Naydenov, A.; Kolev, H.; Holgado, J.P.; Ivanov, G.; Kadinov, G.; Caballero, A. Mechanism of complete n-hexane oxidation on silica supported cobalt and manganese catalysts. *Appl. Catal. A Gen.* **2012**, *413–414*, 43–51. [CrossRef]
53. Markova-Velichkova, M.; Lazarova, T.; Tumbalev, V.; Ivanov, G.; Kovacheva, D.; Stefanov, P.; Naydenov, A. Complete oxidation of hydrocarbons on $YFeO_3$ and $LaFeO_3$ catalysts. *Chem. Eng. J.* **2013**, *231*, 236–244. [CrossRef]
54. Honkanen, M.; Kärkkäinen, M.; Kolli, T.; Heikkinen, O.; Viitanen, V.; Zeng, L.; Jiang, H.; Kallinen, K.; Huuhtanen, M.; Keiski, R.L.; et al. Accelerated deactivation studies of the natural-gas oxidation catalyst–verifying the roleof sulfur and elevated temperature in catalyst aging. *Appl. Catal. B Environ.* **2016**, *182*, 439–448. [CrossRef]
55. Bartholomew, C.H. Mechanisms of catalyst deactivation. *Appl. Catal. A Gen.* **2001**, *212*, 17–60. [CrossRef]
56. Neyestanaki, A.K.; Klingstedt, F.; Salmi, T.; Murzin, D.Y. Deactivation of postcombustion catalysts, a review. *Fuel* **2004**, *83*, 395–408. [CrossRef]
57. Carp, O.; Huisman, C.L.; Reller, A. Photoinduced reactivity of titanium dioxide. *Prog. Solid State Chem.* **2004**, *21*, 33–177. [CrossRef]
58. Hirano, M.; Nakahara, N.; Ota, K.; Tanaike, O.; Inagaki, N. Photoactivity and phase stability of ZrO_2-doped anatase-type TiO_2 directly formed as nanometer-sized particles by hydrolysis under hydrothermal conditions. *J. Solid State Chem.* **2003**, *170*, 39–47. [CrossRef]

59. Kim, J.; Song, K.C.; Foncillas, S. Pratsinis S Dopants for synthesis of stable bimodally porous titania. *J. Eur. Ceram. Soc.* **2001**, *21*, 2863–2872. [CrossRef]
60. Gouma, P.I.; Mills, M.J. Anatase-to-rutile transformation in titania powders. *J. Am. Ceram. Soc.* **2001**, *84*, 619–622. [CrossRef]
61. Hyoung, G.L.; Zuo, J.M. Growth and phase transformation of nanometer-sized titanium oxide powders produced by the precipitation method. *J. Am. Ceram. Soc.* **2004**, *87*, 473–479. [CrossRef]
62. Zhang, Q.; Gao, L.; Guo, J. Effects of calcination on the photocatalytic properties of nanosized TiO_2 powders prepared by $TiCl_4$ hydrolysis. *Appl. Catal. B* **2000**, *26*, 207–215. [CrossRef]
63. Zhang, Z.; Wang, C.; Zakaria, R.; Ying, J. Role of Particle Size in Nanocrystalline TiO_2-Based Photocatalysts. *J. Phys. Chem. B* **1998**, *102*, 10871–10878. [CrossRef]
64. Suzana, M.; Francisco, P.; Mastelaro, V.R. Inhibition of the anatase−rutile phase transformation with addition of CeO_2 to $CuO-TiO_2$ system: raman spectroscopy, X-ray diffraction, and textural studies. *Chem. Mater.* **2002**, *14*, 2514–2518. [CrossRef]
65. Bagheri, S.; Shameli, K.; Abd Hamid, S.B. Synthesis and characterization of anatase titanium dioxide nanoparticles using egg white solution via sol-gel method. *J. Chem.* **2013**, *2013*, 848205. [CrossRef]
66. Kominami, H.; Kalo, J.-I.; Takada, Y.; Doushi, Y.; Ohtani, B.; Nishimoto, S.; Inoue, M.; Inui, T.; Kera, Y. Novel synthesis of microcrystalline titanium(IV) oxide having high thermal stability and ultra-high photocatalytic activity: Thermal decomposition of titanium(IV) alkoxide in organic solvents. *Catal. Lett.* **1997**, *46*, 235–240. [CrossRef]
67. Su, S.C.; Carstens, J.N.; Bell, A.T. A study of the dynamics of Pd oxidation and PdO reduction by H2 and CH4. *J. Catal.* **1998**, *176*, 125–135. [CrossRef]
68. Belfiore, L.A. *Transport Phenomena for Chemical Reactor Design*; John Wiley & Sons, Inc.: Hoboken, NJ, USA, 2003.
69. Nauman, E.B. *Chemical Reactor Design, Optimization, and Scaleup*; McGraw-Hill Companies: New York, NY, USA, 2002.
70. Satterfield, C.N. *Mass Transfer in Heterogeneous. Catalysis, Cambridge, Mass*; MIT Press: Cambridge, UK, 1970.
71. Tomašić, V. Application of the monoliths in DeNOx catalysis. *Catal. Tod.* **2007**, *119*, 106–113. [CrossRef]
72. Tomašić, S.; Gomzi, Z. Experimental and theoretical study of NO decomposition in a catalytic monolith reactor. *Chem. Engin. Proc.* **2004**, *43*, 765–774. [CrossRef]
73. Tomašić, V.; Jović, F. State-of-the-art in the monolithic catalysts/reactors. *Appl. Catal. Gen.* **2006**, *311*, 112–121. [CrossRef]

materials

MDPI

Article

Machine Learning Prediction of the Redox Activity of Quinones

Ilia Kichev [1,2], Lyuben Borislavov [1,*], Alia Tadjer [2,*] and Radostina Stoyanova [1]

1 Institute of General and Inorganic Chemistry, Bulgarian Academy of Sciences, 1113 Sofia, Bulgaria; ikichev@uni-sofia.bg (I.K.); radstoy@svr.igic.bas.bg (R.S.)
2 Faculty of Chemistry and Pharmacy, University of Sofia, 1164 Sofia, Bulgaria
* Correspondence: lborislavov@svr.igic.bas.bg (L.B.); tadjer@chem.uni-sofia.bg (A.T.)

Abstract: The redox properties of quinones underlie their unique characteristics as organic battery components that outperform the conventional inorganic ones. Furthermore, these redox properties could be precisely tuned by using different substituent groups. Machine learning and statistics, on the other hand, have proven to be very powerful approaches for the efficient in silico design of novel materials. Herein, we demonstrated the machine learning approach for the prediction of the redox activity of quinones that potentially can serve as organic battery components. For the needs of the present study, a database of small quinone-derived molecules was created. A large number of quantum chemical and chemometric descriptors were generated for each molecule and, subsequently, different statistical approaches were applied to select the descriptors that most prominently characterized the relationship between the structure and the redox potential. Various machine learning methods for the screening of prospective organic battery electrode materials were deployed to select the most trustworthy strategy for the machine learning-aided design of organic redox materials. It was found that Ridge regression models perform better than Regression decision trees and Decision tree-based ensemble algorithms.

Keywords: quinones; machine learning; ridge regression; decision tree; ensemble methods; density functional theory; organic electrode materials

Citation: Kichev, I.; Borislavov, L.; Tadjer, A.; Stoyanova, R. Machine Learning Prediction of the Redox Activity of Quinones. *Materials* **2023**, *16*, 6687. https://doi.org/10.3390/ma16206687

Academic Editor: Pasquale Fernando Fulvio

Received: 21 September 2023
Revised: 9 October 2023
Accepted: 11 October 2023
Published: 14 October 2023

1. Introduction

In recent years, the global demand for effective energy-storage materials has constantly grown [1]. Traditionally, the widely used electrode materials in metal-ion batteries are inorganic compounds capable of reversible redox transformations [2,3]. Organic electrode materials, on the other hand, have some gainful properties, such as structural diversity and flexibility, synthetic tunability, lower price, and harmless recyclability [4–7]. Among the organic compounds considered for research on battery electrode materials, quinones have engendered the most ubiquitous expectations and extensive investigation. Quinones are a class of organic compounds derived from aromatic diols, whose redox capacity makes them interesting for designing novel organic electrode materials [8]. Quinones with a low molecular weight, such as 1,4-benzoquinone, have a relatively high redox potential [9] and, in case the two-electron redox reaction of benzoquinone takes place, a high capacity could be expected. However, due to the sublimation and dissolution of benzoquinone in the organic electrolyte solvents, a poor capacity is observed in practice [10]. These problems can be overcome by immobilizing benzoquinone on nanoparticles [11], by using various polymers containing benzoquinone fragments [12–14], or by introducing different substituent groups [15]. The redox potential of the quinones is dependent on the substituent type; electron-withdrawing substituents, such as carbonyl, nitro, and carboxylate groups, make quinones stronger oxidants, while electron-donating groups, such as amine, hydroxyl, and alkoxy groups, turn quinones into weaker oxidants [16]. In the present study, a dataset of quinones with electron-withdrawing substituents was constructed, since this class of materials exhibits a fairly high redox potential.

Machine learning and statistics approaches have successfully been applied for capturing the complex relationships between material structures and different properties of interest [17]. This kind of approach has also effectively been employed in the design of novel energy-storage materials: Joshi et al. [18] demonstrated that deep neural networks (DNNs), support vector regression (SVR), and kernel ridge regression (KRR) can be used to predict the redox potential of inorganic electrode materials extracted from the Materials Project Database; Zhang et al. [19] used a Crystal Graph Convolutional Neural Network (CGCNN) to creatively build an interpretable deep learning model that predicts redox potential based on inorganic crystal structures [19]. Machine learning algorithms have also been productively applied in the design of organic electrode materials: Allam et al. [20] developed a pre-screening procedure that relies on the density functional theory to compute both the redox potential of organic electrode materials and molecular descriptors, such as the electron affinity and the gap between the highest occupied molecular orbital (HOMO) and the lowest unoccupied molecular orbital (LUMO), to be used as input features of artificial neural networks (ANNs), gradient boosting regression (GBR) and KRR. A major disadvantage is that the density functional theory, which is comparatively computationally expensive, is used for descriptor computation. Tutte et al. [21] propose a Hammet-like approach to model quinone solubility in organic electrolytes that are typically used in lithium-ion batteries (the organic electrode materials must have low solubility in the battery electrolyte). Machine learning screening has also been applied for the design of quinone electrolytes for redox flow batteries [22]: Wang et al. created a dataset by generating various disubstituted quinones, replacing hydrogens in different quinone backbones with a predefined set of substituents, and, subsequently, utilized the extreme gradient boosting algorithm to build a model for screening the HOMO–LUMO gap and the free energy of solvation. In the current study, different linear and nonlinear regression models were built to predict the electrode potential of substituted quinones.

Dataset construction plays a central role in any data-driven study. In this report, two tactics for the creation of application-specific datasets were combined. Firstly, a top-down approach was used: molecular structures that satisfy some application-specific conditions (i.e., contain a quinone fragment) were extracted from PubChem [23] (a large, publicly available database). Next, a bottom-up approach was applied: the dataset of molecular structures produced in the first step was expanded via inclusion of the systematically generated derivatives of the already-selected species. This strategy guarantees that the final dataset created is structurally consistent.

2. Materials and Methods

2.1. Dataset Construction

2.1.1. Molecular Structure Generation

To construct the dataset, 100 benzoquinone derivatives were extracted from the PubChem database [23] as simplified molecular-input line-entry system (SMILES) strings. The SMILES strings were converted into 3D structures using the OpenBabel software package (version 3.1.1) [24] and, subsequently, the DerGen software (version 0.1) [25] was used to generate all possible derivatives of those compounds containing a -CN or a -C≡CMe group. In total, 494 structures were produced. This dataset construction procedure guarantees that the generated molecules are structurally similar, and hence makes it easier to establish the structure–electrode potential relationship for a quinone series with electron-withdrawing substituents—a group of compounds that is particularly interesting for the design of organic energy-storage materials.

2.1.2. Dataset Splitting

The dataset was shuffled and split into a training set (395 compounds, 80% of the whole dataset) and a test set (99 compounds, 20% of the whole dataset). To avoid data leakage [26], the descriptor selection and hyperparameter optimization were performed on the training set. An average R^2 metric over 5-fold cross-validation was used for model performance assessment during the descriptor selection and hyperparameter optimization.

2.2. Molecular Descriptors

Representing molecular structures in an unambiguous machine-readable format is not a trivial task. Many different molecular representations have been developed [27]. Molecular structures can be represented as the following:

- Strings—for example, the SMILSES representation that contains information about atom types and connectivity [28];
- Connection table formats [29]: tabular formats that provide information about atom counts, atom types, connectivity matrix, bonded pairs of atoms, chirality, etc.; an example for such molecular representation format is the MDL molfile;
- Vectors of features: a molecule can be represented either as a vector of molecular properties (descriptors) such as molecular weight, molecular volume, numbers of certain atom types, topology, etc., [30] or as a molecular fingerprint: a bitstring (can be regarded as vector of ones and zeros) is derived from the molecular structure according to a predefined set of rules [31]—among the most employed fingerprints are the extended-connectivity fingerprints (ECFPs) based on Morgan's algorithm [32], since they are specially designed for establishing structure–property relationships [33];
- Computer-learned representations: in recent years, a large number of machine learning-based molecular representations were developed—those methods rely on convolutional neural networks (CNNs) and/or recurrent neural networks (RNNs) to transform a molecule represented as a SMILES string or as 3D Cartesian atom coordinates to a low dimensional latent space [34,35] that can be used both for property prediction and for the generation of new molecular structures [36].

In the current study, the PaDEL [37] software package was employed to generate a multitude (750 descriptors per molecule) of cheminformatics-based molecular descriptors, and the MOPAC program suite [38] was used to produce semi-empirical descriptors such as HOMO and LUMO energies and the dipole moments of the reducible compounds.

Descriptor Selection

Feature selection is a key step in any data-driven study [39]. The objective of the feature selection procedure is to assort features that have a strong correlation with the target variable. In the current work, the following steps were taken:

- Low-variance descriptors were removed: descriptors whose value equalled the descriptor mode for 60% or more of the molecules in the dataset were discarded;
- Descriptors that had a weak correlation with the target value were discarded. Correlations with covariance between the normalized descriptors and normalized target values of less than 0.25 were considered as weak correlations. The normalization was performed as follows:

  $$V_{norm} = \frac{V - V_{mean}}{\sigma_V},$$

  where V_{norm} is the normalized value, V is the unnormalized value, V_{mean} is the mean of V in the dataset, and σ_V is the standard deviation of V in the dataset.
- Strongly mutually correlated descriptors (covariance between normalized descriptors greater than 0.7) were discarded. After this operation, 52 descriptors were left;
- Backward stepwise regression [40] was used for further descriptor reduction (Figure 1). Finally, 32 descriptors remained (Table 1).

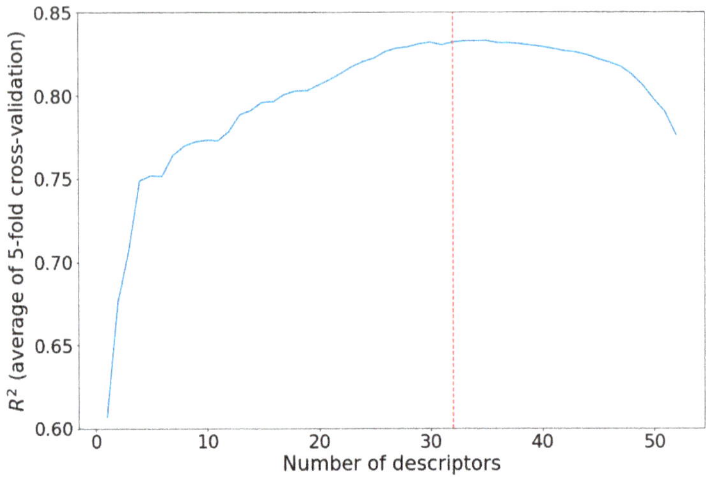

Figure 1. Results of the backward stepwise regression for descriptor selection.

Table 1. Molecular descriptors' names and descriptions.

Description	Descriptor Name
Lowest partial charge weighted BCUTS [41]	BCUTc-1l
Highest partial charge weighted BCUTS [41]	BCUTc-1h
Total number of double bonds (excluding aromatic bonds)	nBondsD2
Triply bound carbon bound to another carbon	C1SP1
Doubly bound carbon bound to three other carbons	C3SP2
A topological descriptor combining distance and adjacency information [42]	ECCEN
Count of atom-type H E-State: H on aaCH, dCH2 or dsCH * [43]	nHother
Count of atom-type E-State: =C< [43]	ndssC
Count of atom-type E-State: aaC- [43]	naasC
Count of atom-type E-State: N≡ [43]	ntN
Sum of E-States for weak hydrogen bond acceptors [43]	SwHBa
Sum of atom-type H E-State: =CH- [43]	SHdsCH
Sum of atom-type H E-State: H on aaCH, dCH2 or dsCH [43]	SHother
Sum of atom-type E-State: =C< [43]	SdssC
Sum of atom-type E-State: aaC- [43]	SaasC
Sum of atom-type E-State: N≡ [43]	StN
Minimum atom-type H E-State: H on aaCH, dCH2 or dsCH [43]	minHother
Minimum atom-type E-State: aaC- [43]	minaasC
Minimum atom-type E-State: =O [43]	mindO
Maximum atom-type H E-State: H on aaCH, dCH2 or dsCH [43]	maxHother
Maximum atom-type E-State: aaC- [43]	maxaasC

Table 1. *Cont.*

Description	Descriptor Name
Mean intrinsic state values I [43]	meanI
Maximum negative intrinsic state difference in the molecule (related to the nucleophilicity of the molecule) [44]	MAXDN2
Maximum positive intrinsic state difference in the molecule (related to the electrophilicity of the molecule) [44]	MAXDP2
Complexity of the system [45]	fragC
Number of rings	nRing
Topological diameter (maximum atom eccentricity)	topoDiameter
Mean topological charge index of order two [46]	JGI2
Topological polar surface area	TopoPSA
Van der Waals volume calculated using the method proposed in Zhao et al. JACS 2003, 68, 7368–7373 [47]	VABC
Molecular weight	MW
Energy of the lowest unoccupied molecular orbital estimated by PM6 [eV]	LUMO

* a = aromatic; s = single; d = double.

2.3. Redox Potential Calculation

The redox potential was calculated with the density functional theory (DFT) for the redox reaction:

i ii

Geometry optimization was performed on all quinone derivatives in the dataset (i) and their respective reduced forms (ii) using the B3LYP functional in combination with the 6-311++G(2df,2p) basis set, as implemented in the Gaussian 16 software package (version EM64L-G16RevB.01, 20 December 2017) [48]. This protocol was chosen as a trade-off between precision and computational time.

The electrode potential was calculated using the Nernst equation:

$$\Delta E = \frac{-\Delta G}{nF}, \quad (1)$$

where n is the number of exchanged electrons, F is the Faraday's constant, and the reaction free energy, ΔG, is calculated as follows:

$$\Delta G = G_{ii} - G_i - 2G_{Li}. \quad (2)$$

G_{ii} and G_i were obtained from the B3LYP/6-311++G(2df,2p) calculation, as follows [49]:

$$H_X = E_0 + ZPE + H_{trans} + H_{rot} + H_{vib} + RT \quad (3a)$$

$$S_X = S_{trans} + S_{rot} + S_{vib} + S_{el} \quad (3b)$$

$$G_X = H_X - TS_X, \quad (3c)$$

where E_0 is the total electronic energy, ZPE is the unscaled zero-point energy, H_{trans}, H_{rot}, and H_{vib} are, correspondingly, the translational, rotational, and vibrational shares in the enthalpy, S_{trans}, S_{rot}, S_{vib}, and S_{el} are, respectively, the rotational, translational, vibrational, and electronic motion contributions to the entropy. RT represents the work term converting the internal energy into enthalpy (T = 298 K). G_{Li} is the free energy of lithium in the gas phase. A comparison of calculated and experimental values of electrode potentials showed that when the free energy change in the redox reaction is estimated as the difference of the free energies of the reacting molecules in the gas phase, then the gas phase free energy for lithium should be considered as well (see Supplementary Information in [50]).

2.4. Machine Learning Methods Used

Different machine learning methods were deployed to investigate the relationship between the molecular structure and the electrode potential.

2.4.1. Ridge Regression

Ridge regression is a method for estimating the coefficients of l2-regularized multiple linear regression models:

$$\boldsymbol{X}\boldsymbol{\beta} = \boldsymbol{y}, \tag{4}$$

where, for a dataset consisting of n molecules, each is represented as an m-dimensional vector; $\boldsymbol{X}$ is an $n_x(m+1)$ matrix of n-dimensional column vectors $\boldsymbol{x}_j$ ($\boldsymbol{x}_1$ is $[1\,1\ldots 1]^{\mathrm{T}}$, while $x_2, x_3, \ldots x_{(m+1)}$ are the values of the corresponding descriptors), known as explanatory variables; $\boldsymbol{\beta}$ is an $(m+1)$ dimensional vector of parameters, where β_1 is the intercept term and $\boldsymbol{y}$ is the vector of the observed values (redox potentials in the current study). The ridge estimator of $\boldsymbol{\beta}$ is given using the following equation [50]:

$$\boldsymbol{\beta} = \left(\boldsymbol{X}^T\boldsymbol{X} + \lambda\boldsymbol{I}\right)^{-1}\boldsymbol{X}^T\boldsymbol{y}, \tag{5}$$

where λ is a regularization coefficient and $\boldsymbol{I}$ is the identity matrix. Ridge regression is known to perform better than linear regression in cases of mutually correlated explanatory variables (molecular descriptors in our case) [51].

2.4.2. Decision Tree

First introduced in 1987 [52], decision trees are hierarchical supervised machine learning models that logically combine a sequence of decisions, based on simple tests, and their possible outcomes. This is achieved by optimizing the simple test condition threshold during the training process [53]. In the course of training, all possible data splits are considered:

$$Q_m^l = \{(x, y) | x_j < t_m\} \tag{6a}$$

$$Q_m^r = Q_m \setminus Q_m^l, \tag{6b}$$

where Q_m is the data at node m, Q_m^l and Q_m^r are the candidate splits, $\boldsymbol{x}$ is the training data vectors, and $\boldsymbol{y}$ is the target variable vector. The threshold condition is optimized by comparing the quality of the splits using an appropriate cost function. For regression decision trees, the mean square error (MSE—Equation (8a)) or the Poisson deviance (Equation (8b)) can be used as cost functions [52]:

$$y_m = \frac{1}{n_m}\sum\nolimits_{y \in Q_m} y \tag{7}$$

$$H(Q_m) = \frac{1}{n_m}\sum\nolimits_{y \in Q_m} (y - y_m)^2 \tag{8a}$$

$$H(Q_m) = \frac{1}{n_m}\sum_{y\in Q_m}\left(ylog\left(\frac{y}{y_m}\right) - y - y_m\right)^2 \tag{8b}$$

This splitting operation is performed for all the features, and the feature split that leads to the largest decrease in the cost function is kept at node m. This allows for the estimation of the feature importance—the more efficiently a feature split decreases the cost function, the more important the feature.

It should be noted that, due to their structure of sequential simple tests, decision trees are able to capture nonlinear dependencies between the explanatory variables and the measured property. Decision trees have been successfully utilized to solve both classification and regression problems [54–56]. There exist numerous algorithms for decision tree construction: ID3, C4.5, CART, MARS, and CHAID [57]. In the present study, the CART (classification and regression tree) algorithm with a mean square error cost function was used.

2.4.3. Random Forest

Random forests are ensemble machine learning algorithms that can be used for classification and regression. Multiple decision trees are constructed using randomly selected explanatory variables (molecular descriptors in our case) and each tree is trained on different bootstrapped samples (sampling, allowing for multiple selection of the same items) of the training set. When a prediction is made, the average result of all trees is returned [58].

2.4.4. Extra Trees

The extra trees algorithm [59] is similar to the random forest algorithm—a multitude of decision trees are used; however, the individual decision trees are trained on subsamples of the training set taken without replacement (in contrast to bootstrapping). Another important difference is that, in the extra trees algorithm, the cut point is selected randomly, while in the random forest algorithm, the optimal split is chosen. These differences generally lead to the reduction of bias and variance. The random choice of a cut point also makes the algorithm faster (in the random forest algorithm, the optimal split is found by computing some impurity metric for all possible splits).

2.4.5. Gradient Boosting

Gradient boosting relies on the fitting of a sequence of weak prediction models (decision trees in this case) on repeatedly altered versions of training data [60]. The predictions of all individual weak predictors are combined as a sum:

$$\hat{y}_i = F_M(\boldsymbol{x}_i) = \sum_{m=1}^{M} h_m(\boldsymbol{x}_i), \tag{9}$$

where $\hat{y}_i$ is the model prediction, $\boldsymbol{x}_i$ is a vector of all features that describes the i-th object in the dataset (in our case, all descriptors used to represent a molecule), M is the number of weak estimators, and $h_m(\boldsymbol{x}_i)$ is the prediction of the m-th weak estimator. From Equation (9), it follows that

$$F_m(\boldsymbol{x}_i) = F_{m-1}(\boldsymbol{x}_i) + h_m(\boldsymbol{x}_i). \tag{10}$$

The weak predictor $h_m(\boldsymbol{x}_i)$ in Equation (10) is fitted to minimize a sum of the cost functions, Cm:

$$h_m = argmin_h(C_m) = argmin_h\left(\sum_{i=1}^{n} c(y_i,\ F_{m-1}(\boldsymbol{x}_i) + h_m(\boldsymbol{x}_i))\right), \tag{11}$$

where n is the number of training entries and $c(y_i, F(\boldsymbol{x}_i))$ is a cost function, such as the mean square error (MSE, Equation (8a)).

Friedman [59] proposed a regularization strategy, based on scaling the contribution of each new weak predictor, based on a learning rate (γ):

$$F_m(\boldsymbol{x}_i) = F_{m-1}(\boldsymbol{x}_i) + \gamma h_m(\boldsymbol{x}_i). \tag{12}$$

It has been demonstrated [61] that, in many cases, gradient boosting outperforms other ensemble methods such as random forests and extra trees.

In the present study, all machine learning algorithms were exploited as implemented in the scikit-learn library [62].

3. Results and Discussion

The redox potential distribution (Figure 2) over the entire dataset (494 compounds) shows that the redox potential spans the range of 0.3–2.8 V. The distribution plot has an asymmetric bell-like shape, with the majority of compounds having potentials between 0.75 V and 1.60 V.

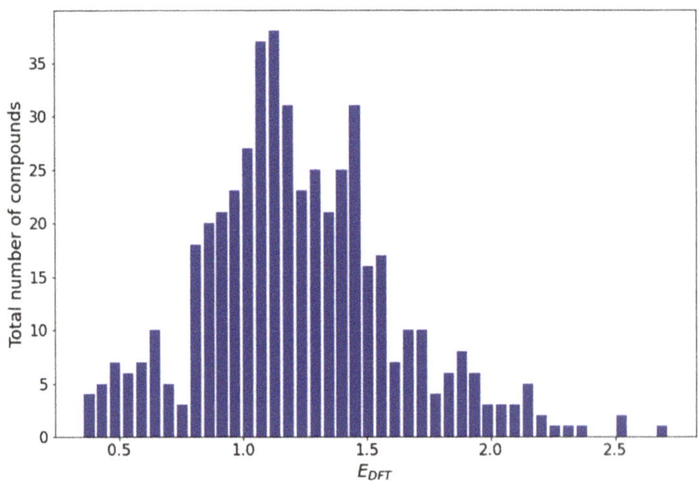

Figure 2. Redox potential distribution histogram.

In order to find an optimal approach for the machine learning modelling of structure–redox potential relationships, the following algorithms were tested: ridge regression, decision tree, random forest, extra trees, and gradient boosting. Artificial neural networks were not considered, since they are prone to overfitting, especially when trained on an insufficient amount of data [63].

In order to attain maximal predictive ability, the hyperparameters (parameters that control the learning process) of each of the machine learning models were optimized using a grid search. The model performance was evaluated based on the averaged coefficient of determination (R^2) [64] of the five-fold cross-validation over the training set. The training R^2 was also taken into account, since the difference between the validation and training R^2 can be used to judge whether the model is overfitted.

The l2-regularization value (λ in Equation (5)) in ridge regression does not have a significant impact on the model performance (Figure 3a); increasing the l2-regularization value leads to a decrease (by an almost equal amount) of the training and validation R^2. It should be noted that the difference between the training and validation R^2 reached a minimum at $\lambda = 0.1$, and hence, this value of lambda results in an optimal (neither underfitted, nor overfitted) ridge regression model.

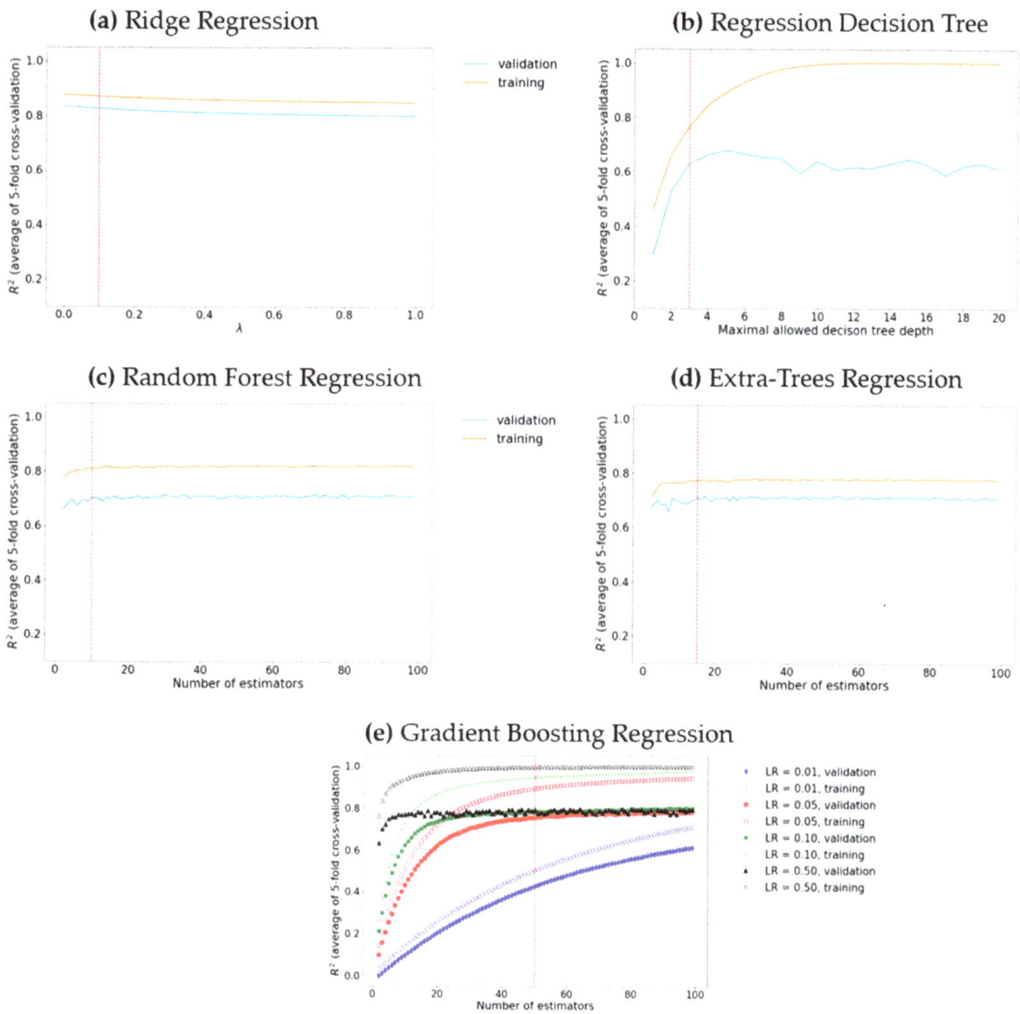

Figure 3. Model tuning via grid search for (**a**) the optimal learning rate in ridge regression; (**b**) the optimal maximal decision tree depth; ((**c**) and (**d**), respectively) the optimal number of decision tree estimators in random forest regression and extra trees regression; (**e**) the optimal number of decision tree estimators and the learning rate (LR) in gradient boosting. The optimal hyperparameter value is depicted by a dotted red line.

The decision tree maximal allowed depth plays a central role in determining whether the decision tree underfits or overfits the training data: a larger maximal allowed depth results in a deeper tree that fits the training data better; however, if a tree is too deep, the noise in the training data is also learned, i.e., the decision tree overfits. In the present work, the maximal tree depth varied from two to twenty (Figure 3b). Optimal algorithm performance was attained when the maximal tree depth was three. A serious advantage of decision trees is their ability to visualize the learning process (Figure 4). Furthermore, the decision tree algorithm enables the examination of the descriptor significance. It was found that the most significant descriptors, MAXDN2, LUMO, SaasC, SHdsCH, and BCUTc-1H (see Table 1), are all related to the electronic structure of the molecules—quinones that contain more CN and C≡C-Me groups (lower LUMO, large MAXDN2 due to CN groups) exhibit larger redox potential.

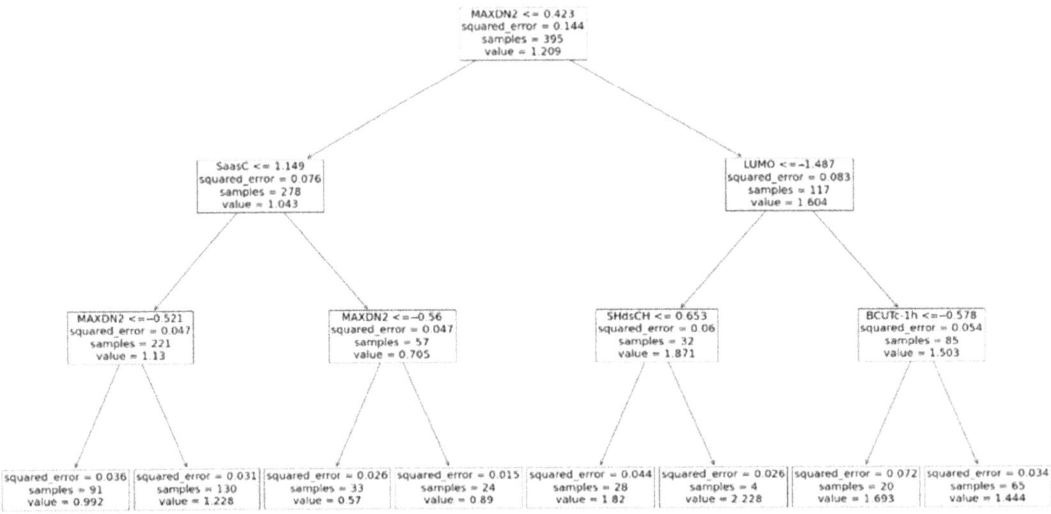

Figure 4. Regression decision tree chart with maximal depth of three.

To examine the predictive ability of maximal random forest regression and extra trees regression, the depth of the decision tree estimator was set to three (since we established that this value of maximal depth ensures maximal learning performance), and the number of estimators was optimized to achieve the maximal coefficient of determination over the validation set (Figure 3c,d): 10 and 15 estimators were chosen for random forest and extra trees, respectively.

It was found that the extra trees algorithm is less prone to overfitting: the R^2 value over the validation set is closer to the R^2 value over the test set. The random forest and extra trees algorithms can also be used to estimate the descriptors' importance—the most significant descriptors for the decision tree (described above) are found among the ten most significant descriptors of both algorithms, which confirms that the descriptors related to the electronic structure, such as the LUMO energy, and descriptors derived from electronegativity, such as SaasC, SHdsCH, MAXDN2, and meanI, are important for the machine learning prediction of the redox potential of organic energy-storage materials. As expected, we found that the gradient boosting regression exhibits a better predictive ability and is less prone to overfitting than the other ensemble methods used (random forest regression and extra trees regression). The learning rate (γ in Equation (12)) and the number of weak predictors values of 0.05 and 50, respectively, were found via grid searching (Figure 3e).

The prediction models' performance, as evaluated based on the average coefficient of determination over a five-fold cross-validation (R^2_{CV}), increases in the following order: regression decision tree (R^2_{CV} = 0.632), random forest regression (R^2_{CV} = 0.705), extra trees regression (R^2_{CV} = 0.715), gradient boosting regression (R^2_{CV} = 0.756), and ridge regression (R^2_{CV} = 0.832).

All machine learning algorithms were evaluated on the test set. To visualize the model performance, scatter plots of the redox potential calculated based on the density functional theory versus the redox potential estimated using the corresponding machine learning algorithms were drawn (Figure 5). Linear regression was implemented to construct a trendline in the (E_{model}, E_{DFT})—space (Figure 5, red lines), and the slope, intercept, and coefficient of determination (R^2) of the trendline were calculated. When the model ideally fits the data, the trendline slope is supposed to have one and zero for the slope and trendline, respectively, and the R^2 value should be close to one. It was found that the models' performance on the test set does not differ significantly from the models' performance observed upon the five-fold cross-validation, which means that the models fit the data fairly well (i.e., the models are not significantly overfitted or underfitted). All models tend to give worse prediction for large voltages: a possible explanation is that, in the training set, there are fewer molecules exhibiting a high redox potential. The dataset and machine models' source code are publicly available: https://github.com/carim2020/org-redox-dataset (accessed on 12 October 2023).

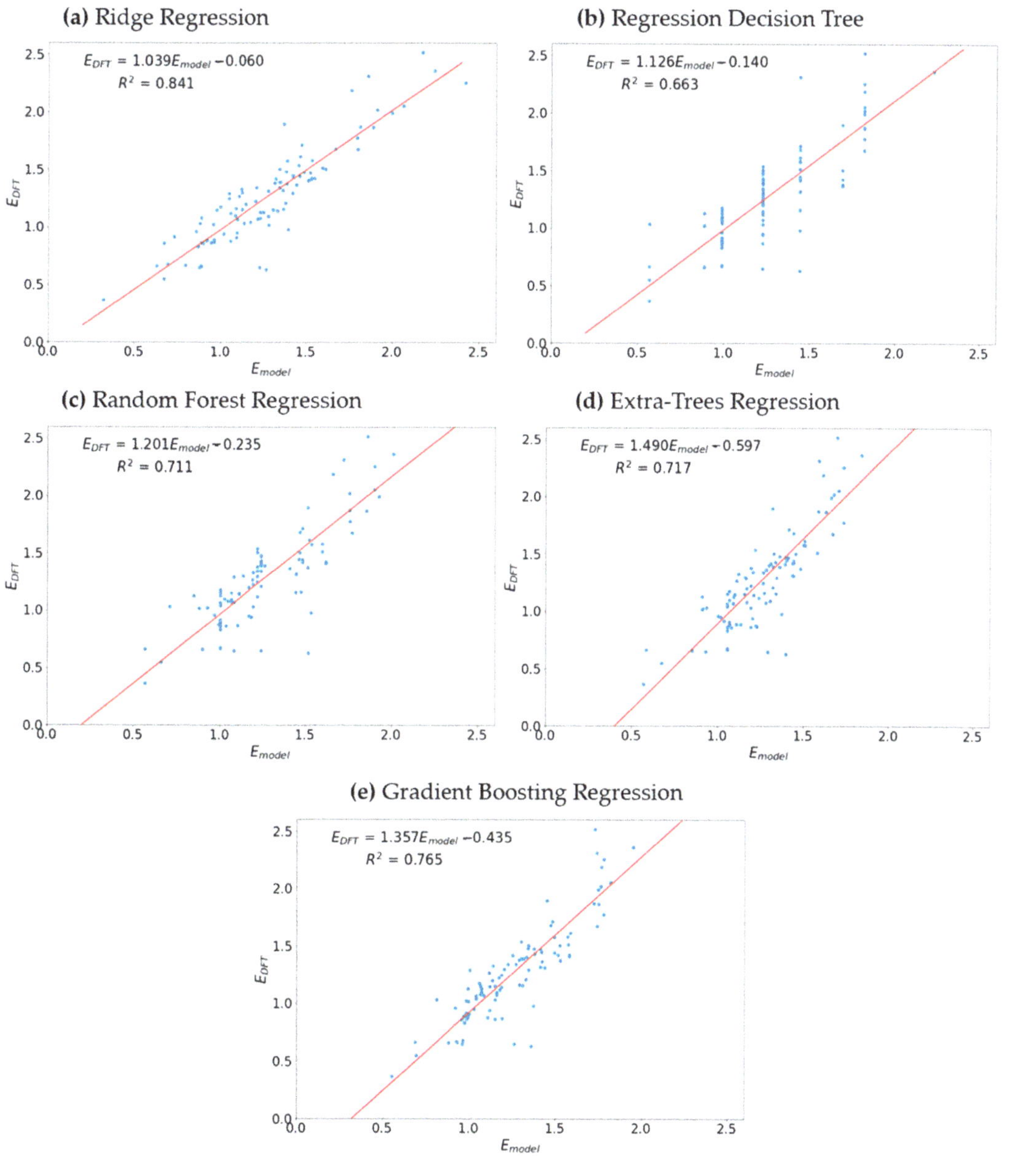

Figure 5. Models' performance on the test set.

4. Conclusions

We have constructed a dataset of 494 potential organic electrode materials through the automated generation of derivatives of 100 quinones, extracted from a general-purpose public database (PubChem). A descriptor selection procedure that combines low-variance descriptor removal with covariance matrix analysis and stepwise linear regression for finding uncorrelated descriptors, on which the redox potential of the molecules in the dataset depends, was devised. Due to the comparatively small dataset size, deep learning approaches were not deployed as inappropriate, since they are prone to overfitting when trained on small amounts of data. Five different supervised machine learning models for regression that tend to give better results for smaller datasets were built. The hyperparameters of all those models were tuned to attain the maximal electrode potential predictive ability. The models' performance was evaluated on a test set containing molecules that are completely unknown to the model. It was established that the model performance increases in the following order: regression decision tree < random forest regression < extra trees regression, gradient boosting regression < ridge regression. It turned out that the linear model, i.e., the ridge regression, outperforms the decision tree-based algorithms, known to be able to capture nonlinear dependencies between the descriptors and the target variable. This is an implication that the relationship between the electrode potential and some chemical properties is most probably linear. In particular, it was found that descriptors related to the electronic structure (LUMO and E-state descriptors) have a large significance. In addition, ridge regression is an excellent method for the screening of databases, as it is a very fast and computationally inexpensive approach, compared to other machine learning algorithms.

Author Contributions: Conceptualization, A.T. and R.S.; methodology, I.K. and L.B.; software, I.K. and L.B.; validation, L.B.; formal analysis, L.B.; resources, L.B.; data curation, I.K.; writing—original draft preparation, I.K. and L.B.; writing—review and editing, A.T.; visualization, L.B.; supervision, R.S. and A.T.; project administration, R.S.; funding acquisition, R.S. All authors have read and agreed to the published version of the manuscript.

Funding: This research was funded by CARIM-VIHREN, grant number КП-06-ДВ-6/2019, and the APC was funded by European Twinning on Materials Chemistry Enabling Clean Technologies (TwinTeam), grant number D01-272/10.2020.

Informed Consent Statement: Not applicable.

Data Availability Statement: The computed data are available from the authors on request.

Acknowledgments: The European Regional Development Fund within the Operational Programme "Science and Education for Smart Growth 2014–2020" under the Project CoE, "National center of mechatronics and clean technologies" (BG05M2OP001-1.001-0008)—for computational facilities; I.K. is grateful to Pascal Friederich for internship hosting; thanks are due to Nina Markova and Yanislav Danchovsky for the initial dataset (100 compounds) construction.

Conflicts of Interest: The authors declare no conflict of interest.

References

1. Poizot, P.; Dolhem, F. Clean energy new deal for a sustainable world: From non-CO_2 generating energy sources to greener electrochemical storage devices. *Energy Environ. Sci.* **2011**, *4*, 2003–2019. [CrossRef]
2. Larcher, D.; Tarascon, J.M. Towards greener and more sustainable batteries for electrical energy storage. *Nat. Chem.* **2015**, *7*, 19–29. [CrossRef]
3. Poizot, P.; Gaubicher, J.; Renault, S.; Dubois, L.; Liang, Y.; Yao, Y. Opportunities and Challenges for Organic Electrodes in Electrochemical Energy Storage. *Chem. Rev.* **2020**, *120*, 6490–6557. [CrossRef]
4. Schon, T.B.; McAllister, B.T.; Li, P.-F.; Seferos, D.S. The rise of organic electrode materials for energy storage. *Chem. Soc. Rev.* **2016**, *45*, 6345–6404. [CrossRef]
5. Lu, Y.; Zhang, Q.; Li, L.; Niu, Z.; Chen, J. Design Strategies toward Enhancing the Performance of Organic Electrode Materials in Metal-Ion Batteries. *Chemistry* **2018**, *4*, 2786–2813. [CrossRef]
6. Lu, Y.; Chen, J. Prospects of organic electrode materials for practical lithium batteries. *Nat. Rev. Chem.* **2020**, *4*, 127–142. [CrossRef] [PubMed]

7. Esser, B.; Dolhem, F.; Becuwe, M.; Poizot, P.; Vlad, A.; Brandell, D. A perspective on organic electrode materials and technologies for next generation batteries. *J. Power Sources* **2021**, *482*, 228814. [CrossRef]
8. Yan, L.; Zhao, C.; Sha, Y.; Li, Z.; Liu, T.; Ling, M.; Zhou, S.; Liang, C. Electrochemical redox behavior of organic quinone compounds in aqueous metal ion electrolytes. *Nano Energy* **2020**, *73*, 10476. [CrossRef]
9. Tobishima, S.; Yamaki, J.; Yamaji, A. Cathode Characteristics of Organic Electron. Acceptors for Lithium Batteries. *J. Electrochem. Soc.* **1984**, *131*, 57–63. [CrossRef]
10. Senoh, H.; Yao, M.; Sakaebe, H.; Yasuda, K.; Siroma, Z. A two-compartment cell for using soluble benzoquinone derivatives as active materials in lithium secondary batteries. *Electrochim. Acta* **2011**, *56*, 10145–10150. [CrossRef]
11. Genorio, B.; Pirnat, K.; Cerc-Korosec, R.; Dominko, R.; Gaberscek, M. Electroactive Organic Molecules Immobilized onto Solid Nanoparticles as a Cathode Material for Lithium-Ion Batteries. *Angew. Chem. Int. Ed.* **2010**, *49*, 7222–7224. [CrossRef]
12. Foos, J.S.; Erker, S.M.; Rembetsy, L.M. Synthesis and Characterization of Semiconductive Poly-l,4-Dirnethoxybenzene and Its Derived Polyquinone. *J. Electrochem. Soc.* **1986**, *133*, 836–840. [CrossRef]
13. Häringer, D.; Novák, P.; Haas, O.; Piro, B.; Pham, M.-C. Poly(5-amino-1,4-naphthoquinone), a Novel Lithium-Inserting Electroactive Polymer with High Specific Charge. *J. Electrochem. Soc.* **1999**, *146*, 2393–2396. [CrossRef]
14. Gall, T.L.; Reiman, K.H.; Grossel, M.C.; Owen, J.R. Poly(2,5-dihydroxy-1,4-benzoquinone-3,6-methylene): A new organic polymer as positive electrode material for rechargeable lithium batteries, Journal of Power Sources. *J. Power Sources* **2003**, *119–121*, 316–320. [CrossRef]
15. Son, E.J.; Kim, J.H.; Kim, K.; Park, C.B. Quinone and its derivatives for energy harvesting and storage materials. *J. Mat. Chem. A* **2016**, *4*, 11179–11202. [CrossRef]
16. Chambers, J.Q. *Quinonoid Compounds*, 1st ed.; John Wiley & Sons Ltd.: Hoboken, NJ, USA, 2010.
17. Mueller, T.; Kusne, A.G.; Ramprasad, R. Machine learning in materials science: Recent progress and emerging applications. In *Reviews in Computational Chemistry*; Parrill, A.L., Lipkowitz, B.K., Eds.; John Wiley & Sons, Inc.: Indianapolis, IN, USA, 2016; Volume 29, pp. 186–273.
18. Joshi, R.P.; Eickholt, L.; Li, L.; Fornari, M.; Barone, V.; Peralta, J.E. Machine Learning the Voltage of Electrode Materials in Metal-ion Batteries. *ACS Appl. Mater. Interfaces* **2019**, *11*, 18494–18503. [CrossRef] [PubMed]
19. Zhang, X.; Zhou, J.; Lu, J.; Shen, L. Interpretable learning of voltage for electrode design of multivalent metal-ion batteries. *NPJ Comput. Mater.* **2022**, *8*, 175. [CrossRef]
20. Allam, O.; Kuramshin, R.; Stoichev, Z.; Cho, B.W.; Lee, S.W.; Jang, S.S. Molecular structure–redox potential relationship for organic electrode materials: Density functional theory–Machine learning approach. *Mater. Today Energy* **2020**, *17*, 100482. [CrossRef]
21. Tuttle, M.R.; Brackman, E.M.; Sorourifar, F.; Paulson, J.; Zhang, S. Predicting the Solubility of Organic Energy Storage Materials Based on Functional Group Identity and Substitution Pattern. *J. Phys. Chem. Lett.* **2023**, *14*, 1318–1325. [CrossRef]
22. Wang, F.; Li, J.; Liu, Z.; Qiu, T.; Wu, J.; Lu, D. Computational design of quinone electrolytes for redox flow batteries using high-throughput machine learning and theoretical calculations. *Front. Chem. Eng.* **2022**, *4*, 1086412. [CrossRef]
23. Kim, S.; Chen, J.; Cheng, T.; Gindulyte, A.; He, J.; He, S.; Li, Q.; Shoemaker, B.A.; Thiessen, P.A.; Yu, B.; et al. PubChem 2023 update. *Nucleic Acids Res.* **2023**, *51*, D1373–D1380. [CrossRef]
24. Open Babel Development Team. Open Babel. 2016. Available online: http://openbabel.org/wiki/Main_Page (accessed on 12 October 2023).
25. Kichev, I.; Borislavov, L.; Tadjer, A. Automated generation of molecular derivatives—DerGen software package. *Mater. Today Proceedings* **2022**, *61*, 1287–1291. [CrossRef]
26. Nayak, S.K.; Ojha, A.C. Data Leakage Detection and Prevention: Review and Research Directions. In *Machine Learning and Information Processing. Advances in Intelligent Systems and Computing*; Swain, D., Pattnaik, P., Gupta, P., Eds.; Springer: Singapore, 2020; Volume 1101, pp. 203–212.
27. Wigh, D.S.; Goodman, J.M.; Lapki, A.A. A review of molecular representation in the age of machine learning. *WIREs Comput. Mol. Sci.* **2022**, *12*, e1603. [CrossRef]
28. Weininger, D. SMILES, a chemical language and information system. 1. Introduction to methodology and encoding rules. *J. Chem. Inf. Comput. Sci.* **1988**, *28*, 31–36. [CrossRef]
29. Dalby, A.; Nourse, J.G.; Hounshell, W.G.; Gushurst, A.K.I.; Grier, D.L.; Leland, B.A.; Laufer, J. Description of several chemical structure file formats used by computer programs developed at Molecular Design Limited. *J. Chem. Inf. Comput. Sci.* **1992**, *32*, 244–255. [CrossRef]
30. Todeschini, R. Consonni Handbook of Molecular Descriptors. In *Methods and Principles in Medicinal Chemistry*; WILEY-VCH Verlag GmbH: Weinheim, Germany, 2000.
31. Cereto-Massagué, A.; Ojeda, M.J.; Valls, C.; Mulero, M.; Garcia-Vallvé, S.; Pujadas, G. Molecular fingerprint similarity search in virtual screening. *Methods* **2015**, *71*, 58–63. [CrossRef] [PubMed]
32. Morgan, H.L. The Generation of a Unique Machine Description for Chemical Structures-A Technique Developed at Chemical Abstracts Service. *J. Chem. Doc.* **1965**, *5*, 107–113. [CrossRef]
33. Rogers, D.; Hahn, M. Extended-Connectivity Fingerprints. *J. Chem. Inf. Model.* **2010**, *50*, 742–754. [CrossRef]
34. Elton, D.C.; Boukouvalas, Z.; Fuge, M.D.; Chung, P.W. Deep learning for molecular design—A review of the state of the art. *Mol. Syst. Des. Eng.* **2019**, *4*, 828–849. [CrossRef]

35. Kuzminykh, D.; Polykovskiy, D.; Kadurin, A.; Zhebrak, A.; Baskov, I.; Nikolenko, S.; Shayakhmetov, R.; Zhavoronkov, A. 3D Molecular Representations Based on the Wave Transform for Convolutional Neural Networks. *Mol. Pharm.* **2018**, *15*, 4378–4385. [CrossRef] [PubMed]
36. Skalic, M.; Jiménez Luna, J.; Sabbadin, D.; De Fabritiis, G. Shape-Based Generative Modeling for de-novo Drug Design. *J. Chem. Inf. Model.* **2019**, *59*, 1205–1214. [CrossRef]
37. Yap, C.W. PaDEL-descriptor: An open source software to calculate molecular descriptors and fingerprints. *J. Comput. Chem.* **2011**, *32*, 1466–1474. [CrossRef]
38. James, J.P. *Stewart, Stewart Computational Chemistry*; MOPAC: Colorado Springs, CO, USA, 2016. Available online: http://OpenMOPAC.net (accessed on 12 October 2023).
39. Pudjihartono, N.; Fadason, T.; Kempa-Liehr, A.W.; O'Sullivan, J.M. A Review of Feature Selection Methods for Machine Learning-Based Disease Risk Prediction. *Front. Bioinform.* **2022**, *2*, 927312. [CrossRef] [PubMed]
40. Hocking, R.R. The Analysis and Selection of Variables in Linear Regression. *Biometrics* **1976**, *32*, 1–49. [CrossRef]
41. Burden, F.R. Molecular identification number for substructure searches. *J. Chem. Inf. Comput. Sci.* **1989**, *29*, 225–227. [CrossRef]
42. Sharma, V.; Goswami, R.; Madan, A.K. Eccentric Connectivity Index: A Novel Highly Discriminating Topological Descriptor for Structure-Property and Structure-Activity Studies. *J. Chem. Inf. Comput. Sci.* **1997**, *37*, 273–282. [CrossRef]
43. Hall, L.H.; Kier, L.B. Electrotopological state indices for atom types: A novel combination of electronic, topological, and valence state information. *J. Chem. Inf. Comput. Sci.* **1995**, *35*, 1039–1045. [CrossRef]
44. Gramatica, P.; Corradi, M.; Consonni, V. Modelling and prediction of soil sorption coefficients of non-ionic organic pesticides by molecular descriptors. *Chemosphere* **2000**, *41*, 763–777. [CrossRef] [PubMed]
45. Nilakantan, R.; Nunn, D.S.; Greenblatt, L.; Walker, G.; Haraki, K.; Mobilio, D. A family of ring system-based structural fragments for use in structure-activity studies: Database mining and recursive partitioning. *J. Chem. Inf. Model.* **2006**, *46*, 1069–1077. [CrossRef] [PubMed]
46. Todeschini, R.; Consonni, V. *Molecular Descriptors for Chemoinformatics*; WILEY-VCH Verlag GmbH: Weinheim, Germany, 2009; pp. 809–812.
47. Zhao, Y.H.; Abraham, M.H.; Zissimos, A.M. Fast Calculation of van der Waals Volume as a Sum of Atomic and Bond Contributions and Its Application to Drug Compounds. *JACS* **2003**, *68*, 7368–7373. [CrossRef]
48. Frisch, M.J.; Trucks, G.W.; Schlegel, H.B.; Scuseria, G.E.; Robb, M.A.; Cheeseman, J.R.; Scalmani, G.; Barone, V.; Petersson, G.A.; Nakatsuji, H.; et al. *Gaussian 16, Revision C.01*; Gaussian, Inc.: Wallingford, CT, USA, 2016.
49. Ochterski, J.W. *Thermochemistry in Gaussian*; Gaussian, Inc.: Wallingford, CT, USA, 2000.
50. Danchovski, Y.; Rasheev, H.; Stoyanova, R.; Tadjer, A. Molecular Engineering of Quinone-Based Nickel Complexes and Polymers for All-Organic Li-Ion Batteries. *Molecules* **2022**, *27*, 6805. [CrossRef] [PubMed]
51. Hoerl, A.E.; Kennard, R.W. Ridge Regression: Biased Estimation for Nonorthogonal Problems. *Technometrics* **1970**, *12*, 55–67. [CrossRef]
52. Breiman, L. *Classification and Regression Trees*, 1st ed.; CHAPMAN & HALL/CRC: Boca Raton, FL, USA, 1984.
53. Kotsiantis, S.B. Decision trees: A recent overview. *Artif. Intell. Rev.* **2013**, *39*, 261–283. [CrossRef]
54. Klekota, J.; Roth, F.P. Chemical substructures that enrich for biological activity. *Bioinformatics* **2008**, *24*, 2518–2525. [CrossRef]
55. Hou, T.; Wang, J.; Li, Y. ADME evaluation in drug discovery. The prediction of human intestinal absorption by a support vector machine. *J. Chem. Inf. Model.* **2007**, *47*, 2408–2415. [CrossRef] [PubMed]
56. Lamanna, C.; Bellini, M.; Padova, A.; Westerberg, G.; Maccari, L. Straightforward recursive partitioning model for discarding insoluble compounds in the drug discovery process. *J. Med. Chem.* **2008**, *51*, 2891–2897. [CrossRef]
57. Patel, H.H.; Prajapati, P. Study and Analysis of Decision Tree Based Classification Algorithms. *Int. J. Comput. Sci. Eng.* **2018**, *6*, 74–78. [CrossRef]
58. Ho, T.K. The random subspace method for constructing decision forests. *IEEE Trans. Pattern Anal. Mach. Intell.* **1998**, *20*, 832–844.
59. Geurts, P.; Ernst, D.; Wehenkel, L. Extremely randomized trees. *Mach. Learn.* **2006**, *63*, 3–42. [CrossRef]
60. Friedman, J.H. Greedy function approximation: A gradient boosting machine. *Ann. Stat.* **2001**, *29*, 1189–1232. [CrossRef]
61. Piryonesi, S.M.; El-Diraby, T.E. Data Analytics in Asset Management: Cost-Effective Prediction of the Pavement Condition Index. *J. Infrastruct. Syst.* **2020**, *26*, 04019036. [CrossRef]
62. Pedregosa, D.; Varoquaux, G.; Gramfort, A.; Michel, V.; Thirion, B.; Grisel, O.; Blondel, M.; Prettenhofer, P.; Weiss, R.; Dubourg, V.; et al. Scikit-learn: Machine Learning in Python. *J. Mach. Learn. Res.* **2011**, *12*, 2825–2830.
63. Bejani, M.M.; Ghatee, M. A systematic review on overfitting control in shallow and deep neural networks. *Artif. Intell. Rev.* **2021**, *54*, 6391–6438. [CrossRef]
64. Wright, S. Correlation and causation. *J. Agric. Res.* **1921**, *20*, 557–585.

materials MDPI

Article

Mg, Zn Substituted Calcium Phosphates—Thermodynamic Modeling, Biomimetic Synthesis in the Presence of Low-Weight Amino Acids and High Temperature Properties

Diana Rabadjieva [1,*], Rumiana Gergulova [1], Kostadinka Sezanova [1], Daniela Kovacheva [1] and Rositsa Titorenkova [2]

1 Institute of General and Inorganic Chemistry, Bulgarian Academy of Sciences, Acad. G. Bonchev Str., bl. 11, 1113 Sofia, Bulgaria; rumigg@yahoo.com (R.G.); ksezanova@abv.bg (K.S.); didka@svr.igic.bas.bg (D.K.)
2 Institute of Mineralogy and Crystallography, Bulgarian Academy of Sciences, Acad. G. Bonchev Str., bl. 107, 1113 Sofia, Bulgaria; rositsatitorenkova@imc.bas.bg
* Correspondence: didiarab@svr.igic.bas.bg

Abstract: The preparation of specially doped calcium phosphates (CaPs) is receiving a great deal of attention from researchers due to CaPs' enhanced capabilities for application in medicine. Complexation and precipitation in a complicated electrolyte system including simulated body fluids that are enriched with Mg^{2+} and Zn^{2+} ions and modified with glycine, alanine and valine were first evaluated using a thermodynamic equilibrium model. The influence of the type and concentration of amino acid on the incorporation degree of Mg and Zn into the solid phases was predicted. Experimental studies, designed on the basis of thermodynamic calculations, confirmed the predictions. Amorphous calcium phosphates double-doped with Mg and Zn were biomimetically precipitated and transformed into Mg, Zn-β—tricalcium phosphates (TCP) upon calcination. The Rietveld refinement confirmed that Mg^{2+} and Zn^{2+} substituted Ca^{2+} only at the octahedral sites of β-TCP, and in some cases, fully displacing the Ca^{2+} from them. The resulting Mg, Zn-β–TCP can serve as a reservoir for Mg and Zn ions when included in the formulation of a biomaterial for bone remodeling. The research conducted reveals the effect of combining mathematical models with experimental studies to pre-evaluate the influence of various additives in the design of materials with predetermined properties.

Keywords: calcium phosphates; thermodynamic modeling; biomimetic synthesis; amino acids; complexation reaction; Rietveld analysis

Citation: Rabadjieva, D.; Gergulova, R.; Sezanova, K.; Kovacheva, D.; Titorenkova, R. Mg, Zn Substituted Calcium Phosphates—Thermodynamic Modeling, Biomimetic Synthesis in the Presence of Low-Weight Amino Acids and High Temperature Properties. *Materials* **2023**, *16*, 6638. https://doi.org/10.3390/ma16206638

Academic Editor: Andreas Taubert

Received: 17 September 2023
Revised: 5 October 2023
Accepted: 9 October 2023
Published: 11 October 2023

1. Introduction

The biomimetic approach is used to obtain materials whose composition, structure and morphology resemble those of hard tissues. Simulation of the biomineralization process is most often reduced to the use of simulated body fluids (SBF) with a composition corresponding to the inorganic component of blood plasma [1,2]. Both the nucleation and growth of calcium phosphate crystals in natural biomineralization processes are controlled by organic molecules, mainly proteins and polysaccharides. Therefore, research on the synthesis of calcium phosphates in the presence of organic molecules is a promising trend for preparation of biomaterials applicable in dental and orthopedic medicine. Since amino acids (AA) are involved in the acidic domains of non-collagen proteins and are found in trace amounts as free amino acids in blood plasma, red blood cells and muscles, they are given special consideration [3]. The majority of studies on the subject focus on how amino acids affect the rate of conversion of primary precipitated amorphous calcium phosphate (ACP) into hydroxyapatite (HA) and the growth of the latter's crystals [4–12].

The synthesis of specifically doped calcium phosphates attracts considerable attention because of the changes in their properties achieved in this way. It was recently shown that Zn-doped hydroxyapatite possesses better anti-bacterial properties and osteoblastic-proliferation activities [13]; Zn- or Ga-doped calcium phosphate coatings decreased corrosiveness of metal implants and enhanced cell affinity [14]; Mg-doped biphasic calcium phosphate nanoparticles containing silver revealed excellent cytocompatibility with a culture of human bone marrow-derived mesenchymal stem cells [15], etc. Mg and Zn are preferred over all other substituents because they are crucial in the formation of the skeleton [16,17]. Their incorporation in a wide concentration range in the composition of calcium phosphate materials does not alter the biocompatibility of the latter [17,18].

There are no data on the effect of amino acids on the composition and morphology of ion-substituted calcium phosphates despite the fact that ion-substituted calcium orthophosphates build up the inorganic component of the hard tissues.

Mathematical models are an effective tool to predict ongoing processes under varying conditions. Their main advantage is that they take into account the simultaneous influence of multiple factors in the process of obtaining materials with predetermined properties and thus to facilitate experimental research or to optimize material applications [19,20]. The models based on the principle of the chemical thermodynamics are usually employed to determine the concentration range at which stable or metastable crystallization of solid phases occurs, and as a result, to predict system behavior with changes of the temperature, component concentrations, etc. [21,22].

Conducting systematic studies on the effect of type and concentration of AAs over a wide range would be a long and complex process. In our previous studies [23,24], we have shown the possibility of using a mathematical approach based on chemical thermodynamics to predict and elucidate the influence of the inorganic composition of the simulated body fluids on the formation and transformation of amorphous calcium phosphate and dicalcium phosphate dihydrate. Based on the results of the thermodynamic calculations, targeted experimental studies were carried out that confirmed the predictions made. An analogous approach was applied in the present work, in which the thermodynamic modeling was complicated by the presence of AAs. This affects the complexation reactions with the metal cations in SBF.

The aim of this work is to predict and experimentally confirm the synthesis of doubly-doped (Mg and Zn) calcium phosphates by thermodynamic modeling of the complexation and precipitation processes in simulated body fluid (SBF) solutions enriched with various low-weight amino acids. To this purpose, the systems MCl_2—AA—SBF– H_2O (M = Ca, Mg, Zn; AA = Gly, Ala, Val; SBF includes Na^+, K^+, Ca^{2+}, Mg^{2+}, HPO_4^{2-}, HCO_3^-, Cl^-, and SO_4^{2-} ions) were modeled. On this basis, experimental studies were carried out with SBFs containing three amino acids, glycine (Gly), alanine (Ala) and valine (Val), using two concentrations in diluted and concentrated ranges. The influence of amino acids on the composition and morphology of freshly precipitated amorphous calcium phosphate was monitored, and its phase transformations during high-temperature treatment (400–1000 °C) were followed by applying chemical, IR, TEM, BET, XRD analyses and Rietveld refinement.

2. Materials and Methods

2.1. *Thermodynamic Calculations*

The MCl_2—AA—SBF– H_2O systems (M = Ca, Mg, Zn; AA = Gly, Ala, Val; SBF includes Na^+, K^+, Ca^{2+}, Mg^{2+}, HPO_4^{2-}, HCO_3^-, Cl^-, SO_4^{2-}) were thermodynamically modeled and species distributions in the solutions together with saturation indices of possible solid phases were calculated.

The computer program PHREEQCI, v. 2.14.3 [25], which is based on an ion-association model, was used. The ion-association model is a chemical equilibrium model which takes into account weak interactions between ions with opposing charges to create ion pairs (associates) in solution or solid phases. Ion association and salt crystallization reactions are specified by a mass-action expression with suitable formation constants, in which the extended Debye–Hückel theory is used to compute the activity coefficients of all potential species.

An extended thermodynamic database with thermodynamic formation constants (logK) from NIST databases [26] was used in the calculations (Table S1, Supplementary Material). Three solubility products and ten complex formation constants were additionally included by us (Table 1).

Table 1. Thermodynamic formation constants included in the used database.

Reaction	log*K*	Source
$Mg_3(PO_4)_2{\cdot}8H_2O = 3Mg^{2+} + 2PO_4^{3-} + 8H_2O$	−25.20	[27]
$Mg_3(PO_4)_2.22H_2O = 3Mg^{2+} + 2PO_4^{3-} + 22H_2O$	−23.30	[27]
$KMgPO_4{\cdot}6H_2O = Mg^{2+} + K^+ + PO_4^{3-} + 6H_2O$	−10.62	[27]
$H^+ + Ala^- = H(Ala)$	9.72	[28]
$2H^+ + Ala^- = H_2(Ala)+$	12.05	[28]
$H^+ + Val^- = H(Val)$	9.54	[28]
$2H^+ + Val^- = H_2(Val)^+$	11.82	[28]
$Ca^{2+} + Ala^- = Ca(Ala)^+$	1.24	[28]
$Mg^{2+} + Ala^- = Mg(Ala)^+$	1.96	[28]
$Zn^{2+} + Ala^- = Zn(Ala)^+$	5.21	[28]
$Ca^{2+} + Val^- = Ca(Val)^+$	1.02	This study
$Mg^{2+} + Val^- = Mg(Val)^+$	1.72	This study
$Zn^{2+} + Val^- = Zn(Val)^+$	5.00	[28]

The logK values for alanine species $MAla^+$ (M = Ca, Mg, Zn) at zero ionic strength and for valine species $ZnVal^+$ at ionic strength of 0.01 (closest to 0) were taken from Sovago et al. [28]. The logK values for $CaVal^+$ and $MgVal^+$ were calculated in this study through translation of line AB (Figure 1) to the values of logK for $MgGly^+$—$MgAla^+$ (line CD) and $CaGly^+$—$CaAla^+$ (line EF), respectively. Our calculations considered only 1:1 complexes because experimental data indicated the formation of these complexes in aqueous solutions [29].

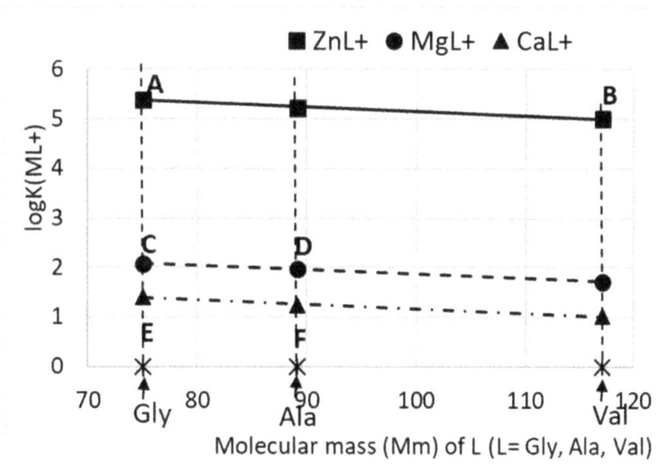

Figure 1. Thermodynamic formation constants (logK) of ML+ complexes. M = Mg, Ca, Zn; L = Gly, Ala, Val. The amino acids were presented with their molecular masses on the abscissa.

The concentrations of ions in Solution 2 and Solution 3 (Table 2), along with varying concentrations of amino acids, were used as input data for calculation of chemical species distribution. The solubility of AAs at 25 °C determines their concentrations range, namely: from 0 to 200 g L^{-1} for Gly (solubility at 25 °C 249.9 g L^{-1} [30]); from 0 to 100 g L^{-1}

for Ala (solubility at 25 °C 166.5 g L^{-1} [30]); from 0 to 50 g L^{-1} for Val (solubility at 25 °C 88.5 g L^{-1} [30])

Table 2. Composition of the conventional (SBFc), modified (solutions 1, 2 and 3) simulated body fluids and of the buffer solution, g L^{-1}.

Components	SBFc [31]	Solution 1	Solution 2	Solution 3	Buffer Solution
Na^+	3.26	3.26	3.26	3.26	2.30
K^+	0.12	19.8	0.12	0.12	
Mg^{2+}	0.04	0.04	1.41	0.04	
Ca^{2+}	0.10		14.7	0.10	
Zn^{2+}				0.86	
Cl^-	5.07	5.07	26.2	6.03	3.46
$SO_4{}^{2-}$	0.05	0.05	0.05	0.05	
$HCO_3{}^-$	0.26	0.78			
$HPO_4{}^{2-}$	0.10	24.1			
Gly/Ala/Val					
Series A		7.5/7.5/7.5	7.5/7.5/7.5	7.5/7.5/7.5	7.5/7.5/7.5
Series B		220/142/60	220/142/60	220/142/60	220/142/60
pH	7.2–7.4	8.0–8.2	8.0–8.2	6.5	8.0

Additionally, Solution 1 and the buffer solution (Table 2), as well as a solution of 0.05 M KOH, were included in the modeling, simulating the precipitation process (as it is described in Section 2.2.2) for calculation of saturation indices of possible solid phases.

The saturation indices (SI) (Equation (1)) were calculated as indicators for possible salts precipitation:

$$SI = \lg (IAP/K) \quad (1)$$

where IAP is an ion activity product, and K is a solubility product.

From 24 solid phases having SI > 0 in the examined systems, only 15 were included in the model, namely $CaCO_3$, $CaHPO_4{\cdot}2H_2O$, $Ca_3(PO_4)$ (amorphous), $Ca_8H_2(PO_4)_6{\cdot}5H_2O$, $Ca_9Mg(HPO_4)(PO_4)_6$, $Ca_5(PO_4)_3(OH)$, $KMgPO_4{\cdot}6H_2O$, $MgHPO_4{\cdot}3H_2O$, $Mg_3(PO_4)_2{\cdot}22H_2O$, $Mg_3(PO_4)_2$, $Mg_3(PO_4)_2{\cdot}8H_2O$, $Zn(OH)_2$, $ZnCO_3$, $ZnCO_3{\cdot}H_2O$ and $Zn_3(PO_4)_2{\cdot}4H_2O$. Phases with a negative SI or with a positive SI but no expected spontaneous crystallization were excluded from the model because they are high-temperature phases (β-$Ca_3(PO_4)_2$, $CaHPO_4$ and $Ca_4(PO_4)_2O$), long-term maturated phases ($CaMg(CO_3)_2$, $CaMg_3(CO_3)_4$, $Zn_2(OH)_3Cl$, $Zn_4(OH)_6SO_4$, $Zn_5(OH)_8Cl_2$), or the result of solution oxidation (ZnO). $Ca_9Mg(HPO_4)(PO_4)_6$ was used as an example of Mg-doped calcium phosphate for which the thermodynamic precipitation constant is known. Double-doped calcium phosphates were not included in the model because of a lack of data.

2.2. *Biomimetic Synthesis of (Mg, Zn)-Doped Calcium Phosphates*

2.2.1. Initial Solutions

The synthesis of calcium phosphates in this work involved modified simulated body fluids. They were all created using a conventional simulated body fluid (SBFc) [26] (Table 2). K_2HPO_4 (Merck, Darmstatd, Germany, A.R.) was dissolved in modified calcium-free conventional simulated body fluid (Solution 1, Table 2), whereas $CaCl_2{\cdot}2H_2O$ (Sigma-Aldrich, St. Louis, MO, USA, A.R.), $MgCl_2{\cdot}H_2O$ (Merck, Darmstatd, Germany, A.R.), and $ZnCl_2{\cdot}2H_2O$ (Merck, Darmstatd, Germany, A.R.) were dissolved in modified phosphorus-free conventional simulated body fluid (Solution 2 and Solution 3, Table 2). In this way, preliminary precipitation was avoided.

The modified simulated bodily fluids (Table 2) were all made by mixing salt solutions that had already been prepared, namely: 1.37 M NaCl (INEOS, A.R., London, UK), 0.04 M $NaHCO_3$ (SOLVAY, A.R., Brussels, Belgium), 0.03 M KCl (Merck, Darmstatd, Germany, A.R.), 0.01 M K_2HPO_4·(Merck, Darmstatd, Germany, A.R.), 0.015 M $MgCl_2{\cdot}6H_2O$,

0.025 M $CaCl_2{\cdot}2H_2O$, 0.005 M Na_2SO_4 (JLC-CHEMIE Hendel GmbH, Wohlen bei Bern, Switzerland, A.R.). The calculated amounts of $K_2HPO_4{\cdot}3H_2O$, $CaCl_2{\cdot}2H_2O$, $MgCl_2{\cdot}6H_2O$ and $ZnCl_2{\cdot}6H_2O$ and the corresponding amounts of Gly (Sigma-Aldrich, St. Louis, MO, USA, A.R.), Ala (Sigma-Aldrich, St. Louis, MO, USA, A.R.) or Val (Sigma-Aldrich, St. Louis, MO, USA, A.R.) were used to prepare calcium- and phosphorus-free simulated body fluids. The concentrations of the AAs were selected to represent diluted solutions (7.5 g L^{-1}) and concentrated solution regions (220 g L^{-1} Gly, 142 g L^{-1} Ala and 60 g L^{-1} Val) of AAs. In the second case, the aim was for the concentrations to be as close as possible to solubility at 25 °C but not to cause solid phase crystallization at room temperature (18–28 °C).

The pH of Solutions 1 and 2 was adjusted to 8.0–8.2 using 0.1 M HCl (Merck, Darmstadt, Germany, A.R.) or 0.05 M tris (hydroxymethyl) aminomethane (Sigma-Aldrich, St. Louis, MO, USA, A.R.). To prevent hydrolysis processes, the pH of Solution 3 was not adjusted.

2.2.2. Precipitation Method

Solution 1, Solution 2 and Solution 3 (Table 2) were appended to matching buffer solutions having Gly, Ala or Val, with a velocity of 3 Ml min^{-1}. pH of 8.0–8.2 was maintained with 0.05 M KOH. Combined apparatus for automatic titration and controlled synthesis (Titrando 907, Methrom AG) was used.

Two sets of experiments were carried out:

Series A—with concentration of amino acids Gly/Ala/Val of 7.5 g L^{-1}, whose products will be hereinafter referred to as CPGly7, CPAla7 and CPVal7, respectively. Series A represents diluted solutions of AA.

Series B—with concentrated solutions of amino acids, namely 220 g L^{-1} of Gly, 142 g L^{-1} of Ala and 60 g L^{-1} of Val, whose products will be hereinafter referred to as CPGly220, CPAla142 and CPVal60, respectively.

Concentrations of Mg^{2+}, Zn^{2+}, Ca^{2+} and $PO_4{}^{3-}$ ions were calculated to produce calcium phosphate precursors with ratios $Mg^{2+}/\Sigma(M^{2+})$ = 7 mol%, $Zn^{2+}/\Sigma(M^{2+})$ = 3 mol% and $\Sigma(M^{2+})/P$ = 1.67 (M = Ca, Mg, Zn), respectively. Then, the concentration of Mg^{2+} was doubled in order to provide the desired percentage of magnesium ions in the structure of the precursor [32]. The concentrations of Mg^{2+} and Zn^{2+} ions were selected on the basis of our previous studies on the biocompatibility of Mg- or Zn-modified tri-calcium phosphates [18].

The suspensions were left in the mother liquor for one hour while being constantly stirred at room temperature. They were then freeze-dried after being cleaned with water numerous times through decantation.

2.3. *Calcination of (Mg, Zn)-Doped Calcium Phosphates*

The dry precursors were calcined at 200, 400, 600, 800 and 1000 °C and atmospheric pressure. The powders were heated at a rate of 3 °C/min up to the desired temperature, which was kept constant for 3 h. A high-temperature furnace, type VP 04/17 of LAC Ltd. Company (Tokyo, Japan), was used.

2.4. *Characterization*

2.4.1. Chemical Analysis

Complexometric analysis using EDTA and the indicator eriochrome black T at pH 10 was used to determine the total amount of Ca^{2+}, Zn^{2+} and Mg^{2+} ions in the liquid and dissolved solid phases. The concentrations of Zn^{2+} and Mg^{2+} ions were determined by ICP-OES (PRODIGY 7, Teledyne, Leeman Labs, Hudson, NH, USA). Spectrophotometer NOVA 60 and Merck Spectroquant test kits were used for determination of the concentrations of $PO_4{}^{3-}$ and Cl^- ions. Each experiment is the result of three parallel samples. Analytical results are the mean values of three parallel measurements (n = 3) of each sample, i.e., RSD = 0.2–0.5% of complexomeric determination, RSD = 1–2% of spectrophotometric determination and RSD = 0.5–1% for ICP-OES measurements.

2.4.2. Infrared Spectroscopy (IR)

Infrared spectra in transmission mode were measured on standard KBr pellets on a FT-IR spectrometer (Tensor 37, Bruker, Karlsruhe, Germany). The spectrometer was a grant to the Institute of mineralogy and crystallography, BAS from the Alexander von Humboldt Foundation (Berlin, Germany) in 2005.The operating parameters were a spectral range of 400–4000 cm^{-1} and a spectral resolution of 4 cm^{-1}. The spectra were evaluated using OPUS 6.5; they were smoothed at 5 points, normalized to the matching maximum intensity, and baseline-corrected using rubber band correction.

2.4.3. X-ray Diffraction Analysis (XRD)

A Bruker D8 Advance diffractometer (Bruker AXS Advanced X-ray Solutions GmbH, Karlsruhe, Germany) was used to perform powder XRD. The X-ray source was a Cu tube (λ = 1.5418 Å). The pattern record was made by a LynxEye detector (Bruker AXS Advanced X-ray Solutions GmbH, Karlsruhe, Germany). The data were gathered in the 10 to 90° 2θ range with a step of 0.03° 2θ and a counting rate of 57 s/step for the phase identification. With the aid of the ICDD-PDF2 (2014) database and Diffracplus EVA software(v. 4, 2014), the phase composition was determined. In the examined samples, whitlockite ($Ca_3(PO_4)_2$—PDF # 09-0169), $Ca_{19}Zn_2(PO_4)_{14}$—PDF # 48-1196, and $Ca_{2.81}Mg_{0.19}(PO_4)_2$—PDF # 70-0682 were the principal calcium phosphate phases.

For the purpose of Rietveld structure refinement, the collection of diffraction patterns has the following parameters: room temperature; 2θ range—5 to 120° 2θ; 2θ step −0.02° 2θ; counting time per step—175 s, sample rotation—15 rpm. The Bruker Topas v. 4.2 program was used to carry out the Rietveld refinement [33]. Yashima et al. [34] crystal structure was used as the initial model for the refining.

The parameters varied in the calculations were as follows: (i) unit cell parameters were refined at zero shift, scale factor; (ii) a sixth order Chebyshev polynomial function was used to refine the background; (iii) occupancies of the Ca(1), Ca(2) and Ca(3) positions were kept fixed because their variation led to values within the experimental error; (iv) the occupancies of the Ca(4) and Ca(5) positions were varied, with the sum of the Ca, Mg and Zn ion occupancy values constrained to 0.5 for the Ca(4) position and to 1 for the Ca(5) position; (v) $Ca_2P_2O_7$ was also included in the refinement.

2.4.4. Transition Electron Microscopy (TEM) Analysis

Transmission electron microscope, apparatus JEOL TEM JEM-2100, Tokyo, Japan was used to analyze the morphology of the acquired precursors. Samples were prepared by dispersing the powders in water and sonicating for 1 min. The suspensions were dropped on standard carbon-copper grids.

2.4.5. Measurement of Specific Surface Area

Low-temperature (77.4 K) N_2 adsorption was applied to obtain the specific surface area (SSA) using a Quantachrome Instruments NOVA 1200e (USA). The Brunauer–Emmett–Teller (BET) theory was applied by the calculations.

3. Results

3.1. Thermodynamic Modeling

Thermodynamic modeling of chemical species distribution (Figure 2) in the studied system shows that only three metal species dominate in all solutions at different concentrations: free M^{2+} ions, MCl^+ and ML^+ species. For low concentrations of Ala and Val, there are also $ZnCl_2^0$ species in amounts of up to 1–2%.

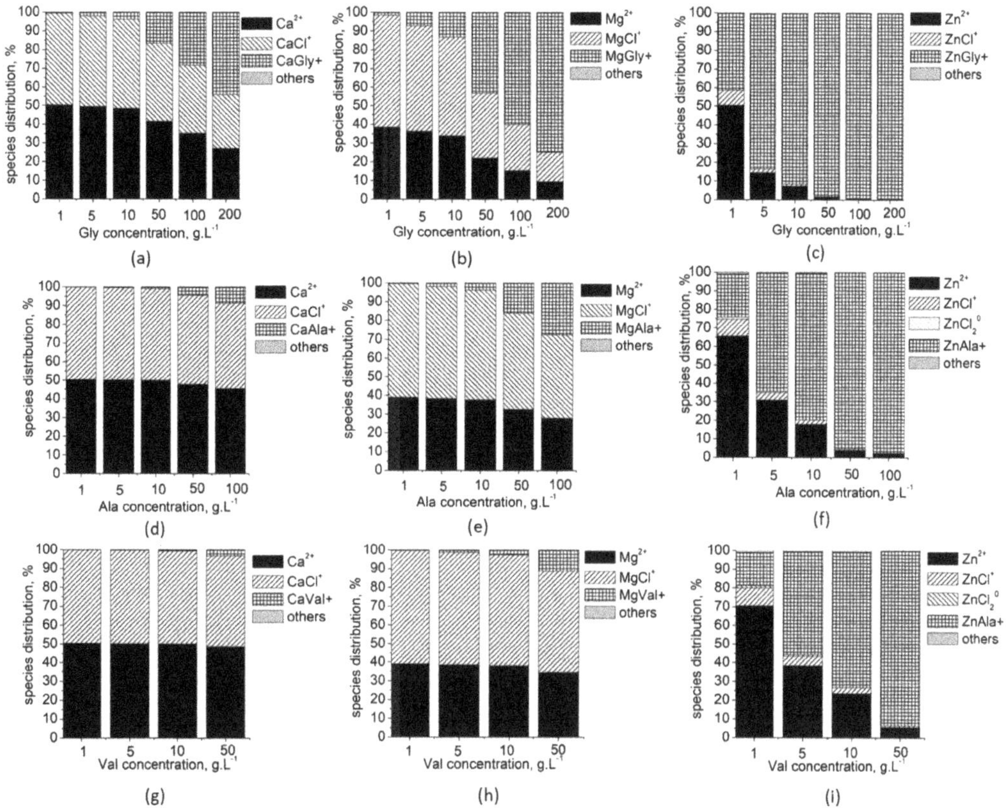

Figure 2. Species distribution of: Ca^{2+} ions in solution with varying concentration of Gly (**a**), Ala (**d**) and Val (**g**); Mg^{2+} ions in solution with varying concentration of Gly (**b**), Ala (**e**) and Val (**h**); and Zn^{2+} ions in solution with varying concentration of Gly (**c**), Ala (**f**) and Val (**i**).

The calculated values of the saturation indices of the solid phases (Figure 3) show that, for the entire concentration range of the three amino acids, they are highest in the calcium phosphate phases (Figure 3a), and lowest in the zinc compounds (Figure 3d), of which only $Zn_3(PO_4)_2{\cdot}4H_2O$ (SI > 0) is expected to precipitate.

The magnesium solid phases take an intermediate position, and only $Mg_3(PO_4)_2{\cdot}8H_2O$ has a positive SI (Figure 3c). $KMgPO_4{\cdot}6H_2O$ (SI < 0) does not co-precipitate in the concentrated solutions of the amino acids, while $Mg_3(PO_4)_2$, $Mg_3(PO_4)_2{\cdot}3H_2O$ and $Mg_3(PO_4)_2{\cdot}22H_2O$ co-precipitate in concentrated solutions of Val and Ala, but not of Gly.

The solutions remain supersaturated (SI > 0) with respect to all phases of calcium phosphate, including $CaCO_3$, in the entire concentration range (Figure 3a,b), which means that, from a thermodynamic point of view, their precipitation is possible in all investigated solutions.

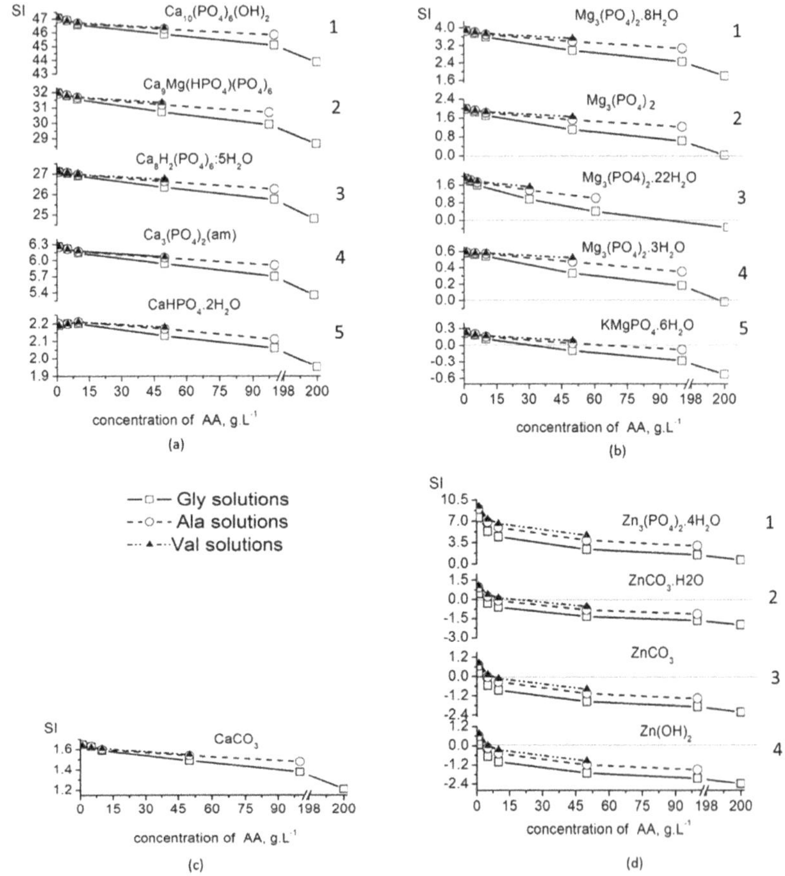

Figure 3. Changes of the saturation indices (SI) of possible-to-crystallize solid phases with increased the amino acid concentrations: (**a**) calcium salts 1—$Ca_{10}(PO_4)_6(OH)_2$; 2—$Ca_9Mg(HPO_4)(PO_4)_6$; 3—$Ca_8H_2(PO_4)_6{\cdot}5H_2O$; 4—$Ca_3(PO_4)_2$(am); 5—$CaHPO_4{\cdot}2H_2O$; (**b**) magnesium salts: 1—$Mg_3(PO_4)_2{\cdot}8H_2O$; 2—$Mg_3(PO_4)_2$; 4—$Mg_3(PO_4)_2{\cdot}3H_2O$; 5—$KMgPO_4{\cdot}6H_2O$ (**c**) $CaCO_3$; and (**d**) zinc salts: 1—$Zn_3(PO_4)_2{\cdot}4H_2O$; 2—$ZnCO_3{\cdot}2H_2O$; 3—$ZnCO_3$; 4—$Zn(OH)_2$. A straight line at SI = 0, separating the diluted from saturated solutions, is added where possible.

3.2. Biomimetic Synthesis

Based on the conclusions drawn from the thermodynamic modeling, two sets of experiments were carried out with different concentrations of amino acids: Series A, with concentration of Gly/Ala/Val of 7.5 g L^{-1}, and Series B, with concentrated solutions of amino acids, namely 220 g L^{-1} of Gly, 142 g L^{-1} of Ala and 60 g L^{-1} of Val.

The results from chemical analysis are shown in Table 3 and reveal that substances with (Ca^{2+} + Mg^{2+} + Zn^{2+})/P ratio between 1.47 and 1.56 and content of Mg^{2+} ions between 5.86 and 8.09 m0l% and Zn^{2+} ions between 0.71 and 2.81 mol% were observed.

Table 3. Composition of the obtained calcium phosphates.

Sample	C_{Mg}	C_{Zn}	$(Ca^{2+} + Mg^{2+} + Zn^{2+})/P$
	mol%		Molar Ratio
	Series A		
CPGly7	7.63	2.61	1.54
CPAla7	7.73	2.72	1.56
CPVal7	7.91	2.82	1.56
	Series B		
CPGly220	5.86	0.71	1.47
CPAla142	6.37	1.73	1.54
CPVal60	8.09	2.63	1.56

Regardless of the type and concentration of amino acids used, amorphous calcium phosphates were precipitated (Figure 4).

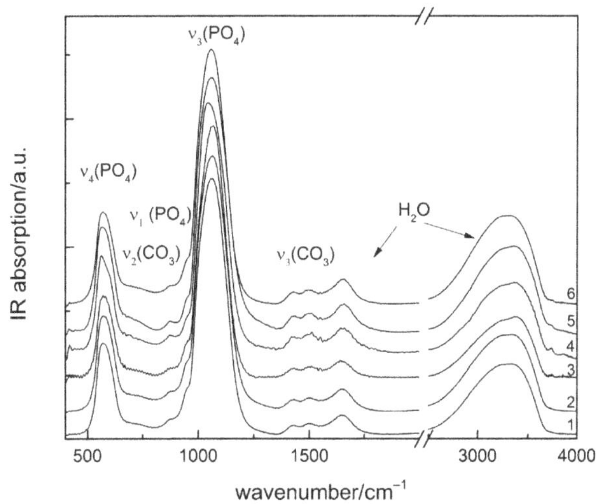

Figure 4. IR spectra of as precipitated calcium phosphates: 1—CPGly7; 2—CPAla7; 3—CPVal7; 4—CPGly220; 5—CPAla142; 6—CPVal60.

All the IR spectra are typical for amorphous calcium phosphate (ACP), with a broad absorption band of phosphate antisymmetric stretching centered near 1060 cm^{-1} and other phosphate peaks at 575 (ν_4) and 950 (ν_1) cm^{-1}. Peaks at 1650 and around 3300 cm^{-1} are due to water molecules. Other weak absorption peaks at 1425, 1495 and 870 cm^{-1} are in the range of carbonate stretching and bending vibrations, respectively.

TEM images (Figure 5) show, the particles of samples from Series A are spherical, with a diameter of about 50 nm in the case of Gly and Ala as additives and about 30 nm in the case of Val (Figure 5a–c). The shape of the particles in Series B that have Val as an additive is preserved, but the diameter increases to 100 nm (Figure 5f). In the case of Gly and Ala, the particles lose their spherical shape and elongate along one axis (Figure 5d,e).

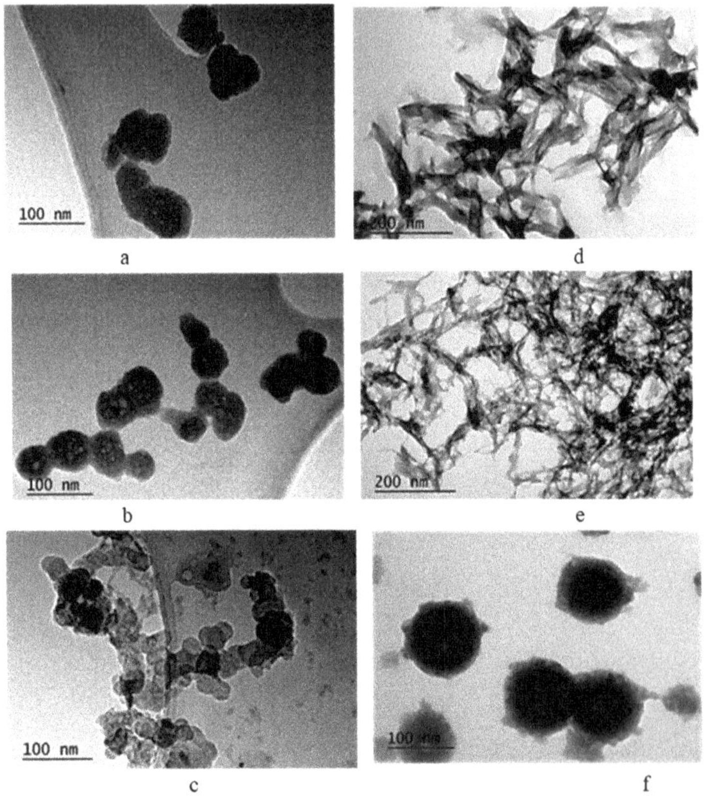

Figure 5. TEM images of CPGly7 (**a**), CPAla7 (**b**), CPVal7 (**c**), CPGly220 (**d**), CPAla142 (**e**) and CPVal60 (**f**).

The radical change in particle shape leads to an eight-fold increase in specific surface area (Table 4).

Table 4. Specific surface area (SSA), $m^2\ g^{-1}$.

Series A		Series B	
CPGly7	39	CPGly220	242
CPAla7	37	CPAla142	222
CPVal7	49	CPVal60	48

3.3. High Temperature Properties

The high-temperature phase transformations of the obtained calcium phosphate precursors were monitored by step-wise heating at 400, 600, 800 and 1000 °C. The IR and XRD analyses of the calcined samples revealed the transformation of the ion-modified amorphous calcium phosphate into (Mg, Zn)-containing β-tricalcium phosphate ((Mg, Zn)-β-TCP). Since all diffraction patterns and IR spectra were of the same type, only the XRD pattern of CPVal7 (Figure 6a) and the IR spectrum of CPGly220 (Figure 6b) are shown as examples.

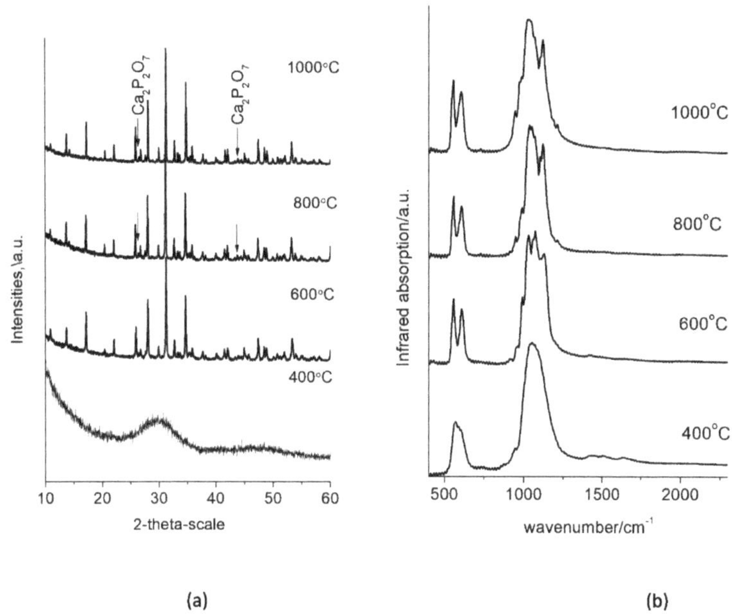

Figure 6. XRD powder patterns and IR spectra of selected calcium phosphate precursors calcined at different temperatures: (**a**) XRD powder patterns of CPVal7 (unmarked peaks—Mg, Zn-β-TCP) and (**b**) IR spectra of CPGly220.

At 400 °C all samples were still amorphous, but after heating at 600 °C for 3 h, crystalline (Mg, Zn)-β-TCP was obtained. At 800 and 1000 °C, the phase $Ca_2P_2O_7$ appeared (Figure 6a). In the IR spectra, it was connected with the peaks at 735 and 1200 cm^{-1} characteristic for a P_2O_7 group (Figure 6b). The formation of $Ca_2P_2O_7$ is a result of the thermal decomposition of the HPO_4^- ions present in the hydrate shell of the amorphous particles.

4. Discussion

A useful tool for calculating the species distribution of the elements in the solution and the type and concentration interval of stable or metastable crystallization of solid phases is thermodynamic modeling, which is based on quantitative relationships between the concentration of the solutions and thermodynamic parameters, such as chemical potential, solubility product, equilibrium complexation constants, etc. In this work, a thermodynamic modeling approach for the systems MCl_2—AA—SBF– H_2O (M = Ca, Mg, Zn; AA = Gly, Ala, Val; SBF includes Na^+, K^+, Ca^{2+}, Mg^{2+}, HPO_4^{2-}, HCO_3^-, Cl^-, and SO_4^{2-} ions) is applied in order to predict incorporation of Mg^{2+} and Zn^{2+} ions in the solids depending on the type and concentration of the AAs used. On this basis, only two series of experimental studies were carried out, with SBFs enriched with Mg^{2+} and Zn^{2+} ions and containing three amino acids, glycine (Gly), alanine (Ala) and valine (Val), and using two concentrations in the diluted (7.5 g L^{-1}) and concentrated (200 g L^{-1} Gly, 100 g L^{-1} Ala and 50 g L^{-1} Val) ranges. The differences in concentrations of the concentrated solutions are due to different solubility of the amino acids used.

For this purpose, the unknown stability constants of MgVal+ and CaVal+ were first determined in this study by approximating the known ones (Figure 1). The obtained values were included in database, and the calculations of Ca, Mg and Zn specie distribution (Figure 2) show that complexation in the solutions increases in the order Ca < Mg < Zn and Val < Ala < Gly, as well as with increasing amino acid concentration. As a result, the amounts of free ions in the solution decrease, which reduces the possibility that precipitation processes will take place and that Mg^{2+} and Zn^{2+} ions will be included in the composition of the solid phase. Thus, it is to be expected that, in concentrated solutions of Gly, the inclusion of Mg^{2+} and Zn^{2+} ions in the precipitates would be minimal and would increase in the concentrated solutions of Ala and Val. The calculated SI (Figure 3) show that, in dilute solutions of AA (up to 10 g L^{-1}), the values of SI are close for all calcium, magnesium and zinc phases, which means that no effect of the type of amino acid on the degree of incorporation of Mg^{2+} and Zn^{2+} ions into the precipitate is to be expected in this case.

Experimental studies were designed on the basis of thermodynamic studies, and the results from the chemical analysis (Table 3) prove the predictions made. In the case of Series A, the concentrations of Mg^{2+} and Zn2+ ions for the three amino acids were similar. A slight trend to increase the content of Mg^{2+} and Zn^{2+} ions was observed in the sequence Gly $\rightarrow$ Ala $\rightarrow$ Val. This trend was more pronounced when concentrated acid solutions (Series B) were used (Table 3). The lowest concentrations of Mg^{2+} and Zn^{2+} ions were determined in the solid phase precipitated in a concentrated solution of Gly, and the highest concentrations were in valine.

The precipitation of amorphous calcium phosphates is in accordance with the dependence we previously discovered [23,24], which states that kinetic conditions, as opposed to thermodynamic ones, control the precipitation of ACP instead of the thermodynamically stable HA. During synthesis, upon addition of a solution containing ${HPO_4}^{2-}$ ions (Solution 1, Table 2), the latter will bind to the free Ca^{2+} ions, forming so-called Posner clusters at pH 8. The latter are the structural units that construct ACP [35]. The highest SI and thermodynamic stability are displayed by hydroxyapatite ($Ca_{10}(PO_4)_6(OH)_2$) followed by whitlockite ($Ca_9Mg(HPO_4)(PO_4)_6$). However, it is to be expected that the salt, whose structural units are already formed in the solution, will be the first to precipitate. Thus, quick kinetics and high pH favor the formation of ACP. As the ionic radii and electrical charge of Mg^{2+} and Zn^{2+} ions are close to those of Ca^{2+} ions, they are easily incorporated in the calcium phosphate structure, especially in dilute solutions where they exist as free ions in relatively high concentrations.

The concentration of amino acid influences the size and shape of primary precipitated particles and thus affects their specific surface area (Figure 5 and Table 4). Transformation from spherical to needle-shaped particles is connected with phase transformation from amorphous to poorly crystalline hydroxyapatite [6,7,9]. In our study, we did not obtain poorly crystalline hydroxyapatite. The samples remained X-ray amorphous even after calcination at 400 °C (Figure 6a,b). This is due to the presence of Mg^{2+} and Zn^{2+} ions that stabilize the amorphous phase and delay transformation to hydroxyapatite. We suppose that organic molecules with a linear carbon chain (Gly and Ala) promote the elongation of the spherical particles (Figure 5d,e), while a branched carbon chain (Val) stimulates their enlargement (Figure 5f).

High temperature treatment reveal formation of double-doped (Mg, Zn)-β-TCP (Figure 6). The Rietveld refinement was used in this study following the cation distribution in the structure of the double-substituted phases (calcined at 1000 °C).

The substitution of Mg^{2+} and Zn^{2+} for Ca^{2+} in the β-TCP structure was confirmed by the decrease in unit cell parameters in comparison with the unsubstituted β-TCP (Table 5). The decrease in the unit cell parameters a and c and in the volume V is proportional to the increase in the sum of concentrations of the substituting ions.

Comparison of the calculated average distances around the five Ca positions shows that distances around Ca(1) and Ca(2) positions remain almost unchanged for all substituted samples obtained at different concentrations of amino acids as compared to unsubstituted β-TCP (Figure 7). A slight decrease in the average distance is observed for Ca(3), while distances around Ca(4) and especially around Ca(5) drastically decrease in comparison with the values for the unsubstituted samples.

Table 5. Unit cell parameters, mean coherent domain size (mean crystallite size), unit cell volume and % of impurity phase for samples of β-TCP synthesized from different amino acid solutions.

Sample	$C_{Mg} + C_{Zn}$ mol,%	*a* [Å]	*c* [Å]	Mean Size [nm]	V, [Å^3]	$Ca_2P_2O_7$ wt%
β—TCP [34]		10.4352 (2)	37.4029 (5)		3482	
			Series A			
CPGly7	10.3	10.3240 (1)	37.2646 (3)	287 (1)	3439	4.34
CPAla7	10.5	10.3234 (1)	37.2623 (6)	280 (2)	3439	3.58
CPVal7	10.7	10.3211 (2)	37.3006 (8)	322 (1)	3441	5.23
			Series B			
CPGly220	6.57	10.3677 (1)	37.2023 (3)	372 (2)	3463	3.87
CPAla142	8.10	10.3503 (1)	37.1326 (3)	309 (5)	3445	3.37
CPVal60	10.7	10.3238 (1)	37.2624 (6)	276 (2)	3439	1.44

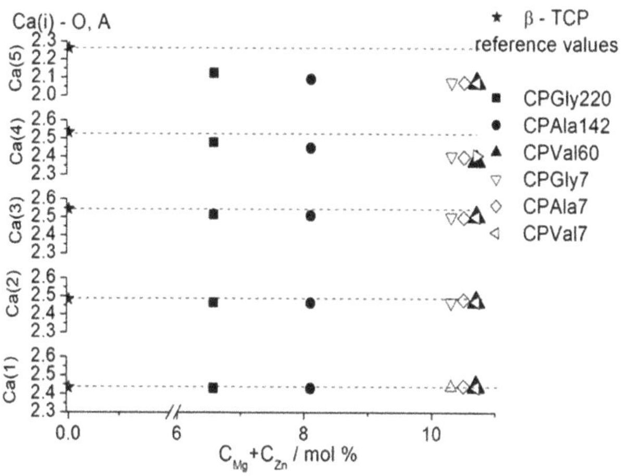

Figure 7. Dependence of the Ca(i)—O (i = 1–5) average distances on the sum of concentrations of substituting ions. Reference data are from Yashima et al. [34].

The calculated occupancy of Ca(4) and Ca(5) positions (Table 6) shows that both Mg and Zn ions have a tendency to predominantly occupy octahedral Ca(5) positions, in some cases fully displacing the Ca ions (sample CPVal7). Ca(4) position seems to be less preferable for small ions, and Ca^{2+} ions in that position are substituted by smaller amounts of Mg^{2+} and Zn^{2+} ions.

Table 6. Occupancies of Ca(4) and Ca(5) in Mg, Zn substituted β-TCP.

Samples	Ca(4) Position			Ca(5) Position		
	Ca	Mg	Zn	Ca	Mg	Zn
			Series A			
CPGly7	0.45	0.05	0.00	0.30	0.66	0.05
CPAla7	0.29	0.07	0.14	0.08	0.76	0.15
CPVal7	0.33	0.08	0.09	0.00	0.86	0.14
			Series B			
CPGly220	0.37	0.10	0.02	0.24	0.62	0.13
CPAla142	0.33	0.13	0.04	0.16	0.65	0.19
CPVal60	0.43	0.07	0.00	0.06	0.76	0.18

5. Conclusions

Mathematical models and experimental studies were combined in this research to pre-evaluate the impact of different additives on the design of materials applicable in medicine with predetermined attributes. Based on thermodynamic modeling of the ongoing processes in the systems MCl_2—AA—SBF—H_2O (M = Ca, Mg, Zn; AA = Gly, Ala, Val; SBF includes Na^+, K^+, Ca^{2+}, Mg^{2+}, HPO_4^{2-}, HCO_3^-, Cl^-, SO_4^{2-}), the degree of inclusion of Mg and Zn in the probable solid phases was predicted depending on the amount of amino acids added. It was calculated that dilute amino acid solutions did not affect the incorporation of Mg and Zn into the solid phases, while concentrated ones did, i.e., the lowest concentrations were predicted in the presence of glycine, but the highest were predicted in valine. To fulfill the calculations, the unknown stability constants of MgVal+ and CaVal+ were first determined in this study by approximating the known ones.

Experimental studies are focused in only the two concentration regions of amino acids that give the most promising results, namely: 7.5 g L^{-1}, representing the range of dilute solutions, and 60 g L^{-1} of Val, 142 g L^{-1} of Ala and 220 g L^{-1} of Gly, closed to saturation at 25 °C solution for each amino acid. The experiments verify the thermodynamic calculations. The content of Mg^{2+} and Zn^{2+} ions in the precipitated amorphous phase varies within narrow limits when the synthesis proceeds in dilute solution of amino acids. A strong tendency toward an increase in the content of Mg^{2+} and Zn^{2+} ions in the sequence Gly → Ala → Val was observed in concentrated solutions of the latter.

Mg^{2+} and Zn^{2+} ions included in the composition of the precipitates stabilize the amorphous phase up to 400 °C, after which they stimulate its transformation into Mg, Zn substituted β-TCP without intermediate phases. Both Mg^{2+} and Zn^{2+} ions preferentially substituted Ca^{2+} ions at the octahedral sites of (Mg, Zn)-β-TCP, and in some cases, fully displaced the Ca^{2+} ions.

The concentration of Mg and Zn in human bones vary up to 2.5 mol% Mg and up to 0.16 mol% Zn [36] which is lower than quantitative ranges of Mg (5.86–8.09 mol%) and Zn (0.71–2.82 mol%) that were prepared in ceramic powders during this study. Used in the composition of a biomaterial for bone remodeling, they can act as a reservoir for Mg and Zn ions to maintain their physiological extracellular concentrations. Our earlier research on the behavior of calcium phosphates substituted with Mg-only or Zn-only demonstrated that they did not alter the materials' biocompatibility over a broad concentration range [18], and that when they came into contact with various simulated body fluids, the ions were gradually released into the solutions [37,38].

Supplementary Materials: The following supporting information can be downloaded at: https://www.mdpi.com/article/10.3390/ma16206638/s1, Table S1. Thermodynamic formation constants of the aqueous species used in calculation [39–41]. Table S2. Thermodynamic formation constants of the solids used in calculation [39,40,42–47].

Author Contributions: Conceptualization, D.R.; Methodology, D.R. and K.S.; Data curation, R.G., K.S., D.K. and R.T.; Formal analysis, D.K.; Investigation, R.G. and R.T.; writing—original draft preparation, D.R., K.S., D.K. and R.T.; writing—review and editing, D.R., K.S., D.K. and R.T.; visualization, D.R., K.S. and R.T. All authors have read and agreed to the published version of the manuscript.

Funding: This research was funded by the BULGARIAN NATIONAL SCIENCE FUND (BNSF) grant numbers KP-06-H49/6 (2020–2023) and D01-272.

Institutional Review Board Statement: Not applicable.

Informed Consent Statement: Not applicable.

Data Availability Statement: The data presented in this study are available in this article and in Supplementary Materials.

Acknowledgments: The authors thank the BULGARIAN NATIONAL SCIENCE FUND (NSF) for the financial support under the project No KP-06-H49/6 (2020–2023). The authors acknowledge the TwinTeam project D01-272 "European Network on Materials for Clean Technologies" for providing the opportunity to present the results at the SizeMat4 conference as well as for the financial publication support.

Conflicts of Interest: The authors declare no conflict of interest.

References

1. Chen, J.; Zhou, Y.; Lin, X.; Li, H. Macrophage Polarization Related to Biomimetic Calcium Phosphate Coatings: A Preliminary Study. *Materials* **2023**, *16*, 332. [CrossRef] [PubMed]
2. Kulhánková, J.; Rohanová, D.; Horkavcová, D.; Bezdička, P.; Boccaccini, A.R. The role of $(HCO_3)^-$ ions in SBF on the interaction with bioactive glass-ceramic scaffold. *Mater. Today Chem.* **2023**, *28*, 101367. [CrossRef]
3. Canepa, A.; Filho, J.C.; Gutierrez, A.; Carrea, A.; Forsberg, A.M.; Nilsson, E.; Verrina, E.; Perfumo, F.; Bergstrom, J. Free amino acids in plasma, red blood cells, polymorphonuclear leukocytes, and muscle in normal and uraemic children. *Nephrol. Dial. Transplant.* **2002**, *17*, 413–421. [CrossRef] [PubMed]
4. Rautaray, D.; Mandal, S.; Sastry, M. Synthesis of Hydroxyapatite Crystals Using Amino Acid-Capped Gold Nanoparticles as a Scaffold. *Langmuir* **2005**, *21*, 5185–5191. [CrossRef] [PubMed]
5. Jack, K.S.; Vizcarra, T.G.; Trau, M. Characterization and Surface Properties of Amino-Acid-Modified Carbonate-Containing Hydroxyapatite Particles. *Langmuir* **2007**, *23*, 12233–12242. [CrossRef]
6. Palazzo, B.; Walsh, D.; Iafisco, M.; Foresti, E.; Bertinetti, L.; Martra, G.; Bianchi, C.L.; Cappelletti, G.; Roveri, N. Amino acid synergetic effect on structure, morphology and surface properties of biomimetic apatite nanocrystals. *Acta Biomater.* **2009**, *5*, 1241–1252. [CrossRef]
7. Tavafoghi, M.; Cerruti, M. The role of amino acids in hydroxyapatite mineralization. *J. R. Soc. Interface* **2016**, *13*, 20160462. [CrossRef]
8. Tavafoghi, M.; Yao, G.; Cerruti, M. The importance of amino acid interactions in the crystallization of hydroxyapatite. *J. R. Soc. Interface* **2013**, *10*, 20120906. [CrossRef]
9. Wang, Z.; Xu, Z.; Zhao, W.; Sahai, N. A potential mechanism for amino acid-controlled crystal growth of hydroxyapatite. *J. Mater. Chem. B* **2015**, *3*, 9157–9167. [CrossRef]
10. Wang, H.; Xiao, Z.; Yang, J.; Lu, D.; Kishen, A.; Li, Y.; Chen, Z.; Zhang, Q.; Deng, X.; Yang, X.; et al. Oriented and Ordered Biomimetic Remineralization of the Surface of Demineralized Dental Enamel Using HAP@ACP Nanoparticles Guided by Glycine. *Sci. Rep.* **2017**, *7*, 40701. [CrossRef]
11. Tao, J.; Pan, H.; Zeng, Y.; Xu, X.; Tang, R. Roles of Amorphous Calcium Phosphate and Biological Additives in the Assembly of Hydroxyapatite Nanoparticles. *J. Phys. Chem. B* **2007**, *111*, 13410–13418. [CrossRef] [PubMed]
12. Gerk, S.A.; Golovanova, O.A.; Odazhiu, V.N. Structural, Morphological, and Resorption Properties of Carbonate Hydroxyapatite Prepared in the Presence of Glycine. *Inorg. Mater.* **2018**, *54*, 305–314. [CrossRef]
13. Okada, M.; Oshita, M.; Kataoka, M.; Azuma, Y.; Furuzono, T. Shareability of antibacterial and osteoblastic-proliferation activities of zinc-doped hydroxyapatite nanoparticles in vitro. *J. Biomed. Mater. Res. B Appl.* **2022**, *110*, 799–805. [CrossRef]
14. Tamay, D.G.; Gokyer, S.; Schmidt, J.; Vladescu, A.; Huri, P.Y.; Hasirci, V.; Hasirci, N. Corrosion Resistance and Cytocompatibility of Magnesium–Calcium Alloys Modified with Zinc- or Gallium-Doped Calcium Phosphate Coatings. *ACS Appl. Mater. Interfaces* **2022**, *14*, 104–122. [CrossRef] [PubMed]
15. Yang, N.; Wang, S.; Ding, P.; Sun, S.; Wei, Q.; Jafari, H.; Wang, L.; Han, Y.; Valentine, O.O.; Wang, T.; et al. Magnesium-doped biphasic calcium phosphate nanoparticles with incorporation of silver: Synthesis, cytotoxic and antibacterial properties. *Mater. Lett.* **2022**, *322*, 132478. [CrossRef]
16. Boanini, E.; Gazzano, M.; Bigi, A. Ionic substitutions in calcium phosphates synthesized at low temperature. *Acta Biomater.* **2010**, *6*, 1882–1894. [CrossRef]
17. Ballouze, R.; Marahat, M.H.; Mohamad, S.; Saidin, N.A.; Kasim, S.R.; Ooi, J.P. Biocompatible magnesium-doped biphasic calcium phosphate for bone regeneration. *J. Biomed. Mater. Res. B Appl. Biomater.* **2021**, *109*, 1426–1435. [CrossRef]
18. Andonova-Lilova, B.; Alexandrova, R.; Rabadjieva, D.; Tepavitcharova, S. Application of cultured murine cells for initial evaluation of the biocompatability of Mg and Zn-modified tri-calcium phosphates. *Compt. Rend. Acad. Bulg. Sci.* **2012**, *65*, 1099–1104.
19. Fu, T.; Hu, X.; Yang, C. Impact response analysis of stiffened sandwich functionally graded porous materials doubly-curved shell with re-entrant honeycomb auxetic core. *Appl. Math. Model.* **2023**, *124*, 553–575. [CrossRef]
20. Yu, Y.; Fu, T. Dynamic response of the hybrid honeycomb sandwich panel with zero Poisson's ratio under low-speed impact. *Mech. Adv. Mater. Struct.* **2023**, 1–15. [CrossRef]
21. Hidayat, T.; Fallah-Mehrjardi, A.; Abdeyazdan, H.; Shishin, H.D.; Shevchenko, M.; Hayes, P.C.; Jak, E. Integrated Experimental and Thermodynamic Modeling Study of the Pb-Fe-O-S-Si System: Effect of Temperature and $p(SO_2)$ on Slag-Matte-Metal-Tridymite Equilibria. *Metall. Mater. Trans. B* **2023**, *54*, 536–549. [CrossRef]

22. López-Porfiri, P.; Gorgojo, P.; González-Miquel, M. Solubility study and thermodynamic modelling of succinic acid and fumaric acid in bio-based solvents. *J. Mol. Liq.* **2023**, *369*, 120836. [CrossRef]
23. Rabadjieva, D.; Gergulova, R.; Titorenkova, R.; Tepavitcharova, S.; Dyulgerova, E.; Balarew, C.; Petrov, O. Biomimetic transformations of amorphous calcium phosphate: Kinetic and thermodynamic studies. *J. Mater. Sci. Mater. Med.* **2010**, *21*, 2501–2509. [CrossRef] [PubMed]
24. Rabadjieva, D.; Sezanova, K.; Gergulova, R.; Titorenkova, R.; Tepavitcharova, S. Precipitation and phase transformation of dicalcium phosphate dihydrate in electrolyte solutions of simulated body fluids: Thermodynamic modeling and kinetic studies. *J. Biomed. Mater. Res. Part A* **2020**, *108*, 1607–1616. [CrossRef] [PubMed]
25. Parkhurst, D.L.; Appelo, C.A.J. Description of Input and Examples for PHREEQC Version 3—A Computer Program for Speciation, Batch-Reaction, One-Dimensional Transport, and Inverse Geochemical Calculations (US Geological Survey Techniques and Methods 2013). Available online: https://pubs.usgs.gov/tm/06/a43/ (accessed on 23 January 2013).
26. Martell, A.E.; Smith, R.M.; Motekaitis, R.J. *NIST Critically Selected Stability Constants of Metal Complexes Database*; Texas A & M University: College Station, TX, USA, 2004.
27. Magalhaes, M.C.; de Aguiar Pereira Marques, P.A.; Correi, R. *Biomineralization-Medical Aspects of Solubility*; Königsberger, E., Königsberger, L.C., Eds.; John Wiley & Sons: Hoboken, NJ, USA, 2006; pp. 71–123.
28. Sovago, I.; Kiss, T.; Gergely, A. Critical survey of the stability constants of complexes of aliphatic amino acids. *Pure Appl. Chem.* **1993**, *65*, 1029–1080. [CrossRef]
29. Golovanova, O.A.; Tomashevsky, I.A. Kinetics and Thermodynamics of the Formation of Compounds of Calcium Ions and Amino Acids: Their Structure and Stability. *Russ. J. Phys. Chem. A* **2019**, *93*, 7–17. [CrossRef]
30. Budavari, S. *The Merck Index: An Encyclopedia of Chemicals, Drugs, and Biologicals*, 11th ed.; Merck: Rahway, NJ, USA, 1989.
31. Kokubo, T. Surface chemistry of bioactive glass-ceramics. *J. Non-Cryst. Solids* **1990**, *120*, 138–151. [CrossRef]
32. Rabadjieva, D.; Tepavitcharova, S.; Gergulova, R.; Sezanova, K.; Titorenkova, R.; Petrov, O.; Dyulgerova, E. Mg- and Zn-modified calcium phosphates prepared by biomimetic precipitation and subsequent treatment at high temperature. *J. Mater. Sci. Mater. Med.* **2011**, *22*, 2187–2196. [CrossRef]
33. Bruker AXS. TOPAS V4 General Profile and Structure Analysis Software for Powder Diffraction Data—User's Manual Bruker AXS Karlsruhe. Germany. 2008. Available online: http//algol.fis.uc.pt/jap/TOPAS%204-2%20Users%20Manual.pdf (accessed on 15 January 2009).
34. Yashima, M.; Sakai, A.; Kamiyama, T.; Hoshikawa, A. Crystal structure analysis of β-tricalcium phosphate $Ca_3(PO_4)_2$ by neutron powder diffraction. *J. Solid State Chem.* **2003**, *175*, 272–277. [CrossRef]
35. Betts, F.; Blumenthal, N.C.; Posner, A.S.; Becker, G.L.; Lehninger, A.L. Atomic structure of intracellular amorphous calcium phosphate deposits. *Proc. Natl. Acad. Sci. USA* **1975**, *72*, 2088–2090. [CrossRef]
36. Zaichick, V. Data for the Reference Man: Skeleton content of chemical elements. *Radiat. Environ. Biophys.* **2013**, *52*, 65–85. [CrossRef] [PubMed]
37. Rabadjieva, D.; Tepavitcharova, S.; Sezanova, K.; Gergulova, R.; Titorenkova, R.; Petrov, O.; Dyulgerova, E.B. Biomimetic modifications of calcium orthophosphates. In *On Biomimetic*, 1st ed.; Pramatarova, L., Ed.; InTech: Rijeka, Croatia, 2011; pp. 135–162.
38. Rabadjieva, D.; Titorenkova, R.; Gergulova, R.; Dyulgerova, E.; Balarew, C. Influence of Zn on the Biomimetic Transformation of Amorphous Calcium Phosphate to Nano-Sized Apatite. *Nanosci. Nanotechnol.* **2009**, *9*, 235–238.
39. Martell, A.E.; Smith, R.M. (Eds.) *NIST Standard Reference Database 46*; Versions, 3, 4, 6, 7, 8; NIST: Gaithersburg, MD, USA, 2019.
40. Plummer, L.N. Busenberg, E. The solubilities of calcite, aragonite and vaterite in CO_2–H_2O solutions between 0 and 90 °C, and an evaluation of the aqueous model for the system $CaCO_3$–CO_2–H_2O. *Geochim. Cosmochim. Acta* **1982**, *46*, 1011–1040. [CrossRef]
41. Turner, D.R.; Whitfield, M.; Dickson, A.G. The equilibrium speciation of dissolved components in fresh water and sea water at 25 °C and 1 atm pressure. *Geochim. Chosmochim. Acta* **1981**, *45*, 855–881. [CrossRef]
42. CEAM. *US-EPA Minteq v. 3.11*; US Environmental Protection Agency, CEAM: Athens, GA, USA, 1991.
43. Ball, J.W.; Nordstrom, D.K. *WATEQ4F-User's Manual*; Open-File Report 90-129; U.S. Geological Survey: Washington, DC, USA, 1991.
44. Tung, M.S.; Eidelman, N.; Sieck, B.; Brown, W.E. Octacalcium Phosphate Solubility Product from 4 to 37 °C. *J. Res. Natl. Inst. Stand. Technol.* **1988**, *93*, 613–624. [CrossRef]
45. McDowell, H.; Gregory, T.M.; Brown, W.E. Solubility of $Ca_5(PO_4)_3OH$ in the system $Ca(OH)_2$-H_3PO_4-H_2O at 5, 15, 25, and 37 °C. *J. Res. Natl. Bur.Stand. Sect. A Phys. Chem.* **1977**, *81A*, 273–281. [CrossRef]
46. Christoffersen, M.R.; Christoffersen, J.; Kibalczyc, W. Apparent solubilities of two amorphous calcium phosphates and of octacalcium phosphate in the temperature range 30–42 °C. *J. Cryst. Growth* **1990**, *106*, 349–354. [CrossRef]
47. Hamad, M.; Heughebaert, J.C. The growth of whitlockite. *J. Cryst. Growth* **1986**, *79*, 192–197. [CrossRef]

materials

MDPI

Article

Polycarboxy/Sulfo Betaine—Calcium Phosphate Hybrid Materials with a Remineralization Potential

Diana Rabadjieva [1,*], Rumiana Gergulova [1], Konstans Ruseva [2], Alexander Bonchev [3], Pavletta Shestakova [4], Marin Simeonov [2], Radosveta Vasileva [3], Dragomir Tatchev [5], Rositsa Titorenkova [6] and Elena Vassileva [2]

1 Institute of General and Inorganic Chemistry, Bulgarian Academy of Sciences, Acad. G. Bonchev Str., bl. 11, 1113 Sofia, Bulgaria; rumigg@yahoo.com
2 Laboratory on Structure and Properties of Polymers, Faculty of Chemistry and Pharmacy, University of Sofia, 1, James Bourchier Blvd., 1164 Sofia, Bulgaria; ohtkr@chem.uni-sofia.bg (K.R.); m.simeonov@chem.uni-sofia.bg (M.S.); ohtev@chem.uni-sofia.bg (E.V.)
3 Faculty of Dental Medicine, Medical University, 1, G. Sofiiski Str., 1431 Sofia, Bulgaria; dralexanderbonchev@abv.bg (A.B.); etienet@abv.bg (R.V.)
4 Institute of Organic Chemistry with Centre of Phytochemistry, Bulgarian Academy of Sciences, Acad. G. Bonchev Str., bl. 9, 1113 Sofia, Bulgaria; pavletta.shestakova@orgchm.bas.bg
5 Rostislaw Kaischew Institute of Physical Chemistry (IPC), Bulgarian Academy of Sciences, Acad. G. Bonchev Str., bl. 11, 1113 Sofia, Bulgaria; dtachev@ipc.bas.bg
6 Institute of Mineralogy and Crystallography, Bulgarian Academy of Sciences, Acad. G. Bonchev Str., bl. 107, 1113 Sofia, Bulgaria; rositsatitorenkova@imc.bas.bg
* Correspondence: didiarab@svr.igic.bas.bg

Citation: Rabadjieva, D.; Gergulova, R.; Ruseva, K.; Bonchev, A.; Shestakova, P.; Simeonov, M.; Vasileva, R.; Tatchev, D.; Titorenkova, R.; Vassileva, E. Polycarboxy/Sulfo Betaine—Calcium Phosphate Hybrid Materials with a Remineralization Potential. *Materials* **2023**, *16*, 6640. https://doi.org/10.3390/ma16206640

Academic Editor: Nikolaos Bouropoulos

Received: 16 September 2023
Revised: 3 October 2023
Accepted: 9 October 2023
Published: 11 October 2023

Correction Statement: This article has been republished with a minor change. The change does not affect the scientific content of the article and further details are available within the backmatter of the website version of this article.

Abstract: Biomacromolecules control mineral formation during the biomineralization process, but the effects of the organic components' functionality on the type of mineral phase is still unclear. The biomimetic precipitation of calcium phosphates in a physiological medium containing either polycarboxybetaine (PCB) or polysulfobetaine (PSB) was investigated in this study. Amorphous calcium phosphate (ACP) or a mixture of octacalcium phosphate (OCP) and dicalcium phosphate dihydrate (DCPD) in different ratios were identified depending on the sequence of initial solution mixing and on the type of the negative functional group of the polymer used. The more acidic character of the sulfo group in PSB than the carboxy one in PCB determines the dominance of the acidic solid phases, namely, an acidic amorphous phase or DCPD. In the presence of PCB, the formation of ACP with acicular particles arranged in bundles with the same orientation was observed. A preliminary study on the remineralization potential of the hybrid material with the participation of PSB and a mixture of OCP and DCPD did not show an increase in enamel density, contrary to the materials based on PCB and ACP. Moreover, the latter showed the creation of a newly formed crystal layer similar to that of the underlying enamel. This defines PCB/ACP as a promising material for enamel remineralization.

Keywords: biomimetic synthesis; de- and remineralization; micro-CT analysis; NMR; pre-nucleation clusters; zwitterionic functionality

1. Introduction

The preparation of inorganic/organic hybrid materials (HMs) via a macromolecule-controlled biomimetic process helps for understanding naturally occurring biomineralization and results in the development of novel hybrid biomaterials with potential applications in dental medicine.

Biomineralization is a process by which inorganic mineral crystals are formed and deposited in an organic matrix. Thus, structures with unique physicochemical and mechanical properties are formed. The organic substances involved in the process control the nucleation, orientation, and bonding of the newly formed nuclei into crystalline formations [1]. The proteins amelogenin, enamelin, ameloblastin, and amelotin together with matrix metalloprotease-20 and kallikrein-4 are responsible for dental enamel formation [1,2], whose mineral part is hydroxyapatite. Amelogenin is a major organic component. Its composition is rich in histidine, glutamine, proline, and leucine amino acid residues, which are responsible for protein–mineral interactions [3,4]. Studying the murine tooth enamel, Beniash et al. [5] reported that, at the beginning of the enamel genesis, mineral particles from amorphous calcium phosphate, suspended in a protein gel, are formed, which furthermore transform into apatite crystals. Tao et al. [4] found that the adsorption of amelogenin onto mineral surfaces depends on the amelogenin structure and affects the induction time for the transformation of the amorphous calcium phosphate to hydroxyapatite. The absence of amelogenin and metalloprotease-20 led to the formation of octacalcium phosphate instead of amorphous calcium phosphate [2]. All studies on the mechanism of enamel formation reveal the leading role of bio-macromolecules, which shows the potential of composite and hybrid organic/inorganic materials as remineralizing agents.

Hybrid materials are intriguing for a variety of applications due to their properties, which are a result of the synergetic effect of the organic and inorganic components' combination. Polymer-based biocompatible materials have attracted a lot of interest due to the possibility of using them to protect human health and improve the quality of life [6]. As inorganic components' different substances like silanized silica particles [7], graphene/carbon nanotubes [8], bioactive ceramics, and glasses or glass ceramics [9], metallic materials [10], etc., have been used depending on the biomaterial application. Calcium phosphate/polymer hybrid materials are preferred for the remineralization of primary tooth enamel defects as they promote fast and noninvasive enamel restoration [11]. Although hydroxyapatite (HA) is considered the closest in composition and structure to the mineral part of enamel, amorphous calcium phosphate (ACP) and α- or β-tricalcium phosphate (TCP) are usually used in hybrid remineralization systems. ACP is a primary solid phase precipitated from calcium phosphate solutions at physiological pH. It is a metastable phase and transforms into thermodynamically stable HA with time. In addition, the solubility of ACP is higher than that of β-TCP and even of α-TCP, which accelerates the dissolution/crystallization/recrystallization processes [12].

Polysaccharides as cellulose [13], different synthetic polymers [14,15] (polylactic acid, poly(lactic-co-glycolic acid), polycaprolactone, etc.), and natural substances [16–18] (collagen, chitosan, gelatine, etc.) have been used as organic components in HMs. Investigations concern mainly the effect of the preparation method and the type of organic compounds on the mechanical properties and the bioactivity of the obtained biomaterials [19,20]. Jee et al. [21] showed that the molecular weight of poly-L-aspartic acid influenced the degree of the mineralization process. Amino acids were also used in a large part of the investigations because they simulate the behavior of the corresponding residues in the protein structure [22–24]. Hybrid materials are usually prepared by in situ precipitation of calcium phosphate (CaP) in the presence of amino acids. The effect of the amino acid side chain charge on the structure, morphology, and surface properties of apatite crystals have been studied.

Synthetic polymers offer a variety of functional groups, which could be identical to the ones possessed by the natural polymers or scaffolds, but at the same time, they offer other parameters that could control the mineralization process, e.g., template architecture, topology, different molecular weights, self-organization, as well as some additional benefits. Polyzwitterions are a unique class of synthetic polymers as they possess both positively and negatively charged functional groups, which are directly covalently bonded to each other and provide a functionality analogous to the naturally occurring one in the betaine form of amino acids, in the polar heads of phospholipids, etc. Polyzwitterions, in particular PSB and PCB, have an expanding popularity in biomedical applications due to their high biocompatibility, "smart" behavior, as they are able to respond to variations in pH, temperature, salt concentration, as well as to their antibiofouling behavior [25–27]. The latter has been exploited recently for dental applications [28], but still the advantages of their zwitterionic structure for controlled calcium phosphates' mineralization has not been clearly explored. PCB and PSB, in their side chains, bear negatively charged acidic end groups, respectively, carboxy or sulfo, which are expected to interact differently with Ca^{2+} ions supplied for the mineralization process. Thus, PSB and PCB could not only control calcium phosphate formation but would also bear antibiofouling activity, which is especially beneficial for dental materials.

Despite the accumulated data, the relationship between the functionality of the organic components used and the resulting hybrid calcium phosphate systems is still unclear.

The aim of the present work is to study the biomimetic precipitation of calcium phosphates in a physiological medium containing polymer with betaine functionality, either polycarboxybetaine (PCB) or polysulfobetaine (PSB), in order to (i) elucidate the effect of betaine moieties and the synthesis route on the phase composition of calcium phosphates and (ii) prepare and characterize new hybrid materials with a potential remineralization effect. Two synthesis routes, differing in the mixing sequence of the starting solutions, were studied in order to elucidate the effect of the initial pH and of the dominant ions on the type of the pre-nucleation clusters and thus on the solid phases formed. The changes in pH and in the concentration of Ca^{2+} ions during the syntheses were followed in three types of systems—a polymer-free system, a system with PSB, and a system with PCB—to explore the effects of betaine zwitterionic functionality and to prepare different types of HMs. The materials, which were prepared for the first time, were characterized by XRD, NMR, TEM, and DTA-TG-MASS analyses. Further, a preliminary study on the remineralization potential of the selected materials was carried out, and the results were proof by micro-CT, IR, and SEM analyses.

2. Materials and Methods

2.1. Synthesis of Polymers

2.1.1. RAFT Polymerization of Carboxybetaine (CB) Monomers

In a 250 mL round-bottom flask, 10 g of CB monomers, obtained according to a procedure described elsewhere [29] (Scheme S1, Supplementary Materials), and 0.1 mol. % 2,2-azobis(2-methylpropionamide) dihydrochloride were dissolved in a 45 mL acetate buffer (pH = 5.2, 0.27 $mol \cdot L^{-1}$ acetic acid and 0.73 $mol \cdot L^{-1}$ sodium acetate). A total of 0.01 mol. % 4-cyano-4-(thiobenzoylthio)pentanoic acid (CTPA, Sigma Aldrich, St. Louis, MO, USA, A.R.) was neutralized in 5 mL of 0.05 M KOH (Sigma Aldrich, St. Louis, MO, USA, A.R.) and then added dropwise to the CB monomer solution. The amount of CTPA was calculated in such a way as to give a linear PCB with a molar mass of 100,000 g/mol (Scheme S1, Supplementary Material). The solution was heated for 5 h at 70 °C and then placed in an ice bath to stop the polymerization process. PCB was dialyzed against distilled water for two weeks using the dialysis membrane 3.5 K MWCO, 16 mm, SnakeSkin™ Dialysis Tubing, Thermo Scientific, Waltham, MA, USA, until no more traces of residual monomers and other reactants were detected in the wastewater (these were monitored with UV/Vis (JAsco V-730, Jasco, Tokyo, Japan)). The resulting polymer was freeze-dried, thus obtaining a white powder. The conversion was determined gravimetrically to be 87%. The successful polymerization and the chemical structure of the polymer was confirmed by ^{1}H NMR spectroscopy (Figure S1 in the Supplementary Materials).

2.1.2. RAFT Polymerization of Sulfobetaine (SB) Monomers

The RAFT polymerization of sulfobetaine monomers was carried out following Scheme S2 in the Supplementary Materials. A total of 5 g of the monomer sulfobetaine methacrylate (SB, Sigma-Aldrich, St. Louis, MO, USA, A.R.) was dissolved in a 20 mL acetate buffer with a pH of 5.2. Initiator 2,2-azobis(2-methylpropionamide) dihydrochloride (Sigma-Aldrich, St. Louis, MO, USA, A.R.) with a concentration of 0.1 mol was added, with respect to the amount of the SB monomer. A total of 14 mg of CTPA was dissolved in 5 mL of 0.05 M NaOH (Sigma-Aldrich, St. Louis, MO, USA, A.R.), and the resulting solution was added dropwise to the reaction mixture. The amount of CTPA was calculated to give a linear PSB with a molar mass of 100,000 g/mol (Scheme S2, Supplementary Materials). The solution was heated for 5 h at 70 °C and then placed in an ice bath to stop the polymerization process. PSB was dialyzed against distilled water for two weeks using the dialysis membrane 3.5 K MWCO, 16 mm, SnakeSkin™ Dialysis Tubing, Thermo Scientific, Waltham, MA, USA, until no more traces of residual monomers and other reagents were detected in the wastewater (these were monitored with UV/Vis (JASCO V-730, Jasco, Japan)). The resulting polymer was freeze-dried, thus obtaining a white powder. The conversion was determined gravimetrically to be 73%. The successful polymerization and the chemical structure of the polymer was confirmed by ^{1}H NMR spectroscopy (Figure S2 in the Supplementary Materials).

2.2. *Biomimetic Precipitation of Calcium Phosphates in a Physiological Medium Containing Polymers with Betaine Functionality*

A combined apparatus for automatic titration and controlled synthesis (Titrando 907, Methrom AG, Herisau, Switzerland) was used to maintain a constant flow rate and to monitor the pH and concentration of free Ca^{2+} ions during the synthesis procedures. The initial substances $CaCl_2.2H_2O$ (Sigma-Aldrich, St. Louis, MO, USA, A.R.) and Na_2HPO_4 (Merck, Darmstadt, Germany, A.R.) were dissolved in a physiological solution (0.9% NaCl, Merck, Darmstadt, Germany, A.R.) to obtain 0.05 M of a Ca solution and 0.03 M of a P solution, respectively. PCB or PSB, with a molar ratio (monomeric unit)/Ca^{2+} of 1, were dissolved in the Ca solution. An experiment without polymers was also performed for the sake of a comparison. Two series of experiments were carried out, namely, the following:

Series A: A total of 130 mL of the Ca solution was added to 130 mL of the P solution. The products of this series will be called further as follows: A-CaP—for a polymer-free system, A-PSB/CaP—for a system with PSB, and A-CaP/PCB—for a system with PCB.
Series B: A total of 130 mL of the P solution was added to 130 mL of the Ca solution. The products of this series will be called further as follows: B-CaP—for a polymer-free system, B-PSB/CaP—for a system with PSB, and B-CaP/PCB—for a system with PCB.

The parameters of the process are an addition rate of 3 mL/min (0.25 mL per step), the room temperature, constant stirring, and a Ca/P molar ratio of 1.67, which is characteristic for hydroxyapatite.

The pH and concentration of the free Ca^{2+} ions were monitored using combined pH (iConneet, Methrom AG, Herisau, Switzerland) and Ca^{2+} polymer membrane ion selective electrodes (Methrom AG, Herisau, Switzerland). A calibration standard curve for calculating free Ca^{2+} ion concentrations was generated by a measurement of the electrical potential of the standard solutions of $CaCl_2$ in 0.9% solution of NaCl. The concentrations of the standard solutions were within the concentration range of the Ca^{2+} ions under the experimental conditions.

The obtained suspension was left to mature for 1 h, washed out from Na^{+} and Cl^{-} residuals using a dialysis tube (3.5 K MWCO, 16 mm, SnakeSkin™ Dialysis Tubing, Thermo Scientific, Waltham, MA, USA), and then freeze-dried.

2.3. *Characterization*

2.3.1. Powder X-ray Diffraction Analysis

A Bruker D8 Advance diffractometer with Cu K radiation and a LynxEye detector (Bruker AXS Advanced X-ray Solutions GmbH, Karlsruhe, Germany) was used to conduct powder X-ray diffraction. The data were gathered in the 10 to 90° 2θ range with a step of 0.03° 2θ and a counting rate of 57 s/step for the primary phase identification. The phase composition was identified with the ICSD database.

2.3.2. Solid State Nuclear Magnetic Resonance (NMR) Analysis

NMR spectra were recorded on a Bruker Avance II+ 600 spectrometer (^{1}H working frequency of 600.01 MHz and 242.94 MHz for ^{31}P), equipped with a 4 mm $^{1}H/^{31}P$-15N solid state CP MAS dual ^{1}H/X probe head. The samples were loaded in 4 mm ZrO_2 rotors and spun at a magic angle spinning (MAS) rate of 10 kHz for all measurements. The quantitative ^{31}P NMR spectra were acquired with a single pulse sequence (Bruker Topspin library), with a 90° pulse length of 3.1 μs, 8 K time domain data points, a spectrum width of 74 kHz, 256 scans, and a relaxation delay of 120 s. The spectra were processed with an exponential window function (line broadening factor 5) and zero-filled to 32 K data points. The $^{1}H \rightarrow ^{31}P$ cross-polarization MAS (CP-MAS) spectra were acquired with the following experimental parameters: a 1H excitation pulse of 3.4 μs, 10 s of a relaxation delay, 256 scans, and a MAS rate of 12 kHz. A series of spectra with a contact time varied from 100 μs up to 5 ms were measured. A ^{1}H SPINAL-64 decoupling scheme was used during the acquisition of the CP experiments. All ^{31}P chemical shifts were referenced against the external solid reference $NH_4H_2PO_4$ (δ 0.9 ppm). The DMfit software (release #20220502)was used for the deconvolution, simulation, and fitting of the experimental NMR data [30]. The ^{1}H NMR spectra of PSB and PCB were measured on the same spectrometer, using a 5 mm direct detection dual probe head (BBO) ^{31}P–$^{109}Ag/^{1}H$ with a gradient coil with a maximum gradient strength of 53 G/cm.

2.3.3. Differential Thermal Analysis with Release Gas Detection (DTA-TG-MASS Analysis)

The thermal characterization of the samples was performed on a LABSYS^{TM} EVO (Setaram, Caluire, France) apparatus with a Pt/Pt-Rh thermocouple in a corundum crucible at a heating rate of 10 °C/minute in a temperature range of 25–900 °C and in an atmosphere of synthetic air. This apparatus was equipped with a quadrupole mass spectrometer (Pfeiffer vacuum OMNISTAR, GSD 301, Zürich, Switzerland), which serves for the analysis of the escaping gasses. In the studied samples, the release of H_2O (m = 18) and CO_2 (m = 44) was determined.

2.3.4. Transmission Electron Microscopy (TEM)

The morphology of the selected samples was examined by transmission electron microscopy (JEOL transmission electron microscope, JEM-2100, Tokyo, Japan), equipped with an EDS detector (X-Max 80T, Oxford Instruments, High Wycombe, UK). Samples were washed several times with water and centrifuged in order to eliminate polymers and secondary crystallizations from the matter solution. The water-dispersed powders were then sonicated for 1 min and dropped onto standard carbon-copper grids.

2.3.5. Preliminary Studies of the Remineralization Potential

The preparation of the enamel samples was in accordance with the methodology described by Bonchev et al. [31].

The erupted third molars were used to create artificial enamel lesions. The radicular part of each tooth was removed, and the coronal part was separated in two halves (buccal and lingual). The buccal halves were polished with SiC paper disks with grit sizes of 320, 600, and 1200 (Shofu, Super-Snap Rainbow Technique Kit, Kyoto, Japan). The teeth samples were covered with two layers of acid-resistant nail varnish (Jerusalem, Israel), but a flat window with size of 4 × 4 mm was left uncovered. Then, the buccal part was separated into two symmetrical pieces. The demineralization was performed by using an aqueous solution of lactic acid (0.1 mM), NaH_2PO_4 (2.2 mM), $CaCl_2$ (2.2 mM), and sodium fluoride (0.2 ppm) with a pH adjusted to 4.5. The solution was renewed every 24 h, and the enamel samples were immersed in it for 6 days, creating carious enamel lesions with a depth of <100 μm.

The demineralized samples were rinsed with distilled water, sonicated for 5 min, and stored at 4 °C in distilled water prior to use. Following this preparation model, the de- and remineralization processes were compared with two halves from one and the same tooth.

The remineralization procedure was performed with freshly prepared suspensions of PCB/CaP and PSB/CaP hybrid materials synthesized in Series A. The powdered hybrid material was mixed with distilled water at a ratio of 0.1:2 g/mL for PCB/CaP and 0.1:1.25 g/mL for PSB/CaP to obtain thick suspensions. The specimens were covered either with the suspension of PCB/CaP or with suspensions PSB/CaP for 6 h, as described above.

After the remineralization, each sample was carefully rinsed with distilled water and then stored in 6 mL of an artificial saliva solution [32] for 18 h. This procedure was repeated daily with a renewed artificial saliva solution for one week.

2.3.6. Micro-CT Scans

The de- and remineralized samples were scanned with the Bruker SkyScan 1272 X-ray scanner. The tube voltage and current were 90 kV and 111 µA, correspondingly. The voxel size was 2 µm. A tomographic reconstruction was performed with a dedicated software NRecon, version number 1.7.4.2. delivered by Bruker. A ring artefact and beam hardening corrections were applied. The beam hardening correction was adjusted by the tooth enamel. The 3D images before and after treatment were 3D-registered by DataViewer, also delivered by Bruker. A down sampling to a voxel size of 16 microns was necessary for this procedure. The region of interest containing the treatment windows was cut and registered again for better accuracy. The voxel size was 8 µm in the ROI registration. Absorption (density) profiles along a line 160 µm thick were taken in the 3D registration view simultaneously from the original ant-treated samples.

2.3.7. Scanning Electron Microscopy (SEM) Analysis

The JEOL JSM 6390 apparatus was used for studying the morphology of the demineralized and remineralized tooth surfaces. The samples were gold-sputtered in vacuum, and SEM images were taken at several different magnifications.

2.3.8. IR Reflection Micro-Spectroscopic Analysis

The non-treated, demineralized, and remineralized dental specimens were studied with the infrared microscope Hyperion 2000, Bruker Optik, Ettlingen, Germany in a spectral range of 600–4000 cm^{-1} with a 20× Schwarzschild objective in reflectance mode after accumulating 264 scans. Micro-infrared spectra were collected from five different areas, with mean size of 100 µm^2 for each sample.

2.3.9. Raman Spectroscopy

Raman spectra of the remineralized samples were collected using the HR LabRam (Horiba, Jobin Yvon, Germany) spectrometer (600 grooves/mm grating) coupled with an Olympus optical microscope and a 50× objective in a range of 100–4000 cm^{-1}. The 632.8 nm line of He-Ne laser was used for sample excitation. The Origin 9 software package was used for spectral evaluation.

3. Results

3.1. Effect of Betaine Functionality and Synthesis Route on the Precipitated Calcium Phosphate Phase Composition

The observed pH and Ca^{2+} ion concentration data in the polymer solutions, where CaP deposition took place, are presented in Figure 1.

The results show significant differences in the behavior of the systems in both series—A and B. However, a similarity was observed in each of the series, between the polymer-free system and the system with PSB (Figure 1). In Series A, it is expressed in a bend in pH and especially in the concentration of Ca^{2+} ions between the added 45 and 65 mL Ca solution. The pH drops from 6.98 to 6.62 (Figure 1a) and from 6.86 to 6.48 (Figure 1c) and continues to decrease gradually to 5.7. At the same time, log[Ca] decreases in two steps from −3.45 to −3.55 and to −3.76 (Figure 1a) and from −3.67 to −3.77 and to −4.02 (Figure 1c). After that, log[Ca] increases smoothly to −2.73 and −2.90, respectively. In Series B, with the addition of the first portions of the P solution, the pH sharply increases by about 0.5 in the system without polymers (Figure 1b) and by almost two units in the system with PSB (Figure 1d). After that, the pH begins to gradually decrease until the addition of about 35 mL of the P solution, followed by a bend and a sharp decrease to about pH 6, at which value it remains relatively constant until the end of the precipitation process. In this series of experiments, the change in the concentration of the free Ca^{2+} ions is more difficult to notice due to the high concentration of the Ca^{2+} ions during the synthesis (the P solution is added to the Ca solution).

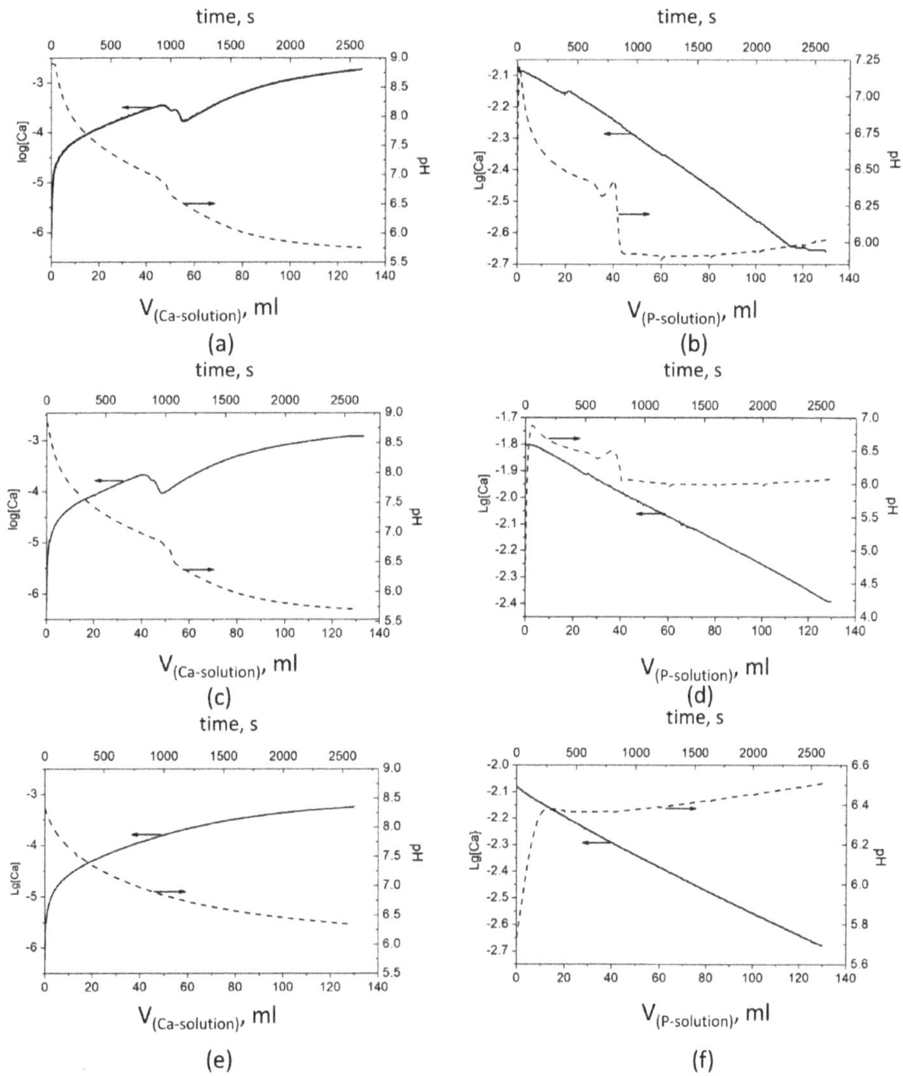

Figure 1. Evolution of pH and concentration of Ca^{2+} ions during the CaP precipitation process: Series A—(**a**) without a polymer, (**c**) with PSB, and (**e**) with PCB; Series B—(**b**) without a polymer, (**d**) with PSB, and (**f**) with PCB The arrows show which ordinate axis the data from the corresponding line belongs to.

In the presence of PCB in Series A, (Figure 1e), the change of both parameters is smooth, with the pH falling from 8.9 to 6.5 and log[Ca] ions increasing from −6 to −3.27. By contrast, in Series B, the pH jumps from 5.7 to 6.4, with the first portion of the P solution, and slowly increases to 6.5 till the end of the experiment.

The results of the XRD analysis (Figure 2) of the final products (after lyophilization) show that, in the polymer-free systems, the calcium phosphate phase is $Ca_8(HPO_4)_2(PO_4)_4.5H_2O$

(OCP, octacalcium phosphate) in Series A (Figure 2(a-1)) but $CaHPO_4.2H_2O$ (DCPD, dicalcium phosphate dihydrate) in Series B (Figure 2(b-1)). In the presence of PSB, a mixture of OCP and $CaHPO_4.2H_2O$ (DCPD, brushite) was obtained in Series A (Figure 2(a-2)), but again only DCPD was obtained in Series B (Figure 2(b-2)). In the presence of PCB, the phase remains XRD-amorphous in both series (Figure 2(a-3,b-3)).

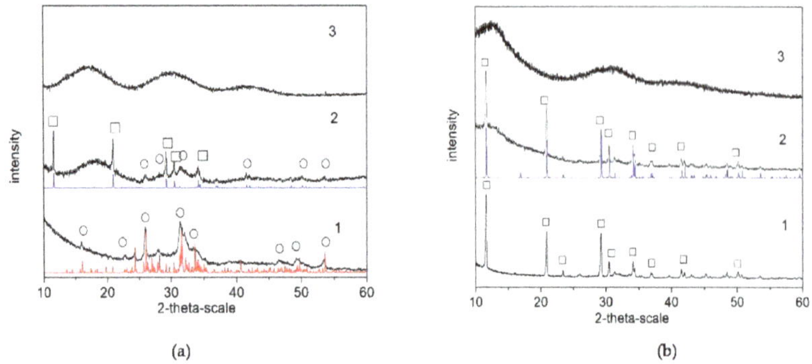

Figure 2. XRD studies of as-prepared biomaterials (after lyophilization) in the system: (**a**) Series A—(1) A-CaP, (2) A-PSB/CaP, and (3) A-PCB/CaP. (**b**) Series B—(1) B-CaP, (2) B-PSB/CaP, and (3) B-PCB/CaP. Red line (o)—powder diffraction data of OCP, ICSD Collection Code 65347; blue line (□)—powder diffraction data of DCPD, ICSD Collection Code 16132.

The single pulse ^{31}P and $^{1}H \rightarrow ^{31}P$ CP-MAS solid state NMR spectra provide a more detailed insight into the structural characteristics and chemical composition of the calcium phosphate phases formed in the studied materials, as a function of the sample preparation procedure, synthesis conditions, and the polymer functional groups. The single pulse ^{31}P spectra give information about all types of CaP phases, while in the $^{1}H \rightarrow ^{31}P$ CP-MAS, the resonances of the P atoms with closely spaced protons, such as hydrogen phosphate species, are specifically enhanced due to the transfer of magnetization to the P atom from the neighboring protons. The relative areas of the characteristic resonances in the single pulse ^{31}P spectra give quantitative information about the relative molar fractions of the different CaP phases. The standard $^{1}H \rightarrow ^{31}P$ CP-MAS spectra, however, cannot be interpreted quantitatively since the effectiveness of cross-polarization transfers depends on the P---H internuclear distance, the relaxation rates, and the local dynamics of the structural fragments, which vary from one chemical environment to another.

Figure 3(Aa–c) shows the single pulse ^{31}P NMR spectra of pure calcium phosphate and the hybrid CaP materials synthesized with the two polymers, PCB and PSB, in Series A. The NMR spectrum of the pure A-CaP (Figure 3(Aa)) indicates the formation of the mixture of several phases. The deconvolution of the spectrum shows the presence of OCP as a main component (signals at −0.4, 2.9, 3.2, and around 4.5 ppm) as well as some amounts of ACP and/or poorly crystalline HA (resonance at around 2.9 ppm) and of DCPD (1.3 ppm). The addition of PCB (sample A-Cap/PCB) results in the formation of an amorphous phase, represented by a broad resonance centered at 2.6 ppm, characteristic for apatite phosphates (Figure 3(Ab)). The ^{31}P spectral pattern of the hybrid material, A-CaP/PSB synthesized in the presence of PSB (Figure 3(Ac)), is very similar to those of the pure A-CaP obtained by the same synthesis procedure; however, in this case, the intensity of the resonance at 1.3 ppm is higher, indicating the formation of a higher amount of DCPD as compared to the pure CaP material. The presence of OCP and DCPD in the pure A-CaP and in the hybrid material, A-Cap/PSB, was further confirmed by the $^{1}H \rightarrow ^{31}P$ CP-MAS spectra (Supplementary Materials, Figure S3a,b). In the $^{1}H \rightarrow ^{31}P$ CP-MAS spectra, the resonances of the P atoms of the hydrogen phosphate groups from OCP at −0.4 ppm and from the DCPD at 1.3 ppm were specifically enhanced due to the transfer of magnetization to the P atom from the neighboring acidic protons of the hydrogen phosphate groups.

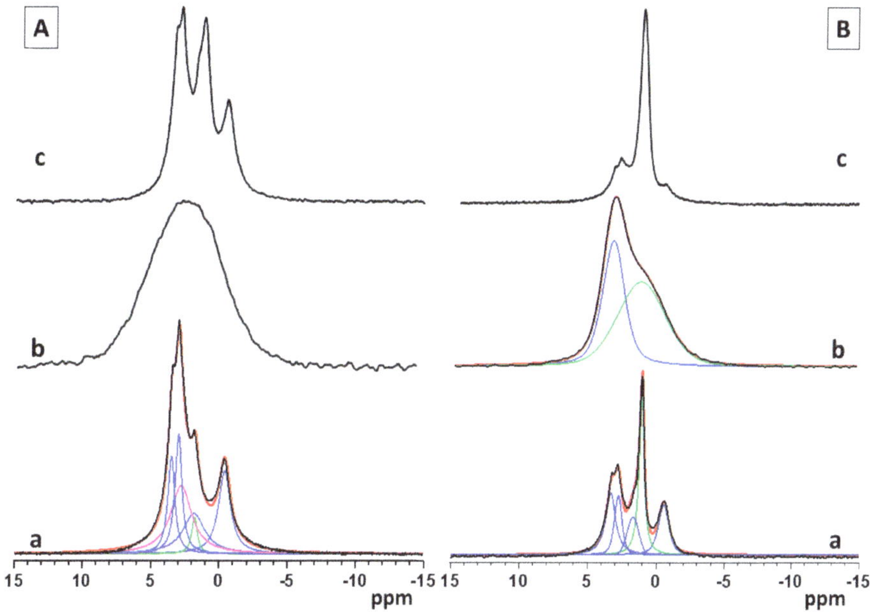

Figure 3. Single pulse ^{31}P NMR spectra of pure and hybrid materials from Series A (**A**)—(**a**) pure A-CaP, (**b**) A-CaP/PCB, and (**c**) A-CaP/PSB—and Series B (**B**): (**a**) pure B-CaP, (**b**) B-CaP/PCB, and (**c**) B-CaP/PSB. The spectra of the pure A-CaP, pure B-CaP, and B-CaP/PCB samples were deconvoluted to identify the different components present in the mixture. Experimental spectra are presented with black lines, while the simulated spectra are given with red lines, and the individual contributions of the different phases obtained by the deconvolution of the spectral patterns are given with colored lines (OCP—blue, ACP/Hap—magenta, DCPD and acidic amorphous phase—green).

The ^{31}P spectra of pure calcium phosphate (B-CaP) and the hybrid materials synthesized in the presence of PCB (B-CaP/PCB) and PSB (B-CaP/PSB) from Series B are presented in Figure 3(Ba–c). Under the used synthesis conditions, the pure B-CaP contains a mixture of OCP and nano-crystalline DCPD, represented by the relatively sharp resonance at 1.3 ppm (Figure 3(Ba)). The ^{31}P spectrum of the hybrid material, B-CaP/PCB, shows a broad non-symmetrical resonance, indicating the formation of at least two different phases (Figure 3(Bb)). Two main peaks could be identified after the deconvolution of the spectrum: a resonance centered at around 2.9 ppm, characteristic of poorly crystalline HA and/or ACP, which is overlapped by a broad second resonance centered at around 1.7 ppm, indicating the formation of a non-stoichiometric amorphous phase of acidic phosphates.

The relative molar ratios of the two resonances were 25:75, respectively. The presence of the acidic amorphous phase was further confirmed by the $^{1}H \rightarrow ^{31}P$ CP-MAS spectra, where a broad symmetric resonance centered at around 1.7 ppm was observed at all mixing times (Supplementary Materials, Figure S4a). The different intensities of the two resonances in the single pulse and the $^{1}H \rightarrow ^{31}P$ C MAS spectra could be explained with the specific enhancement of the signal at 1.7 ppm due to the transfer of magnetization to the P atom from the neighboring protons in the -HPO_4 moiety. The enhancement of the resonance at 2.9 ppm is less pronounced at short mixing times and is clearly visible only at longer mixing times (Supplementary Materials, Figure S4a,b). In the HA structure, the distance between the P atoms and neighboring protons is larger since the protons are not a part of the phosphate groups and, therefore, the transfer of magnetization is less efficient as compared to the acidic phosphate phases. The single pulse ^{31}P (Figure 3(Bc)) and the $^{1}H \rightarrow ^{31}P$ CP-MAS spectra (Figure S4c,d) of the hybrid material, B-CaP/PSB, display similar spectral patterns as observed for the pure B-CaP phase; however, in the presence of PSB, a higher amount of DCPD was detected.

3.2. Characterization and Actions of a Newly Synthesized Hybrid Material

Two hybrid biomaterials were chosen for compositional and morphological characterization and preliminary studies of their remineralization potential, namely, the ones obtained in Series A, due to the absence of the non-stoichiometric amorphous phase of acidic phosphates in A-PCB/CaP and lower amounts of DCPD in A-PSB/CaP than in B-PSB/CaP. The presence of acid phosphates is not characteristic of normal calcification processes.

3.2.1. Composition and Morphology

TEM with EDS and DTA-TG-MASS analyses were performed to characterize the morphology and the composition of the new hybrid materials.

In the A-PSB/CaP material, three types of particles are observed that differ in shape, size, and molar Ca/P ratio, namely: (i) micro-sized single plate-like particles (Figure 4a) with Ca/P = 0.92; (ii) needle shaped particles with Ca/P = 1.02; and (iii) sheet-like particles with Ca/P = 1.14 (Figure 4b).

In contrast, in the A-PCB/CaP materials, only one type of fine needle-like particle with Ca/P = 1.44 is visible in the TEM images (Figure 5). Unlike the randomly scattered particles obtained in the presence of PSB (Figure 4b), these are arranged in bundles of uniformly oriented particles (Figure 5b).

The DTA-TG-MASS analysis of the A-CaP (Figure 6a) shows the decomposition of OCP (T_{max} = 140.9 °C and weight loss of 7.9 mas%), according to the following reaction:

$Ca_8(HPO_4)_2(PO_4)_4.5H_2O \rightarrow 0.5Ca_{10}(PO_4)_6(OH)_2 + 3CaHPO_4 + 4H_2O$ [33].

The next endothermic peaks, accompanied with a weight loss of 10.2%, are due to the decomposition of $CaHPO_4$. The DTA curves of A-PSB/CaP and A-PCB/Ca-P are characterized with strong exothermic effects in a temperature range of 200–600 °C, which can be attributed to the burning of the polymers (Figure 6b,c). They are accompanied with a weight loss of 65 and 65.5%, respectively, and a release of CO_2 and H_2O. The effects of the decomposition of the calcium phosphate compounds in HMs are overlapped by the effects of polymer combustion. This makes it difficult to accurately determine the amount of polymer in the two materials. However, it can be assumed that it is 50–55 mas %, determined by the differences in weight loss of A-PSB/CaP or A-PCB/CaP with A-CaP in the temperature region 200–600 °C.

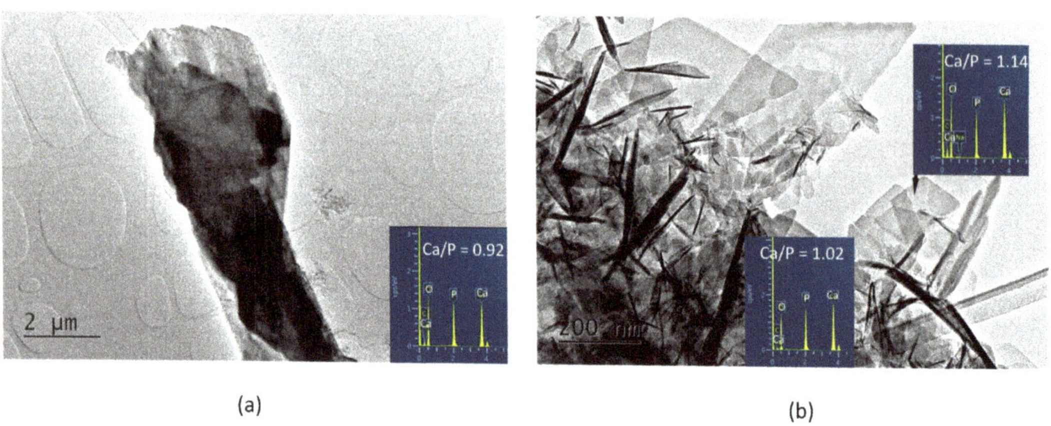

Figure 4. TEM images and EDS analyses of the A-PSB/CaP (**a**) column-like crystal with Ca/P ratio of 0.92; (**b**) needle-shaped crystals with Ca/P ratio of 1.02 and plate-like crystals with Ca/P ratio of 1.14.

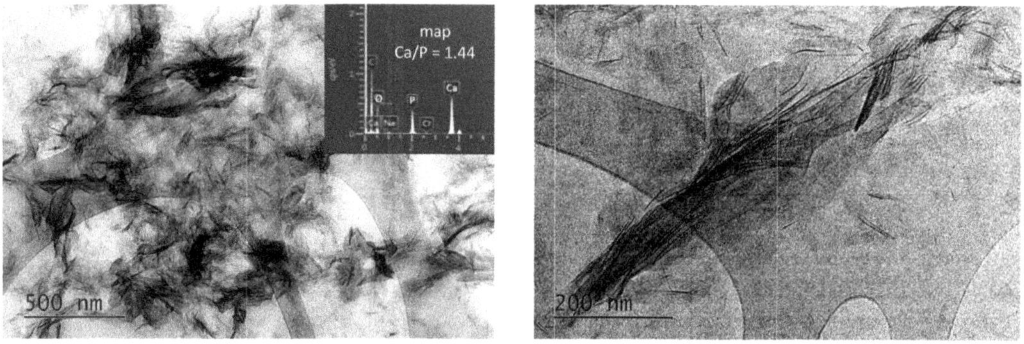

(a) (b)

Figure 5. TEM images at different magnifications and EDS analysis of the particles of A-PCB/CaP. (**a**) magnification 10 kx; (**b**) magnification 30 kx.

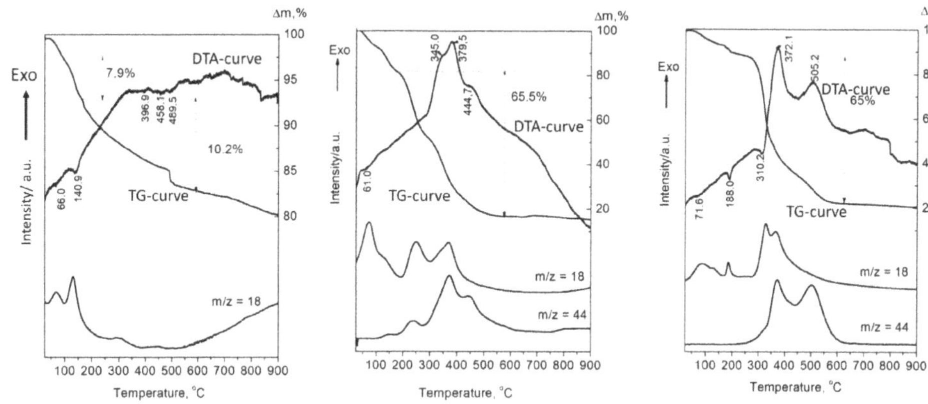

(a) (b) (c)

Figure 6. DTA-TG-MASS analysis of the (**a**) A-CaP, (**b**) A-PSB/CaP, and (**c**) A-PCB/CaP.

3.2.2. Preliminary Studies on the Remineralization Potential of the Selected Hybrid Materials

The mineral loss after the demineralization procedure and the mineral uptake after the remineralization procedure of the enamel were monitored non-destructively using the micro-CT analysis.

The typical 2D micro-CT images of the samples are shown in Figure 7.

In the cross-sectional images, a distinct faint shadow in the enamel (arrows) is observed, which is due to the formed enamel lesion. From the measurements made, it was found that the resulting lesions were located in the superficial enamel layer. The initial lesion depth of the samples was in a range of 30–90 μm.

Grey level profiles of the treated and non-treated samples are shown in Figures 8 and 9. The profiles are scaled so that the background is 0, but the enamel gray level is 100. This allows for the etched sample area in the window to appear as a step. The density (the X-ray absorption ability) increase after treatment is estimated as part of the etched density.

The results show that in the samples treated with A-PSB/CaP, no increase in density was identified (Figure 8a) in the lesions in contrast to those treated with A-PCB/CaP (Figure 8b).

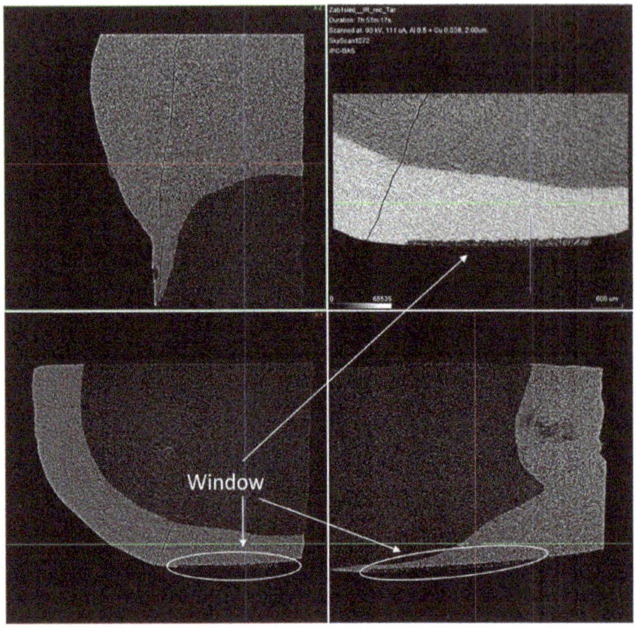

Figure 7. Typical cross-sectional view of the samples with the treatment window enlarged. Voxel size is 2 μm.

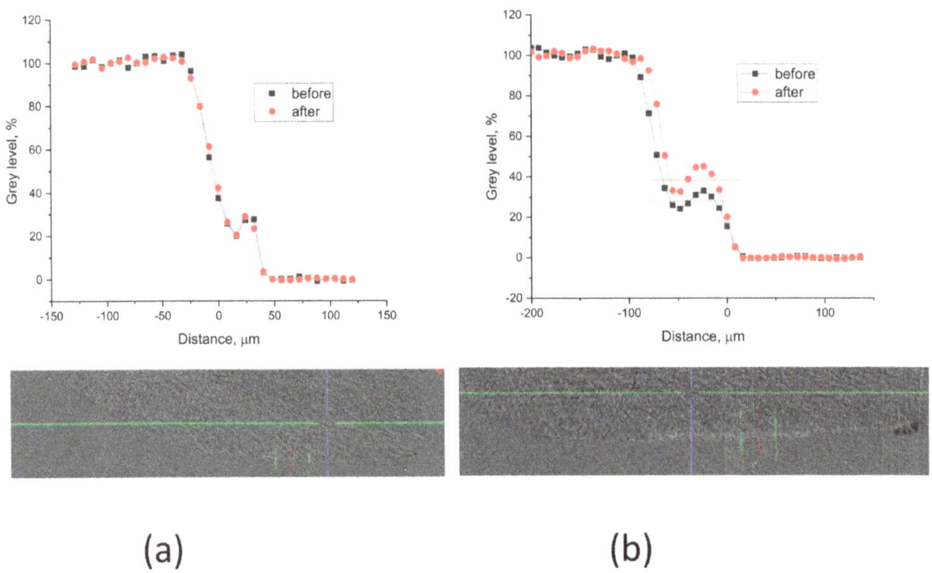

Figure 8. Typical gray level profiles of the non-treated (black line) and treated (red line) specimens with (**a**) A-PSB/CaP—density increase 0%, thickness ~ 40 μm; (**b**) A-PCB/CaP—density increase 14%, thickness ~ 56 μm.

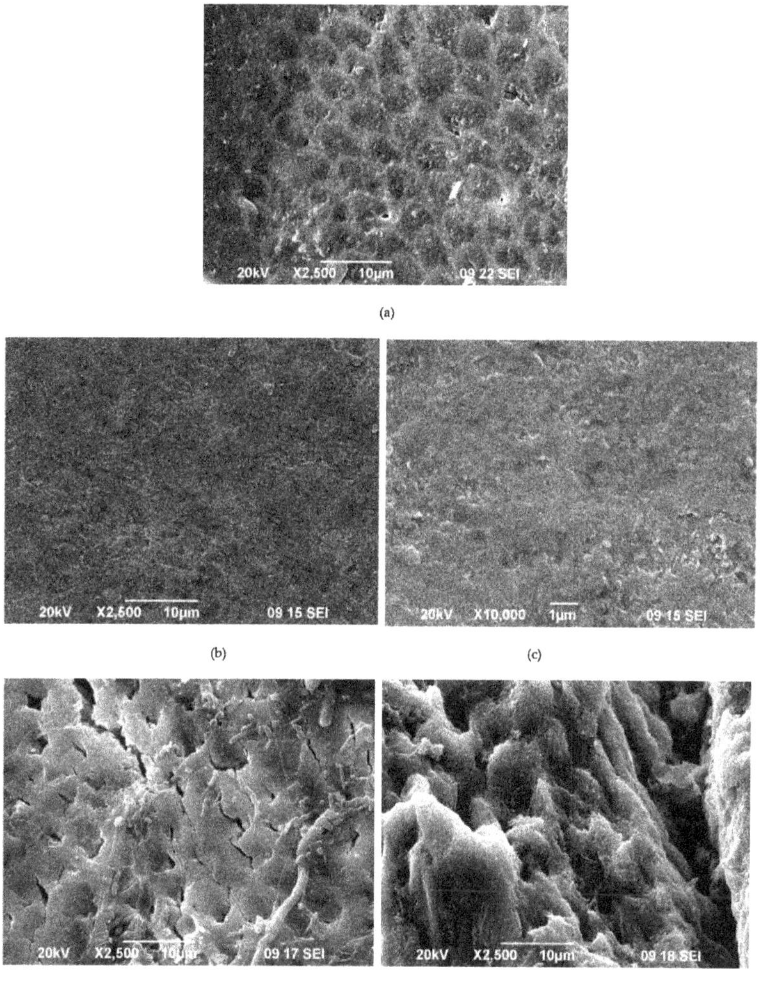

Figure 9. SEM images of demineralized enamel (**a**) after the remineralization procedure with A-PSB/CaP (**b**,**c**) and after the remineralization procedure with A-PCB/CaP (**d**,**e**).

The SEM images of the demineralized enamel (Figure 9a) and after the remineralization procedures (Figure 9b–f) show different behaviors of both materials.

The SEM images of the demineralized enamel showed a dissolution predominantly of the central part of the enamel rods and a widening of the inter-rod zones (Figure 9a). The porous structure of the demineralized enamel changed into a dense flat surface as a result of using A-PSB/CaP (Figure 9b,c). The samples treated with A-PCB/CaP showed a heterogeneous structure due to crystal accumulation and the formation of scattered granular deposits. The newly formed layer is similar to the structure of the underlying enamel (Figure 9d).

A porous net-like substance with a width of the pores up to 2 µm is observed on the surface of the substrates treated with A-PCB/CaP. We hypothesize that the porous structure allows for the penetration of the hydrogel suspension, thus allowing for the deposition of the needle-like crystals of ACP in the demineralized enamel rods (Figure 9e).

The infrared spectra collected in reflectance mode from 100 × 100 micrometer areas of untreated enamel (Figure 10(a-1,b-1)) show the strong characteristic peaks of antisymmetric $\nu_3(PO_4)$ stretching between 1027–1100 cm^{-1}. The weak peak near 957 cm^{-1} results from the symmetric $\nu_1(PO_4)$ stretching in the apatite. The peaks near 870 cm^{-1} and between 1410–1540 cm^{-1} are generated by the out-of-plane ν_2 vibration of CO_3 and the antisymmetric $\nu_3(CO_3)$ stretching impurities in the apatite, respectively [34].

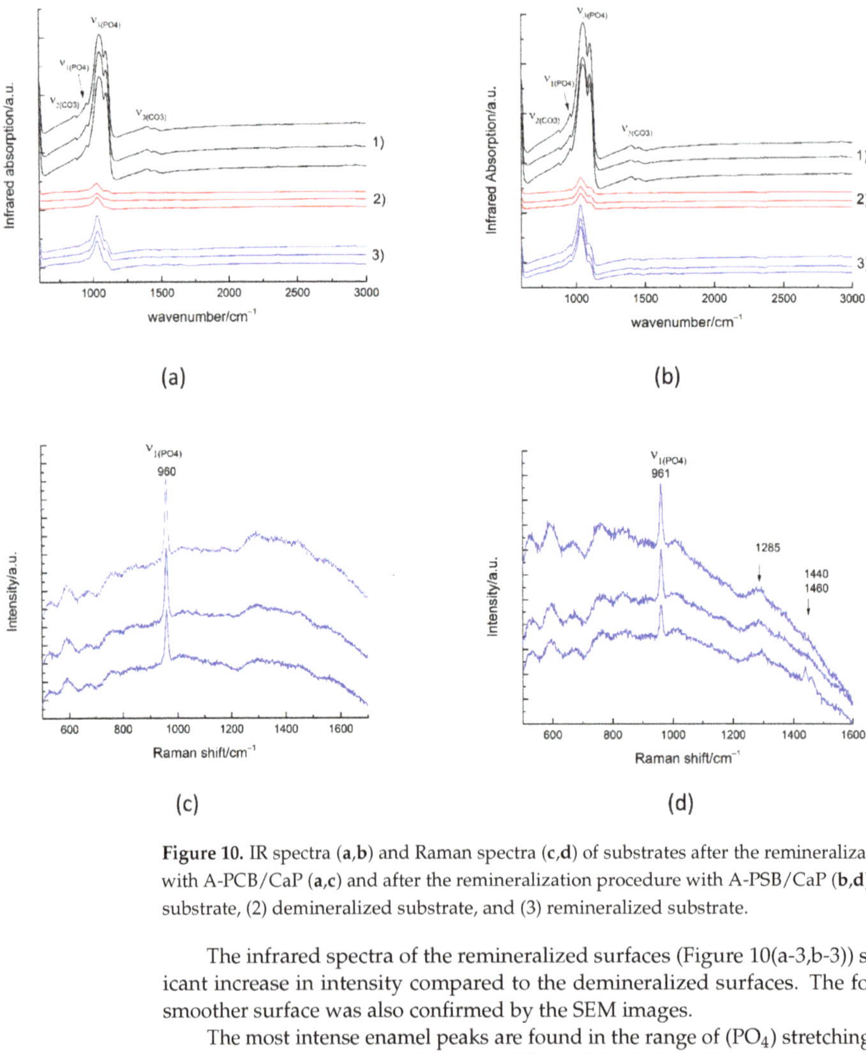

Figure 10. IR spectra (**a**,**b**) and Raman spectra (**c**,**d**) of substrates after the remineralization procedure with A-PCB/CaP (**a**,**c**) and after the remineralization procedure with A-PSB/CaP (**b**,**d**). (1) untreated substrate, (2) demineralized substrate, and (3) remineralized substrate.

The infrared spectra of the remineralized surfaces (Figure 10(a-3,b-3)) show a significant increase in intensity compared to the demineralized surfaces. The formation of a smoother surface was also confirmed by the SEM images.

The most intense enamel peaks are found in the range of (PO_4) stretching. The Raman spectra of the remineralized surfaces (Figure 10c,d) show the most intense enamel peak, generated by the symmetric $\nu_1(PO_4)$ stretching in the apatite and some traces of the polymer matrix. The peak around 1455 cm^{-1} may be due to CH_2 bending.

4. Discussion

4.1. *Effects of Betaine Functionality and Synthesis Routes on the Precipitated CaP Phases*

The effects of the betaine functionalities and the synthesis route on the type of the precipitated calcium phosphates were evaluated during the precipitation process. The varied way of solution mixing has an effect on the initial pH and the microenvironment for

CaP's precipitation during the synthesis and hence on the type of pre-nucleation clusters, according to the non-classical theory of CaP crystallization [35].

In the case of Series A, where the Ca solution was added to the P solution, HPO_4^{2-} ions dominate at the beginning and the formation of $Ca(HPO_4)_3^{4-}$ pre-nucleation clusters is expected to prevail over other ones [36]. The starting pH is high (a pH of 8.9, Figure 1a,c,e) which results in the formation of amorphous calcium phosphate with the common formula $Ca_x(HPO_4)_y(PO_4)_z$ and Ca/P = 1.1–1.6 [37].

In the case of Series B (the P solution was added to the Ca solution), Ca^{2+} ions prevail in the initial solution. The pH was 4.5 for the system with PSB (Figure 1d), 5.7 for the system with PCB (Figure 1f), and 6.3 for the polymer-free system (Figure 1b). Thus, it was more probable that $CaHPO_4^0$ pre-nucleation clusters, which are structural units of DCPD, would be dominant [38]. Further, the composition of the pre-nucleation clusters, and the solid phase formed by them, changed depending on the conditions in the systems.

The different course of the changes in pH and in the concentration of the free Ca^{2+} ions during precipitation in the presence of PSB or PCB in both series (Figure 1) is proof of the influence of the distinct nature of the negatively charged functional group which results in different dipole moments of the formed zwitterionic (betaine) structure. In the case of the polymer-free system or in the presence of PSB, which has a sulfo group that is a stronger acidic group as compared to the carboxylic group, the pH changes to a greater extent (Figure 1a–d). The step change in the concentration of the free Ca^{2+} ions (Figure 1a–d) is an indication of the formation of different pre-nucleation clusters during the synthesis.

The carboxylic group of PCB enters in acid–base interactions, acting as a buffer, which explains the gradual and smaller pH changes observed for the PCB-controlled CaP precipitation, accompanied by a smooth change in the free Ca^{2+} concentration (Figure 1e,f). Also, the polymer macromolecules of PCB interact more actively with the formed pre-nucleation clusters than those of PSB. According to Gebauer [35], the interactions can be either adsorption on the surface or complexation with the Ca^{2+} ions. Thus, the formation of new clusters is hindered, and the formation of amorphous intermediate phases is promoted. In the PSB system, such interactions are suppressed. The dipole moment of the PSB zwitterionic structure is weaker than the PCB one and determines weaker interactions with the surrounding ions and water molecules. Moreover, the smaller size of the $CaHPO_4^0$ clusters than the $Ca(HPO_4)_3^{4-}$ ones inhibits polymer adsorption. In addition, the more acidic character of the sulfo group than the carboxy one determines the lower pH during the synthesis and dominance of the acidic solid phases in the precipitates in the system with PSB (Figures 2 and 3).

4.2. Characterization and Actions of a Newly Synthesized Hybrid Material

Two newly synthesized hybrid materials were chosen to be characterized and for preliminary testing of their remineralization potential, namely, A-PCB/CaP and A-PSB/CaP, due to the absence of an acidic calcium phosphate phase in A-CaP/PCB and the smaller amount of DCPD in A-CaP/PSB (Figure 3). The Acidic phases are not characteristic of normal calcifications in the human body, although DCPD and OCP are considered precursors for obtaining HA, along with ACP [39,40]. The CaP-to-polymer ratio in both materials is 1:1 (Figure 6b,c). The mineral phase, A-PCB/CaP, contains ACP (Figures 2 and 3) with thin acicular, uniformly oriented particles arranged in bundles (Figure 5). A-PSB/CaP contains a mixture of OCP and DCPD (Figures 2 and 3) with larger-sized and differently shaped particles (Figure 4).

Remineralization is the process by which the enamel crystals must regain their size, shape, and strength. Non-fluoride approaches imply the involvement of calcium and phosphate ions from a source external to the tooth to promote ion deposition on the enamel crystals [41]. Such a source is saliva, but the amount of calcium and phosphate ions in saliva is not sufficient for complete remineralization [42]; thus, pre-formed crystals of nano HA or other types of calcium phosphates were used, along with bioactive organic substances, for targeting mineral deposits [43].

The use of a suspension of the hybrid materials, prepared in this study, was intended to induce the penetration of the polymer hydrogel through the thinned enamel layer, which would aid both the introduction and orientation of the preliminarily obtained CaP particles, and the activation of the nucleation centers in the residual enamel. Thus, during the next cycle of immersion in artificial saliva, the processes of crystallization and re-crystallization in HA promoted the faster recovery of the enamel to be stimulated. This strategy needs to be carefully balanced to prevent the clogging of the tubules and the uptake of mineral ions into the lesion.

The comparative study of the remineralization potential of the two materials gives an advantage to A-PCB/CaP. Under the conditions of the experiment, it leads to the creation of a newly formed crystal layer similar to that of the underlying enamel (Figure 9e) and an increase in density in the lesion (Figure 8b). No extraneous mineral phases were found after the remineralization procedure except hydroxyapatite forming the tooth enamel (Figure 10).

The formation of more than one phase during the CaP synthesis within the presence of PSB increases the prospects of the obtained material (A-PSB/CaP) to be successful as a remineralization system because it contains two phases, namely, OCP and DCPD, which are both HA precursors. Moreover, due to their higher solubility, as compared to HA, they are both able to act as ion-releasing systems. According to us, the different solubility of OCP and DCPD [37] would lead to a continuous, gradual release of Ca^{2+} and PO_4^{3-} ions for enhancing the enamel remineralization. The advantage of A-PCB/CaP's performance could be sought rather in the ACP phase, where it prevails, as ACP is known to be unstable, and in an aqueous media, it is easily transformed into stable crystalline phases, such as octacalcium phosphate or apatite, due to the growing of microcrystalline ones [44]. Thus, the type of the CaP phase is the dominant factor defining the success of hybrid materials as remineralization systems.

Mineral phase deposition rather than enamel remineralization was observed as a result of our preliminary investigations. Although further in-depth studies are needed, including a variation of the suspension density and contact time, A-PCB/CaP shows promising properties for a material applicable in dentistry enamel remineralization.

The information gathered from this study can be used to plan future studies to test materials, but its findings cannot be used to guide clinical judgments. The effectiveness of these materials for remineralizing teeth needs to be further investigated, as in a laboratory environment, in clinical conditions too.

5. Conclusions

The effects of betaine zwitterionic functionality and the synthesis route on the biomimetic precipitation of CaP phases were investigated in this study by the monitoring of pH and Ca^{2+} ion concentration during the synthesis along with X-ray powder diffraction and solid-state NMR analyses. Further, preliminary studies on the remineralization potential of the selected new synthesized hybrid materials were carried out, and their effects were characterized by MictoCT, SEM, and IR analyses.

Both polymers used by us, PCB and PSB, have positively charged ammonium groups and negatively charged acidic, carboxy, or sulfo groups covalently bonded to their side chains, respectively. The different chemical behaviors of these groups in the solution determine their different interactions with the surrounding ions and with the pre-nucleation clusters formed during the synthesis. Thus, ACP was obtained to precipitate in the PCB-controlled system, while OCP and DCPD in different ratios were obtained in the PSB or polymer-free systems. The findings of our study demonstrate the significance of organic component functional groups in the biomineralization process.

The preliminary comparative study of the remineralization potential of the two as-prepared hybrid materials gives preference to that consisting of amorphous calcium phosphate and PCB. Under the conditions of the experiment, it leads to the creation of a newly formed crystal layer similar to that of the underlying enamel and an increase in density in the lesion.

Our research is a basis for building a future strategy for a more in-depth study of the remineralization potential of the obtained materials.

Supplementary Materials: The following supporting information can be downloaded at: https://www.mdpi.com/article/10.3390/ma16206640/s1, Scheme S1. RAFT polymerization to obtain PCB, Scheme S2. RAFT polymerization to obtain PSB, Figure S1. ^{1}H spectrum of PCB in D2O, Figure S2. ^{1}H spectrum of PSB in D2O, Figure S3. $^1H \rightarrow ^{31}P$ CP-MAS spectra of pure A-CaP and hybrid A-CaP/PSB materials from Series A. Figure S4. $^1H \rightarrow ^{31}P$ CP-MAS spectra of materials from Series B: pure B-CaP, hybrid B-CaP/PSB material, and hybrid B-CaP/PCB material measured at mixing times of 100 μs and 4 ms; The spectra are discussed in detail in the main text of the paper.

Author Contributions: Conceptualization, D.R., P.S., A.B. and E.V.; methodology, D.R., P.S., R.V. and E.V.; data curation, A.B., D.T., R.T. and M.S.; formal analysis, D.T.; investigation, R.G., K.R., M.S., D.T., R.T. and A.B.; writing—original draft preparation, D.R., P.S., A.B., R.T. and E.V.; writing—review and editing, D.R., P.S., A.B., R.T. and E.V.; visualization, D.R., P.S., R.T. and D.T. All authors have read and agreed to the published version of the manuscript.

Funding: This research was funded by the BULGARIAN NATIONAL SCIENCE FUND (BNSF), grant numbers KP-06-H49/6 (2020–2023) and D01-272.

Institutional Review Board Statement: Study approval was obtained from the Ethics Committee of the Medical University of Sofia, Sofia, Bulgaria, with protocol № 76/8717/18.12.2020.

Informed Consent Statement: Informed consent was obtained from all subjects involved in this study.

Data Availability Statement: The data presented in this study are available in this article and in the Supplementary Materials.

Acknowledgments: The authors thank the BULGARIAN NATIONAL SCIENCE FUND (BNSF) for their financial support, under project No KP-06-H49/6 (2020–2023). The authors acknowledge the TwinTeam project D01-272 "European Network on Materials for Clean Technologies" for providing them with the opportunity to present the results at the SizeMat4 conference as well as for financial publication support. Research equipment of the Distributed Research Infrastructure INFRAMAT, part of the Bulgarian National Roadmap for Research Infrastructures, supported by the Bulgarian Ministry of Education and Science, was used in this investigation.

Conflicts of Interest: The authors declare no conflict of interest.

References

1. Sharma, V.; Srinivasan, A.; Nikolajeff, F.; Kumar, S. Biomineralization process in hard tissues: The interaction complexity within protein and inorganic counterparts. *Acta Biomater.* **2021**, *120*, 20–37. [CrossRef] [PubMed]
2. Bartlett, J.D.; Smith, C.E.; Hu, Y.; Ikeda, A.; Strauss, M.; Liang, T.; Hsu, Y.-H.; Trout, A.H.; McComb, D.W.; Freeman, R.C.; et al. MMP20-generated amelogenin cleavage products prevent formation of fan-shaped enamel malformations. *Sci. Rep.* **2021**, *11*, 10570. [CrossRef] [PubMed]
3. Shaw, W.J.; Tarasevich, B.J.; Buchko, G.W.; Arachchige, R.M.J.; Burton, S.D. Controls of nature: Secondary, tertiary, and quaternary structure of the enamel protein amelogenin in solution and on hydroxyapatite. *J. Struct. Biol.* **2020**, *212*, 107630. [CrossRef] [PubMed]
4. Tao, J.; Hanson, E.; Dohnalkova, A.C.; Buchko, G.W.; Jin, B.; Shaw, W.J.; Tarasevich, B.J. Changes in the Cterminal, N-terminal, and histidine regions of amelogenin reveal the role of oligomer quaternary structure on adsorption and hydroxyapatite mineralization. *Front. Physiol.* **2022**, *13*, 1034662. [CrossRef] [PubMed]
5. Beniash, E.; Metzler, R.A.; Lam, R.S.; Gilbert, P.U. Transient amorphous calcium phosphate in forming enamel. *J. Struct. Biol.* **2009**, *166*, 133–143. [CrossRef]
6. Hamzehlou, S.; Aboudzadeh, M.A. Special Issue on "Multifunctional Hybrid Materials Based on Polymers: Design and Performance. *Processes* **2021**, *9*, 1448. [CrossRef]
7. Zhao, M.; Geng, Y.; Fan, S.; Yao, X.; Zhu, M.; Zhang, Y. 3D-printed strong hybrid materials with low shrinkage for dental restoration. *Compos. Sci. Technol.* **2021**, *213*, 108902. [CrossRef]
8. Prosheva, M.; Aboudzadeh, M.A.; Leal, G.-P.; Gilev, J.B.; Tomovska, R. High-Performance UV Protective Waterborne Polymer Coatings Based on Hybrid Graphene/Carbon Nanotube Radicals Scavenging Filler. *Part. Part. Syst. Charact.* **2019**, *36*, 1800555. [CrossRef]
9. Hench, L.L.; Jones, J.R. Bioactive Glasses: Frontiers and Challenges. *Front. Bioeng. Biotechnol.* **2015**, *3*, 194. [CrossRef]

10. Santos-Coquillat, A.; Martínez-Campos, E.; Sánchez, H.M.; Moreno, L.; Arrabal, R.; Mohedano, M.; Gallardo, A.; Rodríguez-Hernández, J. Matykina, Hybrid functionalized coatings on Metallic Biomaterials for Tissue Engineering. *Surf. Coat. Technol.* **2021**, *422*, 127508. [CrossRef]
11. Xu, J.; Shi, H.; Luo, J.; Yao, H.; Wang, P.; Li, Z.; Wei, J. Advanced materials for enamel remineralization. *Front. Bioeng. Biotechnol.* **2022**, *10*, 985881. [CrossRef]
12. Combes, C.; Rey, C. Amorphous calcium phosphates: Synthesis, properties and uses in biomaterials. *Acta Biomater.* **2010**, *6*, 3362–3378. [CrossRef] [PubMed]
13. Salama, A. Cellulose/calcium phosphate hybrids: New materials for biomedical and environmental applications. *Int. J. Biol. Macromol.* **2019**, *127*, 606–617. [CrossRef] [PubMed]
14. Wang, W.; Zhang, B.; Zhao, L.; Li, M.; Han, Y.; Wang, L.; Zhang, Z.; Li, J.; Zhou, C.; Liu, L. Fabrication and Properties of PLA/Nano-HA Composite Scaffolds with Balanced Mechanical Properties and Biological Functions for Bone Tissue Engineering Application. *Nanotechnol. Rev.* **2021**, *10*, 1359–1373. [CrossRef]
15. Pupilli, F.; Ruffini, A.; Dapporto, M.; Tavoni, M.; Tampieri, A.; Sprio, S. Design Strategies and Biomimetic Approaches for Calcium Phosphate Scaffolds in Bone Tissue Regeneration. *Biomimetics* **2022**, *7*, 112. [CrossRef]
16. Hoshi, M.; Taira, M.; Sawada, T.; Hachinohe, Y.; Hatakeyama, W.; Takafuji, K.; Tekemoto, S.; Kondo, H. Preparation of Collagen/Hydroxyapatite Composites Using the Alternate Immersion Method and Evaluation of the Cranial Bone-Forming Capability of Composites Complexed with Acidic Gelatin and b-FGF. *Materials* **2022**, *15*, 8802. [CrossRef]
17. Bartmanski, M.; Rosciszewska, M.; Wekwejt, M.; Ronowska, A.; Nadolska-Dawidowska, M.; Mielewczyk-Gryn, A. Properties of New Composite Materials Based on Hydroxyapatite Ceramic and Cross-Linked Gelatin for Biomedical Applications. *Int. J. Mol. Sci.* **2022**, *23*, 9083. [CrossRef]
18. Örlygsson, G.; Laxdal, E.H.; Kárason, S.; Dagbjartsson, A.; Gunnarsson, E.; Ng, C.H.; Einarsson, J.M.; Gíslason, J.; Jónsson, H., Jr. Mineralization in a Critical Size Bone-Gap in Sheep Tibia Improved by a Chitosan-Calcium Phosphate-Based Composite as Compared to Predicate Device. *Materials* **2022**, *15*, 838. [CrossRef]
19. Zagho, M.M.; Hussein, E.A.; Elzatahry, Á.A. Recent overviews in functional polymer composites for biomedical applications. *Polymers* **2018**, *10*, 739. [CrossRef]
20. Furko, M.; Balázsi, K.; Balázsi, C. Calcium Phosphate Loaded Biopolymer Composites—A Comprehensive Review on the Most Recent Progress and Promising Trends. *Coatings* **2023**, *13*, 360. [CrossRef]
21. Jee, S.S.; Thula, T.T.; Gower, L.B. Development of bone-like composites via the polymer-induced liquid-precursor (PILP) process. Part 1: Influence of polymer molecular weight. *Acta Biomater.* **2010**, *6*, 3676–3686. [CrossRef] [PubMed]
22. Katti, K.S.; Ambre, A.H.; Peterka, N.; Katti, D.R. Use of unnatural amino acids for design of novel organomodified clays as components of nanocomposite biomaterials. *Philos. Trans. R. Soc. A Math. Phys. Eng. Sci.* **2010**, *368*, 1963–1980. [CrossRef]
23. Palazzo, B.; Walsh, D.; Iafisco, M.; Foresti, E.; Bertinetti, L.; Martra, G.; Bianchi, C.L.; Cappelletti, G.; Roveri, N. Amino acid synergetic effect on structure, morphology and surface properties of biomimetic apatite nanocrystals. *Acta Biomater.* **2009**, *5*, 1241–1252. [CrossRef] [PubMed]
24. Goloshchapov, D.; Kashkarov, V.; Nikitkov, K.; Seredin, P. Investigation of the Effect of Nanocrystalline Calcium Carbonate-Substituted Hydroxyapatite and L-Lysine and L-Arginine Surface Interactions on the Molecular Properties of Dental Biomimetic Composites. *Biomimetics* **2021**, *6*, 70. [CrossRef] [PubMed]
25. Liu, P.; Song, J. Sulfobetaine as a zwitterionic mediator for 3D hydroxyapatite mineralization. *Biomaterials* **2013**, *34*, 2442–2454. [CrossRef] [PubMed]
26. Liu, P.; Emmons, E.; Song, J. A comparative study of zwitterionic ligands-mediated mineralization and the potential of mineralized zwitterionic matrices for bone tissue engineering. *J. Mater. Chem. B Mater. Biol. Med.* **2014**, *2*, 7524–7533. [CrossRef] [PubMed]
27. Słota, D.; Florkiewicz, W.; Piętak, K.; Pluta, K.; Sadlik, J.; Miernik, K.; Sobczak-Kupiec, A. Preparation of PVP and betaine biomaterials enriched with hydroxyapatite and its evaluation as a drug carrier for controlled release of clindamycin. *Ceram. Int.* **2022**, *48*, 35467–35473. [CrossRef]
28. Mangal, U.; Kwon, J.-S.; Choi, S.-H. Bio-Interactive Zwitterionic Dental Biomaterials for Improving Biofilm Resistance: Characteristics and Applications. *Int. J. Mol. Sci.* **2020**, *21*, 9087. [CrossRef]
29. Ruseva, K.; Ivanova, K.; Todorova, K.; Vladov, I.; Nanev, V.; Tzanov, T.; Hinojosa-Caballero, D.; Argirova, M.; Vassileva, E. Antibiofilm poly(carboxybetaine methacrylate) hydrogels for chronic wounds dressings. *Eur. Polym. J.* **2020**, *132*, 109673. [CrossRef]
30. Massiot, D.; Fayon, F.; Capron, M.; King, I.; Le Celve, S.; Alonson, B.; Durand, J.O.; Bujoli, B.; Gan, Z.H.; Hoatson, G. Modelling one- and two-dimensional solid-state NMR spectra. *Magn. Reson. Chem.* **2002**, *40*, 70–76. [CrossRef]
31. Bonchev, A.; Simeonov, M.; Shestakova, P.; Vasileva, R.; Titorenkova, R.; Apostolov, A.; Dyulgerova, E.; Vassileva, E. Bioinspired Remineralization of Artificial Caries Lesions Using PDMAEMA/Carbomer/Calcium Phosphates Hybrid Microgels. *Gels* **2022**, *8*, 681. [CrossRef] [PubMed]
32. Klimek, J.; Hellwig, E.; Ahrens, G. Fluoride taken up by plaque, by the underlying enamel and by clean enamel from three fluoride compounds in vitro. *Caries Res.* **1982**, *16*, 156–161. [CrossRef] [PubMed]
33. Yokoi, T.; Goto, T.; Kato, T.; Takahashi, S.; Nakamura, J.; Sekino, T.; Ohtsuki, C.; Kawashita, M. Hydroxyapatite Formation from Octacalcium Phosphate and Its Related Compounds: A Discussion of the Transformation Mechanism. *Bull. Chem. Soc. Jpn.* **2020**, *93*, 701–707. [CrossRef]

34. Jegova, G.; Titorenkova, R.; Rashkova, M.; Mihailova, B. Raman and IR reflection micro-spectroscopic study of Er: YAG laser treated permanent and deciduous human teeth. *J. Raman Spectrosc.* **2013**, *44*, 1483–1490. [CrossRef]
35. Gebauer, D. How Can Additives Control the Early Stages of Mineralisation? *Minerals* **2018**, *8*, 179. [CrossRef]
36. Habraken, W.; Tao, J.; Brylka, L.; Friedrich, H.; Bertinetti, L.; Schenk, A.S.; Verch, A.; Dmitrovic, V.; Bomans, P.H.H.; Frederik, P.M.; et al. Ion-association complexes unite classical and non-classical theories for the biomimetic nucleation of calcium phosphate. *Nat. Commun.* **2013**, *4*, 1507. [CrossRef]
37. Dorozhkin, S.V. Amorphous calcium (ortho)phosphates. *Acta Biomater.* **2010**, *6*, 4457–4475. [CrossRef]
38. Rabadjieva, D.; Sezanova, K.; Gergulova, R.; Titorenkova, R.; Tepavitcharova, S. Precipitation and phase transformation of dicalcium phosphate dihydrate in electrolyte solutions of simulated body fluids: Thermodynamic modeling and kinetic studies. *J. Biomed. Mater. Res. Part A* **2020**, *108*, 1607–1616. [CrossRef]
39. Lakrat, M.; Jodati, H.; Mejdoubi, E.M.; Evis, Z. Synthesis and characterization of pure and Mg, Cu, Ag, and Sr doped calcium-deficient hydroxyapatite from brushite as precursor using the dissolution-precipitation method. *Powder Technol.* **2023**, *413*, 118026. [CrossRef]
40. Rahmani, F.; Larbi Bouamrane, O.; Ben Bouabdallah, A.; Atanase, L.I.; Hellal, A.; Apintiliesei, A.N. Biomimetic Hydroxyapatite Crystals Growth on Phosphorylated Chitosan Films by In Vitro Mineralization Used as Dental Substitute Materials. *Polymers* **2023**, *15*, 2470. [CrossRef]
41. Cochrane, N.J.; Cai, F.; Huq, N.L.; Burrow, M.F.; Reynolds, E.C. New approaches to enhanced remineralization of tooth enamel. *J. Dent. Res.* **2010**, *89*, 1187–1197. [CrossRef] [PubMed]
42. Reynolds, E.C. Calcium phosphate-based remineralization systems: Scientific evidence? *Aust. Dent. J.* **2008**, *53*, 268–273. [CrossRef] [PubMed]
43. Carey, C.M. Remineralization of Early Enamel Lesions with Apatite-Forming Salt. *Dent. J.* **2023**, *11*, 182. [CrossRef] [PubMed]
44. Zhao, J.; Liu, Y.; Sun, W.B.; Zhang, H. Amorphous calcium phosphate and its application in dentistry. *Chem. Cent. J.* **2011**, *5*, 40. [CrossRef] [PubMed]

materials

MDPI

Article

Ultra-Short Laser-Assisted Micro-Structure Formations on Mg/Zn Double-Doped Calcium Phosphate Ceramics for Enhanced Antimicrobial Activity

Albena Daskalova [1], Kostadinka Sezanova [2], Liliya Angelova [1], Tsvetelina Paunova-Krasteva [3], Rumiana Gergulova [2], Daniela Kovacheva [2] and Diana Rabadjieva [2,*]

1 Institute of Electronics, Bulgarian Academy of Sciences, 1784 Sofia, Bulgaria; albdaskalova@gmail.com (A.D.); lily1986@abv.bg (L.A.)
2 Institute of General and Inorganic Chemistry, Bulgarian Academy of Sciences, 1113 Sofia, Bulgaria; ksezanova@abv.bg (K.S.); rumigg@yahoo.com (R.G.); didka@svr.igic.bas.bg (D.K.)
3 Institute of Microbiology, Bulgarian Academy of Sciences, 1113 Sofia, Bulgaria; pauny@abv.bg
* Correspondence: didiarab@svr.igic.bas.bg

Abstract: Bacterial infection is one of the most common and harmful medical issues following the implantation of materials and devices in the body leading to antibiotic resistance of diverse bacterial strains. In this work, a novel approach is presented combining adopted laser-based patterning method in addition to doping (Mg and Zn) metal ions to prepare calcium phosphate ceramic substrate, applicable in medicine, with enhanced surface antimicrobial characteristics. The preparation of tablets containing Mg (8.44 mol%) and Zn (2.63 mol%) β-tricalcium phosphate involved biomimetic precipitation of amorphous calcium phosphate in media of simulated body fluid enriched with Mg^{2+} and Zn^{2+} ions as well as the presence of valine as an organic additive, followed by step-wise calcination up to 1000 °C. The results from laser processing showed formation of deeper patterns with increased surface roughness (from 4.9 μm to 9.4 μm) as laser power and velocity increase, keeping constant the hatch sizes of 50 μm. The textured surfaces consist of peaks and valleys arrangement that change the morphology of *Escherichia coli* cells and decrease of cell viability. Our study reveals the possibilities of the application of ultra-short laser radiation as a potential alternative therapy for controlling the antimicrobial effect of the ceramic surface.

Keywords: doped tricalcium phosphate; ceramic tablets; femtosecond laser processing; surface modification; antimicrobial characteristics

Citation: Daskalova, A.; Sezanova, K.; Angelova, L.; Paunova-Krasteva, T.; Gergulova, R.; Kovacheva, D.; Rabadjieva, D. Ultra-Short Laser-Assisted Micro-Structure Formations on Mg/Zn Double-Doped Calcium Phosphate Ceramics for Enhanced Antimicrobial Activity. *Materials* **2023**, *16*, 6626. https://doi.org/10.3390/ma16206626

Academic Editor: Csaba Balázsi

Received: 4 September 2023
Revised: 29 September 2023
Accepted: 7 October 2023
Published: 10 October 2023

1. Introduction

The discovery of alternative materials with enhanced antimicrobial characteristics that will help decrease the use of antibiotics has become a priority in developing innovative solutions for fighting the increasing antibiotic resistance of diverse bacterial strains. In the field of orthopedics, implant replacement surgery is becoming more common, especially as society becomes older. One of the most frequent and dangerous medical problems following device implantation is bacterial infection. The ubiquitous use of antibiotics to treat it leads to the development of antibiotic-resistant strains. Thus, the search for new methods or combinations of several methods for more effective antibacterial protection of implants and devices is the subject of increasing research interest.

Calcium orthophosphates are non-toxic biocompatible substances possessing osteoconductive properties, which allow them to be implanted in the body, since they do not cause an adverse immune response [1]. Moreover, their major ionic composition and crystal structure are similar to these of biological apatite. One of the main characteristics of biological apatite is that it contains small amounts of incorporated Na^+, K^+, Mg^{2+}, Zn^{2+}, Sr^{2+}, etc., ions [1–3]. Therefore, interest in the synthesis of ion-substituted calcium phosphate materials approaching the composition of hard tissues has increased. The aspiration of scientists is to obtain materials with improved bioadaptable properties. Zn-doped hydroxyapatite (HA) has been shown to have a better antibacterial effect and osteoblastic proliferation activity [4]. Sulfonated 2 poly(etheretherketone)/(Sr and Ce) co-substituted composite coating of HA possesses increased corrosion resistance and antibacterial activity [5]. In mesenchyme stem cells derived from human bone marrow, Mg-doped biphasic calcium phosphate nanoparticles containing silver showed great cytocompatibility [6]. Zn- [7] or Sr-, Zn- and Mn- [8] containing β-tricalcium phosphate (β-TCP) have been shown to favor the self-repair of the bone. Mg and Zn are preferable as substituents since they are essential for the organism, playing an important role in the formation and growth of the skeleton [9,10]. Mg is essential to living cells such as osteoblasts and osteoclasts [11]. Moreover, deficiency of both Mg and Zn contributes to the development of osteoporosis [11,12].

The main types of calcium-orthophosphate-based biomaterials, differing in preparation techniques and physical–mechanical properties are calcium-phosphate ceramics, cellular and tissue scaffolds, cements, composites and coatings on metal implants improving their surface properties and promoting cellular orientation, adhesion and proliferation. Calcium orthophosphate ceramics, which are already actively used in medicine, are usually based on HA, α- and β-tricalcium-phosphates (TCP) or multiphase formulations based on them [2,3]. One of the most important qualities of ceramic calcium-phosphate-based bone implants is to be resorbable, which requires their rate of dissolution to match the rate of new bone formation [13] to ensure that they will be efficiently absorbed by the body. Solubility diagram of the system $Ca(OH)_2$-H_3PO_4-H_2O at 37 °C, prepared by thermodynamic calculations [14], show that at physiological pH of 7.2–7.4, HA possesses the lowest solubility among all other calcium phosphates, followed by β-TCP (with an order of magnitude difference). The solubility of α-TCP is several orders of magnitude higher. This defines HA ceramics as bioactive, i.e., they stimulate the formation of new tissue on the surface but are not resorbed. From this point of view, β-TCP bioceramics are attractive materials for bone reconstruction and remodeling. Moreover, the degree of dissolution of β-TCP is close to that of bone apatite [15].

The surface topography of calcium phosphate ceramic materials, scaffolds or coatings is of primary importance for osseointegration processes, since they participate in bone-bonding ability and enhance new bone [2,16]. Different techniques were developed for surface treatment to improve its osteogenic and antibacterial properties [16]. The classical method is incorporation of ions with antibacterial effects, such as Ag^+, Cu^{2+} or Zn^{2+} [16,17]. The main disadvantage of this method is the concentration limit, especially for Cu and Zn, above which they can cause harm to the organism. Another type of chemical processing of the surface is with acid–alkali solutions, polymer-containing systems, etc. [17,18]. Surface roughness is a key factor in cell proliferation and growth and as well as in increases in antibacterial activity [16,19,20]. Sandblasting, polishing with SiC papers, acid etching, plasma spring or anodizing is usually used [20,21]. The problem is that more than one processing technique is required, and contamination is possible. As an alternative to conventional methods, the structuration of surfaces by ultrashort-pulsed (femtosecond) lasers introduces contactless patterning of the surfaces, created in a single step, with improved long-term stability of the created structures [22,23]. Successful attempts were performed by Faria et al. [24], by utilizing a hybrid laser-assisted approach for material functionalization of HA-coated zirconia, whereas a surface-patterning technique will prevent the detachment of ceramic coatings. The adopted laser-based approach for producing the functionalized surfaces demonstrates its efficacy in the creation of reproducible micro-textured surfaces coated with the bioactive phase, which assured bioactive retention and non-degradation [24].

Ultra-short laser–matter interactions have become a useful tool for gentle material processing due to the nature of the interaction process in the fs time scale [25,26]. The energy deposition in this regime is highly localized with the capability to ionize a large number of electrons [27]. The surface characteristics of the ceramic biomaterial influence the cell-based interactions, protein adherence, and antimicrobial properties, thus affecting the osseointegration process [28,29]. Therefore, the implant surface topography and chemistry have to be considered when designing the ideal scaffolds. In this study, we will present the combination of laser-induced surface texturing with an examination of doped β-tricalcium phosphate ceramic for the creation of orthopedic implants, to achieve improved biological efficiency and antibacterial activity. Exploring the impact of the laser processing parameters on the efficiency of the obtained patterns has been the focus of numerous studies aimed at minimizing the heat-affected zones and optimizing surface roughness for diverse applications in biomedicine. Research efforts have been made to define the best conditions to process brittle ceramic material with different experimental approaches. Reports on the application of femtosecond laser radiation to process bioceramic material are scarce. However, there are several research efforts that prove the efficacy of the method. For example, Zhang et al. [30] reported successful machining of zirconia (ZrO_2) ceramics utilizing the pump-probe imaging method. They were able to obtain precision processing in the laser drilling of zirconia ceramics. Chen et al. [31] studied the influence of femtosecond laser parameters on laser ablation of Al_2O_3 ceramics. They found that the phase transition develops from solid to vapor without the initiation of melting, thus leading to precise, well-defined micrometer-sized structures.

In this research, a systematic study of diverse patterning designs, showing the dependence of surface characteristics in relation to antimicrobial effects, allows definition of reproducible laser patterning models and eventual achievement of best conditions for microbial biofilm rupture. The aim of this paper is a comparative experimental study of the surface modification of biomimetically synthesized Mg and Zn double-doped β-tricalcium phosphate ceramic, textured by a femtosecond laser to evaluate different processing modes. In addition, a preliminary evaluation of *Escherichia coli* viability and changes in cell morphology were performed on the selected samples to demonstrate that the implementation of laser irradiation for producing surfaces with increased percentage of roughness in addition to doping metal ions could potentially lead to formation of nano- and micro-scale or hybrid surface structures with a potential antimicrobial effect.

2. Materials and Methods

2.1. Synthesis of Mg/Zn Double-Doped β-TCP and Preparation of Ceramic Tablets

Ceramic tablets of Mg/Zn double-doped β-TCP were prepared by a three-step method: (i) biomimetic synthesis of amorphous (Mg, Zn)-modified calcium phosphate; (ii) tablet preparation; and (iii) calcination of the tablets.

A method of continuous precipitation was used to synthesize amorphous (Mg, Zn)-modified calcium phosphate. Calcium- and phosphorus-free conventional simulated body fluids [32], enriched with valine (Sigma-Aldrich, St. Lois, MO, USA, A.R.) were used as solvents for K_2HPO_4, (Merck, Darmstatd, Germany A.R.), $CaCl_2{\cdot}2H_2O$ (Sigma-Aldrich, A.R.), $MgCl_2{\cdot}6H_2O$ (Merck, A.R.) and $ZnCl_2{\cdot}2H_2O$ (Merck, A.R.) (Table 1).

Combined apparatus for automatic titration and controlled synthesis (Titrando 907, Metrohm AG, Herisau, Switzerland) was used. All reagents (Solution 1, Solution 2 and Solution 3, Table 1) were added to a buffer solution (Table 1) at room temperature and continuous stirring at a rate of 3 $mL{\cdot}min^{-1}$ keeping the pH at 8.0–8.2 using 0.05 M KOH. The pH of Solution 3 (Table 1) was not corrected to avoid Zn-hydrolysis products.

Concentration of dopants were selected to achieve Mg/(Ca + Mg + Zn) of 7 mole % and Zn/(Ca + Mg + Zn) of 3 mole %. In the syntheses, the Mg concentration was doubled.

After 2 h of precipitation, the suspension was matured in the mother liquor for 1 h under continuous stirring at room temperature and then washed with water (to remove chlorides) and lyophilized.

Table 1. Composition (mmol·L^{-1}) of the conventional and modified simulated body fluids and the buffer solution.

Components	SBFc [32]	Solution 1	Solution 2	Solution 3	Buffer Solution
Na^+	141.9	141.9	141.9	141.9	100
K^+	3.0	501	3	3	
Mg^{2+}	1.5	0	58.4	1.5	
Ca^{2+}	2.5	0	374	2.5	
Zn^{2+}	0	0	0	12.2	
Cl^-	143	143	1008	167	97.5
$SO_4{}^{2-}$	0.5	0.5	0.5	0.5	
$HCO_3{}^-$	4.2	4.2	4.2	4.2	
$HPO_4{}^{2-}$	1.0	250	0	0	
Valine		512	512	512	512
pH	7.2–7.4	8.0–8.2	8.0–8.2	6.5	8.0–8.2

The lyophilized powder was milled in a ball mill (Fritsch 6) for 0.5 h at 500 rpm, then passed through a 0.1 mm sieve. An amount of 0.65 g of the sieved powder was tableted with a pressure of 6.5 tons for 2 min on a SPECAC GS15011 press (Orpington, GB). Prepared tablets with a diameter of 13 mm were calcined at 200, 400, 600, 800 and 1000 °C and atmospheric pressure in high-temperature furnace (type VP 04/17, LAC Ltd. Company, Rajhrad, Czech Republic). The working regime was heating at a rate of 3 °C·min^{-1} until the desired temperature was reached and then keeping it constant for 3 h.

2.2. Processing with Femtosecond Laser

The patterned rows were acquired with linearly polarized laser pulses on Mg/Zn double-doped β-TCP disks using a Solstice Ace system, delivering laser pulses with a pulse duration of 70 fs, wavelength of λ = 800 nm and repetition frequency of f = 1 KHz at an average output power of the laser system up to 6 W. The focus spot diameter is 25 μm at $1/e^2$ of its intensity maximum at Gaussian-profile intensity distribution. The samples were positioned in air on an XY translation stage orthogonal to the laser beam. The motion stage was then rasterized at diverse scanning speed, generating lines over the sample surface. The energy delivery was controlled via a polarizing beam splitter. The sample stage scanning velocity (V) was varied between 0.2 mm/s and 15 mm/s. Different laser powers (P) and distances between separate rows (d_x), designed in the form of a mesh created by laser raster scanning of the surface in two perpendicular directions, were applied.

2.3. Characterization

2.3.1. Chemical Analysis

The sum of Ca^{2+}, Zn^{2+} and Mg^{2+} ions in the solid samples was determined complexometrically with EDTA at pH 10. The concentrations of Zn^{2+} and Mg^{2+} ions were determined by ICP-OES (PRODIGY 7, Teledyne, Leeman Labs, Hudson, NH, USA). P-$PO_4{}^{3-}$ and Cl^- ions were analyzed spectrophotometrically by NOVA 60 equipment (Darmstatd, Germany) using Merck and Spectroquant test kits.

2.3.2. XRD Analysis

Powder X-ray diffraction (PXRD) was performed using a Bruker D8 Advance diffractometer with Cu Kα radiation and a LynxEye detector (Bruker AXS Advanced X-ray Solutions GmbH, Billerica, MA, USA). For the primary phase identification, the data were collected in the range of 10 to 90° 2θ with step 0.03° 2θ, with counting time 57 s/step. The phase composition was identified using the ICDD-PDF2 (2014) database.

The Rietveld structure refinement was performed by using Bruker Topas v.4.2 program in order to proof substitution of Ca from Mg and Zn in the crystal structure of β-TCP. Powder diffraction patterns were collected at room temperature within the range of 5 to 120° 2θ with a step of 0.02° 2θ and 175 s/step counting time and a sample rotation of 15 rpm.

2.3.3. Confocal Microscopy

For surface and roughness evaluation of the laser-structured or control ceramic pellets, 3D confocal images were obtained by a 3D μsurf explorer confocal microscope (Nano-focus, Oberhausen, Germany). The apparatus provides high-resolution images of the surface. For 3D visualization of the confocal images obtained, ProfilmOnline software Version 3.0.1 (https://www.profilmonline.com) was used (accessed on 30 March 2023). Additional roughness analysis was performed; values of the Sq (the extension of the mean value of the deviations of the surface height from the median line to a surface area) roughness parameter were obtained as average values over 5 separate measurements in accordance with the ISO 4287 standard [33].

2.3.4. Raman Spectroscopy

The effect of the laser processing on the phase of the ceramic material was evaluated by Raman spectroscopy (Raman spectrometer Horiba MicroRaman, Kyoto, Japan). The Raman spectra were obtained using a 532 nm laser as a light source, and all reported spectra were averaged over five measurements, acquired in different points on the same sample.

2.3.5. Scanning Electron Microscopy (SEM) Analysis of Tablet Morphology and Microstructure

The morphology and microstructure of the annealed tablets was investigated using a Lyra I XMU Scanning electron microscope (Tescan, Czech Republic), while the femtosecond laser-treated tablets were investigated using JEOL JSM 6390 (Tokyo, Japan) apparatus. The samples were gold-sputtered (~20 nm Au layer) in vacuum, and SEM images were taken at several different magnifications.

2.3.6. Estimation of Bacterial Viability by Plate Count

The bacterial strain used in the study, *E. coli* 25922, was obtained from the American Type Culture Collection (ATCC). It was stored in 8% DMSO at a temperature of −80 °C. Prior to the experiments, the strain was inoculated in tryptic soy broth (TSB, Sigma-Aldrich, Berlin, Germany) and maintained at 4 °C on tryptic soy agar (TSA, Sigma-Aldrich, Germany) slants.

The ceramic tablets were pre-treated by dry sterilization at a temperature of 200 °C for 30 min. The antimicrobial activity was determined according to a standard evaluation test of antimicrobial surface designs with minor modifications [34]. A volume of 2 mL of overnight 1:100 diluted bacterial inoculum was co-cultivated with sterile ceramic tablets. After a cultivation period of 24 h at 37 °C, bacteria were collected, diluted in phosphate-buffered saline (PBS) and calibrated to 0.5 McFarland units (approximately 1×10^8 colony forming units CFU/mL) by using a Densilameter II (Microlatest). The number of surviving bacteria (CFU) was counted by transferring 100 μL of bacterial suspension onto TSA in a petri dish, followed by incubation for 24 h at 37 °C. The assay was performed in duplicate.

2.3.7. Evaluation of *E. coli* Cell Morphology by Scanning Electron Microscopy (SEM)

For the purpose of evaluating the SEM investigation, the bacteria were incubated overnight (18 h) in TSB at 37 °C. An overnight bacterial inoculum was prepared after 1:100 dilution in TSB. The diluted inoculum was then resuspended and placed in a 12-well plate, with 2 mL of the inoculum added to each well in the presence of sterile ceramic tablets.

Duplicate samples were prepared for each variant. The plate was incubated at 37 °C in a static condition for 24 h. After this interval, the laser-treated tablets were washed and fixed with 4% glutaraldehyde in 0.1M Na cacodylate buffer at 4 °C for 2 h. The tablets were then washed again in cacodylate buffer and post-fixed for 1 h using a solution of 1% OsO_4 at 4 °C. After washing twice with cacodylate buffer, a dehydration procedure was performed in a graded ethanol series (30, 50, 70, 80, 90 and 100%) through 15 min time intervals. Finally, the dehydrated tablets were installed on scanning electron microscopy holders and sputter-coated with a layer of gold using a vacuum evaporator (Edwards, Stansted, UK). The observations of the bacterial morphology were made on Lyra I XMU Scanning electron microscope (Tescan, Czech Republic) with an accelerating voltage of 20 kV.

3. Results

3.1. Synthesis of Mg/Zn Double-Doped β-TCP and Preparation of Ceramic Tablets

The chemical and XRD analysis of biomimetically precipitated powder showed the formation of amorphous calcium phosphate (Figure 1a) with (Ca + Mg + Zn)/P = 1.56 and concentration of Mg 8.44 mole % and Zn 2.63 mole %. In addition, 4.3 mmolCl·g^{-1}, and less than 0.036 mmolNa·g^{-1} and 0.06 mmolK·g^{-1}, were detected. After calcination, the amorphous products were transformed into Mg/Zn doped β-TCP (Figure 1e). The transformation started at 400 °C (Figure 1b) and was fully completed at 600 °C (Figure 1c).

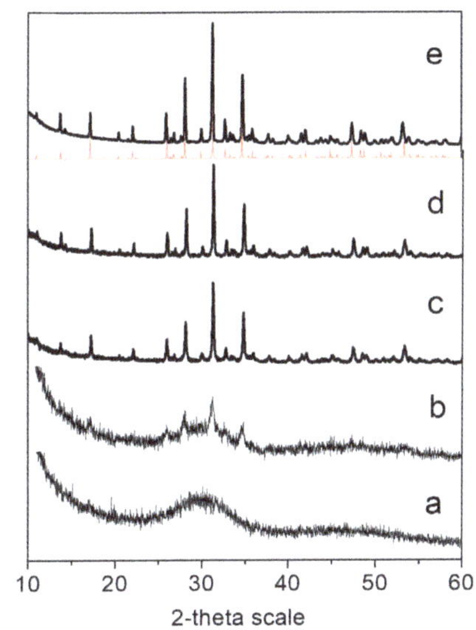

Figure 1. XRD studies of the initial (**a**) and the calcined at different temperature tablets: (**b**) 400 °C; (**c**) 600 °C; (**d**) 800 °C and (**e**) 1000 °C. Red line—$Ca_3(PO_4)_2$—PDF # 09-0169.

In order to determine the crystal structure of the substituted phase, a Rietveld refinement method was used. The best results were obtained under the assumption that Ca is replaced by Zn and Mg in the Ca(5) position in the structure of β-TCP [35]. The calculated unit cell parameters of samples calcined at 1000 °C reveal a decrease in their values in comparison with the unsubstituted β-TCP (Table 2), which is evidence that Mg and Zn are included in the crystal structure of β-TCP.

Table 2. Unit cell parameters for (Mg/Zn)-β-TCP heated at 1000 °C.

Sample	*a* [Å]	*c* [Å]	V, [Å^3]
β-TCP [35]	10.4352 (2)	37.4029 (5)	3482
This study	10.3238 (1)	37.2624 (6)	3439

The calculated geometric density of the amorphous calcium phosphate tablets was 2.21 g·cm^{-3}, which decreased by 32% (to 1.50 g·cm^{-3}) upon calcination. This significant volume contraction leads to the formation of a dense ceramic structure as seen in the SEM images (Figure 2).

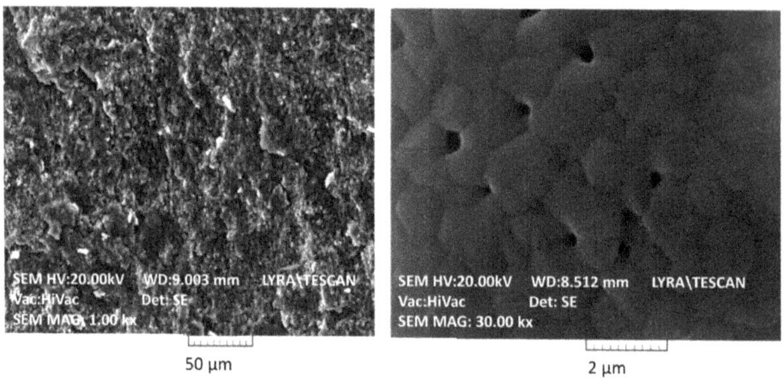

Figure 2. SEM images (at ×1000 [left] and ×30,000 [right] magnification) of ceramic pellets from Mg/Zn-β-TCP calcined at 1000 °C.

3.2. *Processing with Femtosecond Laser*

Ultra-short laser texturing was successfully applied to doped CaP substrates without monitoring changes in the ceramic microstructure. The laser modification induces selective material removal, thus inducing a gentle modification of the irradiated zone of the brittle ceramic material.

In order to evaluate the best texturing conditions, a precise selection of the laser parameters is required to create surface patterns with high quality.

Figure 3 shows the change in surface morphology with changing laser parameters. Obtained results demonstrate a decrease in surface roughness by increasing scanning velocities (V) at constant laser power (P) and distance between separate rows (d_x).

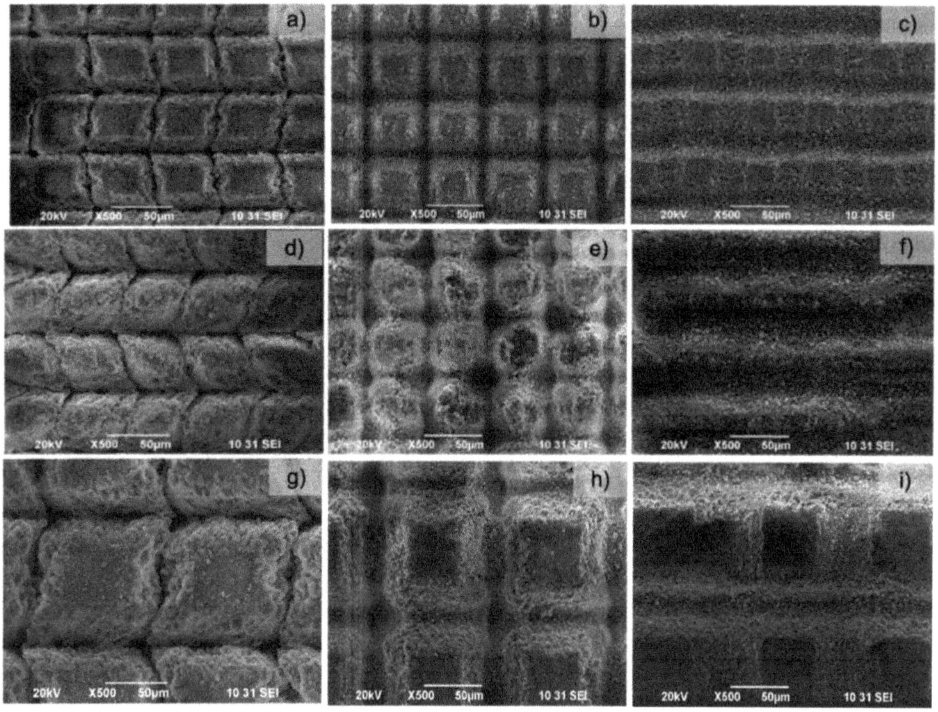

Figure 3. SEM images (at ×500 magnification) of laser-treated pellets with laser power (P) of 20 mW (**a–c**), 60 mW (**d–f**), and 100 mW (**g–i**), with distance between separate rows (d_x) = 50 μm and velocity 0.2 mm/s (**a**,**d**,**g**); 1.5 mm/s (**b**,**e**,**h**); and 7.5 mm/s (**c**,**f**,**i**).

By adjusting the laser scanning parameters, line- and cross-like patterns with diverse row depths were produced via changing the scanning velocities and the applied laser power. We observed a decrease in the pattern depth with scanning velocity increment. The morphological analysis demonstrates the formation of different granular and porous microstructures without evidence of cracks or quality deterioration. The laser-treated and non-treated (Mg/Zn)-β-TCP surfaces presented different surface morphologies (Figure 3a–f). The (Mg/Zn)-β-TCP surface treated with lower laser power (P = 20 mW) (Figure 3a–c) possesses slightly smoother structures. Furthermore, with increasing P to 60 mW (Figure 3d–f), a grain deposition of ablated material resembling on the top of and between the created microstructures was observed for V between 0.2 mm/s and 1.5 mm/s. The 3D confocal analysis (Figure 4) complements the SEM results and estimated relation between scanning speed, laser power and depth of the ablated craters.

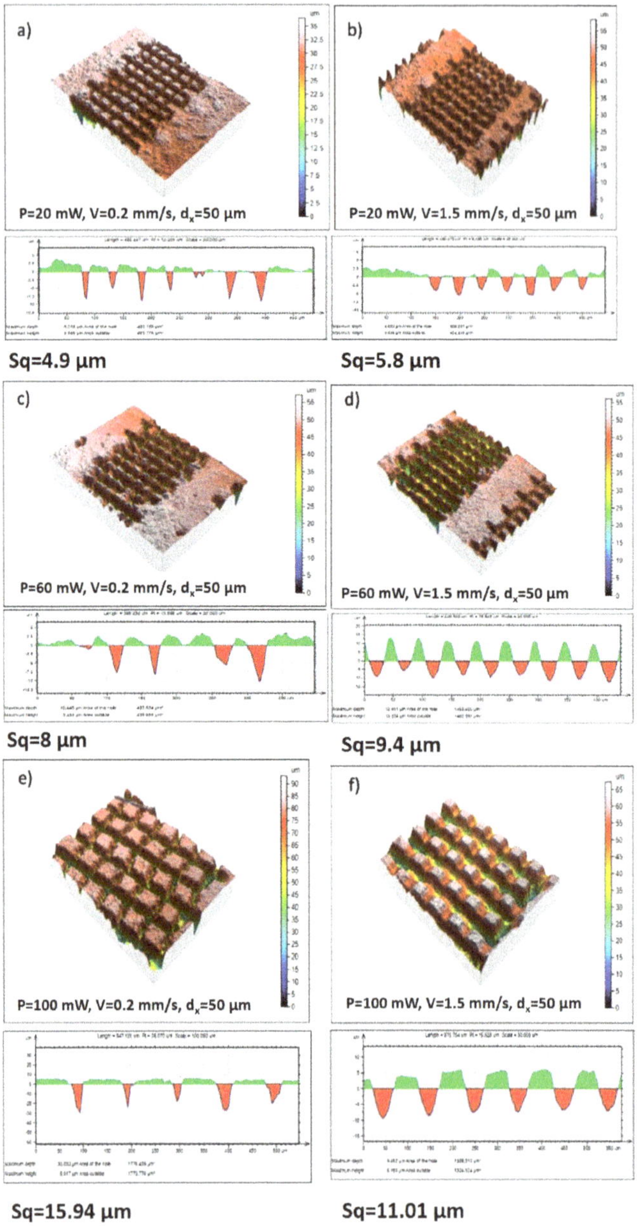

Figure 4. 3D confocal microscope (CLSM) images (at ×20 magnification) of femtosecond laser-treated pellets by different power and velocity: (**a**) P = 20 mW, V = 0.2 mm/s; (**b**) P = 20 mW, V = 1.5 mm/s; (**c**) P = 60 mW, V = 0.2 mm/s; (**d**) P = 60 mW, V = 1.5 mm/s; (**e**) P = 100 mW, V = 0.2 mm/s; (**f**) P = 100 mW, V = 1.5 mm/s.

It was found that increasing the depth of the single pattern was strongly influenced by the laser power. Crater width increased with increasing laser power, while increasing scanning speed to 1.5 mm/s led to a change in crater forms, from V- to U-shaped (Figure 4a,d). The surface roughness was assessed through an evaluation of the arithmetic average roughness (Sq) from the profiles obtained through confocal microscopy. A cross-sectional profile of the created structures is also presented. However, the increase in scanning velocity did not alter the repeatability of the patterns produced.

The textures represent highly ordered rectangle-like, V-shaped micropillars (in the case of V = 0.2 mm/s for P = 20 and 60 mW—Figure 3b,e) due to the laser raster scanning pattern used in two perpendicular directions, so a grid design of the surface is created. When a further increase in the scanning velocity is performed (V = 1.5 mm/s), the depth of the created microchannel around the created rectangles decreases, and the formation of clear pillar-like features diminishes (Figure 3d,f). The grid microporous structure is not well defined, nor are the channel depth and edges; no ejected material in the form of micropillars is observed, due to the increase in the scanning velocity applied and, as a consequence, more gentle laser-material interaction is observed. All of the structures obtained are highly and precisely reproducible and controlled by the precise selection of the applied laser parameters.

Both XRD (Figure 5) and Raman analysis (Figure 6) did not show the presence of α-TCP or other high-temperature calcium-phosphate phases after laser modifications with variation of scanning speed and applied laser power. As can be seen from Figure 5, the presence of β-TCP in all laser-processed and control pellets was confirmed by the reflection in accordance with standard $Ca_3(PO_4)_2$—PDF # 09-0169 (ICDD-PDF2 (2014) database).

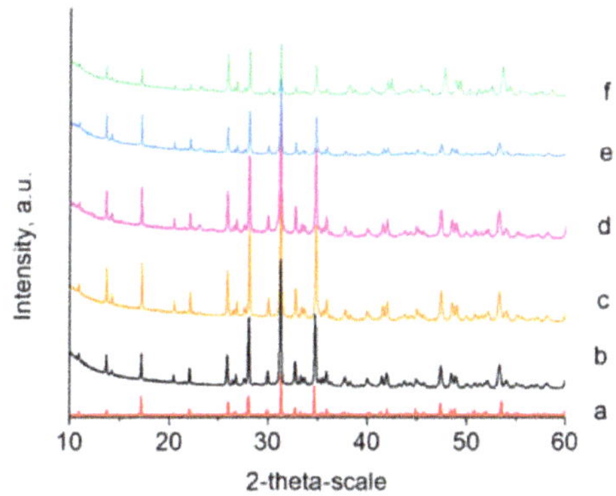

Figure 5. XRD studies of (Mg/Zn)-β-TCP ceramic pellets (**a**) standard $Ca_3(PO_4)_2$—PDF # 09-0169; (**b**) without laser modification; and laser-treated at (**c**) P = 20 mW, V = 1.5 mm/s, d_x = 50 μm; (**d**) P = 60 mW, V = 1.5 mm/s, d_x = 50 μm; (**e**) P = 100 mW, V = 1.5 mm/s, d_x = 100 μm; (**f**) P = 100 mW, 15 mm/s, d_x = 100 μm.

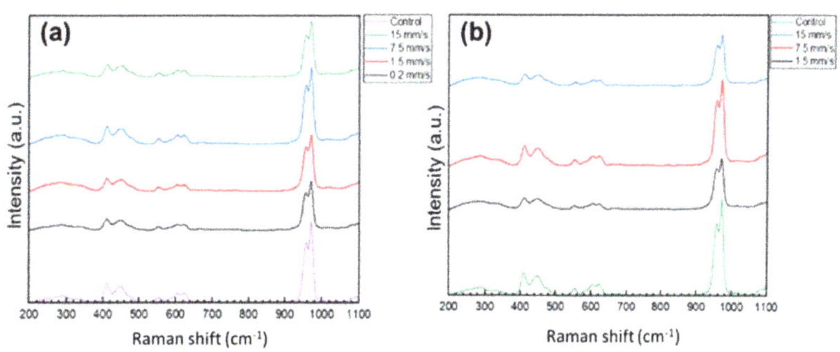

Figure 6. Raman spectra of the (Mg/Zn)-β-TCP ceramic pellets ablated in a grid pattern at (**a**) P = 20 mW, V= 0.2, 1.5, 7.5, 15 mm/s, d_x = 50 μm; and (**b**) P = 60 mW, V = 1.5, 7.5, 15 mm/s, d_x = 50 μm.

However, a decrease in V to 0.2 mm/s, P = 20 mW led to a slight decrease in the intensity of the Raman signal (Figure 6), but no change in the form or position of the peaks was observed after laser ablation of the ceramic material with respect to control surfaces. The complementary results obtained by the XRD and Raman analyses indicate once again that the femtosecond laser surface structuring of the (Mg/Zn)-β-TCP pellets is a contactless, finely controlled method that allows precise tuning of surface micromorphology without significantly affecting the crystallinity and phase of the ceramic material.

3.3. *Analysis of an Antimicrobial Effect*

To evaluate the viability of *E. coli* bacterial cells co-cultured with the sterile ceramic tablets, we applied the classic agar plating method. The plate-counting measurements verify the antimicrobial activity of the treated tablets (Figure 7). The percentage of surviving cells on the tablets treated with 20 mW and 60 mW parameters varied between 66% and 49% compared to control values. The most significant antimicrobial effect was observed during cultivation with the laser-treated tablets with 100 mW, where the viable cells were 37%. These findings are also confirmed by SEM observations, which revealed significant morphological damage to the bacterial cells, as well as alterations in cell size. The experiments were carried out two times, and the resulting data was used to calculate the average values. Mean values are given with standard deviations of $\leq$10%. Statistical analysis was performed using OriginPro 6.1.

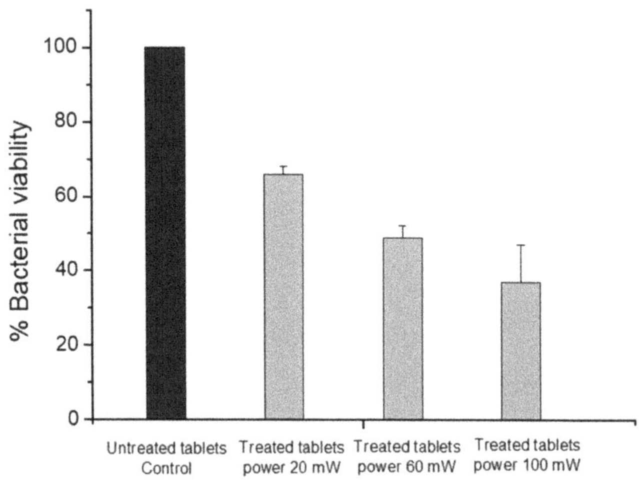

Figure 7. Assessment of the viability of bacterial cells co-cultivated with laser-treated ceramic tablets at three different laser powers (P = 20, 60, 100 mW) and scanning velocity of V = 1.5 mm/s when compared to the control samples (untreated tablets). Colony-forming units (CFUs) were counted and represented as a percentage of viable cells in comparison to colonies obtained from untreated tablets. The control samples were accepted as 100% and the other values were normalized to it.

SEM images of the *E. coli* cell morphology (Figure 8) demonstrate the surface topology and morphological changes in bacterial cells resulting from contact with the treated surfaces. We observed significant changes in the morphology of the bacterial cells after 24 h of cultivation in the presence of laser-treated ceramic tablets (Figure 8).

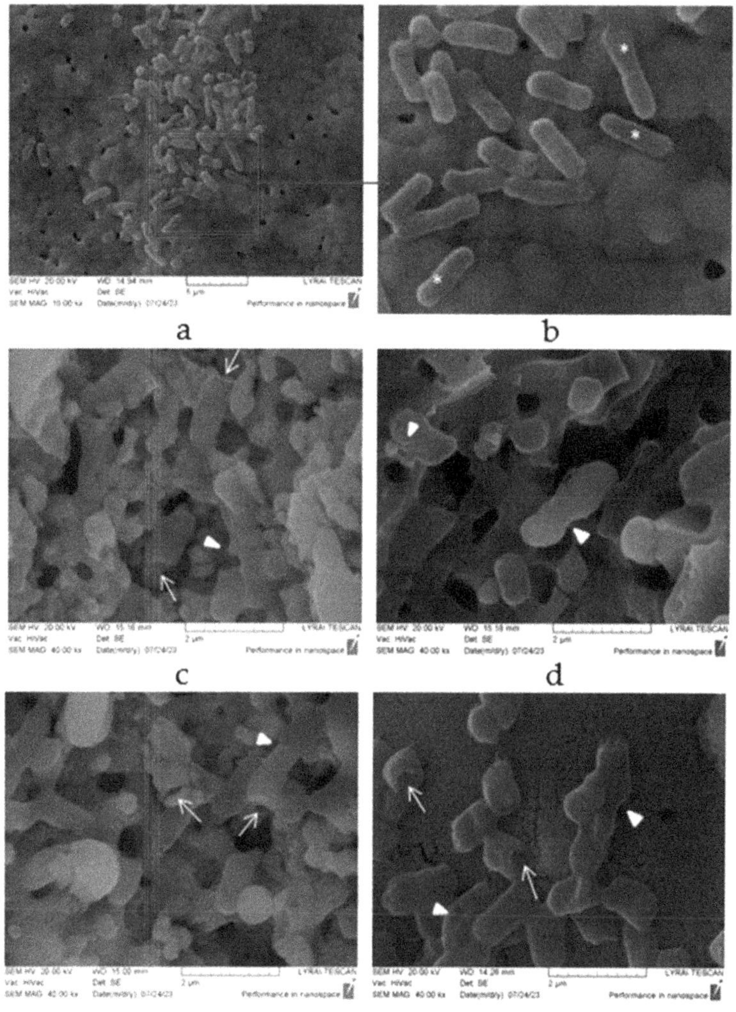

Figure 8. Scanning electron microscopy micrographs of *E. coli* 25922 strain, cultivated on laser-treated ceramic tablets: (**a**,**b**) control group (untreated tablets); (**c**) laser power (P) 20 mW, (**d**) 60 mW and (**e**,**f**) 100 mW and scanning velocity V = 1.5 mm/s. Designations: white stars—normal cells; white arrows—unipolar indentation; white triangles—structural invaginations. Bars = 2–5 μm.

It was shown that in the control probe, where the tablets were not treated with laser, the bacterial cells presented with normal morphology typical of the species. They appeared as rod-shaped cells with rounded ends and intact cell walls. Some cells in the population also showed the formation of bacterial septa, indicating the stage of cell division (Figure 8a,b). However, after the laser treatment, notable morphological changes were observed in the bacteria. The different laser parameters used in the treatment led to the deformation of the cell wall, resulting in the presence of infolds and grooves on the bacterial cell surfaces. These deformations potentially caused a loss of cellular integrity. Additionally, a portion of the bacterial cells became shorter in length, and their surfaces displayed unipolar indentations or the appearance of holes (Figure 8c–f).

4. Discussion

Calcium phosphates that have undergone ion modification are non-stoichiometric compounds in which some of the ions that build up the crystal structure have been replaced by others. The closer the admixture ions are in charge and chemical affinity to the parent ion, the easier the substitution. In this study, Mg/Zn double-doped β-TCP was prepared by a biomimetic approach that includes synthesis of amorphous precursor in a medium of simulated body fluid in the presence of valine as an organic additive. The selected concentrations of Zn (3 mole %) and Mg (7 mole %) are higher than the corresponding concentrations measured for human bones (up to 2.5 mole % Mg and up to 0.16 mole % Zn [36]) in order for the as-prepared material to act as a reservoir for Mg and Zn ions to maintain physiological extracellular concentrations. Our previous studies on the behavior of calcium phosphates substituted with Mg-only or Zn-only showed that they did not change the biocompatibility of the materials in a wide concentration range [37] and in contact with different simulated body fluids, released the ions step by step into the solutions [38,39]. Electrolyte environment of simulated body fluid that is similar to the inorganic part of blood plasma, provides additional inclusion of Cl^-, Na^+ and K^+ ions in the composition analogous to the composition of human bones and teeth [1,2]. Valine was chosen to mimic the organic biomolecules of blood plasma. Included in the composition of the initial modified body fluids (see Section 2.1), the valine forms complexes with Ca^{2+} and Mg^{2+} ions in Solution 2 (Table 1) and with Zn^{2+} ions in Solution 3 (Table 1) reducing the activity of free cations analogous to organic components in the blood plasma [40]. Since the calcium phosphate salts are poorly soluble, HPO_4^- ions from Solution 1 (Table 1) are associated with free metal ions during the synthesis procedure. The precipitation of ion-modified amorphous calcium phosphate leads to stepwise decomposition of the metal-valine complexes in order to obtain new portions of free metal ions and thus to their depletion in the solution. The released valine is washed away with the washing water.

The precipitation of amorphous calcium phosphates (Figure 1a) is consistent with the notion that kinetic rather than thermodynamic conditions determine the precipitation of amorphous calcium phosphate, as opposed to the thermodynamically stable HA, which we found previously [41] in systems with one doping ion and only simulated body fluid as the medium by precipitation. Both Mg and Zn, together (in this study) or separately [41], increase the stability of amorphous calcium phosphate and stimulate its transformation into Mg/Zn-substituted β-TCP (Figure 1). The contraction of unit cell obtained by Rietveld refinement (Table 2) confirms the substitution of Ca by Mg and Zn in the structure of β-tricalcium phosphate.

The results of laser-processing (Mg/Zn)-β-TCP showed the ability of fs laser sources to obtain highly accurate laser–matter structuration that can be implemented to functionalize the surface of brittle ceramic scaffolds by precisely adjusting the patterning geometry. The design of surface structures was influenced by the need to create reproducible surface features, by formation of textured surfaces with an impact on the behavior of bacterial cells.

By creating surface textures with increasing laser power and velocity, keeping constant the hatch sizes of 50 μm, formation of deeper patterns with increased surface roughness was observed (Figure 4d). It can be hypothesized that the bacterial cells would not be able to adhere, since the surface could not offer anchoring possibilities due to the laser-created roughness, which varied from 4.9 μm to 9.4 μm (Figure 4a–d). This would prevent the formation of multilayers in the bacterial population and subsequent biofilm formation, because of the additional patterning since the bacterial cells would not be able to overcome the gaps between the laser-carved areas. The textured surfaces consist of peaks and valleys (Figure 3), which can serve as a trap for bacteria. In the experiments, the maximum ablation depth that was achieved was approximately 10 μm regardless of the applied laser power, which could be explained with the development of partial shielding of the incoming laser pulse from the laser-generated plasma. This feature makes patterning conditions reproducible. From the cross-sectional confocal microscopy images, the ablated zones possess a conical form, which enlarges with increasing scanning velocities and applied laser power.

The data obtained from the analysis of an antimicrobial effect reveal that it increases with increasing laser power (P) (Figures 7 and 8). The lowest cell viability accompanied by deformation of the cell wall was obtained at P = 100 mW and scanning velocity V = 1.5 mm/s, which provide optimal surface roughness. Analysis of the *E. coli* cell viability (Figure 7) and morphology (Figure 8) give significant information on existing antimicrobial activity. Such promising results are a reasonable premise for expanding further antimicrobial research in order to improve the properties of ceramic tablets.

5. Conclusions

Tablets of Mg (8.44 mol %) and Zn (2.63 mol %) double-doped β-TCP were prepared by biomimetic precipitation of amorphous calcium phosphate in media of simulated body fluid enriched with Mg^{2+} and Zn^{2+} ions and in the presence of valine as an organic additive, followed by step-wise calcination up to 1000 °C.

Femtosecond laser structuring was successfully applied to the prepared substrates without monitoring changes in the ceramic microstructure. The laser modification induced selective material removal, thus causing a gentle modification of the irradiated zone of the brittle ceramic material. All structures obtained were highly and precisely reproducible and controlled by the precise selection of the applied laser parameters. The complementary results obtained by the SEM and roughness analyses, as well as by the XRD and Raman spectra, clearly demonstrate that the femtosecond laser surface structuring of the (Mg/Zn)-β-TCP pellets is a contactless, finely controlled method that allows precise surface micromorphology design and structuring without significantly affecting the crystallinity and phase of the ceramic material. Moreover, the laser-based patterning method adopted in this study, in combination with additional doping with Mg or Zn, represents a novel approach that was applied directly to enhance the antimicrobial activity of TCP biomaterial. The results demonstrate the potential of femtosecond laser radiation to tune laser-generated structures by texturing, as an alternative therapy for controllin the antibacterial effect of the ceramic surface.

Author Contributions: Conceptualization, A.D., D.R. and T.P.-K.; Methodology, A.D., D.R., L.A. and D.K.; Data curation, L.A., K.S. and R.G.; Formal analysis, D.K.; Investigation, L.A., K.S., R.G. and T.P.-K.; writing—original draft preparation, A.D., D.R. and T.P.-K.; writing—review and editing, A.D., D.R. and T.P.-K.; visualization, L.A. and K.S.; project administrations, A.D. and D.R. All authors have read and agreed to the published version of the manuscript.

Funding: This research was funded by Bulgarian National Science Fund (BNSF) grant numbers KP-06-H48/6 (2020–2023), KP-06-H49/6 (2020–2023) and TwinTeam project D01-272.

Institutional Review Board Statement: Not applicable.

Informed Consent Statement: Not applicable.

Data Availability Statement: The data presented in this study are available in this article.

Acknowledgments: The authors thank the Bulgarian National Science Fund (NSF) for the financial support of the laser-assisted studies under the project No KP-06-H48/6 (2020–2023), and of the material preparation under the project No KP-06-H49/6 (2020–2023). The authors acknowledge the TwinTeam project D01-272 "European Network on Materials for Clean Technologies" for providing the opportunity to present the results at the SizeMat4 conference as well as for the financial publication support.

Conflicts of Interest: The authors declare no conflict of interest.

References

1. Dorozhkin, S.V. Calcium orthophosphate ($CAPO_4$) scaffolds for bone tissue engineering applications. *J. Biotechnol. Biomed. Sci.* **2018**, *1*, 25–93. [CrossRef]

2. Dorozhkin, S.V. Calcium Orthophosphate ($CaPO_4$)-Based Bioceramics: Preparation, Properties, and Applications. *Coatings* **2022**, *12*, 1380. [CrossRef]
3. Samavedi, S.; Whittington, A.R.; Goldstein, A.S. Calcium phosphate ceramics in bone tissue engineering: A review of properties and their influence on cell behavior. *Acta Biomater.* **2013**, *9*, 8037–8045. [CrossRef] [PubMed]
4. Okada, M.; Oshita, M.; Kataoka, M.; Azuma, Y.; Furuzono, T. Shareability of antibacterial and osteoblastic-proliferation activities of zinc-doped hydroxyapatite nanoparticles in vitro. *J. Biomed. Mater. Res. B Appl.* **2022**, *110*, 799. [CrossRef] [PubMed]
5. Rajeswari, D.; Gopi, D.; Ramya, S.; Kavitha, L. Investigation of anticorrosive, antibacterial and in vitro biological properties of a sulphonated poly(etheretherketone)/strontium, cerium co-substituted hydroxyapatite composite coating developed on surface treated surgical grade stainless steel for orthopedic applications. *RSC Adv.* **2014**, *4*, 61525–61536. [CrossRef]
6. Yang, N.; Wang, S.; Ding, P.; Sun, S.; Wei, Q.; Jafari, H.; Wang, L.; Han, Y.; Valentine, O.O.; Wang, T.; et al. Magnesium-doped biphasic calcium phosphate nanoparticles with incorporation of silver: Synthesis, cytotoxic and antibacterial property. *Mater. Lett.* **2022**, *322*, 132478. [CrossRef]
7. Luo, X.; Barbieri, D.; Davison, N.; Yan, Y.; de Bruijn, J.D.; Yuan, H. Zinc in calcium phosphate mediates bone induction: In vitro and in vivo model. *Acta Biomater.* **2014**, *10*, 477–485. [CrossRef] [PubMed]
8. Ferreira, M.; Brito, A.; Brazete, D.; Pereira, I.; Carrilho, E.; Abrantes, A.; Pires, A.; Aguiar, M.; Carvalho, L.; Botelho, M.; et al. Doping β-TCP as a Strategy for Enhancing the Regenerative Potential of Composite β-TCP—Alkali-Free Bioactive Glass Bone Grafts. Experimental Study in Rats. *Materials* **2019**, *12*, 4. [CrossRef]
9. Boanini, E.; Gazzano, M.; Bigi, A. Ionic substitutions in calcium phosphates synthesized at low temperature. *Acta Biomater.* **2010**, *6*, 1882. [CrossRef]
10. Ballouze, R.; Marahat, M.H.; Mohamad, S.; Saidin, N.A.; Kasim, S.R.; Ooi, J.P. Biocompatible magnesium-doped biphasic calcium phosphate for bone regeneration. *J. Biomed. Mater. Res. B Appl. Biomater.* **2021**, *109*, 1426. [CrossRef]
11. Castiglioni, S.; Cazzaniga, A.; Albisetti, W.; Maier, J.A.M. Magnesium and Osteoporosis: Current State of Knowledge and Future Research Directions. *Nutrients* **2013**, *5*, 3022–3033. [CrossRef] [PubMed]
12. Ceylan, M.N.; Akdas, S.; Yazihan, N. Is Zinc an Important Trace Element on Bone-Related Diseases and Complications? A Meta-analysis and Systematic Review from Serum Level, Dietary Intake, and Supplementation Aspects. *Biol. Trace Elem. Res.* **2021**, *199*, 535–549. [CrossRef] [PubMed]
13. Hsu, P.-H.; Kuo, H.-C.; Syu, M.-L.; Tuan, W.-H.; Lai, P.-L. A head-to-head comparison of the degradation rate of resorbable bioceramics. *Mater. Sci. Eng. C* **2020**, *106*, 110175. [CrossRef] [PubMed]
14. Chow, L.C.; Eanes, E.D. *Octacalcium Phosphate*, 1st ed.; S Karger Ag: Basel, Switzerland, 2001.
15. LeGeros, R.Z. Properties of Osteoconductive Biomaterials: Calcium Phosphates. *Clin. Orthop. Relat. Res.* **2002**, *395*, 81–98. [CrossRef] [PubMed]
16. de Oliveira Rigotti, R.L.; Tardelli, J.D.C.; dos Reis, A.C. Influence of antibacterial surface treatment on dental implants on cell viability: A systematic review. *Heliyon* **2023**, *9*, e13693. [CrossRef] [PubMed]
17. Jemat, A.; Ghazali, M.J.; Razali, M.; Otsuka, Y. Surface Modifications and Their Effects on Titanium Dental Implants. *Biomed Res. Int.* **2015**, *2015*, 791725. [CrossRef] [PubMed]
18. Parsikia, F.; Amini, P.; Asgari, S. Influence of mechanical and chemical surface treatments on the formation of bonelike structure in cpTi for endosseous dental implants. *Appl. Surf. Sci.* **2012**, *259*, 283–287. [CrossRef]
19. Kubo, A.L.; Rausalu, K.; Savest, N.; Žusinaite, E.; Vasiliev, G.; Viirsalu, M.; Plamus, T.; Krumme, A.; Merits, A.; Bondarenko, O. Antibacterial and Antiviral Effects of Ag, Cu and Zn Metals, Respective Nanoparticles and Filter Materials Thereof against Coronavirus SARS-CoV-2 and Influenza A Virus. *Pharmaceutics* **2022**, *14*, 2549. [CrossRef]
20. Xiao, D.; Zhang, J.; Zhang, C.; Barbieri, D.; Yuan, H.; Moroni, L.; Feng, G. The role of calcium phosphate surface structure in osteogenesis and the mechanisms involved. *Acta Biomater.* **2020**, *106*, 22–33. [CrossRef]
21. Deligianni, D.D.; Katsala, N.D.; Koutsoukos, P.G.; Missirlis, Y.F. Effect of surface roughness of hydroxyapatite on human bone marrow cell adhesion, proliferation, differentiation and detachment strength. *Biomaterials* **2000**, *22*, 87–96. [CrossRef]
22. Daskalova, A.; Angelova, L.; Trifonov, A.; Lasgorceix, M.; Hocquet, S.; Minne, M.; Declercq, H.; Leriche, A.; Aceti, D.; Buchvarov, I. Development of Femtosecond Laser-Engineered β-Tricalcium Phosphate (β-TCP) Biomimetic Templates for Orthopaedic Tissue Engineering. *Appl. Sci.* **2021**, *11*, 2565. [CrossRef]
23. Daskalova, A.; Angelova, L.; Carvalho, A.; Trifonov, A.; Nathala, C.; Monteiro, F.; Buchvarov, I. Effect of surface modification by femtosecond laser on zirconia based ceramics for screening of cell-surface interaction. *Appl. Surf. Sci.* **2020**, *513*, 145914. [CrossRef]
24. Faria, D.; Henriques, B.; Souza, A.C.; Silva, F.S.; Carvalho, O. Laser-assisted production of HAp-coated zirconia structured surfaces for biomedical applications. *J. Mech. Behav. Biomed. Mater.* **2020**, *112*, 104049. [CrossRef] [PubMed]
25. Lutey, A.; Gemini, L.; Romoli, L.; Lazzini, G.; Fuso, F.; Faucon, M.; Kling, R. Towards Laser-Textured Antibacterial Surfaces. *Sci. Rep.* **2018**, *8*, 10112. [CrossRef] [PubMed]
26. Dyukin, R.; Martsinovskiy, G.; Sergaeva, O.; Shandybina, G.; Svirina, V.; Yakovlev, E. Interaction of Femtosecond Laser Pulses with Solids: Electron/Phonon/Plasmon Dynamics. In *Laser Pulses—Theory, Technology, and Applications*; Peshko, I., Ed.; InTechOpen: London, UK, 2012.
27. Malinauskas, M.; Žukauskas, A.; Hasegawa, S.; Hayasaki, Y.; Mizeikis, V.; Buividas, R.; Juodkazis, S. Ultrafast laser processing of materials: From science to industry. *Light Sci. Appl.* **2016**, *5*, e1613. [CrossRef] [PubMed]

28. Stanciuc, A.; Flamant, Q.; Sprecher, C.M.; Alini, M.; Anglada, M.; Peroglio, M. Femtosecond laser multi-patterning of zirconia for screening of cell-surface interactions. *J. Eur. Ceram.* **2018**, *38*, 939–948. [CrossRef]
29. Buchberger, G.; Muck, M.; Plamadeala, C.; Heitz, J. Laser Structuring for Biomedical Applications. In *Laser Pulses—Theory, Technology, and Applications*; Peshko, I., Ed.; InTechOpen: London, UK, 2012.
30. Zhang, Y.; Ito, Y.; Sun, H.; Sugita, S. Investigation of multi-timescale processing phenomena in femtosecond laser drilling of zirconia ceramics. *Opt. Express* **2022**, *30*, 37394–37406. [CrossRef]
31. Chen, B.C.; Tsai, Y.; Ho, C.; Chen, C.; Ma, C. Parametric effects on femtosecond laser ablation of Al2O3 ceramics. *Ceram. Int.* **2013**, *39*, S341–S344. [CrossRef]
32. Kokubo, T.J. Surface chemistry of bioactive glass-ceramics. *Non-Cryst. Solids* **1990**, *120*, 138–151. [CrossRef]
33. *ISO 4287:1997*; Geometrical Product Specifications (GPS)—Surface texture: Profile method—Terms, definitions and surface texture parameters. ISO: Geneva, Switzerland, 1997.
34. Sjollema, J.; Zaat, S.A.J.; Fontaine, V.; Ramstedt, M.; Luginbuehl, R.; Thevissen, K.; Li, J.; van der Mei, H.C.; Busscher, H.J. In vitro methods for the evaluation of antimicrobial surface designs. *Acta Biomater.* **2018**, *70*, 12–24. [CrossRef]
35. Yashima, M.; Sakai, A.; Kamiyama, T.; Hoshikawa, A. Crystal Structure Analysis of β-Tricalcium Phosphate $Ca_3(PO_4)_2$ by Neutron Powder Diffraction. *J. Solid State Chem.* **2003**, *175*, 272–277. [CrossRef]
36. Zaichick, V. Data for the Reference Man: Skeleton content of chemical elements. *Radiat. Environ. Biophys.* **2013**, *52*, 65–85. [CrossRef]
37. Andonova-Lilova, B.; Alexandrova, R.; Rabadjieva, D.; Tepavitcharova, S. Application of cultured murine cells for initial evaluation of the biocompatibility of Mg and Zn-modified tri-calcium phosphates. *Compt. Rend. Acad. Bulg. Sci.* **2012**, *65*, 1099.
38. Rabadjieva, D.; Tepavitcharova, S.; Sezanova, K.; Gergulova, R.; Titorenkova, R.; Petrov, O.; Dyulgerova, E.B. Biomimetic modifications of calcium orthophosphates. In *On Biomimetic*, 1st ed.; Pramatarova, L., Ed.; InTech: Rijeka, Croatia, 2011; pp. 135–162.
39. Rabadjieva, D.; Titorenkova, R.; Gergulova, R.; Dyulgerova, E.; Balarew, C. Influence of Zn on the Biomimetic Transformation of Amorphous Calcium Phosphate to Nano-Sized Apatite. *Nanosci. Nanotechnol.* **2009**, *9*, 235–238.
40. Oyane, A.; Kim, H.M.; Furuya, T.; Kokubo, T.; Miyazaki, T.; Nakamura, T. Preparation and assessment of revised simulated body fluids. *J. Biomed. Mater. Res.* **2003**, *65*, 188–195. [CrossRef] [PubMed]
41. Rabadjieva, D.; Tepavitcharova, S.; Gergulova, R.; Sezanova, K.; Titorenkova, R.; Petrov, O.; Dyulgerova, E. Mg- and Zn-modified calcium phosphates prepared by biomimetic precipitation and subsequent treatment at high temperature. *J. Mater. Sci. Mater. Med.* **2011**, *22*, 2187. [CrossRef]

materials

MDPI

Article

Carbon-Based Composites with Mixed Phosphate-Pyrophosphates with Improved Electrochemical Performance at Elevated Temperature

Sonya Harizanova, Trajche Tushev, Violeta Koleva * and Radostina Stoyanova *

Institute of General and Inorganic Chemistry, Bulgarian Academy of Sciences, 1113 Sofia, Bulgaria; sonya@svr.igic.bas.bg (S.H.); tushev@svr.igic.bas.bg (T.T.)
* Correspondence: vkoleva@svr.igic.bas.bg (V.K.); radstoy@svr.igic.bas.bg (R.S.); Tel.: +359-979-3915 (R.S.)

Abstract: Sodium iron phosphate-pyrophosphate, $Na_4Fe_3(PO_4)_2P_2O_7$ (NFPP) emerges as an excellent cathode material for sodium-ion batteries. Because of lower electronic conductivity, its electrochemical performance depends drastically on the synthesis method. Herein, we provide a simple and unified method for synthesis of composites between NFPP and reduced graphene oxide (rGO) and standard carbon black, designed as electrode materials for both sodium- and lithium-ion batteries. The carbon additives affect only the morphology and textural properties of the composites. The performance of composites in sodium and lithium cells is evaluated at elevated temperatures. It is found that NFPP/rGO outperforms NFPP/C in both Na and Li storage due to its hybrid mechanism of energy storage. In sodium half-cells, NFPP/rGO delivers a reversible capacity of 95 mAh/g at 20 °C and 115 mAh/g at 40 °C with a cycling stability of 95% and 88% at a rate of C/2. In lithium half-cells, the capacity reaches a value of 120 mAh/g at 20 and 40 °C, but the cycling stability becomes worse, especially at 40 °C. The electrochemical performance is discussed on the basis of ex situ XRD and microscopic studies. The good Na storage performance of NFPP/rGO at an elevated temperature represents a first step towards its commercialization.

Keywords: sodium-ion batteries; lithium-ion batteries; hybrid metal-ion batteries; $Na_4Fe_3(PO_4)_2P_2O_7$; carbon composites; rGO; synthesis; structure; characterization

Citation: Harizanova, S.; Tushev, T.; Koleva, V.; Stoyanova, R. Carbon-Based Composites with Mixed Phosphate-Pyrophosphates with Improved Electrochemical Performance at Elevated Temperature. *Materials* **2023**, *16*, 6546. https://doi.org/10.3390/ma16196546

Academic Editor: Giovanni Battista Appetecchi

Received: 11 September 2023
Revised: 29 September 2023
Accepted: 2 October 2023
Published: 4 October 2023

1. Introduction

Mixed phosphate-pyrophosphates of sodium and transition metals, $Na_4M_3(PO_4)_2P_2O_7$ (M = Fe, Mn, Co, Ni), emerge as excellent cathode materials for sodium-ion batteries due to their flexible structure and ability to adopt ions in different oxidation states [1–3]. Given the two building polyanion units, namely PO_4^{3-} and $P_2O_7^{4-}$ groups, the manner of their connectivity yields an unique structure matrix which ensures a three-dimensional insertion pathway with a low activation energy barrier for both smaller Li^+ and bigger K^+ via Na^+ [2–8]. In addition, the interconnectivity of PO_4^{3-} and $P_2O_7^{4-}$ groups contributes to high structure stability and easy alkaline ion diffusion owing to a more open framework [5,9].

Among mixed phosphate-pyrophosphates, the iron analogue, $Na_4Fe_3(PO_4)_2P_2O_7$, operating at about 3 V vs. Na^+/Na, is considered to be the most attractive material for large scale applications owing to its low cost, natural abundance and reasonable energy and power density [5,6,10–13]. After the first report of Kang and coworkers [2], numerous studies have been directed to improve the poor cycling and rate performance of $Na_4Fe_3(PO_4)_2P_2O_7$ (hereafter NFPP) mainly through particle design and coating with various carbon materials, including reduced graphene oxide (rGO), in order to increase the material conductivity and overcome the sluggish Na^+ diffusion kinetics [6,10,11,14–16]. Carbon-coated NFPP obtained by a solution combustion synthesis shows a comparatively stable cycling at 1 C rate with a loss of capacity from 95 mAh/g to about 80 mAh/g after 200 cycles and 100% Coulombic efficiency [6]. The research group of Cao et al. has developed effective methods (such as a template approach using triblock copolymer F127 as surfactant and spray drying) for preparation of NFPP/C nanospheres and microspheres NFPP@rGO with a longer cycle life even at high current densities [10,14]. NFPP/C nanospheres (~30 nm) coated with carbon layer (~3 nm) deliver a high discharge capacity of 128.5 mAh/g (near to the theoretical capacity of 129 mAh/g) at C/5 and capacity retention of 63.5% at 10 C after 4000 cycles [10]. The excellent cycling stability is combined with an excellent rate capability, even at such ultrahigh current rates of 80 and 100 C, where the achieved capacities are 90.0 and 79 mAh/g, respectively. The addition of rGO into the initial solution during the spray-drying synthesis procedure results in the formation of 3D graphene-decorated $Na_4Fe_3(PO_4)_2P_2O_7$ microspheres (NFPP@rGO) with high reversible capacity of 128 mAh g^{-1} at C/10, a long cycling life over 6000 cycles at 10 °C with 62.3% capacity retention and superior rate capability of 35 mAh/g at 200 C [14]. A recent study of Chen et al. demonstrates the great potential of $Na_4Fe_3(PO_4)_2P_2O_7$ for commercial application towards SIB [16]. They have offered an economical method to prepare a rust-derived NFPP/C composite with a porous and hollow spherical morphology (HS-NFPP/C) which ensures a shortening in the Na^+ diffusion pathway, facilitates fast electron transport as well as efficiently accommodates the volume changes upon charge/discharge reactions. This material exhibits outstanding cycling performance in both Na half-cell and Na full battery (pouch cell) which has never been reported before for iron-based polyanionic cathodes [16]. Thus, in Na half-cell, it displays a reversible capacity of 69.5 mAh/g at 10 C with a capacity retention of 87.5% after 10,000 cycles as the operating voltage retains above 2.98 V vs. Na/Na^+. The full battery, using hard carbon as the anode, delivers a discharge capacity of 2761 mAh at 0.5 C after 1000 cycles, corresponding to a high-capacity retention of 93.3%. Moreover, HS-NFPP/C is able to fast Na-ion intercalation/deintercalation reactions which is evident by the achieved reversible capacities of 86.5, 68.9, and 57.1 mAh/g at rates of 1, 50, and 100 C, respectively [16].

A limited number of papers report on the operation of carbon composites of NFPP (with carbon black and soot) in hybrid Na/Li electrochemical cells [15,17,18]. When cycling in Li cells, a competitive co-intercalation of Li^+ and Na^+ occurs at the cathode side with formation of the mixed $Na_{4-x}Li_x(PO_4)_2P_2O_7$ compositions [15,17,18]. In these mixed compositions, a part of the Na^+ ions (about 1–1.2 Na^+ per formula unit) always remains in the NFPP structure (the final composition being $Li_3Na(PO_4)_2P_2O_7$) and it is responsible for its stability [2,4,17]. Kosova et al. [17] have established that the initial capacity in the Na cell with $NaClO_4$ electrolyte is higher than that in Li cell with $LiPF_6$ electrolyte: 104 vs. 95 mAh g^{-1} at C/5 rate, but the rate capability and reversibility are better in the Li cell—on going from rate C/10 to 1 C, the decrease in the discharge capacity is 40% vs. 20% in the Na and Li cell, respectively. Furthermore, the increase in the Na concentration in the dual Na/Li electrolyte is found to improve significantly the electrochemical performance of NFPP/C, which is explained by some competitive processes with participation of Na^+ and Li^+ (different conductivities in $LiPF_6$ and $NaPF_6$, higher polarizing ability of Li^+ than Na^+, and higher diffusion coefficient of Li^+ in the solid state than that of Na^+) [18]. The effect of Li electrolyte on the electrochemical behavior of NFPP/C has also been examined and it is shown that the Li^+ intercalation proceeds to a higher extent in the LiTFSI-based electrolyte than in $LiPF_6$ [15]. In sodium cells with an imide-based electrolyte, the cycling stability and rate capability outperform those in lithium ion cells with imide-based electrolytes [15].

The use of an NFPP@rGO cathode results in the improvement of the K^+ storage properties in comparison with in situ carbon-coated NFPP/C [6,7]. An NFPP/C material prepared via the solution combustion method is able to intercalate reversibly K^+ with a capacity of 116 mAh/g at C/20 rate and after 100 cycles at 1 C rate, the capacity is still about 70 mAh/g [6], while NFPP@rGO delivers an initial capacity of 80.3 mAh/g at 2 C current density and retains 82.1% of this capacity after 500 cycles [7].

The benefits of the rGO decoration have been established in other mixed polyanionic cathode materials such as $Na_4Ni_3(PO_4)_2P_2O_7$/rGO [19], $Na_3V_2(PO_4)_2F_3$@C@rGO [20] and $NaFe_2(PO_4)(SO_4)_2$ @rGO [21]. Two groups, Xia et al. [22] and later Cao et al. [23], have focused on another iron-based mixed phosphate–diphosphate with close chemical composition to NFPP, namely $Na_3Fe_2(PO_4)P_2O_7$ and reported results for excellent cycling stability and rate capability of $Na_3Fe_2(PO_4)P_2O_7$/rGO composites obtained via spray drying. At a 20 C rate, the material of Xia et al. [20] shows a capacity of 55 mAh/g after 6400 cycles (89.7% capacity retention), while that of Cao et al. [23]—51 mAh/g after 8000 cycles (72.4% capacity retention). In addition, Cao et al. have studied the electrochemical performances of $Na_3Fe_2(PO_4)P_2O_7$/rGO and $Na_3Fe_2(PO_4)P_2O_7$/C and the comparison shows that the rGO-based composite overperforms the C-based one in terms of achieved reversible capacity at different current densities (from 1 C to 100 C) and cycling stability [23]. Better electrochemical performance in the presence of rGO was explained by the enhanced sodium ion diffusion kinetics in $Na_3Fe_2(PO_4)P_2O_7$/rGO than that of $Na_3Fe_2(PO_4)P_2O_7$/C as revealed by the Na^+ diffusion coefficients within the composite electrodes.

All these findings clearly manifest the effectiveness of the carbon decoration and particularly with rGO for improved cycling and rate performances of $Na_4Fe_3(PO_4)_2P_2O_7$ positive electrodes. For practical application, there is a need to elaborate a simple and reproducible method for coating of NFPP with carbon-based materials. In addition, the operation of NFPP-electrodes at elevated temperatures is also of importance for commercialization. However, these studies are still scarce. To the best of our knowledge, only two papers report on the effect of the operating temperature on the electrochemical performance of $Na_4Fe_3(PO_4)_2P_2O_7$/C composites [11,24]. According to the results of Yang et al. [24], slightly higher discharged capacities and better rate capability (from 0.2 C to 10 C rate) have been achieved at an elevated cell temperature of 55 °C: for instance, a discharge capacity of 78 mAh/g at 25 °C vs. 82 mAh/g at 55 °C at 10 C rate. W. Chen et al. have tested NFPP/C composites at three cell temperatures: −20 °C, 20 °C, and 50 °C (all-climate condition) [11]. Their data show that there is almost no difference between room temperature and 50 °C in respect to achieved capacities and rate performances. At low temperature (−20 °C), the capacity decreases in comparison with that at 50 °C (95 mAh/g vs. 110 mAh/g at C/10), as well as a fast capacity drop being established with increased current density.

Herein, we provide a simple and unified method for the synthesis of composites between NFPP and carbonaceous materials designed as electrode materials for both sodium- and lithium-ion batteries. The synthetic method consists in the ball-milling of NFPP with carbonaceous materials followed by thermal treatment at 400 °C in Ar atmosphere. Two types of carbonaceous materials are selected: rGO and standard carbon black additives. The performance of NFPP/rGO and NFPP/C in sodium and lithium half-cells is evaluated at elevated temperatures in the framework of CV and CDG experiments. Through detailed diffraction, spectroscopic, and microscopic tools, the changes in composites at pristine state and after electrochemical reactions are monitored.

2. Materials and Methods

2.1. Materials and Synthesis

High-purity powder of $NaH_2PO_4 \cdot H_2O$ (ACS, Sigma-Aldrich, Rockford, IL, USA) and $Fe(HCOO)_2 \cdot 2H_2O$ prepared by us were used as starting materials. The crystals of $Fe(HCOO)_2 \cdot 2H_2O$ were synthesized via dissolution of metal iron in 40% solution of formic acid (Merck, Darmstadt, Germany) at about 60–70 °C, followed by the solution concentration and filtration of the crystals [25]. As conductive additives, we utilized Super C65 carbon black (TIMCAL Ltd., Bodio, Switzerland) and reduced graphene oxide (rGO) provided by Graphit Kropfmühl GmbH (Hauzenberg, Germany).

$Na_4Fe_3(PO_4)_2P_2O_7$ was prepared after a thermal treatment at 500 °C of a freeze-dried phosphate-formate precursor containing $NaH_2PO_4 \cdot H_2O$ and $Fe(HCOO)_2$ (4:3 mole ratio). The details for the synthesis can be found elsewhere [15].

Two types of carbon composites were prepared using simple, one-step ball-milling of NFPP with 15 wt.% carbon black and 15 wt.% rGO (labelled NFPP/C and NFPP/rGO, respectively). The ball-milling process was performed via means of planetary mono mill "Pulverisette 6" (Fritsch GmbH, Idar-Oberstein, Germany) with agate balls having ϕ of 10 mm, powder to balls mass ratio is 1:10, the duration being of 4 h at a speed of 300 rpm. Thus, obtained composites have been annealed at 400 °C for 3 h under Ar flow.

2.2. Methods

The structure of NFPP, carbon composites, and cycled cathodes was determined by powder X-ray diffraction (Bruker Advance D8 diffractometer, Karlsruhe, Germany) with LynxEye detector (CuKα radiation). The lattice parameters were calculated by WinPLOTR program. The local structure of anionic groups was analyzed by IR spectroscopy. The FT-IR spectra were recorded on a Nicolet Avatar-320 FTIR spectrometer (Thermo Fisher Scientific, Waltham, MA, USA) in KBr pellets (resolution < 2 cm^{-1}). The specific surface area of the powder samples was measured using the BET method from low-temperature adsorption–desorption isotherms (77.4 K) recorded via NOVA 1200e device (Quantachrome, Boynton Beach, FL, USA). The total pore volume was calculated according to Gurwitsch's rule at p/po = 0.99 and the pore size distribution was estimated according to the Barett–Joyner–Halenda (BJH) method. The pore size distribution for rGO was calculated using the DFT model. The morphology of powder samples and electrodes was studied via scanning electron microscopy (SEM) using JSM 6390 microscope (JEOL, Tokyo, Japan). TEM analysis was performed with a JEOL 2100 microscope (Tokyo, Japan) with a GATAN Orius 832 SC1000 camera (Plesantan, CA, USA). The optical images of the electrodes were obtained using ZEISS Stemi 508 stereo microscope (Carl Zeiss, Jena, Germany).

The electrochemical characterization was carried out in sodium and lithium half-cells in galvanostatic and potentiostatic regimes by means of 32 channel Biologic VMP-3e battery cycler (Seyssinet-Pariset, France) at 20 and 40 °C (KB-53 incubator, Binder GmbH, Tuttlingen, Germany). The positive electrodes, supported using aluminum foil, were made from mixtures containing 80 wt.% composite, 10 wt.% Super C65 carbon and 10 wt.% polyvinylidene fluoride (PVDF) (Sigma-Aldrich, St. Louis, MO, USA). The slurry was then cast with a doctor blade film coater (ZAA 2600.A, Proceq SA, Schwerzenbach, Switzerland) onto the aluminum foil, followed by a drying at 80 °C overnight and the resulting electrode film is cut into 10 mm discs, pressed and dried at 120 °C under vacuum for 10 h. The electrode loading is around 8 mg/cm^2 which corresponds to a thickness of about 300 μm. The negative electrodes consist of sodium and lithium metals. The electrolytes are 1 M $NaPF_6$ in PC (Sigma Aldrich, St. Louis, MO, USA) and 1 M $LiPF_6$ (EC/DMC, 1:1 by volume, Sigma Aldrich, St. Louis, MO, USA). Swagelok type sodium and lithium half-cells were assembled in an argon-filled glovebox (MB-Unilab Pro SP (1500/780), H_2O, and O_2 content < 0.1 ppm, MBraun, Garching, Germany) using Whatman GF/D (Whatman International Ltd., Maidstone, UK) as the separator. The galvanostatic cycling was performed over the range 1.5–4.5 V vs. Na^+/Na and 1.5–4.8 V vs. Li^+/Li at a rate of C/2 (1 C = 129 mA/g), while the cyclic voltammetry measurements were performed with a scanning rate of 1 mV/s in the range of 1.5–5 V vs. Na^+/Na or Li^+/Li. The specific capacity was calculated based on the mass of the active NFPP phase in the electrodes. For ex situ analyses, the electrodes were switched off at 1.5 V and the electrochemical cells were disassembled inside the glove-box; the electrodes were covered with plastic film and then subjected to different analyses.

3. Results and Discussion

3.1. Structure and Morphology Characterization of NFPP/C and NFPP/rGO Composites

Figure 1 compares the XRD patterns of pristine NFPP and its composites with carbon black and rGO. For all composites, the diffraction peaks are indexed in the space group $Pn2_1a$, which is identical to that for pristine NFPP. Moreover, the unit-cell parameters also remain intact (Table 1). This indicates that NFPP in both composites NFPP/rGO and NFPP/C preserves the NASICON-type crystal structure of the pristine NFPP. The XRD peak profiles also appear to be unchanged after the formation of composites, thus evidencing the retention of the crystallinity of NFPP. In addition to the main NASICON phase, low-intensive peaks due to maricite phase, $NaFePO_4$, are also distinguished irrespective of the type of carbon additives. This means that the impurity phase, already formed during the synthesis of pristine NFPP (including freeze-drying reactions followed by thermal treatment at 500 °C), is not affected after the ball milling with carbon additives. It is worth mentioning that, irrespective of the synthesis methods, NFPP formation is always accompanied with impurity phases [2,17,19].

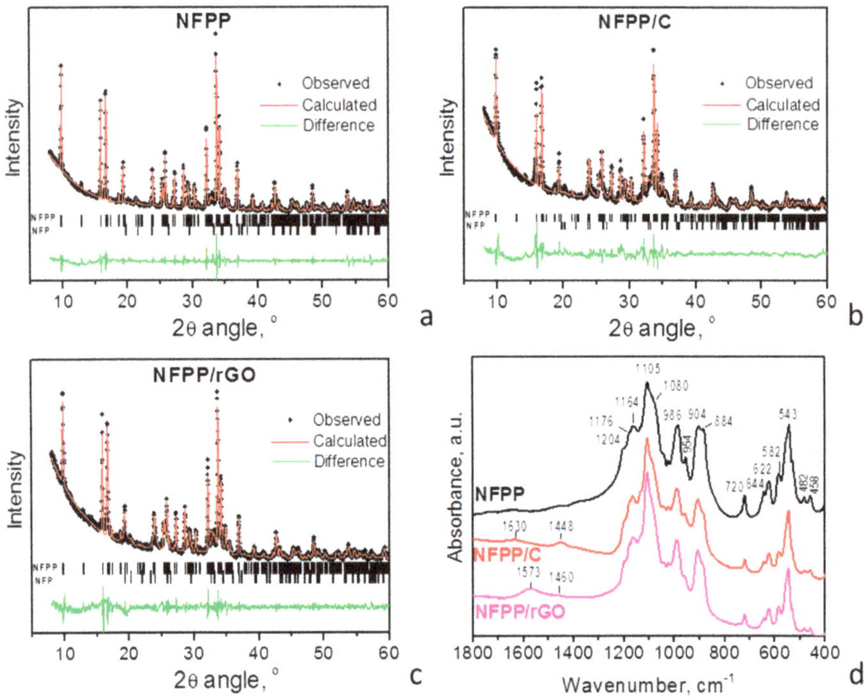

Figure 1. XRD patterns of pristine NFPP and composites NFPP/C and NFPP/rGO (**a**–**c**). "NFP" in the figures denotes the diffraction peaks due to maricite $NaFePO_4$. FT-IR spectra of NFPP, NFPP/C, and NFPP/rGO (**d**).

Table 1. Lattice parameters of powder NFPP, carbon composites NFPP/C and NFPP/rGO (SG $Pn2_1a$), and electrodes cycled in $NaPF_6$/PC electrolyte and stopped at 1.5 V. For comparison, the literature data for NFPP/C and NFPP/rGO obtained using different methods are also provided.

Description	*a*, Å	*b*, Å	*c*, Å	*V*, Å^3
NFPP	18.1018(6)	6.5331(2)	10.6452(3)	1258.93(7)
NFPP/C	18.0863(8)	6.5390(3)	10.6461(4)	1259.09(10)
NFPP/rGO	18.0854(6)	6.5345(6)	10.6467(7)	1259.95(12)
Electrodes cycled in $NaPF_6$/PC electrolyte, analyzed in discharged state (1.5 V)				
NFPP/C (200 cycles)	18.0711	6.5340	10.6626	1259.02
NFPP/rGO (200 cycles)	18.0838	6.5344	10.6428	1257.63
Literature data for NFPP/C and NFPP/rGO obtained by different methods				
NFPP/C [10]: template method	18.0604(6)	6.5354(3)	10.6533(7)	1257.446(6)
NFPP/C [24]: sol-gel	18.1953(5)	6.5639(3)	10.6965(4)	1277.50
NFPP/C [16]: spray-drying	17.938	6.507	10.623	1239.94
NFPP/C [6]: solution combustion	18.038(4)	6.5429(13)	10.6744(17)	1259.8(5)
NFPP/rGO [14]: spray-drying	17.9675	6.5402	10.6672	1253.51

The formation of well-crystallized NFPP in the composites NFPP/C and NFPP/rGO is further supported by the IR spectra (Figure 1d). In the region of 1250–400 cm^{-1}, the IR spectra enable clear differentiation of the orthophosphate and diphosphate groups in the NASCON-type structure. The detailed assignment of the IR bands in NFPP was provided in our previous paper [15]. In this study, we compared the spectroscopic characteristics (i.e., band shape, intensities, and frequency positions) of NFPP and its carbon composites.

The comparison shows that NFPP/C and NFPP/rGO composites display the same vibrational characteristics as those of carbon-free NFPP (Figure 1d). It should be noted that the position deviation between 1 and 2 cm^{-1} falls in the limit of the experimental resolution. The similarity in the spectral features of the sample provides evidence of the identical local structure of PO_4 and P_2O_7 groups in the three samples. Therefore, the formation of composites of NFPP with carbon black and rGO neither induces change in their anion frameworks nor oxidation processes related to the Fe^{2+} ions.

In addition, the IR spectra of NFPP/C and NFPP/rGO composites display bands that are associated with the nature of the carbon additives. The bands at 1573 cm^{-1} for NFPP/rGO and at 1630 cm^{-1} for NFPP/C (less intensive) are due to the in-plane stretching motion of pairs of sp^2 carbon atoms (C=C) [25,26]. This vibration is typical for aromatic and olefinic molecules and always lies in the region of 1500–1630 cm^{-1} [26,27]. The position of this band in our NFPP/rGO is very close to that in ideal graphite (1588 cm^{-1}, E_{1u} mode) [26,27] which is an indication of the more graphitic-like nature of the rGO material than that of carbon black [26–28]. The other IR active band in graphite appears at 868 cm^{-1} (A_{2u} mode), but in our spectra it is probably hidden by the strong absorptions of the P–O–P bridge at 884 cm^{-1} (Figure 1d). The band in the region 1460–1448 cm^{-1} (more intense in NFPP/C) could be attributed to the carbonate-related and C=N species [29,30].

Further information about the carbon structure can be obtained using Raman spectroscopy based on the intensity ratio of the D and G bands (I_D/I_G ratio) that are characteristic for disordered carbonaceous materials [26,31]. The carbonaceous materials used (Super C65 carbon black and rGO (Graphit Kropfmühl GmbH)) are commercial products and they are well-characterized by Raman spectroscopy [32–34]. In our previous papers, we have presented the Raman spectrum of carbon black which exhibits two bands at 1590 and 1326 cm^{-1} assigned to G and D band, respectively [32,33]. The intensity ratio, I_D/I_G, measured from the areas of the bands is 1.75 (measured from the peaks intensity is 1.15) [32,33]. As reported by Ackermann et al., the Raman spectrum of rGO (Graphit Kropfmühl) also displays G and D bands at approximately 1590 and 1326 cm^{-1}, the peak intensity ratio I_D/I_G being 0.41 [34]. The comparison clearly shows that the level of disorder of carbon black is significantly larger than that of rGO. Moreover, we have established that the carbon structure in the composites between phosphates and Super C65 carbon black prepared via our ball-milling treatment remains unchanged [32,33]. Therefore, the conclusion derived from the Raman spectra of the pure carbon additives can be referred to the respective composites, i.e., the level of disorder of carbon black in NFPP/C composite is much higher than that of rGO in NFPP/rGO composite, in agreement with the IR spectra.

Although the structural characteristics of NFPP are unaffected by the ball-milling with carbon additives, the morphology and porosity of NFPP/C and NFPP/rGO undergo some changes (Figure 2).

The morphology of pristine NFPP comprises irregular micrometric aggregates (above 10 μm), where individual particles are hardly visible (Figure 2a). The ball-milling with carbon additives effectively breaks these aggregates: for NFPP/C, nanoparticles prevail, while for NFPP/rGO, the smaller globe-like aggregates with sizes about 200 nm dominate (Figure 2b,c). It should be mentioned that the ball-milling of NFPP without carbonaceous materials is not energetically enough to break NFPP aggregates into small particles, as seen from Figure S1. The SEM image (Figure S1) demonstrates that, even after ball-milling, NFPP comprises mainly irregular aggregates with micrometer sizes (between 1 and 5 μm), but smaller particles or aggregates below 0.2 μm are also visible. The observed agglomeration could also be explained with a high surface tension of NFPP particles, but it can be effectively overcome by ball-milling with carbon additives (Figures 2A and S1).

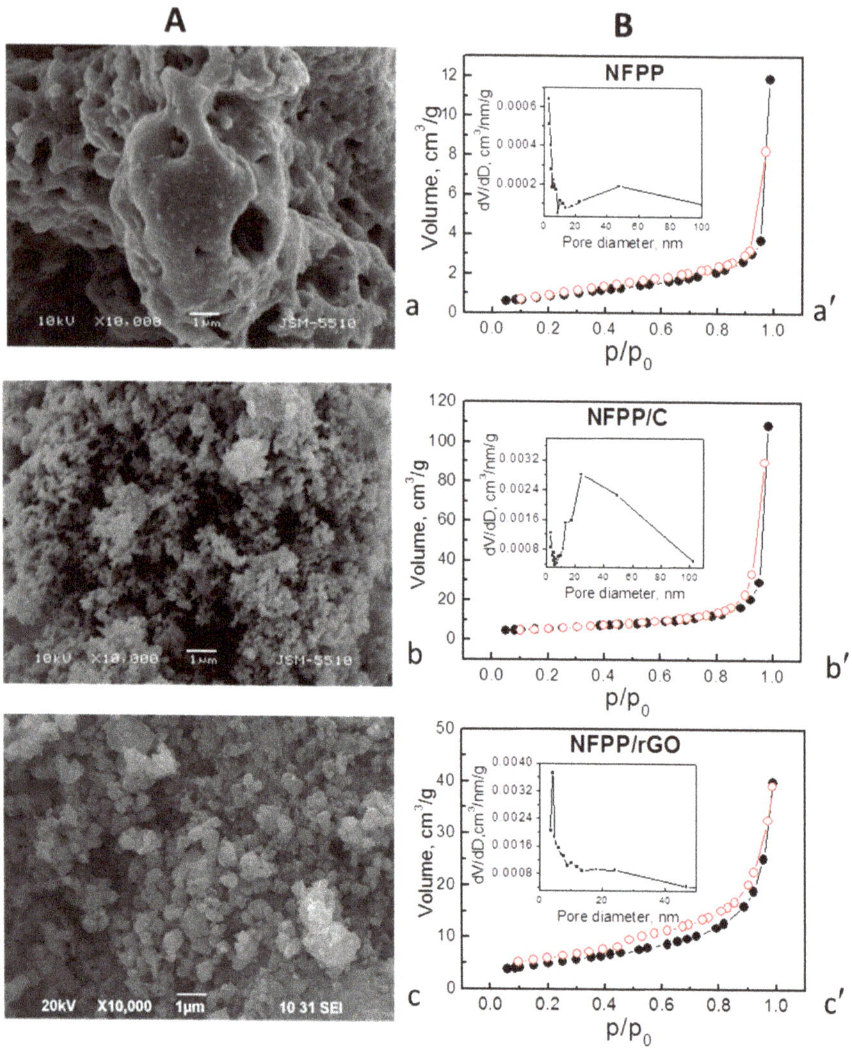

Figure 2. (**A**) SEM images and (**B**) nitrogen adsorption/desorption isotherms (full/open symbols) with pore size distributions (insets) of: NFPP (**a**,**a′**), NFPP/C (**b**,**b′**), and NFPP/rGO (**c**,**c′**).

The co-existence of NFPP and rGO in the composite was confirmed by HR-TEM (Figure 3). The thickness of the rGO shell is inhomogeneous and varies from about 10 to 30 nm. In addition, separate rGO particles are also visible. Inside the composites, the NFPP phase preserves its crystal structure intact (Figure 3), as was established from XRD and IR experiments. HR-TEM image exhibits lattice fringes from the (201) plane and the interplanar space of 0.69 nm coincides very well with that found via the XRD analysis (6.89 Å).

The ball-milling process is concomitant with variations in the textural characteristics together with preservation of the isotherm type: the isotherms are of type II with H3 hysteresis (presence of slit-shaped pores), which is typical of mesoporous materials [35–37]. As can be expected, the composites have considerably higher specific surface areas: 20 and

18 cm^2/g for NFPP/C and NFPP/rGO, respectively, vs. 3 cm^2/g for NFPP (Figure 2B), at least owing to the high specific surface areas of pure carbon additives (49 cm^2/g for carbon black and 363 cm^2/g for rGO, Figure S2). The total pore volumes increase as well: from 0.02 cm^3/g in NFPP to 0.17 and 0.10 cm^3/g in NFPP/C and NFPP/rGO, respectively. On the other hand, the pore volume for NFPP/rGO is rather low (even lower than that for NFPP/C), considering the much larger total pore volume of pure rGO (1.58 cm^3/g, Figure S2). This result implies that during the ball-milling treatment a partial blocking of the slit-shaped mesopores of rGO by the NFPP particles (larger than 100 nm) occurs and information about this can be derived by the pore size distribution curves. For rGO, a homogeneous distribution of the mesopores in a narrow range of sizes between 3.5 and 6.5 nm is clearly visible (Figure S2); meanwhile, for composite NFPP/rGO the distribution appears to be bimodal: there is a predominant part of the pores between 3 and 10 nm and a less expressed part between 13 and 45 nm (Figure 2B). From these data, one could suppose that the small slit-shaped mesopores of rGO with sizes between 3.5 and 6.5 nm are most probably blocked by the NFPP particles to a high extent giving rise to a lowered pore volume and specific surface area of the composite NFPP/rGO. For comparison, carbon black is characterized by a broader pore size distribution with mesopores having diameters between 3 and 50 nm (Figure S2) and the pore blocking by the NFPP particles occurs to a lesser extent. The composite NFPP/C and pristine NFPP also exhibit a monomodal distribution of the pores within a broad range between 10 and 100 nm (Figure 2B).

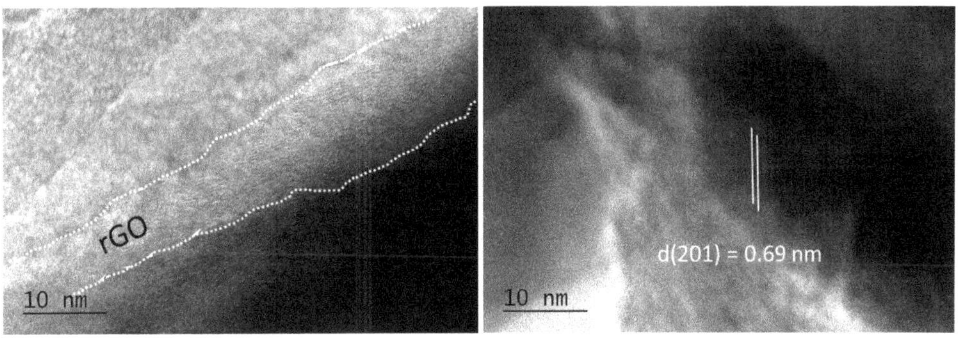

Figure 3. HR-TEM image of NFPP/rGO composite. The rGO layer is marked by white dotted lines.

3.2. Sodium Storage Properties of NFPP/C and NFPP/rGO Composites

The capability of the composites to intercalate Na^+ at elevated temperatures is compared in Figure 4.

In sodium-ion half cells (20 °C), the CV curves of the composites display broad oxidation and reduction peaks at around 3.25 and 2.80 V, respectively. These redox peaks coincide with those previously reported for NFPP and were attributed to the Na^+ intercalation/deintercalation into the NASICON structure of NFPP owing to the Fe^{2+}/Fe^{3+} ionic pair: $Na_4Fe_3(PO_4)_2P_2O_7–3Na^+ \leftrightarrow NaFe_3(PO_4)_2P_2O_7$ [2,4,10,15,17]. The carbon and rGO additives have no measurable effect on the redox peak due to the Na^+ intercalation, which confirms the XRD data on the structure stability of NFPP after the ball-milling treatment (Figure 1). The close inspection of the CV profiles indicates that the redox peak due to Na^+ intercalation is split into several components, the splitting being better expressed for the NFPP/rGO composite. It is worth mentioning that Na^+ deintercalation is a step-wise process due to the availability of several crystallographic sites for sodium [2,4,17,22,24]. The sequence deintercalation of Na^+ from one crystallographic site to another is resolved when sodium electrolytes containing NaTFSI [15] or $NaClO_4$ [2,4,22,24] salts are used, while in $NaPF_6$-based electrolyte, the Na^+ deintercalation seems structureless [15]. It appears that, for the NFPP/rGO composite, the Na^+ deintercalation remains step-wise even in the $NaPF_6$-based electrolyte (Figure 4b). This can be related with the textural characteristics of

NFPP/rGO, where specific bimodal pore-size distribution is observed (Figure 2B). These textural characteristics could be beneficial for better penetration of the electrolyte ions within the bulk of the electrodes, as well as being able to help with the formation of porous electrolyte–electrode interface, thus contributing to the step-wise intercalation of Na^+.

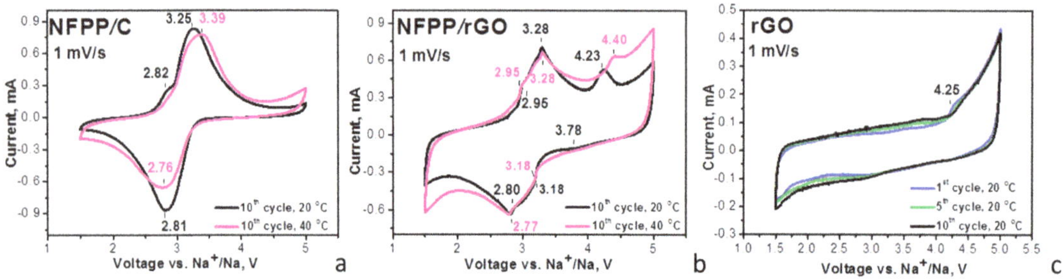

Figure 4. CV curves at a scanning rate of 1 mV/s for NFPP/C (**a**) and NFPP/rGO (**b**) in sodium-half cells. The cells operate firstly at 20 °C (10 cycles) and then at 40 °C (10 cycles). For the sake of comparison, the CV curves of rGO in sodium half-cells are also provided (**c**).

By increasing the operating temperature from 20 to 40 °C, the oxidation and reduction peaks for the NFPP/C composite become broader and the peak-to-peak separation increases from 0.44 to 0.63 V (Figure 4a). This provides evidence that the elevated temperature provokes a suppression in the electron transfer reaction during the Na^+ intercalation into the NFPP/C composite. In contrast to NFPP/C, the composite NFPP/rGO retains unchanged its CV curve profiles at 40 °C (Figure 4b). Again, this can be related to the textural characteristics: while the rGO additive facilitates the formation of a porous electrolyte–electrode interface, the carbon black additive has a negative effect.

Moreover, the CV curve of the NFPP/rGO composite brings some features of capacitive electrochemical storage in addition to the Faradaic ones, while the NFPP/C composite stores energy via the Faradaic reaction only (Figure 4). To understand this observation, Figure 4 also provides the CV curves of rGO used as a positive electrode in sodium half-cell (Figure 4c). The CV curve of rGO displays a rectangular-like profile, which is typical behavior of capacitance electrochemical storage. This is consistent with previous studies on the electrochemical behavior of rGO in sodium-ion devices, where the sodium storage was associated with diffusionless surface reactions [38,39]. Above 4.2 V, the enhancement in the oxidation curve implies the beginning of the interaction between rGO and electrolyte which requires further study. Comparing the CV profiles of the NFPP/rGO and rGO additive, it appears that the composite NFPP/rGO stores energy via a hybrid mechanism: the NFPP constituent contributes to the bulk Faradaic reaction, while the rGO additive initiates the capacitive surface reactions. This mechanism of bulk and surface reactions appears to be facilitated by the textural characteristics of NFPP/rGO composite.

Furthermore, the galvanostatic test of the composites supports the different electrochemical performance of NFFP/C and NFPP/rGO composites (Figure 5). The charge/discharge curves of NFFP/C and NFPP/rGO demonstrate a voltage plateau at around 3.0 V (Figure 5), as was observed in the CV curves (Figure 4). This plateau is related with Na^+ intercalation to the NFPP structure. Irrespective of the similar charge/discharge curve profiles, the specific discharge capacity and the cycling stability at 20 °C of the composite with rGO exceed those of the composite with carbon black additive: 95 mAh/g vs. 80 mAh/g after the 100 cycles at a rate of C/2, the corresponding cycling stability being 95% vs. 84%. It is worth comparing our data on NFPP/rGO with that previously reported on NFPP decorated with rGO: at a rate of C/2, a reversible capacity of 117 mAh/g was reported [14]. We obtained a good storage performance of the composite NFPP/rGO using a simple and fast method of synthesis.

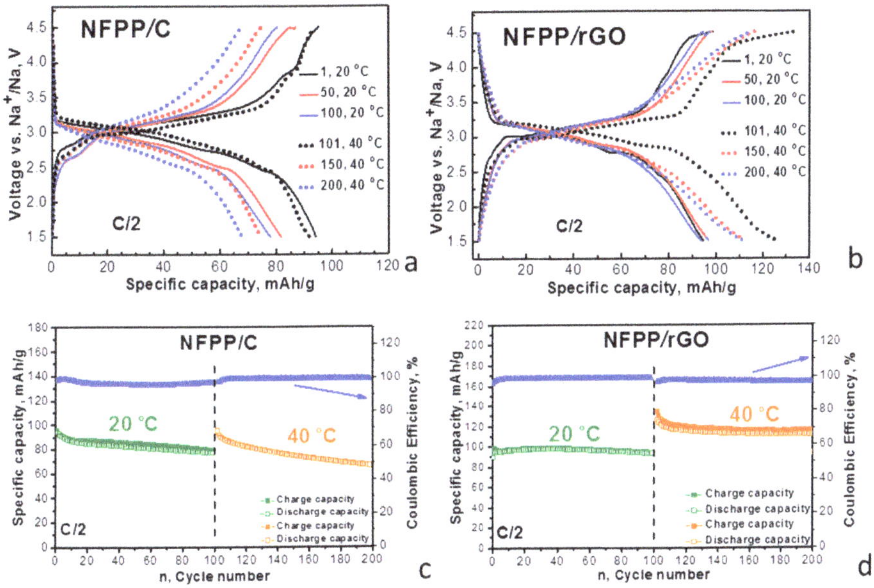

Figure 5. Charge/discharge curves of NFPP/C and NFPP/rGO in sodium half-cells (**a**,**b**). The cells operate firstly at 20 °C (cycles from 1 to 100) and then at 40 °C (cycles from 101 to 200). Cycling stability and Coulombic efficiency at C/2 rate of NFPP/C and NFPP/rGO in sodium half-cells at 20 and 40 °C (**c**,**d**).

The difference in the electrochemical performance of NFPP/rGO and NFPP/C becomes more significant at 40 °C: a discharge capacity of 115 mAh/g vs. 70 mAh/g after the next 100 cycles at a rate of C/2, the corresponding cycling stability being 88% vs. 70% at 20 °C. It is of importance that the Coulombic efficiency varies between 97 and 100% for both NFPP/C and NFPP/rGO composites. The improved storage performance of the composite with rGO can be correlated with its hybrid mechanism of energy storage. In addition, as far as we know, we provide the first data on the sodium storage performance of NFPP/rGO at elevated temperatures.

3.3. Lithium Storage Properties of NFPP/C and NFPP/rGO Composites

In comparison with the Na-ion cell, the energy storage by NFPP in the Li-ion cell is a more complex process [2,15,17,18]. At OCV (i.e., before the electrochemical reaction), NFPP reacts partially with the Li-electrolyte including Na^+/Li^+ exchange reactions. Furthermore, the first charge/discharge cycle gives rise to extraction of the bigger Na^+ and consequent insertion of the smaller Li^+ ions. This reaction is concomitant with a change of Na-to-Li ratio in the electrolyte, but the total alkali content remains 1 M. After the next cycles, Li^+ is preferentially intercalated leading to formation of the $(Na_{3-x}Li_x)NaFe(PO_4)_2P_2O_7$ phase [2,15,17,18].

Lithium intercalation into composites was examined in lithium half-cells with a conventional carbonate-based electrolyte (i.e., 1 M $LiPF_6$ in EC/DMC) (Figure 6).

For the NFFP/C composite, the CV curve at 20 °C displays a couple of broad oxidation/reduction peaks centered at around 3.13 V/2.75 V, which are split into two components at 40 °C: oxidation peaks at 2.95 V and 3.41 V and reduction peaks at 3.01 V and 2.80 V. For the NFPP/rGO composite, a broad oxidation peak centered at 3.58 V with a shoulder at 3.12 V together with a broad reduction peak at 3.06 V are resolved at 20 °C. The elevated operating temperature slightly changes the peak positions, but the curve profile is retained as a whole. Therefore, the operating temperature induces different electrochemical responses of NFPP/C and NFPP/rGO composites in both lithium and sodium ion cells. The textural properties are mainly responsible for the observed temperature-induced phenomenon.

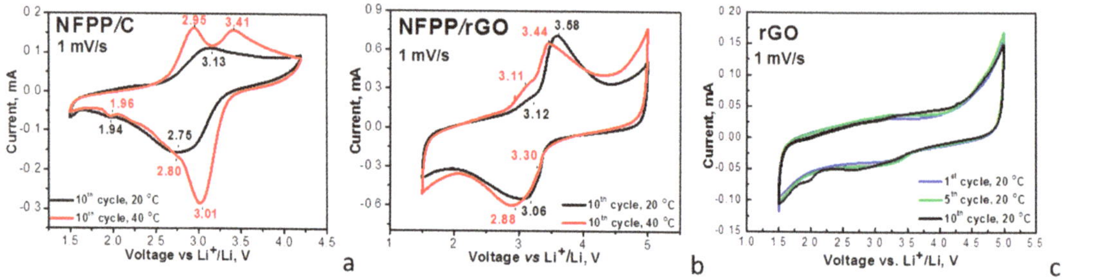

Figure 6. CV curves at a scanning rate of 1 mV/s for NFPP/C (**a**) and NFPP/rGO (**b**) in lithium half-cells. The cells operate firstly at 20 °C (10 cycles) and then at 40 °C (10 cycles). For the sake of comparison, the CV curves of rGO in lithium half-cells are also provided (**c**).

The redox peaks observed for NFPP/C and NFPP/rGO composites coincide with those previously found for NFPP, which are due to the consecutive intercalation of Li^+ into the NASICON structure [15,17]. This means that for the NFPP/C and NFPP/rGO composites, the Li^+ intercalation into NFPP also takes place with the participation of the Fe^{2+}/Fe^{3+} redox couple. As in the case of Na-ion cell, the appearance of a tail in CV curves above 4.2 V implies some side reactions between NFPP composites and the Li electrolyte.

The comparison of the CV curve profiles for NFPP/C and NFPP/rGO discloses the next peculiarity, which is associated with the mechanism of the lithium storage. Similar to sodium storage, the CV curve profile for NFPP/rGO evidences the simultaneous occurrence of capacitive and Faradaic electrochemical reactions, while only Faradaic reactions take place for the NFPP/C composite. This is supported by the observed rectangular-like profile of the CV curve of rGO in lithium-half cells that is characteristic of capacitive reactions (Figure 6c). Therefore, the CV data reveal that the NFPP/rGO composite stores lithium via a hybrid electrochemical mechanism, as we established for sodium storage.

The different mechanism of the lithium storage determines the charge/discharge characteristics of the composites (Figure 7).

At room temperature, the charge/discharge curves of the composites consist of a voltage plateau at 3.2 V vs. Li/Li^+. However, NFFP/C and NFPP/rGO composites deliver completely different specific capacity: 28 mAh/g for NFFP/C and 123 mAh/g for NFPP/rGO. These data are also confirmed by the specific capacity determined from the CV curves: 31 mAh/g for NFFP/C and 107 mAh/g for NFPP/rGO. Furthermore, the cycling stability is also different: about 93% for NFPP/C and 89% for NFPP/rGO after 60 cycles (85% for NFPP/rGO after 100 cycles). However, the Coulombic efficiency tends to 99% for both NFPP/C and NFPP/rGO. We think that the excellent performance of NFPP/rGO as compared to NFPP/C is a result of the versatile role of the rGO additive: (i) rGO ensures a high electrical conductivity; (ii) rGO seems to wrap well NFPP particles thus changing the textural parameters (Figures 2 and 3); and (iii) rGO displays capacitive reactions (Figure 6c).

At an elevated temperature (i.e., 40 °C), the shape of the charge/discharge curves remains constant, but the performance becomes worse for both composites. For NFPP/C, the specific capacity increases at 40 °C from 28 to 48 mAh/g at the expense of worsened cycling stability (i.e., a decrease from 93% to 83%). For NFPP/rGO, the specific discharge capacity at 40 °C is close to that determined at 20 °C (121 mAh/g), but the cycling stability decreases reaching a value of about 71%. An important sign for worsened electrochemical performance of the composites at elevated temperature is the Coulombic efficiency: for NFPP/C, the Coulombic efficiency exceeds 100%, while for NFPP/rGO, the Coulombic efficiency varies around 90%. This provides evidence of the occurrence of side reactions when NFPP/C and NFPP/rGO composites work at elevated temperatures in lithium-ion cells.

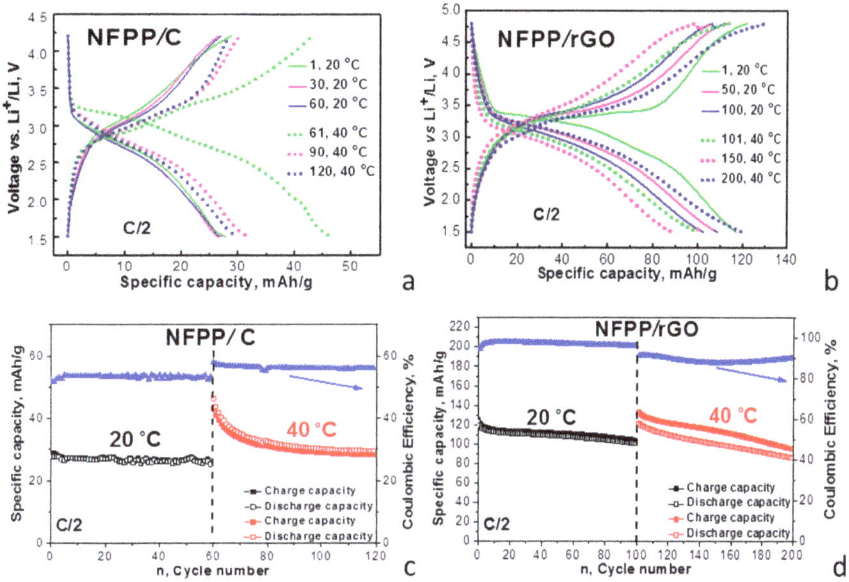

Figure 7. Charge/discharge curves of NFPP/C and NFPP/rGO in lithium half-cells (**a**,**b**). The cells operate firstly at 20 °C, and then at 40 °C. Cycling stability and Coulombic efficiency at C/2 rate of NFPP/C and NFPP/rGO in lithium half-cells at 20 and 40 °C (**c**,**d**).

3.4. Ex Situ Studies

The structural changes during the prolonged electrochemical oxidation/reduction reactions in NFPP/C and NFPP/rGO electrodes in Na- and Li half-cells are followed by ex situ XRD. The electrochemical cells were switched off at 1.5 V (discharged state) after 200 cycles (100 cycles at 20 °C and subsequent 100 cycles at 40 °C) and the electrodes were analyzed. Since the XRD patterns of pristine NFPP/C and NFPP/rGO composites are practically the same, in Figure 8, only the XRD pattern of NFPP/C is given for comparison with these of the cycled electrodes. It can be seen that after 200 cycles there is no substantial change in the XRD patterns of both NFPP/C and NFPP/rGO compared with the pristine state, particularly in the Na half-cells. The diffraction peaks are indexed in the same orthorhombic crystal lattice as pristine NFPP with approximately the same lattice parameters (Table 1). This feature indicates a high structural stability of both NFPP/C and NFPP/rGO electrodes even at prolonged cycling at elevated temperature of 40 °C. The structural stability is an important advantage of NFPP electrode. It has been demonstrated that the Na^+ extraction/insertion into NFPP proceeds topotactically within the framework of one-phase with low volume change (i.e., of about 4%) [2,4,11,17,24].

Contrary to the Na half-cells, in Li half-cells the XRD patterns exhibit some changes in the peak positions and intensities (Figure 8c,d), which is related to the intercalation of smaller Li^+ instead of bigger Na^+. However, the quality of the XRD patterns is not good enough for their indexing. On the other hand, the low quality of the XRD patterns could serve as an indirect sign for the amorphization of NFPP-based composites when they work in Li half-cells, which in turn determines the worse performance of the composites in Li half-cells than those in Na half-cells. This is especially valid for the NFPP/rGO composite.

Furthermore, the surface of the same cycled electrodes was observed using optical microscopy and SEM (Figures 9 and 10). In general, the electrode roughness seems unchanged after the electrode cycling in both Na and Li half-cells (Figure 9). The preservation of the electrode roughness sheds further light on the process of loss of crystallinity of the composites after working in Li half-cells: it appears that the amorphization of the electrode proceeds without any dissolution reactions. At higher magnification, a veil on electrode composites working in Li half-cells becomes visible, especially for NFPP/rGO (Figure 10). This could be tentatively related to the formation of a specific cathode–electrolyte interface, which deserves further studies. In comparison with Li half-cells, the electrodes working in Na half-cells seem to retain their morphology characteristics unchanged (Figures 9 and 10).

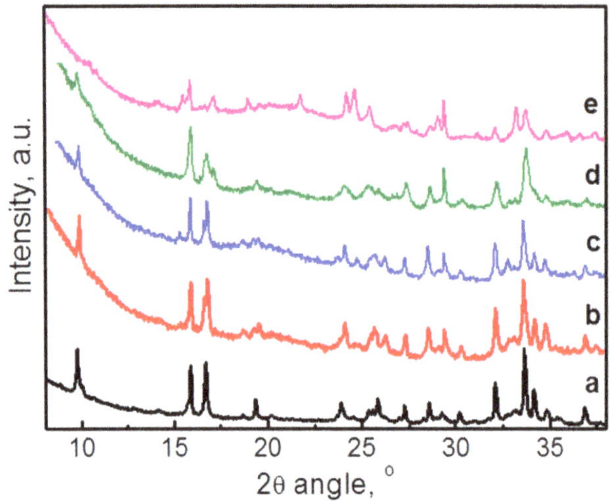

Figure 8. Ex situ XRD patterns of NFPP/C (**a**) and electrodes NFPP/C and NFPP/rGO cycled in Na half-cells (**b**,**c**), NFPP/C and NFPP/rGO cycled in Li half-cells (**d**,**e**) for 200 cycles (100 cycles at 20 °C and subsequent 100 cycles at 40 °C).

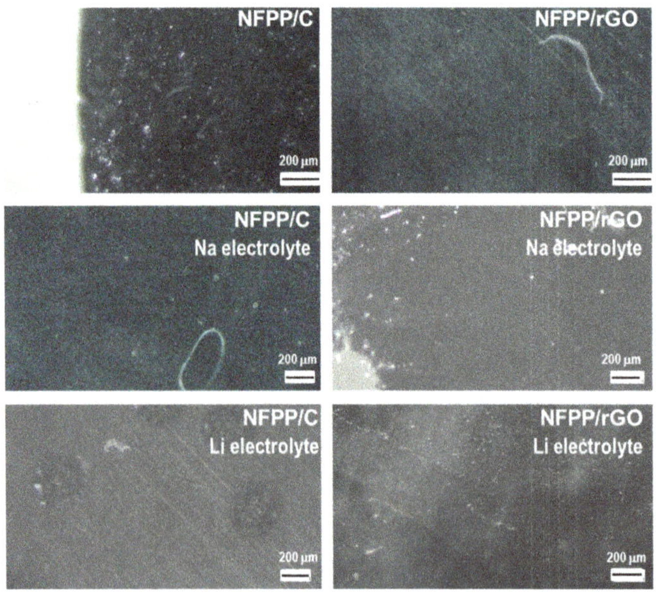

Figure 9. Optical images of initial NFPP/C and NFPP/rGO electrodes and after cycling in Na half-cells and Li half-cells for 200 cycles (100 cycles at 20 °C and subsequent 100 cycles at 40 °C).

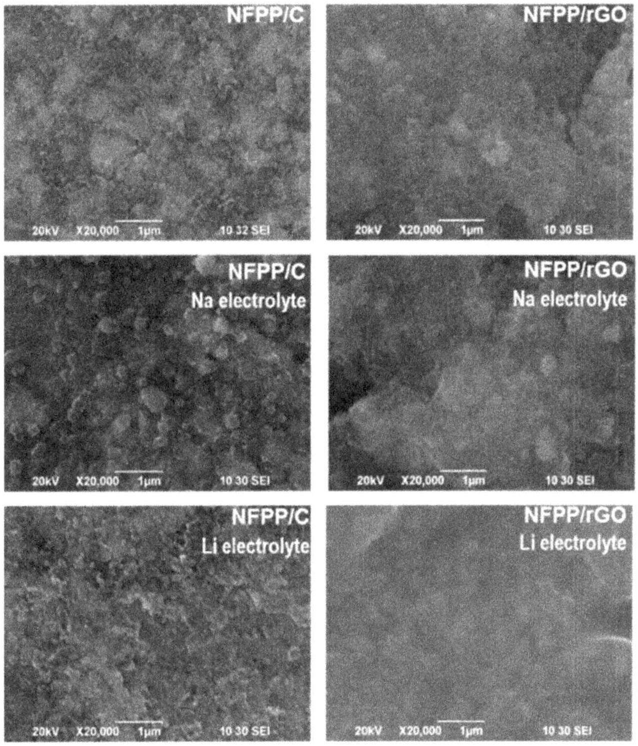

Figure 10. SEM images of initial NFPP/C and NFPP/rGO electrodes and after cycling in Na half-cells and Li half-cells for 200 cycles (100 cycles at 20 °C and subsequent 100 cycles at 40 °C).

4. Conclusions

Ball milling followed by thermal treatment in Ar atmosphere at 400 °C yields composites of NFPP with carbon black and rGO. In these composites, NFPP retains its NASICON-type structure unchanged. The carbon additives affect the morphology and texture of the composites. The carbon black enables the exposure of nanoparticles of NFPP, while rGO causes the formation of globe-like aggregates with sizes of about 200 nm. The NFPP/C composite exhibits a monomodal pore size distribution between 10 and 100 nm, while a bimodal pore size distribution in the ranges of 3–10 nm and 13–45 nm occurs for NFPP/rGO.

In sodium half-cells, the composite of NFPP with rGO outperforms that with carbon black: at 20 °C, the reversible capacity is 95 mAh/g with a cycling stability of 95% after 100 cycles at a rate of C/2; meanwhile, at 40 °C, the reversible capacity increases to about 115 mAh/g at the expense of the lowered cycling stability to 88%. The improved performance of the NFPP/rGO composite is a result of the hybrid mechanism of Na-storage: although NFPP ensures the Na-storage via bulk Faradaic reactions, rGO gives rise to capacitive surface reactions. The hybrid mechanism of Na storage could also be correlated with the specific textural characteristics of NFPP/rGO composite.

In lithium half-cells, the composite NFPP/rGO stores Li via a hybrid mechanism, while NFPP/C operates via standard Li^+ intercalation reactions. As a result, a higher reversible capacity is achieved for the composite with rGO, but the cycling stability of the composite with carbon black is better. However, at elevated temperatures, the performance of both NFPP/C and NFPP/rGO becomes worse due to the initiation of side reactions with the electrolytes. Further studies are needed to suppress these reactions, which will be of importance for improving the performance of the composites in Li ion cells at elevated temperatures.

In general, this study reports the first data on Na storage performance of composites of NFPP with rGO at elevated temperatures, thus revealing the path for their commercialization.

Supplementary Materials: The following supporting information can be downloaded at: https://www.mdpi.com/article/10.3390/ma16196546/s1, Figure S1: SEM image of ball-milled NFPP without carbonaceous additives; Figure S2: Nitrogen adsorption/desorption isotherms (full/open symbols) with pore size distributions (insets) of carbon black and rGO.

Author Contributions: Conceptualization, R.S. and V.K.; methodology, V.K., R.S. and S.H.; formal analysis and investigation, S.H. and T.T.; writing—review and editing, V.K. and R.S.; visualization, T.T., S.H., V.K. and R.S.; funding acquisition, R.S.; project administration, R.S. All authors have read and agreed to the published version of the manuscript.

Funding: This research was funded by TwinTeam project (grant number Д01-272).

Institutional Review Board Statement: Not applicable.

Informed Consent Statement: Not applicable.

Data Availability Statement: Data available on request.

Acknowledgments: The authors acknowledge the TwinTeam project Д01-272 "European Network on Materials for Clean Technologies" for providing the opportunity to present the results at the SizeMat4 conference as well as for the financial publication support. Our thanks are also due to the project "Master" (КП-06-Д002/3 dated 18.05.2023) and project "Energy Storage and Hydrogen Energy" (NSI-ESHER, № 01-161/28.07.2022) for incubators used. The authors are sincerely grateful to Robert Feher from Graphit Kropfmühl GmbH (Hauzenberg, Germany) for providing rGO additive and Pavel Markov for TEM measurements.

Conflicts of Interest: The authors declare no conflict of interest.

References

1. Sanz, F.; Parada, C.; Rojo, J.M.; Ruíz-Valero, C. Synthesis, structural characterization, magnetic properties, and ionic conductivity of $Na_4M^{II}_3(PO_4)_2(P_2O_7)$ (M^{II} = Mn, Co, Ni). *Chem. Mater.* **2001**, *13*, 1334–1340. [CrossRef]
2. Kim, H.; Park, I.; Seo, D.-H.; Lee, S.; Kim, S.-W.; Kwon, W.J.; Park, Y.-U.; Kim, C.S.; Jeon, S.; Kang, K. New iron-based mixed-polyanionc cathodes for lithium and sodium rechargeable batteries: Combined first principles calculations and experimental study. *J. Am. Chem. Soc.* **2012**, *134*, 10369–10372. [CrossRef] [PubMed]
3. Nose, M.; Shiotani, S.; Nakayama, H.; Nobuhara, K.; Nakanishi, S.; Iba, H. $Na_4Co_{2.4}Mn_{0.3}Ni_{0.3}(PO_4)_2P_2O_7$: High potential and high capacity electrode material for sodium-ion batteries. *Electrochem. Commun.* **2013**, *34*, 266–269. [CrossRef]
4. Kim, H.; Park, I.; Lee, S.; Kim, H.; Park, K.-Y.; Park, Y.-U.; Kim, H.; Kim, J.; Lim, H.-D.; Yoon, W.-S.; et al. Understanding the electrochemical mechanism of the new iron-based mixed-phosphate $Na_4Fe_3(PO_4)_2(P_2O_7)$ in a Na rechargeable battery. *Chem. Mater.* **2013**, *25*, 3614–3622. [CrossRef]
5. Niu, Y.; Zhang, Y.; Xu, M. A review on pyrophosphate framework cathode materials for sodium-ion batteries. *J. Mater. Chem. A* **2019**, *7*, 15006–15025. [CrossRef]
6. Senthilkumar, B.; Murugesan, C.; Sada, K.; Barpanda, P. Electrochemical insertion of potassium ions in $Na_4Fe_3(PO_4)_2P_2O_7$ mixed phosphate. *J. Power Sources* **2020**, *480*, 228794. [CrossRef]
7. Shi, K.; Yang, W.; Wu, Q.; Yang, X.; Zhao, R.; She, Z.; Xie, Q.; Ruan, Y. Boosting the fast electrochemical kinetics of $Na_4Fe_3(PO_4)_2(P_2O_7)$ via a 3D graphene network as a cathode material for potassium-ion batteries. *New J. Chem.* **2023**, *47*, 10153–10161. [CrossRef]
8. Wood, S.M.; Eames, C.; Kendrick, E.; Islam, M.S. Sodium ion diffusion and voltage trends in phosphates $Na_4M_3(PO_4)_2P_2O_7$ (M=Fe, Mn, Co, Ni) for possible high-rate cathodes. *J. Phys. Chem. C* **2015**, *119*, 15935–15941. [CrossRef]
9. Zhang, H.; Hasa, I.; Buchholz, D.; Qin, B.; Geige, D.; Jeong, S.; Kaiser, U.; Passerini, S. Exploring the Ni redox activity in polyanionic compounds as conceivable high potential cathodes for Na rechargeable batteries. *NPG Asia Mater.* **2017**, *9*, e370. [CrossRef]
10. Pu, X.; Wang, H.; Yuan, T.; Cao, S.; Liu, S.; Xu, L.; Yang, H.; Ai, X.; Chen, Z.; Cao, Y. $Na_4Fe_3(PO_4)_2P_2O_7$/C nanospheres as low-cost, high-performance cathode material for sodium-ion batteries. *Energy Storage Mater.* **2019**, *22*, 330–336. [CrossRef]
11. Chen, M.; Hua, W.; Xiao, J.; Cortie, D.; Chen, W.; Wang, E.; Hu, Z.; Gu, Q.; Wang, X.; Indris, S.; et al. NASICON-type air-stable and all-climate cathode for sodium-ion batteries with low cost and high-power density. *Nat. Commun.* **2019**, *10*, 1480. [CrossRef] [PubMed]
12. Fernández-Ropero, A.J.; Zarrabeitia, M.; Reynaud, M.; Rojo, T.; Casas-Cabanas, M. Toward safe and sustainable batteries: $Na_4Fe_3(PO_4)_2P_2O_7$ as a low-cost cathode for rechargeable aqueous Na-ion batteries. *J. Phys. Chem. C* **2018**, *122*, 133–142. [CrossRef]

13. Senthilkumar, B.; Rambabu, A.; Murugesan, C.; Krupanidhi, S.B.; Barpanda, P. Iron-based mixed phosphate $Na_4Fe_3(PO_4)_2P_2O_7$ thin films for sodium-ion microbatteries. *ACS Omega* **2020**, *5*, 7219–7224. [CrossRef] [PubMed]
14. Yuan, T.; Wang, Y.; Zhang, J.; Pu, X.; Ai, X.; Chen, Z.; Yanga, H.; Cao, Y. 3D graphene decorated $Na_4Fe_3(PO_4)_2(P_2O_7)$ microspheres as low-cost and high-performance cathode materials for sodium-ion batteries. *Nano Energy* **2019**, *56*, 160–168. [CrossRef]
15. Boyadzhieva, T.J.; Koleva, V.G.; Kukeva, R.R.; Stoyanova, R.K. Mechanochemically desodiated $Na_4Fe_3(PO_4)_2P_2O_7$ as a lithium and sodium storage material. *ACS Appl. Energy Mater.* **2021**, *4*, 7182–7189. [CrossRef]
16. Chen, Y.; Dong, C.; Chen, L.; Fu, C.; Zeng, Y.; Wang, Q.; Cao, Y.; Chen, Z. "One stone two birds" design for hollow spherical $Na_4Fe_3(PO_4)_2P_2O_7$/C cathode enabled high-performance sodium-ion batteries from iron rust. *EcoMat* **2023**, e12393. [CrossRef]
17. Kosova, N.V.; Belotserkovsky, V.A. Sodium and mixed sodium/lithium iron ortho-pyrophosphates: Synthesis, structure and electrochemical properties. *Electrochim. Acta* **2018**, *278*, 182–195. [CrossRef]
18. Kosova, N.V.; Shindrov, A.A. Effect of mixed Li^+/Na^+-ion electrolyte on electrochemical performance of $Na_4Fe_3(PO_4)_2P_2O_7$ in hybrid batteries. *Batteries* **2019**, *5*, 39. [CrossRef]
19. Kumar, P.R.; Yahia, H.B.; Belharouak, I.; Sougrati, M.T.; Passerini, S.; Amin, R.; Essehli, R. Electrochemical investigations of high voltage $Na_4Ni_3(PO_4)_2P_2O_7$ cathode for sodium ion batteries. *J. Solid State Electrochem.* **2020**, *24*, 17–24. [CrossRef]
20. Jiang, X.; Du, P.; Mi, K.; Hu, F.; Wang, D.; Zheng, X. Hierarchical hollow microspheres $Na_3V_2(PO_4)_2F_3$@C@rGO as high-performance cathode materials for sodium ion batteries. *New J. Chem.* **2020**, *44*, 12985–12992.
21. Li, S.-F.; Hou, X.-K.; Gu, Z.-Y.; Meng, Y.-F.; Zhao, C.-D.; Zhang, H.-X.; Wu, X.-L. Sponge-like $NaFe_2(PO_4)(SO_4)_2$ as a high-performance cathode material for sodium-ion batteries. *New J. Chem.* **2021**, *45*, 4854–4859. [CrossRef]
22. Cao, Y.; Yang, C.; Liu, Y.; Xia, X.; Zhao, D.; Cao, Y.; Yang, H.; Zhang, J.; Lu, J.; Xia, Y. A green and scalable synthesis of $Na_3Fe_2(PO_4)P_2O_7$/rGO cathode for high-rate and long-life sodium-ion batteries. *ACS Energy Lett.* **2020**, *5*, 3788–3796. [CrossRef]
23. Wang, H.; Pan, Z.; Zhang, H.; Dong, C.; Ding, Y.; Cao, Y.; Chen, Z. A green and scalable synthesis of $Na_3Fe_2(PO_4)P_2O_7$/rGO cathode for high-rate and long-life sodium-ion batteries. *Small Methods* **2021**, *5*, 2100372. [CrossRef] [PubMed]
24. Wu, X.; Zhong, G.; Yang, Y. Sol-gel synthesis of $Na_4Fe_3(PO_4)_2(P_2O_7)$/C nanocomposite for Sodium ion batteries and new insights into microstructural evolution during sodium extraction. *J. Power Sources* **2016**, *327*, 666–674. [CrossRef]
25. Koleva, V.; Zhecheva, E.; Stoyanova, R. A new phosphate-formate precursor method for the preparation of carbon coated nano-crystalline $LiFePO_4$. *J. Alloys Compd.* **2009**, *476*, 950–957. [CrossRef]
26. Ferrari, A.C.; Robertson, J. Interpretation of Raman spectra of disordered and amorphous carbon. *Phys. Rev. B* **2000**, *61*, 14095–14107. [CrossRef]
27. Eklund, C.; Molden, J.M.; Jishi, R.A. Vibrational modes of carbon nanotubes: Spectroscopy and theory. *Carbon* **1995**, *33*, 959–972. [CrossRef]
28. Strankowski, M.; Włodarczyk, D.; Piszczyk, A.; Strankowska, J. Polyurethane nanocomposites containing reduced graphene oxide, FTIR, Raman, and XRD studies. *J. Spectrosc.* **2016**, *2016*, 7520741. [CrossRef]
29. Coluccia, S.; Marchese, L.; Martra, G. Characterisation of microporous and mesoporous materials by the adsorption of molecular probes: FTIR and UV–Vis studies. *Microporous Mesoporous Mater.* **1999**, *30*, 43–56. [CrossRef]
30. Li, Y.; Ye, K.; Cheng, K.; Cao, D.; Pan, Y.; Kong, S.; Zhang, X.; Wang, G. Anchoring CuO nanoparticles on nitrogen-doped reduced graphene oxide nanosheets as electrode material for supercapacitors. *J. Electroanal. Chem.* **2014**, *727*, 154–162. [CrossRef]
31. Tuinstra, F.; Koenig, J.L. Raman Spectrum of Graphite. *J. Chem. Phys.* **1970**, *53*, 1126–1130. [CrossRef]
32. Boyadzhieva, T.; Koleva, V.; Zhecheva, E.; Nihtianova, D.; Mihaylov, L.; Stoyanova, R. Competitive lithium and sodium intercalation into sodium manganese phospho-olivine $NaMnPO_4$ covered with carbon black. *RSC Adv.* **2015**, *5*, 87694–87705. [CrossRef]
33. Boyadzhieva, T.; Koleva, V.; Kukeva, R.; Nihtianova, D.; Harizanova, S.; Stoyanova, R. Storage performance of Mg^{2+} substituted $NaMnPO_4$ with an olivine structure. *RSC Adv.* **2020**, *10*, 29051–29060. [CrossRef] [PubMed]
34. Ackermann, A.C.; Carosella, S.; Rettenmayr, M.; Fox, B.L.; Middendorf, P. Rheology, dispersion, and cure kinetics of epoxy filled with amine- and non-functionalized reduced graphene oxide for composite manufacturing. *J. Appl. Polym. Sci.* **2022**, *139*, e51664. [CrossRef]
35. Broekhoff, J.C.P. Mesopore determination from nitrogen sorption isotherms: Fundamentals, scope, limitations. *Stud. Surf. Sci. Catal.* **1979**, *3*, 663–684.
36. Thommes, M.; Kaneko, K.; Neimark, A.; Olivier, J.P.; Rodriguez-Reinoso, F.; Rouquerol, J.; Sing, K.S.W. Physisorption of gases, with special reference to the evaluation of surface area and pore size distribution (IUPAC Technical Report). *Pure Appl. Chem.* **2015**, *87*, 1051–1069. [CrossRef]
37. Gauden, P.A.; Terzyk, A.P.; Jaroniec, M.; Kowalczyk, P. Bimodal pore size distributions for carbons: Experimental results and computational studies. *J. Colloid Interface Sci.* **2007**, *310*, 205–216. [CrossRef]
38. Ali, G.; Mehmood, A.; Yong Ha, H.; Kim, J.; Chung, K.Y. Reduced graphene oxide as a stable and high-capacity cathode material for Na-ion batteries. *Sci. Rep.* **2017**, *7*, 40910. [CrossRef]
39. Lee, S.W.; Yabuuchi, N.; Gallant, B.M.; Chen, S.; Kim, B.-S.; Hammond, P.T.; Shao-Horn, Y. High-power lithium batteries from functionalized carbon-nanotube electrodes. *Nat. Nanotechnol.* **2010**, *5*, 531–537. [CrossRef]

materials

Article

Influence of the Chemical Composition of Ceria Conversion Coatings, Sealed in Solution of NaH_2PO_4 and $Ca(NO_3)_2$, on the Corrosion Behavior of Aluminum

Aleksandar Tsanev [1,*], Reni Andreeva [2] and Dimitar Stoychev [2]

1 Institute of General and Inorganic Chemistry, Bulgarian Academy of Sciences, "Acad. G. Bonchev" Street, bl. 11, 1113 Sofia, Bulgaria
2 "Rostislaw Kaischew" Institute of Physical Chemistry, Bulgarian Academy of Sciences, "Acad. G. Bonchev" Street, bl. 11, 1113 Sofia, Bulgaria; randreeva@ipc.bas.bg (R.A.); stoychev@ipc.bas.bg (D.S.)
* Correspondence: tsanew@abv.bg

Abstract: The corrosion-protective influence of eco-friendly ceria conversion coatings deposited on Al-1050 alloy, additionally treated in mixed NaH_2PO_4 and $Ca(NO_3)_2$ solution, was studied. The main aim of this work was to investigate how the obtained mixed systems of coatings eliminates the negative role of cracks and pores on the surface formed after deposition only of ceria coating. For this purpose, the growth structure, main components and corrosion resistance of the so formed protective systems were investigated by SEM, EDS, XRD, XPS and electrochemical (PDP, Rp, etc.) methods. The results obtained show that the basic components of the conversion layers (before and after exposure in model corrosion media) are characterized by Al_2O_3, $Al(OH)_3$, $CePO_4$ and $Ca_5(PO_4)_3(OH)$. Based on these results, the optimal conditions of immersion treatment(s) of Al substrate are established. At these regimes, the relationship of co-deposited Ce^{3+}, $PO_4{}^{3+}$ and Ca^{2+}-containing components of the conversion layers determine the maximum values of their polarization resistance—a basic criterion for corrosion protection of Al. This effect is related to the formation of fill out of the defects of the conversion coatings and additional $Ca_5(PO_4)_3(OH)$, $CePO_4$ $AlPO_4$ and $Al(OH)_3$ deposits, leading to the decrease of the corrosion rate.

Keywords: aluminum; ceria; calcium; phosphate(s); conversion coatings; corrosion

Citation: Tsanev, A.; Andreeva, R.; Stoychev, D. Influence of the Chemical Composition of Ceria Conversion Coatings, Sealed in Solution of NaH_2PO_4 and $Ca(NO_3)_2$, on the Corrosion Behavior of Aluminum. *Materials* **2023**, *16*, 6499. https://doi.org/10.3390/ma16196499

Academic Editor: Sergei A. Kulinich

Received: 5 September 2023
Revised: 26 September 2023
Accepted: 27 September 2023
Published: 29 September 2023

1. Introduction

The requirements and norms of modern legislation [1,2], regarding the application of "green" technologies in the deposition of conversion protective and decorative coatings on aluminum and its alloys (Al), strictly exclude the widely used until recently highly efficient chromate conversion coatings based on Cr^{6+} [3–8]. A wide range of intensive research carried out in this aspect ranks among the favorites of ecofriendly technologies those related to the formation of protective oxide layers of lanthanides and, in particular, cerium oxides [9–23]. However, upon application of the cerium salts, elaborated on the basis of these compositions and regimens for ceria conversion treatment of Al, the high degree of protection of the chromate layers has not yet been achieved. In this connection, systematic investigations continue and they are associated with finding options for promoting their corrosion-protective ability by the inclusion of additional processing of the formed ceria conversion coatings (CCOC). Encouraging results in this respect have been obtained by the inclusion of additional finishing sealing of the already deposited CCOC in solutions of various phosphate compounds, leading to the formation of a thin phosphate layer [5,15,24–28]. Such an approach could lead to the formation of insoluble Ce-phosphates, as well as to a possible synergistic protective effect between ceria layers and the combined metal-phosphate layer formed on it, where the metal could be Ce, Al or Al + Ce [15,29].

It is necessary to note, however, one characteristic disadvantage of these conversion coatings, which is related to the formation of micro-cracked pores [9,30,31]. They can reach the Al substrates and cause corrosion damage when in contact with aggressive environments.

In light of the above, the task of the present study was to establish and juxtapose corrosion behavior, with respect to the efficiency of deposited-on-Al-1050 thin, conversion cerium-oxide coatings (CCOC) [32–35]. They are subjected to additional sealing post-treatment in mixed solution of sodium phosphate mono basic and calcium nitrate. Sodium phosphate mono basic and calcium nitrate can fill and reduce possible crack and pore defects of the CCOC, as well as increase their corrosion-protective influence and ability in corrosion medium.

Our motivation for including and investigating the impact of such additional processing on CCOC is the results of recent systematic studies of H.S. Lee and co-authors [36–39]. They found a strong positive effect of sodium phosphate and calcium nitrate sealing treatment on the corrosion resistance and ability of wire arc sprayed aluminum protective coatings (deposited on steel substrate), the surface of which is characterized by a rough and porous microstructure.

2. Materials and Methods

The studied conversion layers (produced by electroless treatments) were deposited on substrates of "technically pure" Al 1050 (containing 0.40% Fe, 0.25% Si, 0.05% Mn, 0.05% Cu, 0.07% Zn, 0.05% Mg). The studied specimens/substrates of dimensions 3 × 3 cm were cut out of rolled Al sheets (not polished additionally) with thickness 0.1 cm. Their pre-treatment (in 1.5 M NaOH) is described in [32]; The short name for these pre-treated substrates is S1. The formation of thin conversion ceria oxide coatings on S1 was implemented in a solution containing 0.5 M $CeCl_3 \times 7H_2O + 1 \times 10^{-5}$ M $CuCl_2 \times 2H_2O$ (at a temperature 25 °C and with a time of immersion 2 h). No H_2O_2, or other type of oxidizing agent, was added [33–35]. For these systems, the short name is S2. The thickness of the deposited CCOCs was 110 nm (as measured from XPS in-depth profiles, displaying in detail the change of the ratio between the elements Ce, Al and O in the depth of CCOC [35]). After formation of CCOC, the specimens were rinsed fully in distilled water, dried and post-treated in mixed 0.5 M NaH_2PO_4 + 0.1 M $Ca(NO_3)_2$ solution, as proposed and applied for formation of phosphate protection layers (marked in the text below as PhCa) on aluminum by Lee and co-authors [36]. The temperature of the mixed solution was 85 °C and we chose three different times of immersion treatment: (I) 5 min; (II) 3 × 5 min and (III) 3 × 5 min + 72 h exposure of the so treated sample in humid atmosphere over a distillated water (in thermo-stated glass cell at 50 °C). In procedure (I), after 5 min immersion treatment, the sample(s) were rinsed fully in distilled water and finally dried at room temperature. In procedure (II), the sample(s) were under immersion treatment, rinsing and drying three times after every 5 min of immersion treatment in the mixed solution. In procedure (III), the sample(s) were under immersion treatment, rinsed and dried three times after every 5 min immersion treatment (as in procedure II), after which the sample(s) were left in a humidity chamber [37–39] for 72 h at 50 °C and 95% relative humidity atmosphere for formation of natural oxides and submitted to a final drying, prior to exposure and investigation in model corrosion medium of 0.1 M NaCl (CM). The thickness of the so deposited phosphate layers was ≈0.9 μm (measured by SEM observations of cross sections of the samples). The short names of these prepared samples: for sample (I)–S3; for sample (II)–S4 and for sample (III)–S5 (Table 1).

Table 1. Treatment conditions of the studied samples.

Abbreviation of the Samples	Type of Treatment of the Samples		
	in NaOH (1.5 M; 60 °C; 1 min) Solution	in 0.5 M $CeCl_3 \times 7H_2O$ + 1×10^{-5} M $CuCl_2 \times 2H_2O$ (25 °C; 2 h)	in 0.5 M NaH_2PO_4 + 0.1 M $Ca(NO_3)_2$ Solution (85 °C)
S1	YES	NO	NO
S2	YES	YES	NO
S3	YES	YES	5 min.
S4	YES	YES	3 × 5 min.
S5	YES	YES	3 × 5 min + 72 h in humidity atmosphere

The surface morphology, structure and chemical composition in the bulk of the studied systems were investigated by electron microscopy (JEOL JSM 6390, Tokyo, Japan) under the conditions of secondary electron image (SEM), back-scattered electrons (BEI) and characteristic energy dispersive X-rays analysis (EDS).

The formed phases on the S1 substrate(s) covered by CCOC, after their post-treatment in NaH_2PO_4 and $Ca(NO_3)_2$ solutions, were observed by X-ray diffraction. XRD patterns were recorded on the multipurpose system Empyrean, manufactured by PANalytical, Malvern, UK. The system was equipped with LFHR X-ray tube Cu Kα radiation (λ = 1.54184 Å) generated at 45 kV and 40 mA. The volume fraction (in %) of each phase formed on the samples was analyzed by HighScorePlus 4.5 software and the ICSD database (Inorganic Crystal Structure Database) provided by FIZ Karlsruhe, Eggenstein-Leopoldshafen, Germany.

The X-ray photoelectron spectroscopy (XPS) was carried on an AXIS Supra electron-spectrometer (Kratos Analitycal Ltd., Manchester, UK) using achromatic AlKα radiation with photon energy of 1486.6 eV. The analyzed area was 0.3 × 0.75 mm^2. The energy calibration was performed by fixing the C1s line of adsorbed adventitious hydrocarbons to 284.6 eV. The binding energies (BE) were measured with an accuracy of ±0.1 eV. The changes in composition and chemical surrounding of the chemical elements were determined monitoring the areas and binding energies of the photoelectron peaks of the appropriate elements, that is, C1s, O1s, Ce3d, etc. The chemical composition (in at. %) and the state of the elements of cerium-based CCOC on Al 1050 were examined and determined for the as-deposited samples and after their post-treatment in mixed 0.5 M NaH_2PO_4 + 0.1 M $Ca(NO_3)_2$ solution, as well as after long-time (up to 168 h) exposure in model corrosion media.

The corrosion behavior of the samples was studied in a standard three-electrode cell in open-to-air 0.1 M NaCl ("p.a." Merck, Darmstadt, Germany) CM at 25 °C. A platinum electrode was used as the counter electrode, having dimensions of 10 × 10 × 0.6 mm, while the reference electrode was saturated calomel electrode (SCE) (E_{SCE} = +0.240 V vs. SHE). All the potentials in this study are compared to SCE (for the realization of OCP-Time plots, Chronoamperometry (at E_{pitt}) and Rp measurements, the reference electrode was Ag/AgCl). The anodic and cathode polarization curves were obtained by means of a potentiostate/galvanostate Gamry Interface 1000, the obtained results being processed with the help of specialized software. The potentiodynamic polarization (PDP) curves were recorded at a sweeping rate of the potential 1 mV.s^{-1} within the range of potentials–2500–+2500 mV (starting in cathodic and anodic directions from the open circuit potential (E_{OCP}) of S1 = −0.660 V vs. E_{SCE}).

The polarization resistance investigations of studied samples (as-deposited and after definite time of exposure in CM) were carried out on Gamry Interface 1000 (Software and Frequency Response Analyzer EIS300, Warminster, PA, USA). The scan range was ±15 mV relative to corrosion potential (E_{corr}). The initial delay was 15 min, and the temperature was 25 °C ± 0.5° C. The polarization resistance (Rp) of the coatings was calculated by the Stern-Geary equation [40], which is based on the fact that a higher Rp value (in Ω cm^2) corresponds to higher corrosion resistance and to lower corrosion current (i_{cor}) as well, i.e., $Rp \sim 1/i_{cor}$. The results obtained with this method are highly relevant to standard measurements conducted in natural corrosion conditions. In order to check the reproducibility, the PDP and Rp measurements were repeated at least 3 times.

3. Results

Conducting the research presented in this work, we were guided by the following considerations. During the immersion processing of aluminum, the deposited thin CCOC are characterized by cracks and pores that may reach the Al substrate [41]. Their formation is mainly associated with the occurrence of processes of dehydration of CCOC [20]. It is considered that these areas are one of the main reasons that the corrosion protection of CCOC does not always satisfy enough. Such cases eventually require additional post-treatment operations of CCOC [31,42]. In the present study, we applied post-treatment in mixed 0.5 M NaH_2PO_4 + 0.1 M $Ca(NO_3)_2$ solution, because during this treatment, according to [37,38] and our XRD and XPS investigations, the following reactions may occur:

$$2Al + 6NaH_2PO_4 + 2Ca(NO_3)_2 + 14H_2O \rightarrow 2CaHPO_4(H_2O)_2 + 2Na_3Al(OH)(HPO_4)(PO_4) + 4NH_4OH + 8O_2 + 3H_2 \quad (1)$$

The products obtained during Reaction (1) $Ca(HPO_4)(H_2O)_2)$ and $(Na_3Al(OH)(HPO_4)(PO_4)$ can interact with water according to the following equation [39]:

$$CaHPO_4(H_2O)_2 + Na_3Al(OH)(HPO_4)(PO_4) + 5H_2O \rightarrow Al(OH)_3 + 3NaOH + Ca(OH)_2 + 3H_3PO_4 \quad (2)$$

in which the formed $Al(OH)_3$ (characterized and proven by XRD analyses [39]) is mainly in the form of its thermodynamically stable phase bayerite–α-$Al(OH)_3$.

Along with that, however, the H_3PO_4 formed in Reaction (2) can cause Reaction (3) to occur as well [43]

$$10Ca(OH)_2 + 6H_3PO_4 \rightarrow Ca_{10}(PO_4)_6(OH)_2 + 18H_2O \quad (3)$$

which will lead to the deposition of the practically water-insoluble hydroxylapatite ($Ca_5(PO_4)_3(OH)$) (labeled also as calcium hydroxylapatite ($Ca_{10}(PO_4)_6(OH)_2$) on the S2 system (including on the "bared" areas of the Al substrate—the bottoms of the cracks and pores of the CCOC reaching the Al surface).

We also consider that H_3PO_4 formed during Reaction (2) can interact both with the dense regions of CCOC (made up of a mixture of easily soluble Ce_2O_3 and hardly soluble CeO_2 [32,35,42]), as well as with the exposed Al surface (in the areas of microcracks and pores), according to Reactions (4) and (5).

$$Ce_2O_3 + 2\,H_3PO_4 \rightarrow 2CePO_4 + 3H_2 \quad (4)$$

$$2Al + 2\,H_3PO_4 \rightarrow 2AlPO_4 + 3H_2. \quad (5)$$

which will lead to the formation of the water-insoluble $CePO_4$ and $AlPO_4$ on these surfaces. Confirming this surmise are the XPS studies conducted and presented below on the chemical state of the elements Ce, Al, P and Ca, proving the presence of $CePO_4$, $AlPO_4$ and $Ca_5(PO_4)_3(OH)$ on the surface of the examined samples (see Section 3.3. XPS Studies, below).

3.1. SEM/EDS Studies

Figure 1 illustrates SEM investigations of the surface of obtained conversion layers, deposited on Al 1050 (after its pre-treatment-S1), and its immersion treatment in solution(s) of 0.5 M $CeCl_3 \times 7H_2O$ + 1×10^{-5} M $CuCl_2 \times 2H_2O$ followed by different consecutive sealing treatments in mixed 0.5 M NaH_2PO_4 + 0.1 M $Ca(NO_3)_2$. From the presented results, it appears that the formed CCOCs (Figure 1a) on the S1 (the system S2) are built of spherical agglomerates, which are relatively uniformly distributed over the entire surface, reproducing the profile of the Al substrates. Therefore, the roughness and the rigidity determined by the mechanical treatment of the rolled Al 1050 sheet influence in a specific way the formation of the chemically deposited CCOC [35].

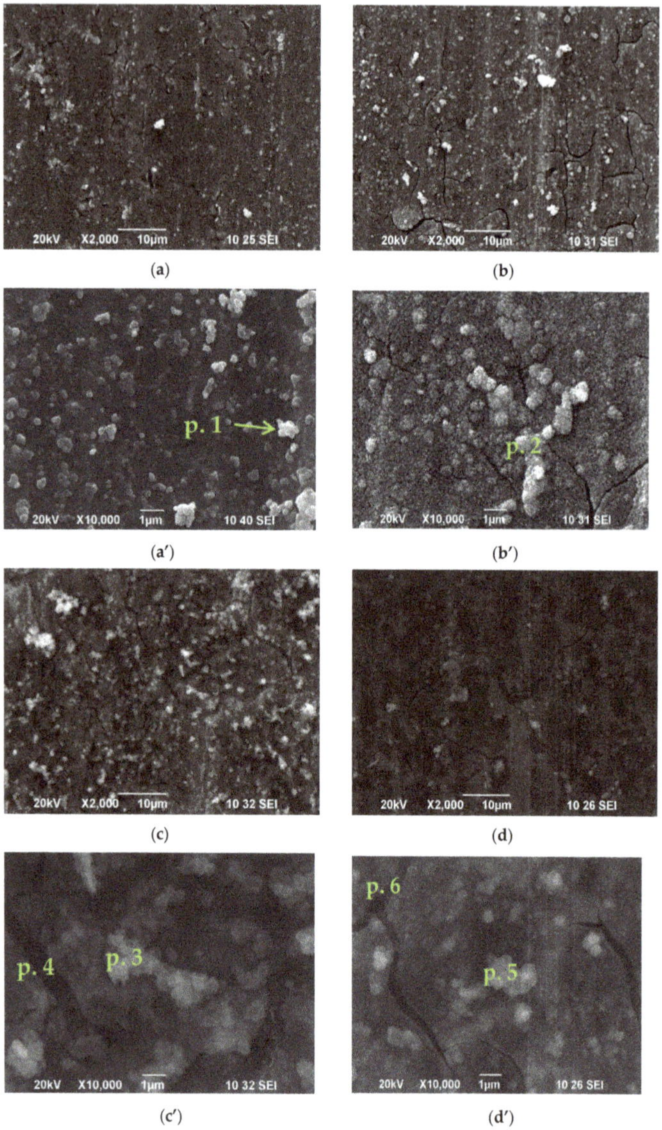

Figure 1. SEM images of studied samples: (**a**) S2; (**b**) S3; (**c**) S4; (**d**) S5 at magnification ×2000; (**a′**) S2; (**b′**) S3; (**c′**) S4. In point 3: Al–27.81%; O–64.56%; P–4.75%; Ca–0.52%; Cu–0; Ce–1.69%; Fe–0.66%. In point 4: Al–90.27%; O–7.66%; P–2.07%; Ca–0%; Cu–0%; Ce–0%; Fe–0%; (**d′**) S5 at magnification ×10,000. In point 5: Al–34.66%; O–34.04%; P–9.37%; Ca–2.13%; Cu–1.03%; Ce–17.88%; Fe–0.89%; In point 6: Al–90.05%; O–8.39%; P–1.56%; Ca–0 %; Cu–0 %; Ce–0 %; Fe–0%.

Pores and cracks are also observed, their width reaching 0.2–0.4 μm (Figure 1a). The integral EDS analysis (Table 2) registered the presence of Al (86.09%), O (10.40%), Fe (0.37%)

and Ce (3.14%). At the same time, the performed EDS analyses at a point (at magnification 300,000×) in the bright areas indicated by arrows (in Figure 1a') characterizing the areas of CCOC deposited on Al_3Fe-rich phase agglomerates [35,41] in the aluminum matrix of Al 1050 (Figure 1a') register the presence of Al (47.25%), O (30.95%), Fe (5.13%) and Ce (16.66%).

Table 2. Chemical composition of the studied systems determined by EDS analysis before and after their exposure in CM for 168 h.

Sample	Al, wt.%	O, wt.%	Ce, wt.%	Ca, wt.%	P, wt.%	Cu, wt.%	Fe, wt.%	Cl, wt.%
S2 as-deposited	86.09	10.4	3.14				0.37	
after 168 h in 0.1 M NaCl	93.68	4.68	1.64				0	
S3 as-deposited	80.83	12.34	6.55	0.28	0	0	0	
after 168 h in 0.1 M NaCl	88.87	8.67	1.33	0	1.13	0	0	0
S4 as-deposited	63.97	25.57	2.58	0.95	6.48	0.45	0	
after 168 h in 0.1 M NaCl	91.38	7.73	0	0	0.89	0	0	0
S5 as-deposited	68.94	22.64	2.12	0.84	5.1	0	0.35	0.37
after 168 h in 0.1 M NaCl	94.48	5.15	0	0	0	0	0	

Figure 1b presents the changes in the structure and composition of the samples subjected to an additional 5-min immersion treatment of the S2 system in mixed 0.5 M NaH_2PO_4 + 0.1 M $Ca(NO_3)_2$ solution (the system S3). The surface morphology of this system is similar to that on Figure 1a. One is to note additionally, the increase in the number and size of the spheroidal agglomerates forming the surface morphology, as well as the changes in the shape and size of the formed micro-cracks, is impressive. At the same time, the integral EDS analysis (Table 2) registered a two-fold increased concentration of Ce (6.55%) and the presence of Ca (0.28%), and the concentrations of Al and O, respectively, were 80.83 and 12.34%. Pronounced to a higher degree are also the sections of the Al_3Fe phase, on which the conversion layers are predominantly formed (Figure 1b', marked as Point 2) [34,35,41]. The analyzes, at a point (300,000×) in them, recorded a significantly higher concentration of Ce (26.31%), P (2.88%), Ca (1.51%), Fe (0.80%) and Cu (0.67%) (Table 2) [34,35,41].

Figure 1c characterizes the changes in the structure and composition of samples subjected to three additional 5-min (3 × 5 min) immersion treatment of the S2 system in mixed 0.5 M NaH_2PO_4 + 0.1 M $Ca(NO_3)_2$ solution (the system S4). The surface morphology of this system is quite different than that on Figure 1b. The micro-profile of the surface of the samples is more uniform and the number of micro-cracks is reduced (mostly the recorded traces of cracks (shown in Figure 1b) are practically filled with conversion coating—Figure 1c). The results of the integral EDS analysis for this system are presented in Table 2. They indicate that the application of three times longer exposure time in mixed 0.5 M NaH_2PO_4 + 0.1 M $Ca(NO_3)_2$ solution causes the inclusion of P in about an eightfold higher concentration (6.48 vs. 0.81%) in such a modified system of conversion coatings. At the same time, phosphates and dihydrogen phosphates (see the XPS studies) are also registered at the bottom of the pores (2.07%), where Ca, Cu and Ce are absent (Figure 1c'). That is, the post-treatment in mixed 0.5 M NaH_2PO_4 + 0.1 M $Ca(NO_3)_2$ solution causes direct interaction with the Al substrate in the "defective" areas (cracks), on which, however, CCOC and insoluble calcium-phosphate compounds (see Section 3.2 XRD Studies) are not registered.

Figure 1d characterizes the changes in the structure and composition of samples subjected to three additional 5-min (3 × 5 min) immersion treatments of the S2 system in mixed 0.5 M NaH_2PO_4 + 0.1 M $Ca(NO_3)_2$ solution and subsequent exposure for 72 h in 95% humidity atmosphere (the system S5). The surface morphology and micro profile of this system to a certain extent are similar to that on Figure 1c. The same is seen for the results of the integral EDS analysis of its chemical composition, but nevertheless, they register certain differences when comparing the results obtained in the EDS analyses at a point. While the surface concentrations of the aggregates forming the conversion coating (Figure 1d′) are Ce–17.88%, Ca–2.13% and P–9.37%, for the system presented in Figure 1c′, these concentrations are significantly lower—respectively, Ce–1.69%, Ca–0.52% and P–4.75%. This significant difference in the concentration of Ce, Ca and P, determined by the exposure for 72 h in 95% humidity atmosphere, is most likely related to the occurrence of Reactions (1)–(4), which lead to the formation on the surface and in the cracks of CCOC of the practically insoluble in the corrosion medium $Ca_5(PO_4)_3(OH)$, $CePO_4$ and $AlPO_4$. The XPS analyses on the surface of the investigated samples carried out in the present study (see P. 3.3. XPS Studies, below), as well as those registered for similar systems earlier [42], give reasons to assume that to the favorable protective effect of the above-mentioned phosphate compounds can be added the effect due to the formed Al_2O_3 and AlOOH [42].

The quantitative characterization of the concentration of the characteristic elements (Ce, Ca, P) found in the volume of the formed conversion layers is presented in Table 2. According to these data, their arrangement in descending order, for each of the studied systems (or their concentration in the chemical compounds formed by them), is the following:

Ce: S3 > S2 > S4 > S5;

Ca: S4 > S5 > S3;

P: S4 > S5 > S3.

These results indicate that when the exposure time of the system S2 in mixed 0.5 M NaH_2PO_4 + 0.1 M $Ca(NO_3)_2$ increases, the concentration of Ce decreases. At the same time, the concentration of Ca and P rises. The concentration of Ca and P at 3 × 5 min exposure in the solution is the highest.

3.2. XRD Studies

The deposited and registered phases on the S5 system (see Section 3.4, below) are shown in Figure 2. Based on the results of the polarization investigations and Rp tests, this sample is characterized by the highest anticorrosion characteristics. It can be seen from the figure that, along with the most salient and prominent peaks (in 2θ scale) of the Al substrate (ICDD PDF #01-073-2661), peaks for the $Ca_5(PO_4)_3(OH)$ (ICDD PDF # 01-070-0798) phase are clearly registered [44,45], while those for the α-$Al(OH)_3$ phase are not observed. The processing of the XPS data (see Section 3.3, below), however, indicates the presence of $Al(OH)_3$. These results are also in good agreement with the results obtained by EDS analysis of the same samples (see Section 3.1, above), characterizing the presence of elements characteristic for $Ca_5(PO_4)_3(OH)$—Ca and P (Table 2). These results correspond and confirm the possibility of Reactions (1)–(3), with respect to the formation of $Ca_5(PO_4)_3(OH)$ and $Al(OH)_3$, which are difficult to dissolve in the corrosion medium. The latter can inhibit the corrosion process, respectively, its speed, both in the areas of exposed Al substrate (filling out the defects of the CCOC), and thanks to thickening and stabilization of the CCOC, caused by the formation of water-insoluble $CePO_4$ and/or $AlPO_4$.

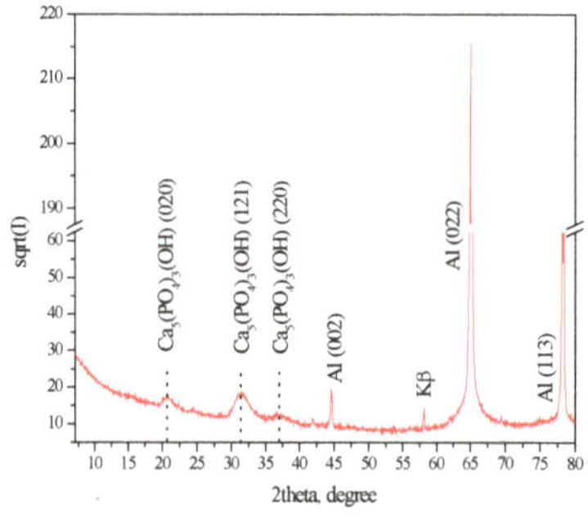

Figure 2. XRD pattern of the system S5 (showed the best corrosion-protective effect) at 2θ = 10°–90°.

3.3. XPS Studies

3.3.1. XPS of As-Deposited Samples

Surface XPS studies of the above-described conversion coating systems formed on the aluminum substrate were performed. The surface of the obtained layers was examined both immediately after receiving the samples and after they were subjected to various electrochemical and corrosion tests. The purpose of these studies was to establish the chemical composition of the deposited conversion coatings, the chemical state of the characteristic elements in them before and after being subjected to corrosion attack in CM.

Figure 3a–c shows XPS–core photoelectron spectra of Al2p, P2p and Ca2p of the as-deposited samples. C1s, O1s, Ce3d, Na1s, N1s and Cu2p spectra are not shown here because they do not provide such essential information about the processes taking place on the surface. The Al2p spectra are shown in Figure 3a. The spectra are deconvoluted into two essential groups of peaks. The first one for samples S2 and S4 at 74.2 eV is attributed to Al_2O_3 [46]. This peak for sample S3 is slightly shifted at 74.5 eV to higher BE. Such a shift we can attribute to the presence of $AlPO_4$ [47] formed during the immersion treatment. The peak at 74.7 eV for sample S4 is related to the existence of $Al(OH)_3$ [48]. The second group around 76.1 eV we can attribute to the presence of some forms of amorphous AlO(OH) with a non-stoichiometric amount of crystalline H_2O groups [49,50].

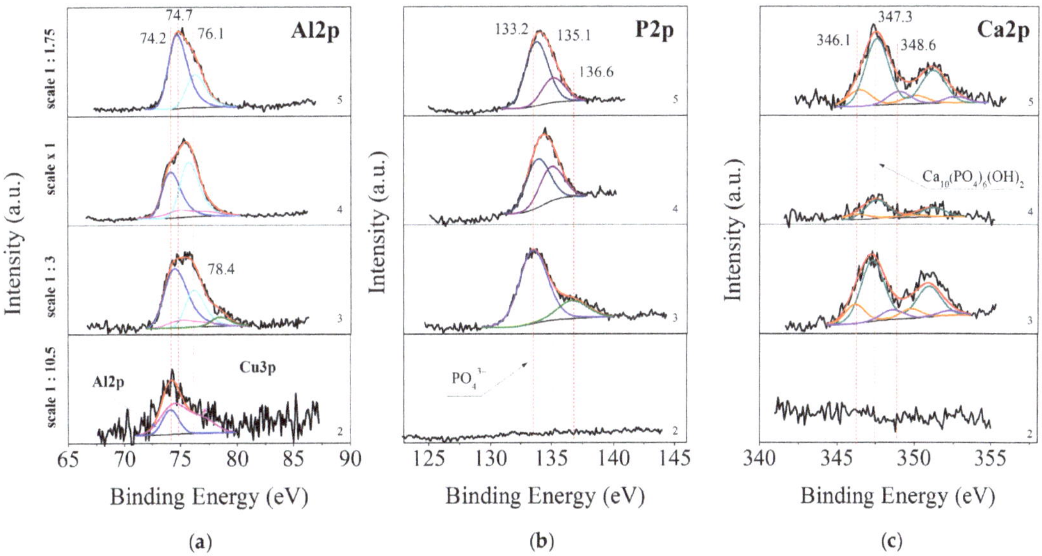

Figure 3. Peak fitting of Al2p (**a**), P2p (**b**) and Ca2p (**c**) XPS spectra for as-deposited samples. From bottom to top: 2—S2, 3—S3, 4—S4 and 5—S5. The sum of peaks is given with a red line, the black line labels the experimental data with the background, and the other color lines represent peak fitting curves.

In Figure 3b are shown XPS spectra of the other chemical element in the conversion coating–P2p for as-deposited samples. The spectra for sample S3 were deconvoluted into two peaks—at 133.2 eV and around 136.6 eV. This peak at 133.2 eV is a characteristic peak for the presence of PO_4^{3-} groups [51]. These groups can originate from the phosphate precursor or can be $CePO_4$ or $AlPO_4$ obtained during the immersion processing (Equations (4) and (5)).

Taking into account the concentration of Na–ions, we can conclude that where the concentration of sodium is higher (Table 3), in sample S4 (134.0 eV) [52], this shift is larger compared to the S5 sample, where this shift is smaller (133.8 eV). From this we can conclude that the amount of unbound phosphate groups in sample S4 is greater than in sample S5. Because NaH_2PO_4 is soluble, this may also be a reason for the lower corrosion resistance of this sample, comparable to that of uncoated aluminum. Also a reason for this lower resistance is the lack of a cerium-oxide layer on the surface. In the XPS analysis, this element was not detected. Looking at the P2p spectrum of the sample S3 and comparing it with the chronoamperometric curves, we can note its higher corrosion resistance. Here, the position of the P2p peak is characteristic of the presence of PO_4^{3-} groups, while approximately 4 at.% Ce^{3+} have been calculated on the surface. Kozhukharov and co-authors assert that when Ce^{3+} is involved in the immersion deposition, $CePO_4$ is formed on the Al surface [29]. Considering the areas of Al and P peaks corresponding to Al-PO_4 bond and P-O bonds (within phosphate groups), the ratio between them should be 1:1, which is the stoichiometry in $AlPO_4$. The established ratio, however, is 1:1.5, an excess of P. From here we can conclude that about 33% of the phosphate groups are connected to cerium ions in the form of insoluble $CePO_4$. However, the relatively low concentration of cerium phosphate formed is not able to sufficiently block the corrosion centers. The samples S5 and S2 have the highest corrosion resistance. In sample S5, NaH_2PO_4 was also registered. Despite the good solubility of NaH_2PO_4, the increased corrosion resistance of this sample can be explained by the protective role of Al_2O_3 and AlOOH, as well as with the presence of the poorly soluble $AlPO_4$ (Table 3). We deduce the presence of $AlPO_4$, not the presence of NaH_2PO_4, from the smaller shift (133.8 eV) of the phosphate groups to higher binding energy values. In sample S2, on which no phosphate coating was deposited, the relatively good corrosion protection ability can be related to the presence of both Ce_2O_3 and Al_2O_3.

Table 3. Chemical composition of the elements of the studied systems determined by XPS analysis for the following samples: as-deposited(the first row); after Rp measurements for 168 h (the second row), after 168 h Chronoamperometry (the third row) and after 168 h OCP (the fourth row).

Sample	Al, at.%	O, at.%	Ce, at.%	Ca, at.%	P, at.%	Cu, at.%	Na, at.%	N, at.%
S2 as-deposited	11.0	70.5	15.8			2.7		
after 168 h Rp	20.5	76.7	2.4			0.4		
after 168 h Chronoamperometry	26.0	74.0						
after 168 h OCP	16.2	78.1	5.7					
S3 as-deposited	17.1	61.8	3.84	2.3	14.3	0.6		
after 168 h Rp	27.7	71.2			0.7		0.4	
after 168 h Chronoamperometry	30.5	69.5						
S4 as-deposited	34.7	53.4		0.4	8.8	1.4	0.7	0.6
after 168 h Rp	29.9	62.6			4.4			3.0
after 168 h OCP	24.2	65.2	0.5	0.2	6.8			3.0
S5 as-deposited	19.6	65.1	0.5	1.7	11.6		0.6	1.0
after 168 h Rp	28.2	68.4			3.5			
after 168 h OCP	24.0	63.4	1.1	0.6	8.2			2.7

Figure 3c shows the deconvoluted core photoelectron spectra of Ca2p. They are separated into three doublets. The first peak of Ca2$p_{3/2}$, positioned at 346.1 eV, corresponds to $Ca(OH)_2$ [53], which is a product of the chemical interactions taking place on the surface (reaction 2). The second peak, positioned at 347.3 eV, corresponds to the presence of insoluble corrosion-protective $Ca_{10}PO_{46}(OH)_2$ [54,55], also a product of reaction (3), and the third one-to $Ca(NO_3)_2$ [56], which is an element of the precursor system.

The analysis of the Al2p, P2p and Ca2p spectra for the as-deposited samples are confirmed by the peak fitting of the O1s spectra for the same samples (spectra not given in the publication). The spectra can be divided into four groups of peaks. The first group of peaks, positioned at 529.6–531 eV, we can associate to the presence of chemical bounds of oxygen with metal ions. It can be noted that in the sample order S2 (positioned at

529.6 eV) → S3 (530.4 eV) → S4 (530.8 eV) → S5 (531.0 eV), the binding energy of these peaks gradually increases, which we can associate with the decrease of the concentration of cerium oxide and the increase of the concentration of aluminum oxide [46] on the surface. The second group of peaks positioned at 531.4 are related to the presence of P–O–P–bonds within phosphate groups [51]. The third group, peaks shifted relative to it by about one electronvolt, at 532.2–532.6 eV, we can associate with the presence of NaH_2PO_4 and OH^- groups. The fourth group of peaks, at 535.3–535.5 eV, can be associated with the presence of OOH groups and adsorbed water [49,50].

3.3.2. XPS of Studied Samples after Their Long-Term Exposure in CM

In Figure 4 are presented the results of the XPS analyses of samples identical to the ones discussed above. These samples are studied after long-term exposure in conditions of Rp, OCP and Chronoamperometry tests. The XPS analyses here give us the quantitative and qualitative changes in the chemical composition of the mixed protective conversion layers after these corrosion tests.

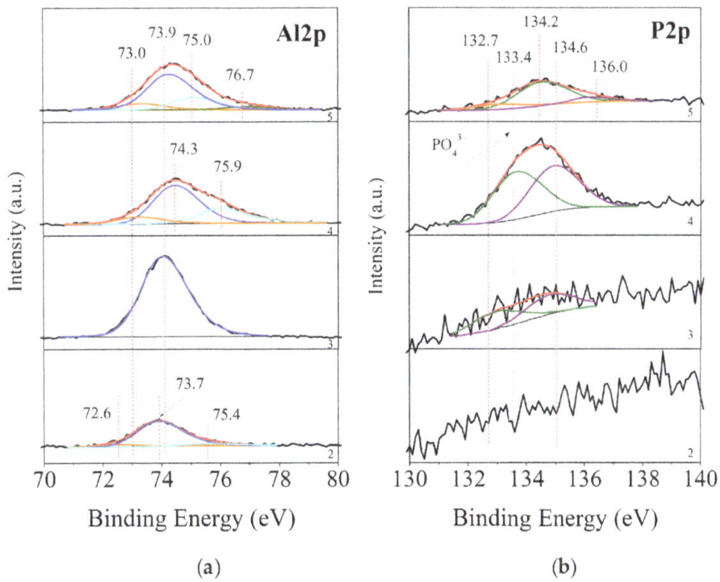

Figure 4. XPS curve peak fitting of Al2p (**a**) and P2p (**b**)–spectra for samples processed in model corrosion medium. From bottom to top: 2—S2, 3—S3, 4—S4 and 5—S5. With a red line is marked the sum of the peaks, the black line labels the originate spectra and the background, and other color lines represent peak fitting curves.

XPS of Studied Samples after Rp Tests

Figure 4a,b show the deconvoluted spectra of Al2p and P2p for samples S2–S5. In contrast to the results for the as-deposited layers (Figure 3), for sample S4, there is a low intensity peak at 72.6 eV, corresponding to the presence of metal aluminum [57]. Calculating the peak area, we determined that 8.6% of the surface has been completely exposed by the corrosion test, which also accounted for the worse results in (Section 3.4.4. Investigations of Rp, Figure 8) of this sample after the 48th h. The following peak at 73.9 eV corresponds to Al_2O_3 [58]. For sample S3, the surface is characterized almost entirely by aluminum oxide. The aluminum peaks for samples S4 and S5 are positioned at 74.1 (sample S5)–74.3 eV (sample S4). A binding energy of 74.1 eV is typical for Al_2O_3, and its shift to higher values of 74.3 eV in sample S4 is a sign of the formation of Al-PO_4 bonds [59]. In the deconvolution of the Al2p peak of samples S4 and S5, another group of peaks is formed at 75.0 eV (sample S5), characteristic of $Al(OH)_3$, and at 75.9 eV [60], characteristic of amorphous non-stoichiometric $AlO(OH) + H_2O$.

By comparing the P2p spectra for these samples, they can be separated into two peaks. For the S3 sample, they are located at binding energies of 132.7 and 134.6 eV. Both peaks are with a low-intensity, which corresponds to the low phosphorus concentration of 0.7 at.% on the surface after the corrosion test. The peak at 132.7 eV can be associated with the presence of phosphorus within the Na_2HPO_4–compound [61], and the peak at 134.6 eV corresponds to Na–P–O–bonds in $NaPO_3$ [62], which also corresponds to the low concentrations of 0.4 at% of Na found in the analysis. The presence of a certain amount of $AlPO_4$ is also possible. Therefore, oxidation-reduction processes of conversion of the phosphorus-containing component of the precursor into its more soluble compounds took place on the surface of this sample.

The concentration of phosphorus on the surface of sample S4 is the highest compared to the other samples subjected to Rp tests–4.4 at. % (Section 3.4.4. Investigations of Rp, Figure 8). The photoelectron peak is split again into two peaks, at 133.4 and 134.7 eV. The first peak positioned at 133.4 eV can be associated with the presence of phosphate groups. The second peak can be associated with the presence of PO_3 groups, probably $AlPO_3$. The phosphorus concentration for sample S5 is lower–3.5 at.%. The spectrum of P2p is divided into three peaks–132.7, 134.2 and 136.0 eV. The first of them corresponds to the presence of hydrogen phosphate groups, the second to -PO_3 bonds [63] and the third is probably due to poorly adsorbed phosphate groups dispersed on the Al surface.

Here again, the analysis of the spectra of the Al2p, P2p and Ca2p Rp samples are confirmed by the peak fitting of the O1s spectra for the same samples (spectra not given in the publication). Here, the O1s spectra can be divided into three groups of peaks. The first group, at 530.1 eV, corresponds to the presence of metal oxides [46]. The second, at 530.7 eV, we can associate to the presence of a mixture of phosphates and hydrogen phosphates [51]. The third group, at 532.8 eV, we can associate with the presence of OH^- and OOH^- groups [49,50].

Analyzing the spectra and concentrations of the chemical elements on the surface, we can conclude that in the Rp tests, the best resistance will be shown by the layers with the presence of phosphate groups present on the surface. Sample S4, which has the highest concentration of phosphate groups (4.4 at. %), however, does not exhibit the best corrosion resistance, which also agrees well with the chronoamperometric studies. For sample S4, the presence of the more soluble $AlPO_3$ and the absence of a cerium conversion layer are combined. In this regard, it can be assumed that the most important role for the protective ability is played by Al_2O_3, $Al(OH)_3$, and amorphous $AlO(OH) + H_2O$, respectively, in samples S3 and S5. It should be noted that after the corrosion tests of the S2 sample, 2.4 at. % Ce_2O_3 remained on its surface, which from the Ce3d spectrum analysis shows to be in the form of Ce^{3+}, the more soluble form of cerium phosphate [64]; it is, therefore, washed away after 48 h.

XPS of Studied Samples after OCP Tests

XPS-surface studies of samples subjected to corrosion tests at OCP were performed. The spectra of Al2p spectra can be divided into three groups of peaks—at 72.9 eV, 73.7–74.6 eV and 76.1 eV (they are not included in this paper). They correspond to the presence of metallic aluminum [65], a mixture of Al_2O_3 and $Al(PO)_4$ and AlOOH [49,50]. In the Al2p-spectrum of the S2 sample, no peak associated with the presence of pure metallic aluminum is observed, indicating that the surface of this sample was not exposed as a result of the corrosion test. The largest amount of Ce was also found on the surface of this sample—5.7 at.%. In this case, it is in the form of Ce^{3+}. In the Al2p spectra of samples S4 and S5, the peaks associated with the presence of metallic aluminum are with a low intensity. They are with the areas corresponding to 13–13.7% of the total amount of Al. This fact indicates that in these samples, this surface baring as a result of the corrosion tests was minimal. In the S2 sample, the first of the second group of peaks is positioned at 73.7 eV. The lower binding energy compared to that of Al_2O_3 (74.1 eV) and the high concentration of Ce on the surface of this sample may be an indication of the formation of a mixture of Al_2O_3 and $CeAlO_3$ [66]. In samples S4 and S5, the binding energy of the peaks is 74.4–74.6 eV, typical of aluminum phosphate. The ratio between Al and phosphorus is 1.8:1. The stoichiometric ratio of Al to P in $AlPO_4$ is 1:1. This excludes the possibility the entire amount of Al being covered with phosphates. For this reason, the peaks at 74.4–74.6 eV in the Al2p spectrum can be attributed to a mixture of Al_2O_3 and $AlPO_4$, although the binding energy is typical only for $AlPO_4$.

The P2p spectra for the studied samples can be divided into two peaks positioned at 133.3 and 137.0 eV. The first peak, already mentioned above in the text, corresponds to the presence of P–O–bonds within aluminum phosphates. The second peak we can again associate with the presence of phosphate agglomerates located on the surface and accordingly, with a different charge.

The Ca established on the surface of samples S4 and S5 Ca is in minimal concentrations—0.2–0.6 at.%. Despite the noisy spectra, their deconvolution was possible and for sample S4, the doublet was split into a single peak positioned at 347.3 eV. This peak corresponds to $Ca_{10}PO_{46}(OH)_2$ [54,55]. For sample S5, its spectrum can be divided into three peaks, the maxima of which are at 346.3, 347.3 and 348.5 eV. Just as with the as-deposited samples, they correspond to the presence of $Ca(OH)_2$, a product of chemical reactions occurring on the surface (reaction (2)), insoluble $Ca_{10}PO_{46}(OH)_2$ [54,55], also a product of Reaction (3) and $Ca(NO_3)_2$, which was part of the precursors. The percentage ratio between the three peaks is 31.0:50.3:19.6%. From this we can conclude that half of the calcium is in the form of insoluble $Ca_{10}PO_{46}(OH)_2$.

XPS of Studied Samples after Chronoamperometric Tests

The XPS spectra of the samples characterized by stable corrosion behavior after 168 h of exposure in CM (see Section 3.4.3. Chronoamperometric Investigations) were taken—samples S2 and S3 (the spectra are not given in this work). Only aluminum and oxygen were found on their surface. The Al2p spectra of the two samples can be divided into three main groups of peaks—72.7 (sample S2), 73.2 eV (sample S3), 74.1 eV and 75.3 eV. The peak positioned at a binding energy of 72.7 eV for the S3 sample is typical for the presence of metallic aluminum [57]. Its area represents only 6% of the total amount of aluminum, i.e., only 6% of the surface is bared in the chronoamperometric test. For the S3 sample, the peak with the lowest binding energy is positioned at 73.0 eV. This energy is 0.3 eV higher than typical for metallic aluminum, and therefore, we can associate it with the presence of slightly oxidized aluminum. In this sample, the peak is more intense and represents 11.4% of the total area, or of the concentration of aluminum. The peak positioned at 74.1 eV, present in both samples, is typical for the presence of Al_2O_3 [46]. For the S2 sample, it is 78% of the total area, and for the S3 sample, it is 73 % of the total area of aluminum. The peak positioned at 75.3 eV, again present in both samples, is characteristic of the presence of AlOOH [42].

3.4. *Electrochemical Investigations*

3.4.1. Polarization Curves

The anodic and cathode potentiodynamic polarization curves of the studied systems are presented in Figure 5. As was marked above (see in paragraph 2), these measurements were started in cathode and anodic directions from the E_{OCP} of S1. They characterize the electrochemical behavior of the S1 (bar) substrate as well as consecutive obtained conversion coatings formed on it in model corrosion medium 0.1 M NaCl. On Figure 5a and Table 4, respectively, it is seen that the polarization of the studied samples in the cathodic direction causes the formation of a wide range of potentials (~1000 mV), in which the value of the current of passivity does not depend on the change in the value of the potential. This effect illustrates and characterizes this interval of potentials as passive layer area (E_{PLA}), after which started the processes of breakdown of the passive layers. At the same time, depending on the type and duration of the immersion post-treatment of the samples in PhCa, a shift of E_{corr} of the studied samples in a positive direction is observed. This shift is most pronounced (by ~200 mV, compared to E_{corr} of S1 substrate) for the system S2. It is less—by ~ 160 mV—for the systems S3 and S4, while for the system S5, E_{corr} is ~25 mV more positive than E_{corr} of S1. At the same time, the decrease of i_{corr} for the studied systems follows the relationship: i_{corr} of S1 > i_{corr} of S2 > i_{corr} of S3 > i_{corr} of S4 > i_{corr} of S5–Table 4.

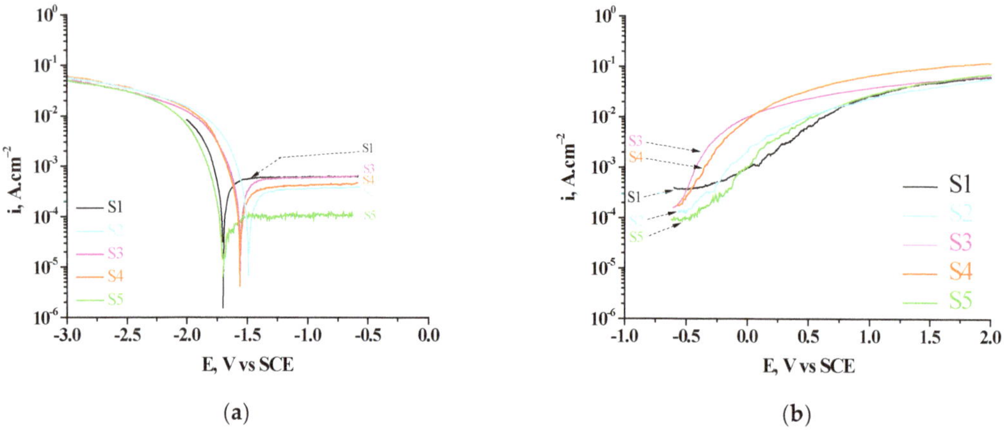

Figure 5. Potentiodynamic polarization curves of the studied systems: S1; S2; S3; S4 and S5 in cathode (**a**) and anodic (**b**) directions.

Table 4. Electrochemical parameters-corrosion potential (E_{cor}), corrosion current (i_{cor}), current of passivity (i_{pass}) and range of passive state ($E_{r.p.s.}$) of the studied systems, determined on the base of polarization investigations.

Sample	E_{cor}, V	i_{cor}, A.cm^{-2}	i_{pass}, A.cm^{-2}	$E_{r.p.s.}$, V
S1	−1.705	9.8×10^{-5}	5.8×10^{-4}	−1.425 ÷ −0.587
S2	−1.472	8.1×10^{-5}	3.4×10^{-4}	−1.263 ÷ −0.480
S3	−1.550	5.7×10^{-5}	5.5×10^{-4}	−1.360 ÷ −0.585
S4	−1.553	4.5×10^{-5}	3.6×10^{-4}	−1.388 ÷ −0.593
S5	−1.682	1.7×10^{-5}	1.0×10^{-4}	−1.537 ÷ −0.470

The comparison of the obtained polarization curves establishes a significant change in the width of the passive state zone ($E_{r.p.s.}$, V), as well as in the current (at the plateau) of the passive state (i_{pass}, A.cm^{-2}) of the studied systems. From the data systematized in Table 4, it can be seen that: (1) the most positive value of E_{corr} for the studied samples is observed for the system S2 (−1.472 V); (2) the lowest value of the corrosion current i_{corr} (1.7×10^{-5} A.cm^{-2}) was registered for system S5; (3) the best corrosion indicators are established for system S5 both in terms of the lowest value of the current of passive state (i_{pass}) (1.0×10^{-4} A.cm^{-2}) and the widest range of passive state (E_{rps}) (−1.537 ÷ −0.470 V)—Table 4.

The registered specific course of the polarization curves (Figure 5a), in the region from potentials −0.680 V to −1.705 V, characterizes and describes the changes in the anodic behavior of the aluminum substrate due to its modification with effective cathodic coatings. The presence of a well-defined, wide passive region in the course of the obtained polarization curves is observed, falling in the region of cathodic potentials characteristic of the modifying cerium oxide layers. These results are related to the following phenomenon: according to the data from a previous study [32–35] and the present one (see Sections 3.1–3.3), both the S1 substrate and the conversion coatings formed on it are characterized by thin surface layers that differ in composition and structure. The potentiodynamic unfolding of the potential (1 mV.s^{-1}) in the interval from -0.660 V to -1.5 V for the different systems, although in a negative direction, will lead to partial or complete dissolution of ceria in components of the deposited conversion layers. This is due to the fact that the E_{corr} of the S1 substrate is more electronegative than the E_{corr} of the studied protective, conversion systems. That is, the area more positive than -1.705 V is an anodic region for the deposited (on S1) protective layers, in which their dissolution is possible after reaching: -1.682 V for system S5; -1.553 V for system S4; -1.550 V for the S3 system; -1.472 V for the S2 system.

The results presented above correspond to the anodic polarization curves obtained for the same samples-Figure 5b. On Figure 5b and Table 4, respectively, it can be seen that the polarization of the studied samples (beginning from E_{OCP}) in the anodic direction maintains the passive state of the S1 substrate up to ~-0.350 V (curve 1); of the S2 system (curve 2)—up to ~-0.480 V; of the system S3 (curve 3)—up to ~-0.585 V; of the system S4 (curve 4)—to ~-0.593 V, and for the system S5 (curve 5)—up to ~-0.470 V. After reaching these values, the course of the anodic polarization curves (in the interval ~-0.500 V–0 V) is characterized as a transition to overpassivation of the samples (transpassive region), and after its further increase (in the interval ~ 0 V–+0.250 V), this stage of the polarization curves more probably is connected with the passive layer breakdown [67].

Briefly, we are led to conclude that this part of cathode curves (started in negative direction from the E_{OCP}-after reach of E_{corr}, respectively) for all of the studied samples correspond to the process of reduction of oxygen. At the same time (as it is shown above), the part of cathode curves obtained in the area of potentials $E_{r.p.s.}$-E_{OCP} and the anode curves (obtained in positive direction, after E_{OCP} of the studied samples) correspond to the process of passivation. They contain all typical steps (local corrosion due to passive layer breakdown; region of passive state; transpassive zone; area of oxygen evolution) which are representative for the all investigated samples. These results, obtained under conditions of external potentiodynamic polarization inform about the kinetics of the related corrosion process reactions, but do not characterize them fully since the real corrosion processes occur at the open circuit potential (OCP). At the same time, it is known that the change in OCP (E_{OCP}) as a function of immersion time can be used to monitor the chemical stability and corrosion process of the Al alloys [68]. Although E_{OCP} does not provide any direct information on the corrosion kinetics, it suggests the corrosion susceptibility [69].

This is the reason why when organizing our experimental studies, along with these considerations, we considered the possibility of the formation of a set of phosphate complexes (obtained at different PhCa-post-treatments of the CCOC layers) with different solubility [36–39] during corrosion attack of Cl^-. In connection with this, our research was conducted both immediately after receiving (as-deposited) the studied systems and after 168 h of their exposure in the CM.

To gain the needed additional insights, we conducted investigations of self-occurring corrosion processes in the systems of interest at: the change in OCP values monitored E_{OCP} vs. time plot; the changes in anodic current (i_a) transients at pitting potential (E_{pit}); the changes of polarization resistance (Rp) as a function of samples exposure time in CM (in OCP conditions).

3.4.2. E_{ocp} vs. Time Plotcharacterization

As it is known, the change in OCP as a function of the immersion time can be used to monitor the chemical stability and corrosion process of the studied samples, i.e., to assess their corrosion susceptibility. In this relation and in light of the marked above, we realized systematically investigations of the studied systems illustrating the change of their OCP vs.time in CM. Figure 6 shows the results obtained for the studied as-deposited (Figure 6a) samples and samples identical of them after 168 h of their exposure in the CM (Figure 6b). It was established on Figure 6a that during the immersion for 3600 s, the course and the changes in E_{OCP} for the bare (S1) sample and system S5 practically coincide at

~−0.660 V. For the systems S3 and S4, both coincide at ~−0.680 V. While for the system S2, this change of E_{OCP} goes in a positive direction till ~−0.630 V. The big fluctuations observed for practically all of the studied systems point to dissimilarities in the potential of different intermetallic inclusions, metastable pit formation and re-passivation in the CM. The fluctuation in the OCP could be due as well to the partial incorporation of the aqua ion and formation of Al oxides/hydroxides. When the samples are exposed to the aggressive medium, the corrosive ions penetrate through the active regions and lead to a more negative OCP. In total, the average positive change of E_{OCP} is at its maximum under the action of ceria conversion treatment (S2) is of ~0.030 V. This positive shift (in the range of E_{OCP} from ~−0.660 to ~−0.630 V) compared to the S1 substrate indicates the effective incorporation and also the stability of the ceria-modified Al substrate. Obviously, at the as-deposited systems S3, S4 and S5, the maximum protective effect is not reached, the maximum positive change of E_{OCP}, respectively, characteristic for the system S2.

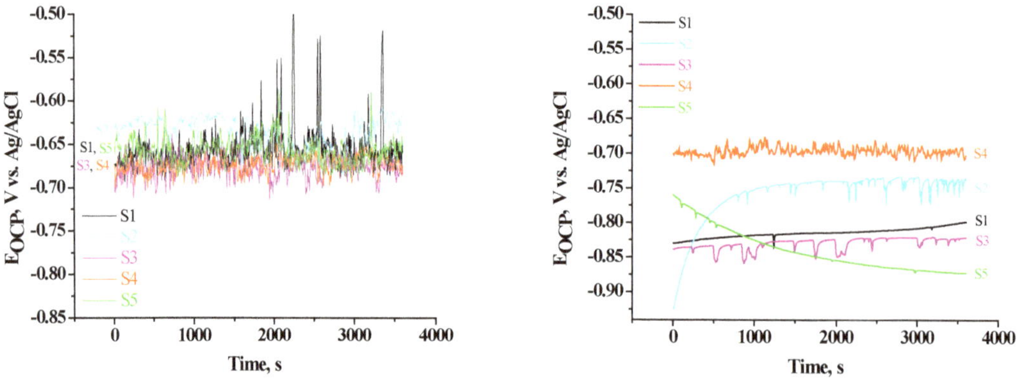

(a) (b)

Figure 6. OCP vs. time plots during immersion in a 0.1 M NaCl solution for as-deposited samples (**a**) and after 168 h exposure of the same samples in CM (**b**): Bare S1 substrate; S2; S3; S4; S5.

The analogous measurements realized after exposure of the as-deposited samples in the CM for 168 h are shown in Figure 6b. It is seen from the course of changes of the E_{OCP} that the tendency of the change of E_{OCP} in this case is quite different. (1) All E_{OCP} transients are shifted from the area −0.680–0.630 V to the area −0.840–0.700 V. At the same time, relative to the E_{OCP} of the Al substrate (E_{OCP} of S1 ~ −0.815 V), with the most positive value for E_{OCP} characterizing system S4 (E_{OCP} ~ −0.700 V), followed by system S2 (E_{OCP} ~ 0.760 V), while systems S3 and S5 are characterized by more negative values for E_{OCP} (−0.830 and −0.840 V, respectively). These results are in good agreement with the data from EDS and XPS studies for the same samples, presented in Tables 2 and 3. From the results, therein, it can be seen that there is a direct relationship between the disposition of E_{OCP} transients (Figure 6) and correspondingly determined values for volume (Table 2) and surface (Table 3) concentrations of the main characteristic elements (Ce, P, Ca) of the formed conversion layers. From this, a conclusion can be drawn that the corrosion-protective ability (of these essentially cathodic coatings) in the as-deposited samples is dominated by the CCOC layers. While after prolonged exposure (168 h) in CM, the corrosion protection of S1 is mainly based on the layers formed after the PhCa-sealing post-treatment of system S2. According to the results of the XPS analyses, these are mainly the additionally formed agglomerates and/or layers of: $Ca_5(PO_4)_3(OH)$, $CePO_4$ and $AlPO_4$ and AlOOH and $Al(OH)_3$, as given by reactions 1–5. Complementing these results and conclusions, the above-mentioned (Section 3.4.1. Polarization Curves) conclusions can be added that the manifestation of zones of passive state (Figure 5), before reaching the values of the registered E_{corr} of the delayed conversion coatings, is completely natural. They indicate the change in the character of the corrosion process, as a result of the movement of the corrosion potentials from the zone of active dissolution to the zone of passivity of the aluminum substrate.

3.4.3. Chronoamperometric Investigations

In these investigations, by polarizing the samples anodically at E_{pit}, we aimed to approach to a maximal extent the actual corrosion process, respectively, to characterize corrosion in view of pitting corrosion, which is a basic characteristic of aluminum and its alloys in Cl^--containing CM [26,70]. Based on the course of the registered curves, we could judge the character of the corrosion attack and the appearance of pitting damages. Figure 7 presents the results for the studied samples.

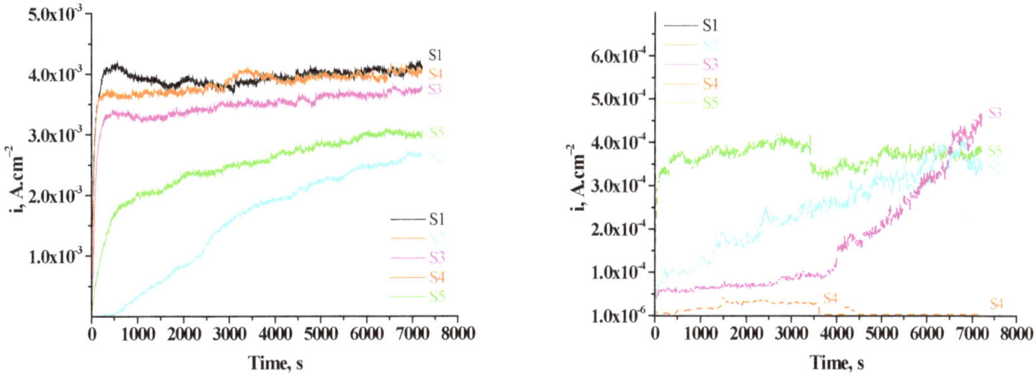

(**a**) (**b**)

Figure 7. Chronoamperometric i_a-Time transients: (**a**) for as-deposited samples S1; S2; S3; S4, S5, recorded in 0.1 M NaCl at the pitting potential of Al 1050 (E_{pit} = –0.550 V vs. Ag/AgCl), and (**b**) after exposure of the same samples for 168 h in 0.1 M NaCl and following analogous chronoamperometric investigation during 7200 s.

As it is observed in Figure 7a, for the S1 (unprotected Al substrate), after its immersion in CM at E_{pit} (–0.550 V vs. Ag/AgCl [70]), the i_a density increases sharply (until the ~350-th s of exposure) up to values ~4.22 × 10^{-3} A.cm^{-2}, whereupon the surface film on the Al is disrupted (Figure 7a), transient S1), which is a prerequisite for the appearance and development of pitting corrosion during the interaction with the CM. After breaking through the passive film, there starts a process of local corrosion characterized by values of the anodic current (i_a) ~3.84 × 10^{-3} A.cm^{-2} (at 3600 s) and the appearance of current oscillations specific for it owing to the unstable pits that are repassivated/activated. Similar behavior is also observed for the current transients of the S4 and S3 systems (Figure 7a) for which a process of local corrosion is characterized by very near values of the i_a –4.05 × 10^{-3} A.cm^{-2} and 3.51 × 10^{-3} A.cm^{-2}, respectively.

The course of the current transients is essentially different for the samples protected by $CCOC_{(Ce+Cu)}$ and sealed in PhCa $_{(3 \times 5\ min + 72\ h\ exp.)}$ solutions (Figure 7a)—transients 2 and 5, respectively. Current fluctuations characteristic for localized breakdown of the passive film decrease, which would lead to the decrease of initiation and growth of corrosion pits in the CM. The transients have a process course characteristic of general corrosion, the i_a of which are 2.54 × 10^{-3} and 1.84 × 10^{-3} A.cm^{-2}.(at 3600 s) for system S5 and S2, respectively.

The comparison of the current transients in the case of a consecutive protected by $CCOC_{(Ce+Cu)}$ and sealed at different PhCa regimes of the S1 shows (Figure 7a) that the order of stabilities of the systems towards the appearance and development of pitting corrosion is the following: (S1) < S4< S3< S5 < S2.

At the same time, the transients shown in Figure 7b, which characterize analogous to Figure 7a samples, after their 168 hexposure in CM, reveal quite different results. The current transients are characterized by a very fast establishment (till 10-th-100-rds) of constant values of i_a, which are an order of magnitude lower (2.59×10^{-4}–2.72×10^{-5} A.cm^{-2}) than those obtained for the as-deposited samples (Figure 7b). After the 3600th s of exposure in the corrosion medium, they are characterized by values for i_a, respectively, for: transient 2–2.59×10^{-4} A.cm^{-2}; transient 3–9.40×10^{-4} A.cm^{-2}; transient 4–2.79×10^{-5} A.cm^{-2}; transient 5–3.19×10^{-4} A.cm^{-2}. In the interval of times (3600–7200 s) of exposure in CM, the values for i_a reached at 3600 s start to rise, reaching ~3–4.5×10^{-3} A.cm^{-2} for systems S2, S5 and S3. At system 4, however, a unique decrease in i_a is registered, maintaining a constant value (in the interval 3600–7200 s) of 1×10^{-6} A.cm^{-2}. The explanation of this effect will be the subject of our next study.

3.4.4. Investigations of Rp vs.Time of Exposure in CM

Figure 8 shows the results reflecting the changes in Rp of the studied systems as a function of the exposure time (0.25–168 h) in the CM. The analysis of the results obtained showed that after the first hour of exposure of S1 in CM, its polarization resistance (Rp) is characterized by a value of ~350 kΩ.cm^2. In the subsequent interval of exposure times, 24–168 h, it decreased more than threefold, varying in the interval 75–115 kΩ.cm^2.

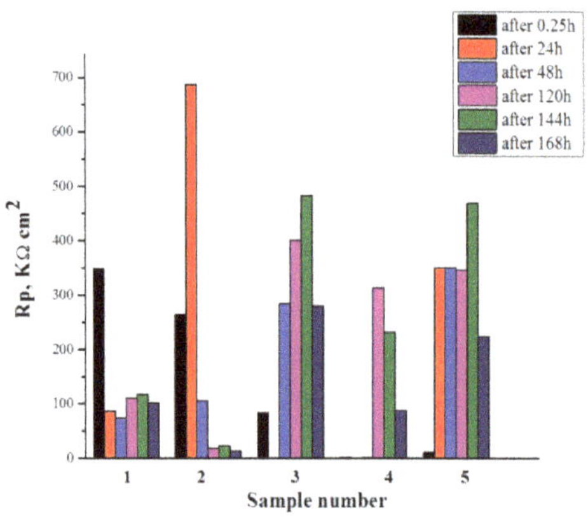

Figure 8. Changes in Rp of the studied systems as a function of the exposure time 0.25 (the initial delay of 15 min before starting of measurements of Rp)—168 h in the CM (1—S1; 2—S2; 3—S3; 4—S4, 5—S5).

The formation of a $CCOC_{(Ce+Cu)}$ layer on S1 (the S2 system) is characterized by values for Rp~ 700 kΩ.cm^2, after its exposure in CM for 24 h. At the subsequent exposure times (48–168 h) in CM, however, this value drastically decreases to ~20 kΩ.cm^2. This effect can be related to the strong decrease in the concentration of Ce_2O_3 (15.8% vs. 2.4%—Table 3) due to the dissolution of CCOC, which will lead to the exposure and activation of certain (containing Fe and Cu) areas of the Al substrate [34].

The sealing post-treatment of system S2 in NaH_2PO_4 and $Ca(NO_3)_2$ solution, leading to the formation of system S3, causes a significant increase in Rp—up to ~480 kΩ.cm^2 after the 144th h of exposure in CM. After the 168th h of CM exposure, however, this value drops to ~280 kΩ.cm^2. This change is most likely due to the strong decrease in volume (Table 2) and surface (Table 3) concentrations of Ce and P after the 168th h of exposure in the CM.

In system S4, characterized by three-fold sealing post-treatment, the values of Rp change are significantly lower (up to ~5 kΩ.cm^2 after 48th h of exposure), then after the 120th h, they rise to ~310 kΩ.cm^2, and after the 168th, they decrease again to ~85 kΩ.cm^2.

In this case, the recorded strong decrease in Rp after the 168th h of exposure in the CM is probably related to the strong decrease in bulk (0 wt.% Ce and 0.89 wt.% P-Table 2) and surface (0 at.% Ce, 4.4 at.% P and 3 at.% N-Table 3) concentrations of Ce, P and N after the 168th h of exposure in CM. At the same time, as indicated in the XPS analysis paragraph, in this sample, apart from the absence of cerium oxide and/or phosphate, phosphorus is in the form of the more soluble $AlPO_3$.

For system S5, which included an additional 72 h treatment at 50 °C in an atmosphere with 95% relative humidity, however, the values for Rp increase again, reaching and exceeding those for system S3. Moreover, they are higher (up to ~350 kΩ.cm^2) and unchanged (up to ~144 h of exposure in CM) compared to system S4. The high values registered for Rp after the 168th h of exposure in CM are in good agreement with the results of XPS analyses of the changes in the surface concentration of the elements characteristic of the deposited conversion layers (Table 3). They describe the presence of cerium-phosphate layers, the concentration of the elements in which it is 28.2 at.% Al and 3.5 at.% P (the rest up to 100 at.% is oxygen). In these cerium-phosphate layers, aluminum is in the form of insoluble Al_2O_3, $Al(OH)_3$, and amorphous $AlO(OH) + H_2O$, which determine its relatively high corrosion resistance.

4. Conclusions

Ceria-based conversion coatings, formed on technically pure Al-1050, were post-treated by sealing in mixed 0.5 M NaH_2PO_4 + 0.1 M $Ca(NO_3)_2$ solution. We paid extra attention to the influence of regimes of this post-treatment operation. Of as-prepared samples, we have characterized the chemical composition in the volume and on the surface by XRD and XPS. The results allow to ascertain the following conclusions:

1. There is substantial influence of the time and type of sealing post-treatment on the chemical composition and chemical state of the elements in the obtained and investigated systems. It is established there is a strong decrease of the concentration of Al_2O_3 and Ce_2O_3 components in the as-deposited CCOCs at the expense of the formation of insoluble phases: $Ca_5(PO_4)_3(OH)$; $AlPO_4$; $CePO_4$ and AlOOH (transformed in maximal stage of increase in $Al(OH)_3$), as well as PO_3^-, compounds with Al and Ce, after their sealing post-treatment in mixed sodium phosphate and calcium nitrate solution;
2. Based on the analyzed specific course of the potentiodynamic polarization dependences, changes in the anodic behavior of the aluminum substrate have been established, which are due to its modification with effective cathodic coatings;
3. The comparison of these results with the changes of the concentrations of Ce^{3+}, Al and P (also their respective oxides and phosphates) before and after exposure of the samples in CM show that the concentrations and chemical state of the Al, Ca, Ce and P (and their form as protective compounds) on the surface of the studied samples are directly related. In this case, it was found that the combination of oxide-phosphate compounds forming the conversion layers of system S5 has the best protective effect;
4. Polarization investigations simultaneously showed that the combination of studied phosphate and ceria conversion layers are not only cathodic barrier coatings, but they also change the kinetics of the conjugated electrochemical reactions characterizing the corrosion process in Cl^--containing media, i.e., they determine the electrochemical protection of Al substrates.

The established protective effect of the mixed conversion coatings on Al at long exposure in CM can be related to the beneficial transformation of the chemical composition of CCOC, formed on Al substrates after sealing post-treatment processing. This effect, as well as the formation of different types of corrosion products (at a long time of exposure in CM) on the surface of Al/CCOC/PhCa systems provide an effective barrier to the diffusion of Cl^- toward to the Al surface, which leads to the corresponding positive and beneficial changes of the i_a and Rp for the studied systems.

Author Contributions: A.T.: Experimental design, writing guide and design picture, R.A.: Experimental design, data processing, writing guide and design picture. D.S.: Conceptualization, Experimental design, Data curation, Writing original draft, Writing–review and editing. All authors have read and agreed to the published version of the manuscript.

Funding: The authors want to thank Project Д01-92/06.2022-"European Network on Materials for Clean Technologies", funded by the Ministry of Education and Science under the National Program "European Scientific Networks".

Institutional Review Board Statement: Not applicable.

Informed Consent Statement: Not applicable.

Data Availability Statement: Not applicable.

Conflicts of Interest: The authors declare no conflict of interest.

References

1. European Commission. *Commission Directive (EU) 2017/2096 of 15 November 2017 Amending Annex II to Directive 2000/53/EC of the European Parliament and of the Council on End-of Life Vehicles*; European Commission: Brussels, Belgium, 2017.
2. European Commission. *Commission Implementing Decision (EU) 2019/2193 of 17 December 2019 Laying down Rules for the Calculation, Verification and Reporting of Data and Establishing Data Formats for the Purposes of Directive 2012/19/EU of the European Parliament and of the Council on Waste Electrical and Electronic Equipment (WEEE)*; European Commission: Brussels, Belgium, 2019.
3. Wernick, S.; Pinner, R. *Surface Treatment and Finishing of Aluminium and Its Alloys*; Robert Draper Ltd.: Sevenoaks, UK, 1956.
4. Wernick, S.; Pinner, R.; Sheasby, P. *The Surface Treatments of Aluminium and Its Alloys*, 5th ed.; ASM International and Finishing Publications Ltd.: Metals Park, OH, USA, 1987.
5. Harvey, T.G. Cerium-Based Conversion Coatings on Aluminium Alloys: A Process Review. *Corros. Eng. Sci. Technol.* **2013**, *48*, 248–269. [CrossRef]
6. Twite, R.L.; Bierwagen, G.P. Review of Alternatives to Chromate for Corrosion Protection of Aluminum Aerospace Alloys. *Prog. Org. Coatings* **1998**, *33*, 91–100. [CrossRef]
7. Hinton, B. 2009 Frank Newman Speller Award Lecture: Prevention and Control of Corrosion in Aircraft Components—Changes over Four Decades. *Corrosion* **2010**, *66*, 0850011–08500115. [CrossRef]
8. Buchheit, R.G.; Hughes, A.E. Chromate and Chromate-Free Conversion Coatings. In *ASM Handbook Vol. 13A: Corrosion: Fundamentals, Testing and Protection*; ASM International: Detroit, MI, USA, 2003; pp. 720–735.
9. Hinton, B.R.W.; Arnott, D.R.; Ryan, N.E. Cerium Conversion Coatings for the Corrosion Protection of Aluminium. *Mater. Forum* **1986**, *9*, 162–173.
10. Hughes, A.E.; Taylor, R.J.; Hinton, B.R.W.; Wilson, L. XPS and SEM Characterization of Hydrated Cerium Oxide Conversion Coatings. *Surf. Interface Anal.* **1995**, *23*, 540–550. [CrossRef]
11. Paussa, L.; Andreatta, F.; De Felicis, D.; Bemporad, E.; Fedrizzi, L. Investigation of AA2024-T3 Surfaces Modified by Cerium Compounds: A Localized Approach. *Corros. Sci.* **2014**, *78*, 215–222. [CrossRef]
12. Eslami, M.; Fedel, M.; Speranza, G.; Deflorian, F.; Andersson, N.E.; Zanella, C. Study of Selective Deposition Mechanism of Cerium-Based Conversion Coating on Rheo-HPDC Aluminium-Silicon Alloys. *Electrochim. Acta* **2017**, *255*, 449–462. [CrossRef]
13. Eslami, M.; Fedel, M.; Speranza, G.; Deflorian, F.; Zanella, C. Deposition and Characterization of Cerium-Based Conversion Coating on HPDC Low Si Content Aluminum Alloy. *J. Electrochem. Soc.* **2017**, *164*, C581. [CrossRef]
14. Hassannejad, H.; Moghaddasi, M.; Saebnoori, E.; Baboukani, A.R. Microstructure, Deposition Mechanism and Corrosion Behavior of Nanostructured Cerium Oxide Conversion Coating Modified with Chitosan on AA2024 Aluminum Alloy. *J. Alloys Compd.* **2017**, *725*, 968–975. [CrossRef]
15. Johnson, B.Y.; Edington, J.; Williams, A.; O'Keefe, M.J. Microstructural Characteristics of Cerium Oxide Conversion Coatings Obtained by Various Aqueous Deposition Methods. *Mater. Charact.* **2005**, *54*, 41–48. [CrossRef]
16. Aldykewicz, A.J.; Isaacs, H.S.; Davenport, A.J. The Investigation of Cerium as a Cathodic Inhibitor for Aluminum-Copper Alloys. *J. Electrochem. Soc.* **1995**, *142*, 3342–3350. [CrossRef]
17. Dabalà, M.; Armelao, L.; Buchberger, A.; Calliari, I. Cerium-Based Conversion Layers on Aluminum Alloys. *Appl. Surf. Sci.* **2001**, *172*, 312–322. [CrossRef]
18. Arenas, M.A.; Bethencourt, M.; Botana, F.J.; De Damborenea, J.; Marcos, M. Inhibition of 5083 Aluminium Alloy and Galvanised Steel by Lanthanide Salts. *Corros. Sci.* **2001**, *43*, 157–170. [CrossRef]
19. Aballe, A.; Bethencourt, M.; Botana, F.J.; Cano, M.J.; Marcos, M. On the Mixed Nature of Cerium Conversion Coatings. *Mater. Corros.* **2002**, *53*, 176–184. [CrossRef]
20. Hughes, A.E.; Gorman, J.D.; Miller, P.R.; Sexton, B.A.; Paterson, P.J.K.; Taylor, R.J. Development of Cerium-Based Conversion Coatings on 2024-T3 Al Alloy after Rare-Earth Desmutting. *Surf. Interface Anal.* **2004**, *36*, 290–303. [CrossRef]
21. Campestrini, P.; Terryn, H.; Hovestad, A.; Wit, J.H.W. de Formation of a Cerium-Based Conversion Coating on AA2024: Relationship with the Microstructure. *Surf. Coat. Technol.* **2004**, *176*, 365–381. [CrossRef]

22. Davó, B.; De Damborenea, J.J. Use of Rare Earth Salts as Electrochemical Corrosion Inhibitors for an Al-Li-Cu (8090) Alloy in 3.56% NaCl. *Electrochim. Acta* **2004**, *49*, 4957–4965. [CrossRef]
23. Danaee, I.; Zamanizadeh, H.R.; Fallahi, M.; Lotfi, B. The Effect of Surface Pre-Treatments on Corrosion Behavior of Cerium-Based Conversion Coatings on Al 7075-T6. *Mater. Corros.* **2014**, *65*, 815–819. [CrossRef]
24. Johnson, B.Y.; Edington, J.; O'Keefe, M.J. Effect of Coating Parameters on the Microstructure of Cerium Oxide Conversion Coatings. *Mater. Sci. Eng. A* **2003**, *361*, 225–231. [CrossRef]
25. You, S.; Jones, P.; Padwal, A.; Yu, P.; O'Keefe, M.; Fahrenholtz, W.; O'Keefe, T. Response of Nanocrystalline Cerium-Based Conversion Coatings on Al 2024-T3 to Chloride Environments. *Mater. Lett.* **2007**, *61*, 3778–3782. [CrossRef]
26. Zhang, H.; Zuo, Y. The Improvement of Corrosion Resistance of Ce Conversion Films on Aluminum Alloy by Phosphate Post-Treatment. *Appl. Surf. Sci.* **2008**, *254*, 4930–4935. [CrossRef]
27. Kiyota, S.; Valdez, B.; Stoytcheva, M.; Zlatev, R.; Schorr, M. Electrochemical Study of Corrosion Behavior of Rare Earth Based Chemical Conversion Coating on Aerospace Aluminum Alloy. *ECS Trans.* **2009**, *19*, 115–123. [CrossRef]
28. Andreeva, R.; Stoyanova, E.; Tsanev, A.; Stoychev, D. Influence of the Processes of Additional Phosphate Post-Treatment of Ceria Conversion Coatings Deposited on Al 1050 on Their Corrosion Protective Behavior. *J. Phys. Conf. Ser.* **2020**, *1492*, 012019. [CrossRef]
29. Girginov, C.; Avramova, I.; Kozhukharov, S. Addition of Transition Elements to the Phosphate Sealing of Cerium Oxide Protective Primer Deposited on AA2024-T3 Alloy. *J. Chem. Technol. Metall.* **2016**, *51*, 705–714.
30. Heller, D.K.; Fahrenholtz, W.G.; O'Keefe, M.J. Effect of Phosphate Source on Post-Treatment of Cerium-Based Conversion Coatings on Al 2024-T3. *J. Electrochem. Soc.* **2009**, *156*, C400–C406. [CrossRef]
31. Heller, D.K. Phosphate Post-Treatment of Cerium-Based Conversion Coatings on Al 2024-T3. Ph.D. Thesis, Missouri University of Science and Technology, Rolla, MO, USA, 2010.
32. Andreeva, R.; Stoyanova, E.; Tsanev, A.; Stoychev, D. Influence of the Surface Pre-Treatment of Aluminum on the Processes of Formation of Cerium Oxides Protective Films. *J. Phys. Conf. Ser.* **2016**, *700*, 012049. [CrossRef]
33. Andreeva, R.; Stoyanova, E.; Stoychev, D. Study of the Chemical Formation of Cerium Oxide Conversion Layers on Aluminium AD-3. *J. Int. Sci. Publ. Mater. Methods Technol.* **2014**, *8*, 751–759.
34. Stoyanova, E.; Stoychev, D. Electrochemical Aspects of the Immersion Treatment of Aluminium. *J. Appl. Electrochem.* **1997**, *27*, 685–690. [CrossRef]
35. Andreeva, R.; Stoyanova, E.; Tsanev, A.; Datcheva, M.; Stoychev, D. On the Role of Pre-Treatment of Aluminum Substrate on Deposition of Cerium Based Conversion Layers and Their Corrosion-Protective Ability. *Int. J. Electrochem. Sci.* **2018**, *13*, 5333–5351. [CrossRef]
36. Lee, H.S.; Kumar, A.; Mandal, S.; Singh, J.K.; Aslam, F.; Alyousef, R.; Albduljabbar, H. Effect of Sodium Phosphate and Calcium Nitrate Sealing Treatment on Microstructure and Corrosion Resistance of Wire Arc Sprayed Aluminum Coatings. *Coatings* **2020**, *10*, 33. [CrossRef]
37. Lee, H.S.; Singh, J.K.; Ismail, M.A. An Effective and Novel Pore Sealing Agent to Enhance the Corrosion Resistance Performance of Al Coating in Artificial Ocean Water. *Sci. Rep.* **2017**, *7*, 41935. [CrossRef]
38. Lee, H.S.; Singh, J.K. Influence of Calcium Nitrate on Morphology and Corrosion Characteristics of Ammonium Phosphate Treated Aluminum Coating Deposited by Arc Thermal Spraying Process. *Corros. Sci.* **2019**, *146*, 254–268. [CrossRef]
39. Jeong, H.R.; Lee, H.S.; Jalalzai, P.; Kwon, S.J.; Singh, J.K.; Hussain, R.R.; Alyousef, R.; Alabduljabbar, H.; Aslam, F. Sodium Phosphate Post-Treatment on Al Coating: Morphological and Corrosion Study. *J. Therm. Spray Technol.* **2019**, *28*, 1511–1531. [CrossRef]
40. Stern, M.; Geary, A.L. Electrochemical Polarization: I. A Theoretical Analysis of the Shape of Polarization Curves. *J. Electrochem. Soc.* **1957**, *104*, 56–63. [CrossRef]
41. Lukanova, R.; Stoyanova, E.; Damyanov, M.; Stoychev, D. Formation of Protective Films on Al in Electrolytes Containing No Cr6+ Ions. *Bulg. Chem. Commun.* **2008**, *40*, 340–347.
42. Andreeva, R.; Stoyanova, E.; Tsanev, A.; Stoychev, D. XPS Characterisation of the Influence of Phosphate Post-Treatment of Chemically Deposited Ceria Protective Layers on Aluminum. *Comptes Rendus L'Academie Bulg. Sci.* **2019**, *72*, 1336–1342. [CrossRef]
43. Tagai, H.; Aoki, H. *Preparation of Synthetic Hydroxyapatite and Sintering of Apatite Ceramics: Mechanical Properties of Biomaterials*; Hastings, G.W., Williams, D.F., Eds.; John Wiley & Sons Ltd.: Hoboken, NJ, USA, 1987; p. 213.
44. Bouyer, E.; Gitzhofer, F.; Boulos, M.I. Morphological Study of Hydroxyapatite Nanocrystal Suspension. *J. Mater. Sci. Mater. Med.* **2000**, *11*, 523–531. [CrossRef] [PubMed]
45. Akhtar, M.; Khajuria, A.; Sahu, J.K.; Swaminathan, J.; Kumar, R.; Bedi, R.; Albert, S.K. Phase Transformations and Numerical Modelling in Simulated HAZ of Nanostructured P91B Steel for High Temperature Applications. *Appl. Nanosci.* **2018**, *8*, 1669–1685. [CrossRef]
46. Dalouji, V.; Rahimi, N.; Elahi, S.H. The Influence of Post-Annealing Temperatures on XPS, XZ Height of Nanoparticles, and Optical Properties of Cu–Al Doped ZnO Films. *Mol. Cryst. Liq. Cryst.* **2023**, *758*, 11–24. [CrossRef]
47. Wang, Y.; Yu, W.; Zhao, L.; Wu, A.; Li, A.; Dong, X.; Huang, H. AlPO4-Li3PO4 Dual Shell for Enhancing Interfacial Stability of Co-Free Li-Rich Mn-Based Cathode. *Electrochim. Acta* **2023**, *462*, 142664. [CrossRef]
48. Sun, F.; Gao, W. Effect of Hydrochloric Acid on the Thermal Stability of Alpha-Aluminum Hydride. *Int. J. Hydrog. Energy* **2023**, *48*, 28998–29010. [CrossRef]

49. Lie, J.; Shuwanto, H.; Abdullah, H.; Ismadji, S.; Warmadewanthi, I.D.A.A.; Soetaredjo, F.E. Fly Ash Electrodes Fabricated by an Acid-Assisted Subcritical Water Extraction Method for Supercapacitor Applications. *New J. Chem.* **2023**, *47*, 3802–3809. [CrossRef]
50. Ivvala, J.; Arora, H.S.; Grewal, H.S. Towards Development of Sustainable Metallic Superhydrophobic Materials. *Colloids Surf. A Physicochem. Eng. Asp.* **2023**, *663*, 131047. [CrossRef]
51. Jiang, H.; Zhang, S.; Yan, L.; Xing, Y.; Zhang, Z.; Zheng, Q.; Shen, J.; Zhao, X.; Wang, L. Stress-Dispersed Superstructure of $Sn_3(PO_4)_2$@PC Derived from Programmable Assembly of Metal–Organic Framework as Long-Life Potassium/Sodium-Ion Batteries Anodes. *Adv. Sci.* **2023**, *10*, 2206587. [CrossRef] [PubMed]
52. Yang, X.; Lu, X.; Zhou, Y.; Xie, Y.; Yang, J.; Wang, F. Formation of Protective Conversion Coating on Mg Surface by Inorganic Inhibitor. *Corros. Sci.* **2023**, *215*, 111044. [CrossRef]
53. Yadav, P.; Rao, R.P.; Azeem, P.A. Optical and Structural Properties of Cost-Effective Nanostructured Calcium Titanate Blue Phosphor. *Ceram. Int.* **2023**, *49*, 6314–6323. [CrossRef]
54. Eighmy, T.T.; Kinner, A.E.; Shaw, E.L.; Eusden, J.D.; Francis, C.A. Hydroxylapatite ($Ca_5(PO_4)_3OH$) Characterization by XPS: An Environmentally Important Secondary Mineral. *Surf. Sci. Spectra* **1999**, *6*, 193–201. [CrossRef]
55. McDowell, H.; Gregory, T.M.; Brown, W.E. Solubility of $Ca_5(PO_4)_3OH$ in the System $Ca(OH)_2$-H_3PO_4-H_2O at 5, 15, 25, and 37 °C. *J. Res. Natl. Bur. Stand. Sect. A Phys. Chem.* **1977**, *81*, 273–281. [CrossRef]
56. Sharma, S.; Medpelli, D.; Chen, S.; Seo, D.K. Calcium-Modified Hierarchically Porous Aluminosilicate Geopolymer as a Highly Efficient Regenerable Catalyst for Biodiesel Production. *RSC Adv.* **2015**, *5*, 65454–65461. [CrossRef]
57. Xiao, X.; Guo, J.; Gao, Z.; Zhai, D.; Liu, R.; Qin, S.; Alam, M.K.; Sun, Z. Understanding the Complementary Resistive Switching in Egg Albumen-Based Single Sandwich Structure with Non-Inert Al Electrode. *Mater. Res. Express* **2023**, *10*, 056301. [CrossRef]
58. Takakura, R.; Murakami, S.; Takigawa, R. Effect of Ar Fast Atom Beam Irradiation on Alpha-Al_2O_3 for Surface Activated Room Temperature Bonding. *Jpn. J. Appl. Phys.* **2023**, *62*, SG1046. [CrossRef]
59. Zhao, R.; Liang, J.; Huang, J.; Zeng, R.; Zhang, J.; Chen, H.; Shi, G. Improving the Ni-Rich LiNi0.5Co0.2Mn0.3O2 Cathode Properties at High Operating Voltage by Double Coating Layer of Al_2O_3 and $AlPO_4$. *J. Alloys Compd.* **2017**, *724*, 1109–1116. [CrossRef]
60. Chubar, N. The Influence of Sulfate on Selenate Sorption on Mg-Al-CO 3 Layered Double Hydroxides Prepared by Fine Inorganic Sol-Gel Synthesis Studied by X-Ray Photoelectron Spectroscopy. *Appl. Surf. Sci.* **2018**, *459*, 281–291. [CrossRef]
61. Yu, H.; Dong, Q.; Jiao, Z.; Wang, T.; Ma, J.; Lu, G.; Bi, Y. Ion Exchange Synthesis of PAN/Ag3PO4 Core-Shell Nanofibers with Enhanced Photocatalytic Properties. *J. Mater. Chem. A* **2014**, *2*, 1668–1671. [CrossRef]
62. Lee, Y.; Kang, J.; Ahn, J.; Ko, W.; Park, H.; Yoo, J.K.; Yoon, W.S.; Kim, J. Enhanced Conversion Reaction of Na-Cu-PO_3 via Amorpholization and Carbon-Coating for Large Na Storage. *Mater. Today Energy* **2023**, *35*, 101325. [CrossRef]
63. Anantharaj, S.; Kundu, S. Enhanced Water Oxidation with Improved Stability by Aggregated RuO2-NaPO3 Core-Shell Nanostructures in Acidic Medium. *Curr. Nanosci.* **2017**, *13*, 333–341. [CrossRef]
64. Plakhova, T.V.; Romanchuk, A.Y.; Yakunin, S.N.; Dumas, T.; Demir, S.; Wang, S.; Minasian, S.G.; Shuh, D.K.; Tyliszczak, T.; Shiryaev, A.A.; et al. Solubility of Nanocrystalline Cerium Dioxide: Experimental Data and Thermodynamic Modeling. *J. Phys. Chem. C* **2016**, *120*, 22615–22626. [CrossRef]
65. Bolli, E.; Kaciulis, S.; Mezzi, A.; Montanari, R.; Varone, A. XPS Investigation of 5N Purity Al Thin Foils for MEMS Devices. *Surf. Interface Anal.* **2023**, *55*, 466–473. [CrossRef]
66. Avramova, I.; Suzer, S.; Guergova, D.; Stoychev, D.; Stefanov, P. CeOx/Al_2O_3 Thin Films on Stainless Steel Substrate—Dynamical X-Ray Photoelectron Spectroscopy Investigations. *Thin Solid Film* **2013**, *536*, 63–67. [CrossRef]
67. Zhuk, M.P. *Theory of Corrosion and Protection of Metals*; Metallurgy Publishing House: Moscow, Russia, 1976. (In Russian)
68. Danilidis, I.; Hunter, J.; Scamans, G.M.; Sykes, J.M. Effects of Inorganic Additions on the Performance of Manganese-Based Conversion Treatments. *Corros. Sci.* **2007**, *49*, 1559–1569. [CrossRef]
69. Zhang, H.; Yao, G.; Wang, S.; Liu, Y.; Luo, H. A Chrome-Free Conversion Coating for Magnesium–Lithium Alloy by a Phosphate–Permanganate Solution. *Surf. Coatings Technol.* **2008**, *202*, 1825–1830. [CrossRef]
70. Szklarska-Smialowska, Z. Pitting Corrosion of Aluminum. *Corros. Sci.* **1999**, *41*, 1743–1767. [CrossRef]

materials

Article

Formation of Oriented Nanowires from Mixed Metal Oxides

Anna Dikovska [1], Genoveva Atanasova [2,*], Rumen Nikov [1], Georgi Avdeev [3], Zara Cherkezova-Zheleva [4], Daniela Paneva [4] and Nikolay Nedyalkov [1]

1 Institute of Electronics, Bulgarian Academy of Sciences, 72 Tsarigradsko Chaussee, 1784 Sofia, Bulgaria; dikovska@ie.bas.bg (A.D.); rumen_nikov24@abv.bg (R.N.); nned@ie.bas.bg (N.N.)
2 Institute of General and Inorganic Chemistry, Bulgarian Academy of Sciences, Acad. G. Bonchev Str., Bl. 11, 1113 Sofia, Bulgaria
3 Rostislaw Kaischew Institute of Physical Chemistry, Bulgarian Academy of Sciences, Acad. G. Bonchev Str., Bl.11, 1113 Sofia, Bulgaria; g_avdeev@ipc.bas.bg
4 Institute of Catalysis, Bulgarian Academy of Sciences, Acad. G. Bonchev Str., Block 11, 1113 Sofia, Bulgaria; zzhel@ic.bas.bg (Z.C.-Z.); daniela@ic.bas.bg (D.P.)
* Correspondence: genoveva@svr.igic.bas.bg

Abstract: In this study, we present a physical method for the fabrication of oriented nanowires composed of mixed metal oxides. Pulsed laser deposition carried out in the air under atmospheric pressure was used for the production of samples. Two sets of experiments were performed by applying nanosecond and picosecond laser ablation, respectively. The depositions were performed using the laser ablation of mixed targets from iron oxide and zinc oxide as the initial materials in different ratios. The experiments were carried out in a magnetic field, which allowed us to control the morphology of nanostructures. The structure, microstructure, morphology, and composition of the structures obtained were studied in relation to the sample composition and laser ablation regime applied. The morphological analysis revealed that the structure of the samples consisted mainly of nanowire-like features reaching tens of micrometers in length. These nanowires were composed of nanoparticles and oriented predominantly in parallel to magnetic field lines. Nanoparticles produced using *ps* ablation were, on average, smaller than those obtained by *ns* ablation of the same target. Using ablation with *ps* laser pulses, we were able to produce new composite materials or materials containing unstable phases.

Keywords: PLD in open air; *ns* ablation; *ps* ablation; metal oxide composites

Citation: Dikovska, A.; Atanasova, G.; Nikov, R.; Avdeev, G.; Cherkezova-Zheleva, Z.; Paneva, D.; Nedyalkov, N. Formation of Oriented Nanowires from Mixed Metal Oxides. *Materials* **2023**, *16*, 6446. https://doi.org/10.3390/ma16196446

Academic Editor: Antonio Polimeni

Received: 25 August 2023
Revised: 21 September 2023
Accepted: 26 September 2023
Published: 27 September 2023

1. Introduction

Iron oxide-containing nanoparticles and nanostructures have long held the attention of researchers and technologists because of their unique physical properties, which provide opportunities for practical applications [1,2]. Magnetite and maghemite are among the most interesting and promising oxides because, at the nanoscale, they exhibit unique magnetic properties, biocompatibility, and biodegradability, and these characteristics make them suitable for applications in medicine [3–8]. They can be used as contrast-enhancing elements for biomedical diagnostics [7], in targeted drug delivery [9], for the magnetic isolation and separation of labeled cells [10], etc. Such uses usually require a specific size distribution and the desired shape of the nanoparticles or nanoparticle ensembles since their efficiency strongly depends on their magnetic properties. Producing ordered nanoparticle ensembles, especially magnetic nanowires, has attracted special attention due to their enhanced magnetic properties: shape anisotropy, high magnetic moment, coercivity, and remanent magnetization [11–13]. Such nanosized objects should ideally be fabricated through precise size- and shape-controlled synthesis without the use of additional toxic chemicals. However, conventional fabrication methods of iron oxide nanoparticles are chemically based and involve chemical reduction [8], chemical vapor condensation [14], co-precipitation [15], sol-gel [16], etc. Thus, the fabrication of nanoparticles or nanoparticle ensembles in an environmentally friendly manner using a simple and flexible method and conventional low-cost equipment is still a challenge.

A novel and promising physical method for the fabrication of oriented nanostructures of magnetic materials is pulsed laser deposition (PLD) in the air at an atmospheric pressure (PLD in the open air or atmospheric PLD), which is applied in the presence of a magnetic field [17–19]. This method allows the production of nanowires composed of nanoparticles arranged along the magnetic force lines. Additionally, the proposed method can be easily modified to fabricate more complicated systems, including composite nanostructures made of magnetic and non-magnetic materials [20,21]. This technology was first developed for *fs*-laser pulses [17]. Nedyalkov and coworkers demonstrated the fabrication of nanoparticle-composed nanowires of magnetic materials such as steel and Ni. This technology was later adapted for *ns* ablation and developed into commercial technology for nanostructure fabrication [18,19]. The main difference between these two techniques is that, in the case of *ns*-laser pulses, the nanoparticles that build the nanowires are formed by the condensation of ablated material [18,19], while ultra-short laser pulses lead to the direct ejection of nanoparticles due to the specific mechanisms of material removal [17].

Industrial applications of *fs* lasers are still limited because of their high price, expensive maintenance, and very specific requirements in the working environment. At the same time, picosecond laser pulse-generating systems with a duration of up to tens of picoseconds are of considerable scientific and commercial interest. In their interaction with the matter, these impulses retain the main characteristics of *fs*-laser pulses: very limited thermal effects, the lack of a broad thermal-affected area around the impact area, and non-thermal mechanisms of phase transformations, such as phase explosion and homogeneous melting. These laser systems are simpler and more stable than *fs* laser systems, leading to their lower cost, which is comparable to that of popular and widely used nanosecond systems. Such lasers could be easily applied in industries. However, due to the specific interaction of radiation with the matter at these pulse durations, the fundamental physical picture of the processes involved is still unclear and needs detailed investigation.

The aim of this work was to fabricate complex ordered nanostructures of iron-containing metal oxides by implementing a physical method, namely PLD, in the open air and the presence of a magnetic field based on ablation using *ns* and *ps* laser pulses. We emphasize the characteristics of this process and the outcomes of *ns* and *ps* ablation employed for the fabrication of nanoparticle-composed nanowires from multicomponent targets. To the best of our knowledge, composite nanowires produced by *ps*-PLD are reported here for the first time. The possibility of obtaining a new composite material or materials containing unstable phases is discussed in view of the different mechanisms of laser–matter interactions using short and ultra-short laser pulses.

2. Materials and Methods

2.1. Sample Fabrication

Oriented nanowires composed of mixed metal oxides were obtained by PLD in the presence of a magnetic field, as described elsewhere [18,19,21]. The experimental setup was a modification of the classical PLD configuration with a permanent magnet (B = 0.4 T) placed on the back side of the substrate. In such a configuration, the magnetic field lines were parallel to the substrate's surface [18,19,21]. Two sets of experiments were carried out based on nanosecond (*ns*) and picosecond (*ps*) laser ablation, respectively. The *ns* ablation was performed using 15-*ns* laser pulses delivered by a Nd:YAG (LS-2147, Lotis TII, Minsk, Belarus) laser system at a repetition rate of 10 Hz. The *ps* ablation process was carried out using a picosecond Nd:YAG laser (PS-A1-1064, CNL laser, Changchun, China) with a pulse duration of 10 ps and a repetition rate of 1 kHz. In both cases, the fundamental wavelength of these lasers at 1064 nm was used. The ablation targets were produced by mixing the initial compound of Fe_2O_3 (ChemPur, CAS No. 1309-37-1) with ZnO (Merck, CAS No. 1314-13-2) at a different weight percent, as follows: target N0—pure Fe_2O_3; target N1—5 wt% ZnO in Fe_2O_3; target N2—10 wt% ZnO in Fe_2O_3; target N3—25 wt% ZnO in

Fe_2O_3; target N4—50 wt% ZnO in Fe_2O_3, and target N5—75 wt% ZnO in Fe_2O_3. The mixed powders were then homogenized, cold pressed at 5 MPa into tablets and synthesized at 900 °C for 4 h. The laser fluence applied on the targets for *ns* and *ps* ablation was 4 and 0.4 J/cm^2, respectively. The material ablated from the targets was deposited on quartz or silicon substrates. The distance between the target and substrate was 25 mm. The depositions were carried out in the air under atmospheric pressure.

2.2. Sample Characterization

The morphology of the samples was analyzed using scanning electron microscopy (SEM) and a LYRA I XMU system (Tescan, Brno, Czech Republic). Their microstructure was studied via transmission electron microscope (TEM) images taken by a JEOL JEM 2100 system (Akishima-Shi, Tokyo, Japan). Using an Empyrean diffractometer (PANalytical), X-ray diffraction (XRD) was employed to examine the sample crystalline structure and phase composition. The crystalline phases were identified through PAN-ICSD and COD database cards. Transmission Mössbauer spectra were recorded using an electromechanical spectrometer Wissenschaftliche Elektronik (Starnberg, Germany) at a constant acceleration mode and ^{57}Co/Rh source (10 mCi). The velocity was calibrated by the α-Fe standard. The parameters of hyperfine interactions in the obtained spectral components (isomer shift ((IS)), quadruple shift ((2ε))/quadruple splitting ((QS)), hyperfine effective field ((B_{hf})), line width ((FWHM)) and partial area ((A)) were determined using the WinNormos program. Computer fitting was based on the least square method. Using an AXIS Supra electron spectrometer (Kratos Analytical Ltd., Manchester, UK) and AlKα radiation with a photon energy of 1486.6 eV, X-ray photoelectron spectroscopy (XPS) was used to determine the surface composition of the samples. The energy calibration was performed by normalizing the C1 line of adsorbed adventitious hydrocarbons to 284.8 eV.

3. Results

Figure 1 displays the SEM images of samples deposited in a magnetic field via the *ns* laser ablation of the pure Fe_2O_3 target (Figure 1a) and mixed Fe_2O_3-ZnO targets with different ratios of the initial compounds (Figure 1b–f). The morphological analysis revealed that the structure of these samples consisted mainly of nanowire-like features (hereafter called nanowires) reaching tens of micrometers in length. These nanowires tend to be oriented parallel to the external magnetic field lines, with this orientation diminishing with an increase in the ZnO content in the target. It should be noted that the oriented nanowires formed bundles, as seen in Figure 1a–d, and these bundles disappeared with the loss of nanowire orientation. In addition, some randomly distributed droplets, typical for the *ns*-PLD technology, were also observed; their numbers increased along with the ZnO content in the target.

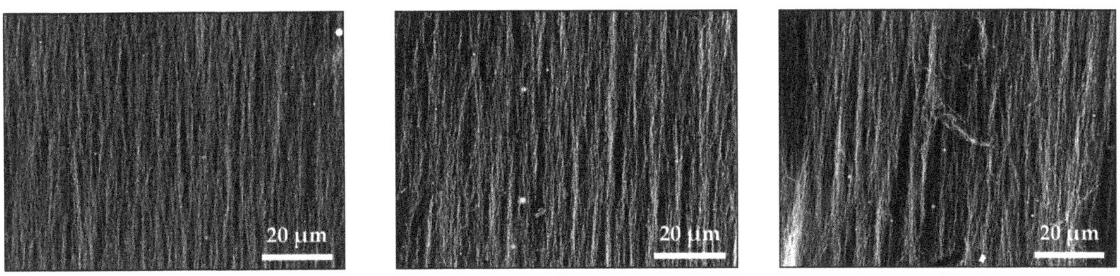

(a) (b) (c)

Figure 1. *Cont.*

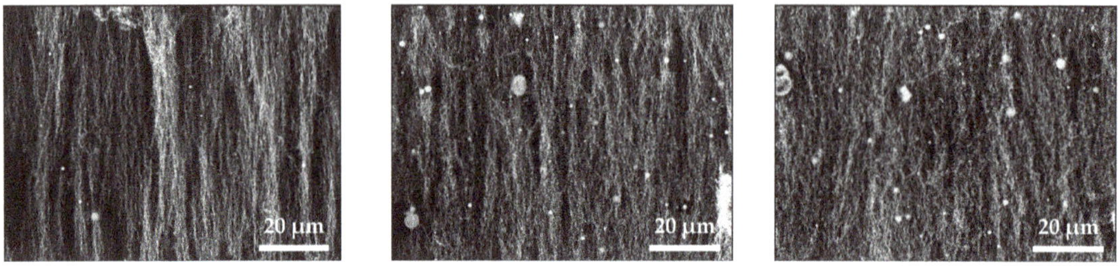

(**d**) (**e**) (**f**)

Figure 1. SEM images of the samples produced by *ns* laser ablation of (**a**) Pure Fe_2O_3 (N0) and mixed (**b**) N1, (**c**) N2, (**d**) N3, (**e**) N4, and (**f**) N5 targets.

Figure 2 shows the SEM images of samples deposited from Fe_2O_3 and mixed with Fe_2O_3-ZnO targets using *ps* laser ablation. The presence of vertically oriented features parallel to magnetic field lines is evident in all the images. The nanowire length is in the range of microns. The insets show the detailed structure of the nanowires. They are composed of mostly spherical nanoparticles of different sizes. Particles with diameters in the range of 150–450 nm and even larger can be clearly distinguished; their number increased with the increase in ZnO content in the targets (Figure 2d–f). It is also evident that the total amount of material deposited on the substrate decreased with the increase in ZnO in the targets. However, bundles of oriented nanowires can still be observed even though the target with the highest ZnO content ablated (Figure 2f).

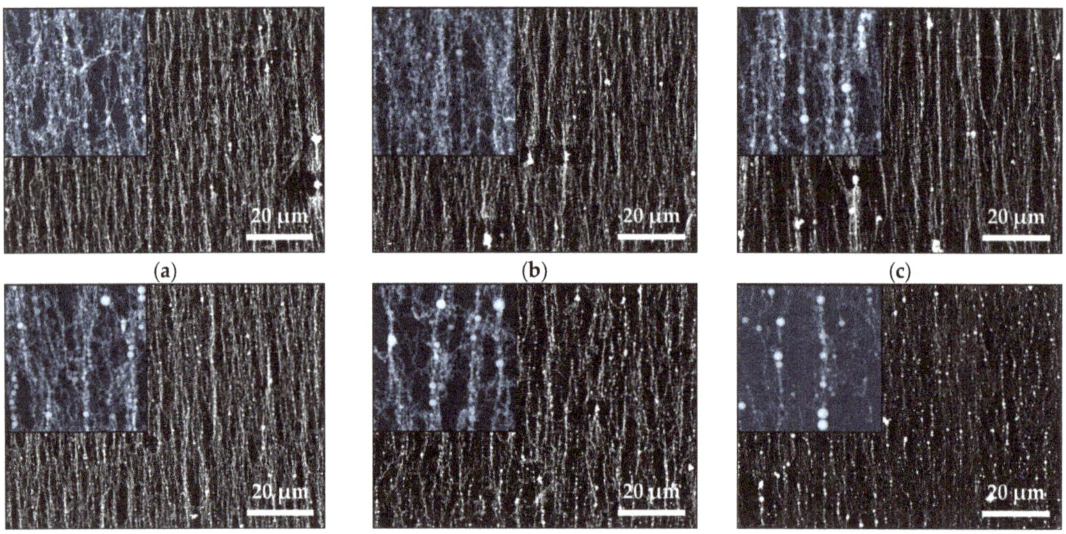

(**d**) (**e**) (**f**)

Figure 2. SEM images of the samples produced using *ps* laser ablation of (**a**) pure Fe_2O_3 (N0) and mixed (**b**) N1, (**c**) N2, (**d**) N3, (**e**) N4, and (**f**) N5 targets. Insets are SEM images at a higher magnification with the size 10×10 µm.

A TEM image of a sample deposited using *ns* ablation from target N4 is shown in Figure 3a. The sample microstructure can be seen as polygon-shaped nanoparticles with different sizes. The size distribution is also presented in Figure 3a; nanoparticle size was in the range of 5–40 nm with a mean diameter of 18 nm. SAED image demonstrated that the nanoparticles were crystalline. The TEM images of samples deposited from other targets display microstructures resembling those in Figure 3a, i.e., with a similar shape and size distribution of nanoparticles. Figure 3b shows a TEM image of a sample deposited using *ps* ablation from target N4. The microstructure of the sample consists of spherical and polygon-shaped nanoparticles. Particles with sizes from approximately 40 nm to 120 nm, as well as smaller ones with sizes in the range of 2–20 nm, can be clearly distinguished (see the size distribution in Figure 3b). The nanoparticles are crystalline with a mean diameter estimated at 14 nm (size distribution in Figure 3b). The microstructure of the samples deposited by the *ps* ablation of the other targets did not differ significantly from that of the sample deposited from target N4.

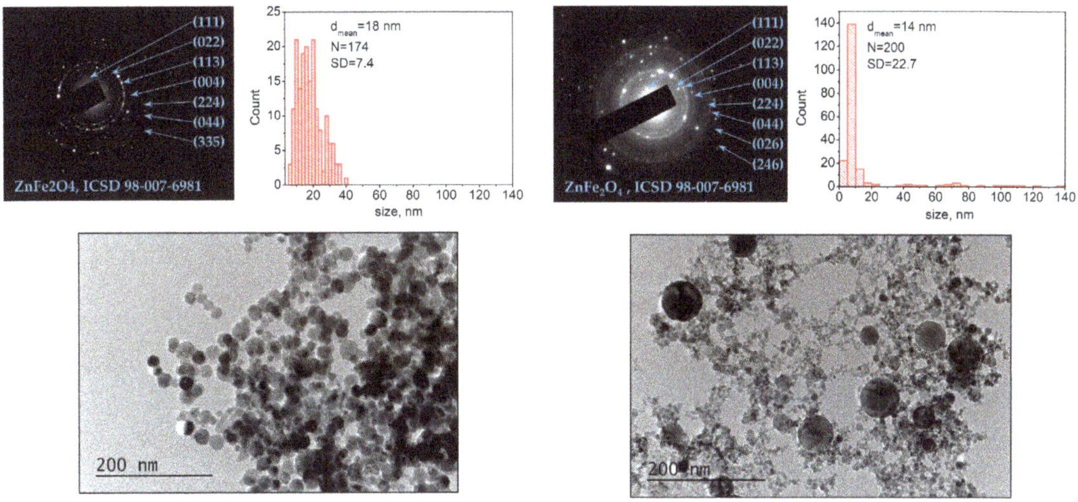

(a) (b)

Figure 3. TEM image with corresponding SAED pattern and size distribution of the sample deposited by (**a**) *ns* ablation and (**b**) *ps* ablation from target N4.

The XRD patterns of the samples (nanowires) deposited using the *ns* laser ablation of Fe_2O_3 and mixed targets are presented in Figure 4a. The phase composition of the sample deposited from Fe_2O_3 is a combination of iron oxides, i.e., magnetite (Fe_3O_4, ICSD 98-015-8741), hematite (α-Fe_2O_3, ICSD 98-005-6372) and wüstite ($Fe_{1-x}O$, ICSD 98-002-7237), as previously reported [15,16]. The phase composition of the sample deposited from target N1 (with the smallest ZnO content in the initial material) did not differ from the composition of the sample deposited from the pure Fe_2O_3 target; the presence of ZnO or other Zn-containing compounds in the sample was not identified. Increasing the Zn content into the target led to the deposition of samples with a slightly higher magnetic phase content, i.e., magnetite, and a lower content of other, non-magnetic, iron oxides (sample deposited from target N2). This tendency continued with the further Zn content increase in the target (deposition from target N3). The presence of a Zn-containing phase was clearly expressed in the sample deposited from target N4. The phase composition of the sample was identified as zinc ferrite ($ZnFe_2O_4$, ICSD 98-007-6981). Lastly, ablating the target with the highest Zn content led to the deposition of composite samples consisting of $ZnFe_2O_4$ and ZnO. The lattice parameters of the main phase of samples deposited from mixed targets are presented in Table 1.

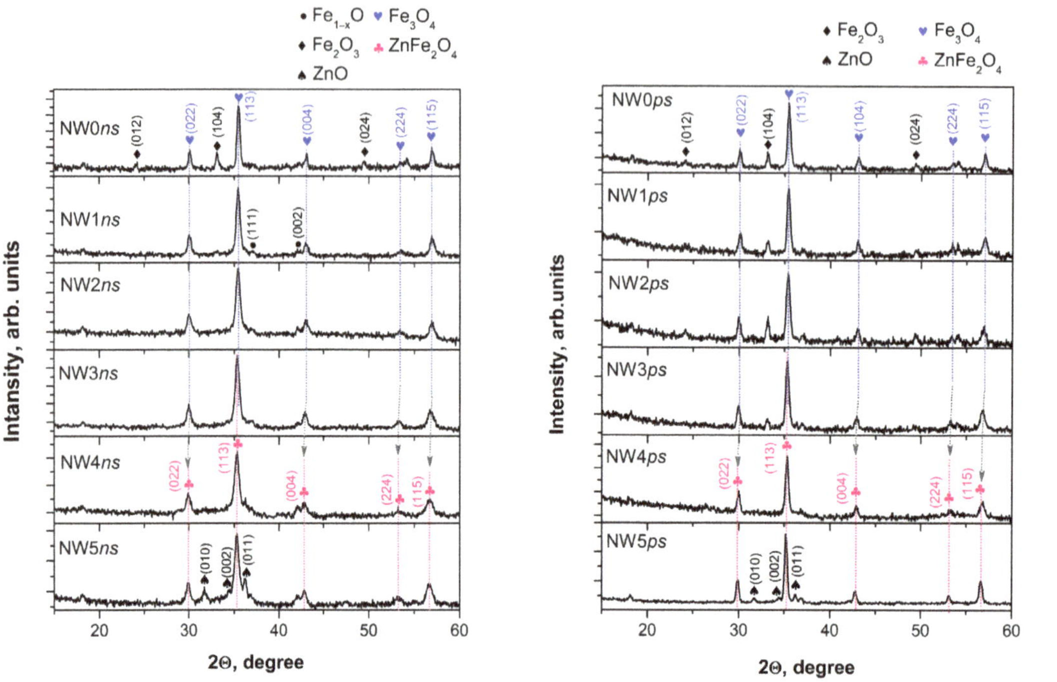

(**a**) (**b**)

Figure 4. XRD patterns of samples produced by the (**a**) *ns* and (**b**) *ps* laser ablation of different targets. The samples (nanowires) deposited from target N0 are named NW0; from N1 as NW1; from N2 as NW2; from N3 as NW3; from N4 as NW4, and from N5 as NW5.

Table 1. Lattice parameter of the main phase of samples deposited from different targets by *ns* and *ps* laser ablation.

Sample Deposited from Target		Fe_3O_4, ICSD 98-015-8741	N0	N1	N2	N3	N4	N5	$ZnFe_2O_4$, ICSD 98-007-6981
Lattice parameter, Å	*ns* ablation	8.3860	8.396 (3)	8.407 (4)	8.415 (4)	8.414 (8)	8.423 (8)	8.432 (8)	8.4220
	ps ablation		8.388 (2)	8.389 (2)	8.394 (2)	8.417 (2)	8.416 (3)	8.437 (2)	

The reference values of magnetite and zinc ferrite are also present. As seen, increasing the Zn content in the targets also increased the lattice parameter of the main phase of the sample. Figure 4b reports the XRD patterns of samples deposited by the *ps* laser ablation of Fe_2O_3 and mixed targets. The *ps* laser ablation of the Fe_2O_3 target led to the deposition of samples consisting of magnetite and hematite phases of iron oxides, with the predominant phase being magnetite. The content of the magnetic phase in the samples increased with the increasing ZnO in the targets (deposition from targets N1, N2, and N3) at the expense of the non-magnetic phase of hematite. Meanwhile, the lattice parameter of the samples' main phase increased, as seen in Table 1. The Zn-containing phase clearly appeared in the sample deposited from target N4; the phase composition of the sample was identified as zinc ferrite. The further increase in the Zn content into the target (depositions from target N5) led to the deposition of composite samples of $ZnFe_2O_4$ and ZnO.

Table 2 summarizes the data calculated after analyzing the decomposition of the Mössbauer spectra of samples deposited using *ns* ablation. The recorded spectra at room temperature of the investigated materials revealed the presence of sextet spectra components superimposed with doublet ones. The appearance of several magnetically split components (sextets) can be resolved in the spectra of samples with different iron/Zn concentrations and samples deposited from the targets N0, N1, and N2. The major contribution of the sextet Mössbauer components is characteristic of the presence of Fe_3O_4 predominantly and α-Fe_2O_3 in studied materials (Table 2) [22,23]. The presence of doublet patterns could be assigned to the contribution of Fe-bearing phases in non-stoichiometric $Fe_{1-x}O$ and $ZnFe_2O_4$ [24,25].

Table 2. Calculated parameters after fitting the Mössbauer spectra of samples deposited by *ns* ablation from different targets. The parameters are as follows: isomer shift (IS), quadruple shift (2ε)/quadruple splitting (QS), hyperfine effective field (B_{hf}), line width (FWHM) and partial area (A).

Sample	Components	IS,	2ε/QS,	B_{hf},	FWHM,	A,
		mm/s	mm/s	T	mm/s	%
NW0*ns*	Sx1-α-Fe_2O_3	0.36	−0.17	51.2	0.25	7
	Sx2-$Fe_{3-x}O_4$	0.27	0.02	48.7	0.39	34
	Sx3-$Fe_{3-x}O_4$	0.64	0.03	44.9	0.58	43
	Db1-$Fe_{1-x}O$	0.98	1.02	-	0.50	16
NW1*ns*	Sx1-α-Fe_2O_3	0.36	−0.22	51.6	0.30	8
	Sx2-$Fe_{3-x}O_4$	0.29	0.01	48.9	0.35	26
	Sx3-$Fe_{3-x}O_4$	0.61	0.03	45.1	0.67	46
	Db1-$Fe_{1-x}O$	0.90	0.78	-	0.68	14
	Db2-$ZnFe_2O_4$	0.35	0.42	-	0.50	6
NW2*ns*	Sx1-$Fe_{3-x}O_4$	0.28	0.01	47.9	0.30	15
	Sx2-$Fe_{3-x}O_4$	0.66	0.01	45.8	0.52	30
	Db1-$Fe_{1-x}O$	1.10	0.41	-	0.40	19
	Db2-$ZnFe_2O_4$	0.32	0.38	-	0.44	36
NW3*ns*	Db1-$Fe_{1-x}O$	1.11	0.40	-	0.30	18
	Db2-$ZnFe_2O_4$	0.32	0.41	-	0.34	82
NW4*ns*	Db1-$Fe_{1-x}O$	1.12	0.09	-	0.50	12
	Db2-$ZnFe_2O_4$	0.32	0.48	-	0.40	88
NW5*ns*	Db-$ZnFe_2O_4$	0.32	0.45	-	0.46	100

Table 3 shows the data estimated after the decomposition of Mössbauer spectra in the samples obtained using *ps* ablation. The Mössbauer patterns reveal the different superpositions of sextets and doublet spectra determined by the target used for ablation. The major contribution of sextet components could be assigned to the presence of Fe_3O_4 predominantly and α-Fe_2O_3 in samples deposited from targets N0–N3. The presence of doublet patterns is characteristic of the contribution of $ZnFe_2O_4$, as this phase presents in the spectra of all samples deposited from the target containing ZnO as an initial material. Also, the presence of γ-Fe_2O_3 is clearly recognized in the spectra of samples deposited from targets N4 and N5.

Table 3. Calculated parameters after fitting the Mössbauer spectra of samples deposited via *ps* ablation from different targets. The parameters are as follows: isomer shift (IS), quadruple shift (2ε)/quadruple splitting (QS), hyperfine effective field (B_{hf}), line width (FWHM) and partial area (A).

Sample	Components	IS, mm/s	2ε/QS, mm/s	B_{hf}, T	FWHM, mm/s	A, %
NW0*ps*	Sx1-α-Fe_2O_3	0.36	−0.21	51.5	0.26	26
	Sx2-$Fe_{3-x}O_4$	0.28	0.01	48.9	0.34	31
	Sx3-$Fe_{3-x}O_4$	0.64	0.00	45.7	0.45	36
	Db-Fe^{3+}	0.32	0.49	-	0.35	7
NW1*ps*	Sx1-α-Fe_2O_3	0.36	−0.20	51.5	0.29	45
	Sx2-$Fe_{3-x}O_4$	0.28	0.02	48.9	0.32	17
	Sx3-$Fe_{3-x}O_4$	0.61	0.02	45.1	0.51	29
	Db-$ZnFe_2O_4$	0.33	0.43	-	0.35	9
NW2*ps*	Sx1-α-Fe_2O_3	0.34	−0.19	52.1	0.26	15
	Sx2-$Fe_{3-x}O_4$	0.27	0.01	49.3	0.35	28
	Sx3-$Fe_{3-x}O_4$	0.57	0.02	45.8	0.66	48
	Db-$ZnFe_2O_4$	0.33	0.42	-	0.45	9
NW3*ps*	Sx1-α-Fe_2O_3	0.37	−0.19	51.8	0.45	5
	Sx2-$Fe_{3-x}O_4$	0.29	0.01	49.4	0.50	27
	Sx3-$Fe_{3-x}O_4$	0.62	0.02	45.8	0.71	50
	Db-$ZnFe_2O_4$	0.34	0.43	-	0.40	18
NW4*ps*	Sx-γ-Fe_2O_3	0.34	0.02	47.9	0.40	7
	Db-$ZnFe_2O_4$	0.34	0.36	-	0.32	93
NW5*ps*	Sx-γ-Fe_2O_3	0.34	0.03	47.3	0.30	23
	Db-$ZnFe_2O_4$	0.33	0.44	-	0.67	77

XPS analysis of the surface of samples deposited from target N4 via ns and ps ablation is presented in Figure 5. As seen, the samples' binding energies for $Fe2p_{3/2}$ and $Fe2p_{1/2}$ were 711 eV and 724.5 eV, respectively. The $Fe2p_{3/2}$ peak had an associated satellite peak situated at 719.6 eV. We can conclude that Fe atoms were in the Fe^{3+} oxidation state due to the estimated spin-orbit splitting of 13.6 eV and the binding energy positions, as well as the presence of a satellite peak at 719.6 eV [26,27]. No peak typical of Fe^{2+} (~53.7 eV) was detected [27]. The Zn $2p_{3/2}$ and Zn $2p_{1/2}$ peaks were seen at 1021.5 eV and 1044.6 eV, respectively, with a spin-orbit splitting of 23.1 eV. The width of the Zn 2p peaks, the binding energy positions, and the spin-orbit splitting all pointed to Zn atoms being in the Zn^{2+} oxidation state [26]. The asymmetric O 1s peak (not shown) implied the presence of different oxygen-containing species. This peak could be deconvoluted into two components: at nearly 529.9 eV, attributed to lattice O^{2-} ions in the metal oxides, and 532.7 eV, attributed to the presence of adsorbed hydroxyl, carbonate, or O_2 species [27–29]. No substantial difference is seen between the spectra of the samples NW4ns and NW4ps presented in Figure 5.

Table 4 summarizes the Fe/Zn ratio on the sample surface against the respective mixed targets used for depositions. It can be concluded that increasing the Zn content in the targets reduced the Fe/Zn ratio on the sample surface.

Table 4. Fe/Zn ratio on the surface of samples deposited using *ns* and *ps* ablation from different targets.

Sample Deposited from Target		N1	N2	N3	N4	N5
Sample Fe/Zn ratio	*ns* ablation	10.3	6.7	2.3	1.4	1.1
	ps ablation	13.4	6.3	3.1	1.9	1.3

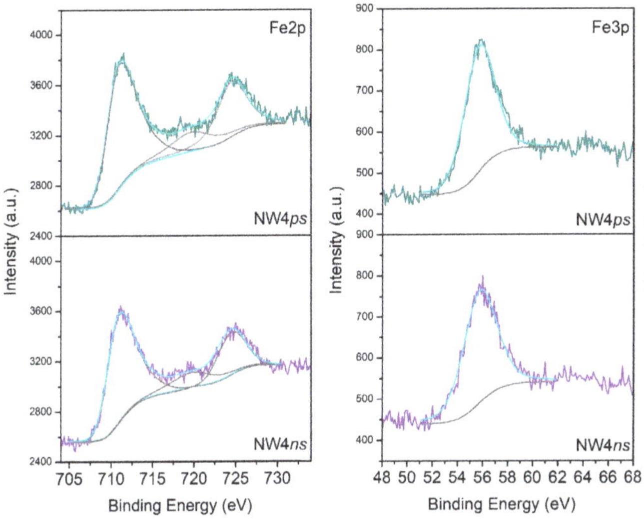

Figure 5. XPS spectra of samples deposited via ns and ps ablation from target N4.

4. Discussion

The phase composition of mixed targets is a combination of hematite and zinc ferrite, as shown in Figure S1 and Table S1, depending on the initial content of ZnO. The predominant hematite phase decreases and disappears in favor of $ZnFe_2O_4$ with the increase in ZnO's weight percentage in the targets. Target N4, obtained from the equal weight percentage of Fe_2O_3 and ZnO, is identified as a zinc ferrite phase. It is not surprising that the structure of the sample deposited from target N4, regardless of the ablation regime, consisted of $ZnFe_2O_4$ because the main advantage of the pulsed laser deposition process is the potential for the stoichiometric transfer of material from the target to the substrate. The predominant phase of samples deposited from targets with a lower ZnO content is magnetite, and no Zn-containing compounds could be identified in the XRD patterns (Figure 4). However, the presence of Zn and/or Zn-containing compounds was confirmed by XPS as well as by Mössbauer analyses. The results from XPS analyses show that the decrease in the Fe/Zn ratio on the sample surface correlates with the rise in the ZnO content in the targets (Table 4). It is worth noting that the sample fabrication technology applied here was a kind of sputtering process, i.e., a physical vapor deposition process, which implies that the composition analyzed on the sample surface is highly unlikely to change in-depth if the target used is homogeneous. In this light, it could be concluded that with the increase in the zinc content in the target, Zn^{2+} tends to substitute Fe^{2+} in Fe_3O_4, producing zinc ferrite. This tendency is already evident in samples from targets with a low ZnO concentration (N1–N3) and persists to the highest concentrations since $ZnFe_2O_4$ is identified as the main phase in the sample deposited from target N5: a target with a higher initial content of ZnO than Fe_2O_3. The change in the lattice parameter of magnetite with the increase in the target ZnO content (Table 1) supports this interpretation. Considering the XRD patterns shown in Figure 4, the main difference in the crystal structure of the samples deposited via *ns* and *ps* ablation is the presence/absence of $Fe_{1-x}O$. The non-stoichiometric wüstite phase is present in the samples prepared using *ns* ablation but cannot be found in the samples deposited via *ps* ablation from the same targets. We attribute this result to the mechanism of the laser–matter interaction during the so-called short (*ns*) and ultra-short (<10 ps) laser pulses and the further evolution of ablated material in the air at atmospheric pressure [30]. In the classical *ns* ablation process, a laser pulse rapidly heats the target. Thus, during *ns* ablation, a significant part of α-Fe_2O_3 is reduced to FeO, which further participates in the formation of the Fe_3O_4 phase ($FeO{\cdot}Fe_2O_3$) [18]. The presence of Fe-containing material allows the potential use of Mössbauer spectroscopy for the precise identification of the sample phase composition. The Mössbauer results for the samples deposited via *ns* ablation are in good agreement with XRD analysis (Table 2 and Figure 4a). The exceptions observed can be related to the distinctive features of the Mössbauer method to register amorphous phases together with crystalline ones, revealing only iron-containing compounds. This latter point means that only the $ZnFe_2O_4$ phase could be registered in the sample deposited from target N5 via *ns* ablation. The possibility of the presence of an amorphous phase is ruled out by the microstructural analysis (Figure 3a), which demonstrates the crystalline nature of the sample. Furthermore, no amorphous halo was observed in any of the XRD patterns shown in Figure 4a. Hence, on the one hand, we refer to the negligible differences in sample composition as obtained by XRD and Mössbauer spectroscopy to the error in quantifying the phase composition and the fabrication method repeatability on the other. In the case of *ps* ablation, the microstructure of the sample is polycrystalline, as revealed from the SAED pattern (Figure 3b). However, the TEM image and the corresponding size distribution of the sample show a significant presence of nanoparticles with very small sizes, in the range of 1–4 nm. The XRD method does not recognize such small particles as crystalline, and they are registered as an amorphous halo in the XRD pattern (Figure 4b). In this regard, we assumed that the differences between XRD and Mossbauer results concerning the sample phase composition are due to the presence of small-sized nanoparticles obtained via *ps* ablation. It should be noted that the precise identification of sample phase composition using Mössbauer spectroscopy shows the presence of maghemite (γ-Fe_2O_3). The maghemite phase is a polymorph form of hematite with a spinel structure-like magnetite [14]. Since this phase is unstable, it transforms into hematite at temperatures between 250 and 750 °C. In this way, using ablation via *ps* laser pulses due to the shorter laser–matter interaction time compared to the *ns* ablation process, we were able to produce an iron oxide phase such as the maghemite, which is difficult to obtain via the sputtering process. Further, it should be noted that the relative weights of the $ZnFe_2O_4$ phase in the samples deposited using the *ps* laser seem not to correspond to the increase in Zn-containing components in the targets (Tables 3 and S1). This indicates that a part of $ZnFe_2O_4$ is decomposed to magnetite or maghemite. It should be emphasized that zinc ferrite is a mixed metal oxide, which, in its bulk form, typically has a paramagnetic behavior. However, prepared at the nanometer scale, $ZnFe_2O_4$ exhibits ferromagnetic/ferrimagnetic properties [31]. It can be summarized that the material deposited by the open-air ablation of mixed targets, regardless of the initial content of Fe_2O_3 and ZnO powders in the target, is a composite with a predominant magnetic phase—magnetite, maghemite, or zinc ferrite.

After the initial ejection of material from the target via *ns* or *ps* ablation and plasma plume formation, the ablated material further evolves in the air at atmospheric pressure. In the case of *ns* ablation, the process taking place in the open air results in the formation of nanoparticles/nanoparticle aggregates in the plasma plume due to the ablated material condensation [18,19]. In the case of *ps* ablation, ultrashort laser pulses result in the direct formation of nanoparticles due to their fragmentation and/or phase explosion, leading to material removal [17]. As a consequence, regardless of the ablation regime applied, the "building blocks" of the samples deposited on the substrate are nanoparticles and/or nanoaggregates with different sizes and shapes, as is evident from the TEM images presented in Figure 3.

Since the atmospheric pressure (as a high surrounding pressure) confines the plume and limits its size to a few millimeters, the ablated material can undergo phase-composition transformations and nanoparticle agglomeration since the nanoparticles formed rapidly decelerate. The small-sized plasma plume means that the evolution of ablated material occurs close to the target; consequently, the ablation process in the open air is an inefficient process. As was previously reported, no material deposited on the substrate is observed or is barely observed for target–substrate distances larger than 5 mm [18]. However, the amount of deposited material significantly increases when a magnetic component is present in the plasma plume and a suitable external magnetic field is applied [18,19]. The external magnet placed at the backside of the substrate attracts the magnetic nanoparticles and nanoaggregates, arranges them in parallel to the magnetic field lines, and collects them on the substrate [18–21]. However, for the cases reported here, XRD and Mössbauer results confirm the presence of non-magnetic particles in the structure of the samples (Figure 4, Tables 2 and 3). Presumably, due to the strong confinement of the plasma plume, the magnetic nanoparticles carry away or coalesce with non-magnetic ones and transport them to the substrate [20,21]. The presence of a magnetic phase is clearly confirmed by the SEM images shown in Figure 1. Micron-sized nanowires were present in all the samples produced via *ns* and *ps* laser ablation. A detailed look at the morphology of the samples deposited from the same target using *ns* and *ps* laser ablation shows differences, which we can attribute to their different microstructure (Figure 3). The nanowires produced via *ns* ablation are composed of nanoparticles with similar sizes (Figure 3a), and separate nanoparticles are not clearly distinguished in the nanowires (Figure 1). By contrast, in *ps* ablation, it is clearly seen that distinct nanoparticles are arranged in nanowires (insets of Figure 2). We can visually recognize nanoparticles of different sizes arranged into nanowires, reflecting the broad range of nanoparticle size distribution (SD = 22.7) shown in Figure 3b. The presence of small nanoparticles also indicates that the magnetic moment induced by the external magnetic field is small since it is proportional to the size (or, more precisely, the volume) of the nanoparticles [17]. Subsequently, the smaller nanoparticles produced via *ps* ablation are weakly or not at all attracted by the magnetic field compared to the nanoparticles deposited via *ns* ablation from the same target. This explains the smaller amount of material deposited by *ps* ablation (Figures 1 and 2).

Although the phase composition of the samples was identified as a combination of different iron-containing oxides, the surface analysis showed that Fe atoms were in the Fe^{3+} oxidation state. We assumed that due to passivation on the surface in open air, the iron oxide nanoparticles were oxidized to Fe_2O_3 [32].

To summarize, oriented nanowires composed of magnetic metal oxides can be easily prepared by pulsed laser deposition in the presence of a magnetic field [33]. The various process parameters, such as laser fluence, target–substrate distance, ambient atmosphere and pressure, and magnetic field, all affect the structure and morphology of the nanostructures produced, as has been previously studied and reported [18,19]. Similar effects of process parameters on the samples' structural and morphological properties also take place when a multicomponent target (targets in our work) is used for ablation. The use of multicomponent targets composed of different and/or mixed oxides additionally widens the possibility of fabricating new composite materials or materials composed of unstable phases. However, it should be noted that when ultrashort lasers are used for ablation, a critical parameter for the fabrication of such nanowires is the size of the laser spot. When the area of the laser spot is a few hundred μm^2 (typical for ultrashort lasers) [34], the quantity of ablated material is too small for efficient interaction in the plasma plume, and consequently, only large particles are deposited on the substrate. Such material deficiency could be easily overcome by increasing the pulse repetition rate, as demonstrated in our study.

5. Conclusions

Micron-sized composite nanowires were produced via *ns* and *ps* laser ablation in the air at atmospheric pressure in the presence of an external magnetic field. Nanowires fabricated using *ns* ablation were composed of nanoparticles with similar sizes, and separate nanoparticles were not clearly distinguished, while nanoparticles of different sizes arranged into nanowires could be visibly recognized when *ps* ablation was applied. We can explain the different sample morphology with the different microstructures of the samples deposited using *ns* and *ps* laser ablation. The laser ablation of multicomponent targets composed of iron oxide and zinc oxide in different ratios as initial materials led to the fabrication of composite materials consisting of Fe-oxides and Zn-ferrite. Based on the Mössbauer results, it can be summarized that the $Fe_{1-x}O$ phase is identified in the phase composition of the samples deposited via *ns* ablation, while *ps* ablation leads to the presence of γ-Fe_2O_3 in the sample composition. Using ablation with *ps* laser pulses, due to the shorter laser–matter interaction time compared to the *ns* ablation process, we were able to produce the maghemite iron oxide phase, which is hard to obtain through a sputtering process. Furthermore, using the *ns* ablation of a multicomponent target, the stoichiometric transfer of material from the target to the substrate can be implemented. Using the *ps* ablation of the same target, this stoichiometric transfer of material from the target to the substrate was no longer possible since part of $ZnFe_2O_4$ was decomposed to magnetite or maghemite. Fabricating chemically clean magnetic nanowires with desirable composition using a simple and flexible method and conventional low-cost equipment makes their application possible in the field of biomedicine.

Supplementary Materials: The following supporting information can be downloaded at: https://www.mdpi.com/article/10.3390/ma16196446/s1, Figure S1: XRD patterns of the targets used for ablation: (a) N0 (Fe_2O_3) and mixed: (b) N1, (c) N2, (d) N3, (e) N4, (f) N5 target. Table S1: Phase composition of the targets used for ablation.

Author Contributions: Conceptualization, A.D.; methodology, A.D. and N.N.; formal analysis, G.A. (Genoveva Atanasova), G.A. (Georgi Avdeev), Z.C.-Z. and D.P.; investigation, A.D. and R.N.; data curation, G.A. (Genoveva Atanasova), R.N., G.A. (Georgi Avdeev), Z.C.-Z. and D.P.; writing—original draft preparation, A.D.; writing—review and editing, N.N.; visualization, (Genoveva Atanasova); project administration, A.D. All authors have read and agreed to the published version of the manuscript.

Funding: The authors wish to thank the financial support of project KP-06-N37/20 "Formation and physical properties of composite nanostructures of metal oxides and noble metals" under the "Competition for financial support of basic research projects—2019" Program of the Bulgarian National Science Fund. The authors acknowledge the TwinTeam project Д01-272 "European Network on Materials for Clean Technologies" for providing the opportunity to present the results at the SizeMat4 conference as well as for financial publication support.

Institutional Review Board Statement: Not applicable.

Informed Consent Statement: Not applicable.

Data Availability Statement: Not available.

Conflicts of Interest: The authors declare no conflict of interest.

References

1. Ali, A.; Zafar, H.; Zia, M.; ul Haq, I.; Phull, A.R.; Ali, J.S.; Hussain, A. Synthesis, characterization, applications, and challenges of iron oxide nanoparticles. *Nanotechnol. Sci. Appl.* **2016**, *9*, 49–67. [CrossRef]
2. Materón, E.M.; Miyazaki, C.M.; Carr, O.; Joshi, N.; Picciani, P.H.S.; Dalmaschio, C.J.; Davis, F.; Shimizu, F.M. Magnetic nanoparticles in biomedical applications: A review. *Appl. Surf. Sci. Adv.* **2021**, *6*, 100163. [CrossRef]
3. Li, F.; Shao, H.; Zhou, G.; Wang, B.; Xu, Y.; Liang, W.; Chen, L. The recent applications of nanotechnology in the diagnosis and treatment of common cardiovascular diseases. *Vascul. Pharmacol.* **2023**, *152*, 107200. [CrossRef]
4. Monteserín, M.; Larumbe, S.; Martínez, A.V.; Burgui, S.; Francisco Martín, L. Recent Advances in the Development of Magnetic Nanoparticles for Biomedical Applications. *J. Nanosci. Nanotechnol.* **2021**, *21*, 2705–2741. [CrossRef]
5. Laurent, S.; Bridot, J.-L.; Elst, L.V.; Muller, R.N. Magnetic iron oxide nanoparticles for biomedical applications. *Future Med. Chem.* **2010**, *2*, 427–449. [CrossRef]
6. Akbarzadeh, A.; Samiei, M.; Davaran, S. Magnetic nanoparticles: Preparation, physical properties, and applications in biomedicine. *Nanoscale Res. Lett.* **2012**, *7*, 144. [CrossRef]
7. Huang, S.H.; Juang, R.S. Biochemical and biomedical applications of multifunctional magnetic nanoparticles: A review. *J. Nanopart Res.* **2011**, *13*, 4411–4430. [CrossRef]
8. Ling, D.; Hyeon, T. Chemical Design of Biocompatible Iron Oxide Nanoparticles for Medical Applications. *Small* **2012**, *9*, 1450–1466. [CrossRef]
9. Chertok, B.; Moffat, B.A.; David, A.E.; Yu, F.; Bergemann, C.; Ross, B.D.; Yanga, V.C. Iron oxide nanoparticles as a drug delivery vehicle for MRI monitored magnetic targeting of brain tumors. *Biomaterials* **2008**, *29*, 487–496. [CrossRef]
10. Kim, S.J.; Lewis, B.; Steiner, M.-S.; Bissa, U.V.; Dose, C.; Frank, J.A. Superparamagnetic Iron Oxide Nanoparticles for Direct Labeling of Stem Cells and In Vivo MRI Tracking. *Contrast Media Mol Imaging* **2016**, *11*, 55–64. [CrossRef]

11. Ivanov, Y.P.; Vázquez, M.; Chubykalo-Fesenko, O. Magnetic reversal modes in cylindrical nanowires. *J. Phys. D Appl. Phys.* **2013**, *46*, 485001. [CrossRef]
12. Nana, A.B.A.; Marimuthu, T.; Kondiah, P.P.D.; Choonara, Y.E.; Du Toit, L.C.; Pillay, V. Multifunctional magnetic nanowires: Design, fabrication, and future prospects as cancer therapeutics. *Cancers* **2019**, *11*, 1956. [CrossRef] [PubMed]
13. Andrade, R.G.D.; Veloso, S.R.S.; Castanheira, E.M.S. Shape Anisotropic Iron Oxide-Based Magnetic Nanoparticles: Synthesis and Biomedical Applications. *Int. J. Mol. Sci.* **2020**, *21*, 2455. [CrossRef] [PubMed]
14. Teja, A.S.; Koh, P.-Y. Synthesis, properties, and applications of magnetic iron oxide nanoparticles. *Prog. Cryst. Growth Charact. Mater.* **2009**, *55*, 22–45. [CrossRef]
15. Jiang, W.Q.; Yang, H.C.; Yang, S.Y.; Horng, H.E.; Hung, J.C.; Chen, Y.C.; Hong, C.-Y. Preparation and properties of superparamagnetic nanoparticles with narrow size distribution and biocompatible. *J. Magn. Magn. Mater.* **2004**, *283*, 210–214. [CrossRef]
16. Laurent, S.; Forge, D.; Port, M.; Roch, A.; Robic, C.; Elst, L.V.; Muller, R.N. Magnetic iron oxide nanoparticles: Synthesis, stabilization, vectorization, physicochemical characterizations, and biological applications. *Chem. Rev.* **2008**, *108*, 2064–2110. [CrossRef]
17. Nedyalkov, N.; Nakajima, Y.; Terakawa, M. Magnetic nanoparticle composed nanowires fabricated by ultrashort laser ablation in air. *Appl. Phys. Lett.* **2016**, *108*, 043107. [CrossRef]
18. Nikov, R.G.; Dikovska, A.O.; Atanasova, G.B.; Avdeev, G.V.; Nedyalkov, N.N. Magnetic field-assisted formation of oriented nanowires produced by PLD in open air. *Appl. Surf. Sci.* **2018**, *458*, 273–280. [CrossRef]
19. Nikov, R.G.; Dikovska, A.O.; Avdeev, G.V.; Amoruso, S.; Ausanio, G.; Nedyalkov, N.N. PLD fabrication of oriented nanowires in magnetic field. *Appl. Surf. Sci.* **2019**, *471*, 368–374. [CrossRef]
20. Nikov, R.G.; Dikovska, A.O.; Avdeev, G.V.; Atanasova, G.B.; Nedyalkov, N.N. Composite magnetic and non-magnetic oxide nanostructures fabricated by a laser-based technique. *Appl. Surf. Sci.* **2021**, *549*, 149204. [CrossRef]
21. Nikov, R.G.; Dikovska, A.O.; Avdeev, G.V.; Atanasova, G.B.; Karashanova, D.B.; Amoruso, S.; Ausanio, G.; Nedyalkov, N.N. Single-step fabrication of oriented composite nanowires by pulsed laser deposition in magnetic field. *Mater. Today Commun.* **2021**, *26*, 101717. [CrossRef]
22. Winsett, J.; Moilanen, A.; Paudel, K.; Kamali, S.; Ding, K.; Cribb, W.; Seifu, D.; Neupane, S. Quantitative determination of magnetite and maghemite in iron oxide nanoparticles using Mössbauer spectroscopy. *SN Appl. Sci.* **2019**, *1*, 1636. [CrossRef]
23. Wareppam, B.; Kuzmann, E.; Garg, V.K.; Singh, L.H. Mössbauer spectroscopic investigations on iron oxides and modified nanostructures: A review. *J. Mater. Res.* **2023**, *38*, 937–957. [CrossRef]
24. McCammon, C.A.; Price, D.C. Mössbauer spectra of Fe_xO (x > 0.95). *Phys. Chem. Miner.* **1985**, *11*, 250–254. [CrossRef]
25. Kopcewicz, M.; Grabias, A.; Kuryliszyn-Kudelska, I.; Dobrowolski, W. Mössbauer Effect Study of Superparamagnetic Behavior of $ZnFe_2O_4$ Nanoparticles Formed in ZnO Doped with Fe_2O_3. *Phys. Status Solidi B* **2019**, *256*, 1800223. [CrossRef]
26. Zhang, J.; Song, J.-M.; Niu, H.-L.; Mao, C.-J.; Zhang, S.-Y.; Shen, Y.-H. $ZnFe_2O_4$ nanoparticles: Synthesis, characterization, and enhanced gas sensing property for acetone. *Sens. Act. B* **2015**, *221*, 55–62. [CrossRef]
27. Yamashita, T.; Hayes, P. Analysis of XPS spectra of Fe^{2+} and Fe^{3+} ions in oxide materials. *Appl. Surf. Sci.* **2008**, *254*, 2441–2449. [CrossRef]
28. Xu, X.; Chen, D.; Yi, Z.; Jiang, M.; Wang, L.; Zhou, Z.; Fan, X.; Wang, Y.; Hui, D. Antimicrobial Mechanism Based on H_2O_2 Generation at Oxygen Vacancies in ZnO Crystals. *Langmuir* **2013**, *29*, 5573–5580. [CrossRef]
29. Zhang, L.; Yin, Y. Large-scale synthesis of flower-like ZnO nanorods via a wet-chemical route and the defect-enhanced ethanol-sensing properties. *Sens. Act. B* **2013**, *183*, 110–116. [CrossRef]
30. Kolasinski, K.W.; Gupta, M.C.; Zhigilei, L.V. Plume and Nanoparticle Formation During Laser Ablation. In *Encyclopedia of Interfacial Chemistry: Surface Science and Electrochemistry*; Wandelt, K., Ed.; Elsevier: Oxford, UK, 2018; Volume 2, pp. 594–603. [CrossRef]
31. Cobos, M.A.; de la Presa, P.; Llorente, I.; García-Escorial, A.; Hernando, A.; Jimenez, J.A. Effect of preparation methods on magnetic properties of stoichiometric zinc ferrite. *J. Alloys Compd.* **2020**, *849*, 156353. [CrossRef]
32. Schwaminger, S.P.; Bauer, D.; Fraga-García, P.; Wagner, F.E.; Berensmeier, S. Oxidation of magnetite nanoparticles: Impact on surface and crystal properties. *Cryst. Eng. Comm.* **2017**, *19*, 246–255. [CrossRef]
33. Lu, Y.; Yang, C.; Wang, H.; Ma, L.; Xu, M.; Xi, L. Structure, principle, and application of magnetic field-assisted pulsed laser deposition: An overview. *Vacuum* **2023**, *211*, 111912. [CrossRef]
34. Huotari, J.; Kekkonen, V.; Haapalainen, T.; Leidinger, M.; Sauerwald, T.; Puustinen, J.; Liimatainen, J.; Lappalainen, J. Pulsed laser deposition of metal oxide nanostructures for highly sensitive gas sensor applications. *Sens. Act B* **2016**, *236*, 978–987. [CrossRef]

materials

MDPI

Article

Effect of B_2O_3 on the Structure, Properties and Antibacterial Abilities of Sol-Gel-Derived $TiO_2/TeO_2/B_2O_3$ Powders

Albena Bachvarova-Nedelcheva [1,*], Reni Iordanova [1], Angelina Stoyanova [2], Nelly Georgieva [3], Veronica Nemska [3] and Tsvetelina Foteva [3]

[1] Institute of General and Inorganic Chemistry, Bulgarian Academy of Sciences, Acad. G. Bonchev Str., bld. 11, 1113 Sofia, Bulgaria; reni@svr.igic.bas.bg

[2] Department Chemistry and Biochemistry, Faculty of Pharmacy, Medical University—Pleven, Kl. Ohridski Str., 1, 5800 Pleven, Bulgaria; angelina.stoyanova@mu-pleven.bg

[3] Department Biotechnology, Faculty of Chemical and Systems Engineering, University of Chemical Technology and Metallurgy, Kl. Ohridski Blvd, 8, 1756 Sofia, Bulgaria; nelly.georgieva@yahoo.com (N.G.); vnemska@uctm.edu (V.N.); tsvetelina_angelova@uctm.edu (T.F.)

* Correspondence: albenadb@svr.igic.bas.bg

Abstract: This paper studies the influence of B_2O_3 on the structure, properties and antibacterial abilities of sol-gel-derived $TiO_2/TeO_2/B_2O_3$ powders. Titanium(IV) butoxide, telluric(VI) acid and boric acid were used as precursors. Differences were observed in the degree of decomposition of Ti butoxide in the presence of H_3BO_3 and H_6TeO_6 acids. The phase transformations of the obtained gels in the temperature range of 200–700 °C were investigated by XRD. Composite materials containing an amorphous phase and different crystalline phases (metallic Te, α-TeO_2, anatase, rutile and $TiTe_3O_8$) were prepared. Heating at 400 °C indicated a crystalline-to-amorphous-phase ratio of approximately 3:1. The scanning electron microscopy (SEM) analysis showed the preparation of plate-like TiO_2 nanoparticles. The IR results showed that the short-range order of the amorphous phases that are part of the composite materials consists of TiO_6, BO_3, BO_4 and TeO_4 structural units. Free B_2O_3 was not detected in the investigated compositions which could be related to the better connectivity between the building units as compared to binary TiO_2/B_2O_3 compositions. The UV-Vis spectra of the investigated gels exhibited a red shift of the cut-off due to the presence of boron and tellurium units. The binary sample achieved the maximum photodegradation efficiency (94%) toward Malachite green dye under UV irradiation, whereas the ternary sample photoactivity was very low. The compositions exhibited promising antibacterial activity against *E. coli* NBIMCC K12 407.

Keywords: sol-gel; IR; UV-Vis spectra; photocatalysis; antibacterial properties

Citation: Bachvarova-Nedelcheva, A.; Iordanova, R.; Stoyanova, A.; Georgieva, N.; Nemska, V.; Foteva, T. Effect of B_2O_3 on the Structure, Properties and Antibacterial Abilities of Sol-Gel-Derived $TiO_2/TeO_2/B_2O_3$ Powders. *Materials* **2023**, *16*, 6400. https://doi.org/10.3390/ma16196400

Academic Editor: Serena Esposito

Received: 2 September 2023
Revised: 18 September 2023
Accepted: 23 September 2023
Published: 25 September 2023

1. Introduction

Boron oxide (B_2O_3) is a well-known classical glass former, and the sol-gel method is not very often used for the synthesis of amorphous borate materials [1]. Within the last two decades, there has been increased interest in B_2O_3-containing compositions due to their unique properties and great promise for engineering applications [2–4]. For example, borate glasses are less chemically durable due to their bonding and structure and are therefore more bioactive. It is well known that numerous investigations focused mainly on the sol-gel processes for silica-based compositions have been published. Up to now, only a limited number of studies on borate glasses created through the sol-gel process [4] have appeared. Recently, borate compositions obtained by using the sol-gel method have been intensively studied due to their promising bioactive properties [4].

The first gel-derived borate materials are reported in the systems: B_2O_3-Li_2O [5] and B_2O_3-BaO [6]. Borosilicate glasses obtained by using the sol-gel method with a more homogeneous network compared to the traditional melting route were reported [7–11].

Some authors found that the short-range order of the alkaline borate gels and multicomponent borosilicate gels closely resembles that of the bulk melt-quenching glasses [12,13]. The positive role of boron ions was proven in the bioactivity of the sol-gel prepared $SiO_2/P_2O_5/CaO/B_2O_3$ samples [14]. The multi-component hybrid materials have been fabricated, and it is shown that the nature of the organic group plays a major role in the design of new sol-gel materials [15]. The as-prepared by sol-gel technology $B_2O_3/Na_2O/TiO_2$ gel region corresponds fairly well to the twin-roller glass region. Up to now, the connectivity between borate and titanate units is questionable, and as it is known, boro-titanate glasses are not obtained from melts. The incorporation of boron into the TiO_2 bulk extends its visible light absorption [16,17]. The photoactivity of TiO_2-B_2O_3 and B_2O_3-SiO_2/TiO_2 catalysts was improved by the boron content [18,19].

As an intermediate oxide, TiO_2 does not form glass by conventional quenching but successfully is used to produce amorphous and crystalline materials via the sol-gel process [20,21]. It was found that the addition of TiO_2 to TeO_2 allows for enhancing the glass-forming ability and obtaining homogeneous glasses [22,23]. Moreover, it has been found that the positive role of TiO_2 is to preserve the glass network of pure TeO_2 which contributes to the optical properties of these materials [24]. On the other hand, sol-gel-derived titania products are extensively used as catalysts, supports, protective coatings, self-cleaning surfaces and as an active ingredient of sunscreen cosmetic products [20].

It is well known that tellurite glasses possess a low melting point and the absence of hygroscopic properties [25,26]. Thus, it is useful to search for new reproducible routes for synthesis in order to extend their applications. The sol-gel technique is a low-temperature alternative that has been developed and is suitable for the investigation of new tellurite compositions.

Sol-gel obtaining of different binary, ternary and multicomponent composite powders containing TiO_2 is one of the main tasks developed by our team. Up to now, we have determined the gel formation regions in several binary and ternary systems [27–31]. Investigations on the structure, thermal, optical, photocatalytic and antibacterial properties of selected compositions have been performed [32–35]. Moreover, the obtained results on the photocatalytic and antibacterial properties of some nanosized powders were promising for their environmental applications [32–35]. The present study extends our investigations on the sol-gel obtaining of TiO_2 nanocomposite powders by applying Te(VI) acid and Ti(IV) n-butoxide as a combination of precursors. The novelty of this work was studying the influence of B_2O_3 on the structure, properties and antibacterial abilities of sol-gel-derived $TiO_2/TeO_2/B_2O_3$ powders as well as verifying their biocidal and environmental applications. The lack of information in the literature concerning this problem additionally emphasizes its recency.

2. Materials and Methods

2.1. *Gelling and Drying*

Based on our previous findings on the gel formation in the ternary TiO_2-TeO_2-B_2O_3 system, compositions containing higher TiO_2 content were selected—$50TiO_2.25TeO_2.25B_2O_3$, $80TiO_2.10TeO_2.10B_2O_3$ and $80TiO_2.20B_2O_3$, denoted as samples A, B and C, respectively. They were subjected to detailed investigations. The gels were prepared using a combination of Te(VI) acid (Aldrich, St. Louis, MO, USA) along with Ti butoxide (Fluka AG, Everett, WA, USA) and H_3BO_3 (Merck, Rahway, NJ, USA) as precursors dissolved in ethylene glycol ($C_2H_6O_2$) (99% Aldrich). Telluric acid (H_6TeO_6) was selected to overcome the problem of the high hydrolysis rate of tellurium(VI) alkoxide that has been analyzed in several papers [36,37]. The precursor solutions were subjected to 5–10 min intensive stirring at room temperature to achieve complete dissolution. No direct addition of water was made to the precursor solutions. Sol-gel hydrolysis reaction was acquired from absorbed atmospheric moisture. The measured pH was 4–5 depending on composition. The gelation time for the investigated compositions occurred immediately. For completing the hydrolysis, the aging of gels was performed in air for several days. The obtained gels were subjected to stepwise heating from 200 to 700 °C for one hour of exposure time in the air. Aiming to verify the phase and structural transformations of the gels, the heating in the range of 200–700 °C until obtaining powders was performed. The selection of the temperatures was made based on our previous investigations [27–31].

2.2. Sample Characterization

Powder XRD patterns were registered at room temperature with a Bruker D8 Advance (Berlin, Germany) X-ray powder diffractometer with a Cu Ka radiation (k = 1.54056 Å) and a LynxEye solid position sensitive detector with X-ray tube operated at 40 kV and 40 mA. X-ray diffraction patterns were recorded in the range of 5.3–80° 2 h with a step of 0.02° 2 h. The differential thermal analysis (LABSYSTM EVO apparatus, Setaram, Lyon, France) with Pt-Pt/Rh thermocouple at a heating rate of 10 K/min in air flow and Al_2O_3 as a reference material was used to study the decomposition process of the gels. The accuracy of the temperature was ±5 °C and the heating of the samples was limited up to 700 °C. Gases evolved (EGA) during the thermal treatments were analyzed by mass spectrometry (MS) with a Pfeiffer OmniStarTM mass spectrometer (Pfeiffer Vacuum Technology AG, Wetzlar, Germany). Mass spectra recorded for investigated samples show the *m/z* = 14, 18 and 44 signals, attributed to CH_2, H_2O and CO_2, respectively. The infrared spectra were made in the range of 1600–400 cm^{-1} using the KBr pellet technique on a Nicolet-320 FTIR spectrometer (Madison, WI, USA) with 64 scans and a resolution of ±1 cm^{-1}. The UV–VIS diffused reflectance Spectrophotometer Evolution 300 (Thermo Electron Corporation, Madison, WI, USA) with a magnesium oxide reflectance standard as the baseline was used for recording the optical absorption spectra of the powdered samples in the wavelength range 200–800 nm. Planck's equation was applied for the absorption edge and optical bandgap (Eg) determination [31–33]. The as-obtained samples were imaged by using a scanning electron microscope (SEM) JSM-5510 (JEOL, Akishima shi, Japan), operated at 10 kV of acceleration voltage. The investigated samples were coated with gold by using JFC-1200 fine coater (JEOL) before observation. The specific surface areas (BETs) were determined by low-temperature (77.4 K) nitrogen adsorption in NOVA (Osaka, Japan) 1200e surface area & pore analyzer at relative pressures p/p0 = 0.1–0.3 using the BET equation.

2.2.1. Photocatalysis

The photocatalytic activity of the investigated pure and composite titania powders was evaluated by UV-light induced photobleaching of Malachite green (MG) dye solution, used as a representative dye pollutant. The MG working solutions of concentration 5 ppm were prepared by serial dilution of 0.1 mol/L stock solution of the dye with bi-distilled water. The procedure was as follows: 100 mg sample was immersed in 150 mL dye solution. The resulting suspension was stirred in the dark for 30 min to obtain MG adsorption equilibrium before irradiation. A black-light blue UV lamp (Sylvania BLB 50 Hz 8W T5) was used as a source of UV light with a main emission wavelength of 365 nm. The lamp was positioned at a distance of 10 cm from the solution surface. All photocatalytic tests were performed at room temperature and constant stirring (450 rpm). At certain time intervals of illumination, 3 mL of the suspensions was withdrawn and centrifuged at 5000 rpm for 10 min to separate the supernatant (containing MG dye with a certain concentration) from the solid particles. Then, the concentration changes of MG dye were monitored by measuring the absorbance of clear aliquots using a Jenway 6505 UV-Vis spectrophotometer (designed and manufactured in Chelmsford, Essex, CM1 3UP, Chelmsford, UK) at 618 nm, the maximum absorption wavelength for MG. As stated in Beer's law, the absorbance at 618 nm increases linearly with increasing concentration of MG, and the degradation ratio of the dye, C/C_o, is represented by the ratio A/A_o, where A_o is the absorbance of MG dye solution at time zero, and A is the solution absorbance at a time t. The photocatalytic properties of the composite sample powders were compared with those of a pure TiO_2 synthesized from Ti(IV) butoxide.

2.2.2. Antibacterial Properties

Test Microorganisms, Media and Culture Conditions

The bacterial strain Escherichia coli NBIMCC K12 407 was supplied by the National Bank for Industrial Microorganisms and Cell Cultures (NBIMCC, Sofia, Bulgaria) and was cultured in Luria-Bertani (LB) broth in a shaker-incubator ES-20/60 (Biosan, Riga, Latvia, 120 rpm) at 37 °C for 24 h.

Bactericidal Effect of the Investigated Materials

The potential ability of materials to exhibit growth-inhibiting effects on *E. coli* K12 407 was examined using an agar-well diffusion test. This method consists of the formation of a clear bacterial zone around the materials which are in the form of powder. For this purpose, 100 µL of bacterial strains with a concentration of 1×10^7 CFU/mL, corresponding to 0.5 McFarland, was spread in LB agar plates. After 30 min, the culture was absorbed into the agar plates, and onto the marked wells was placed 100 mg of the investigated materials. The petri dishes were incubated at 37 °C for 24 h. The bactericidal effect of the samples was evaluated by measuring the size of the formed inhibition zones [38]. Three replicates were made from each sample, and the results show the mean values.

In addition, the percent of the bacteria cell reduction by tested materials was calculated. For this aim, 100 µL bacterial suspension with a turbidity of 0.5 McFarland was poured in flasks with 100 mL LB broth and 10 mg of each material. For control, a flask, which contained only bacteria and LB broth, was used. The incubation conditions were shaking at 120 rpm and 37 °C for 24 h. On the next day, 100 µL diluted suspension of each flask was spread on agar plates, and after incubation, the grown colonies were counted, and the percent reduction was calculated according to Bachvarova-Nedelcheva et al. [34]. All tests were performed in triplicates, and the results obtained showed the mean values.

3. Results and Discussion

3.1. Phase Formation Studies and Thermal Stability of the Gels

The investigated compositions were situated in the gel formation region in the TiO_2-TeO_2-B_2O_3 ternary system, which has been investigated previously and published elsewhere [28]. All gels were transparent. The X-ray diffraction patterns of all heat-treated gels are shown in Figure 1a–c. Although samples A and B contained different TiO_2 amounts, their behavior was very similar up to 400 °C. As can be seen, after heat treatment at lower temperatures (200–300 °C), predominantly an amorphous phase and formation of metallic tellurium only (JCPDS 78-2312) were registered. The amorphous phase amount gradually decreased with increasing the temperature. Tellurium was fully oxidized to TeO_2 at about 500 °C. First TiO_2 (anatase) crystals (JCPDS 78-2486) for samples A and B appeared at 400 °C. At this temperature, the crystalline-to-amorphous-phase ratio was approximately 3:1. A small amount was converted to rutile (JCPDS 21-1276) in sample B (containing 80 mol% TiO_2) at 500 °C, and in the other ternary sample A, rutile was registered at 600 °C. A new crystalline-phase $TiTe_3O_8$ (JCPDS 50-0250) was found at 500 °C for sample A (50 mol% TiO_2), while in sample B, it appeared at 600 °C. At 500 °C, the amount of the amorphous phase strongly decreased, and it was below 10%. The heat treatment at 400 and 500 °C revealed several crystalline phases which were detected and quantified (Figure 1d,e). Additional heating was performed at 700 °C to complete the phase formation. The coexistence of TiO_2 (rutile) and $TiTe_3O_8$ was observed in the prepared powders. The behavior of the binary sample C ($80TiO_2.20B_2O_3$) was different. The amorphous state was preserved up to 300 °C, and further heating of the sample showed the presence of the TiO_2 (anatase) crystalline phase only. The first TiO_2 (rutile) crystals appeared at 700 °C. It has to be noted that in all investigated compositions, H_3BO_3 was not detected. At 400 °C, the average crystallite size (calculated using Sherrer's equation) of TiO_2 (anatase) in the powdered sample C was about 10 nm (Figure 1). The obtained results on the phase formation are similar to those reported by other authors [39,40].

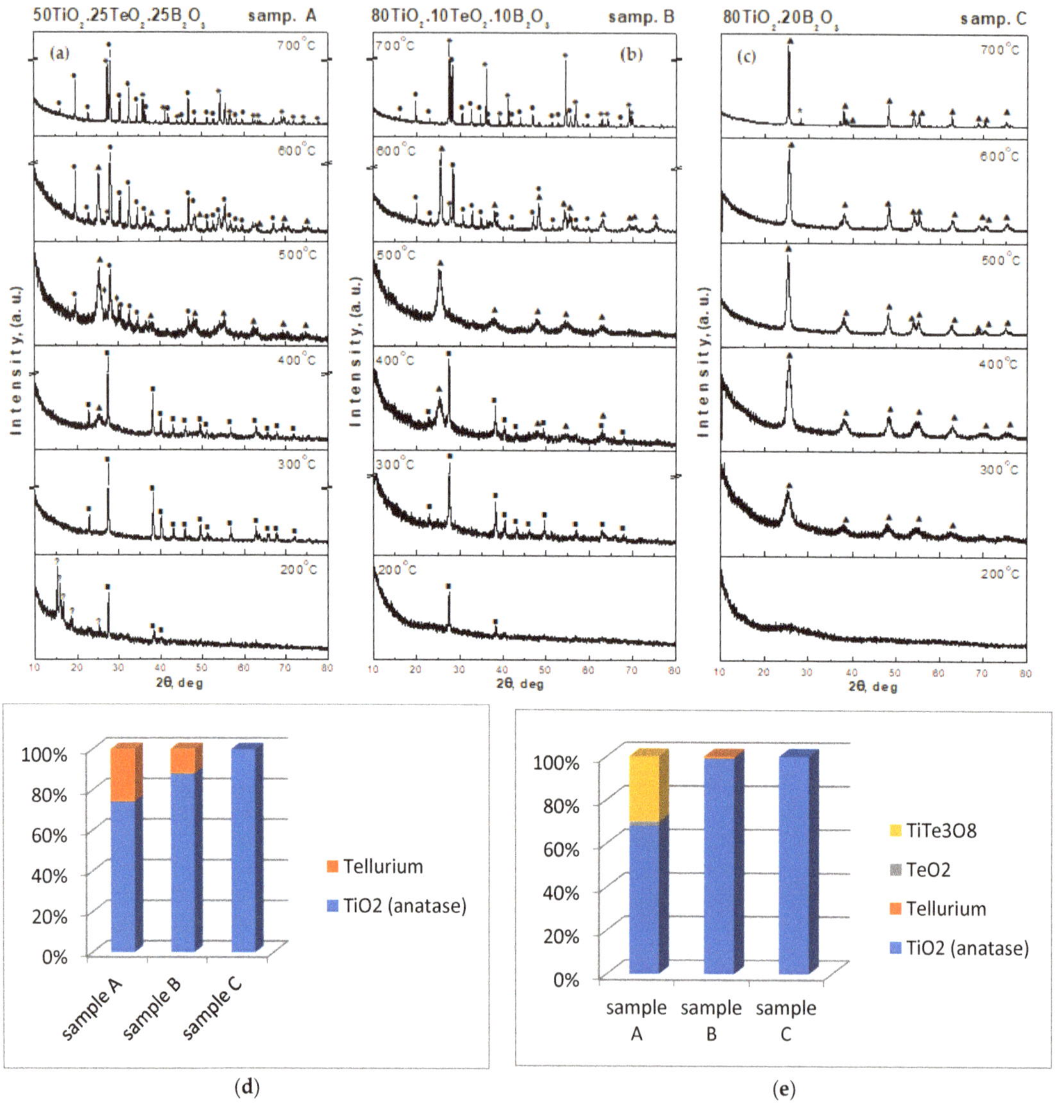

Figure 1. XRD patterns of the investigated samples heat-treated at different temperatures: (■) Te, (♦) α-TeO_2, (▲) TiO_2-anatase, (*) TiO_2-rutile, and (•) $TiTe_3O_8$ (**a**–**c**). Crystalline phase weight content of the studied samples at 400 (**d**) and 500 °C (**e**).

The thermal decomposition data in the air of the investigated gels are illustrated in Figure 2a–f. Several decomposition steps were observed in the DTA/TG curves of the samples. For the interpretation of these results, DTA data for pure Ti(IV) n-butoxide discussed in the literature [41,42] were taken into consideration. The DTA curves of all gels were characterized by an endothermic peak in the range of 100–200 °C while the binary composition (sample C, Figure 2e) exhibited an additional endothermic effect of about 80 °C. The weight loss was between 30 and 37% depending on composition, and it was related to the elimination of water bonded with the material surface as well as to the evaporation of the organic solvent. In the temperature range of 200–600 °C, several exothermic effects could be distinguished. The first one was detected at about 290–300 °C for all gels, but it was observed that this peak was sharper and more intensive for the binary sample. This could be related to the intense decomposition mainly of Ti(IV) butoxide. In all cases, this peak was accompanied by a weight loss of about 10–15% which according to the TG curve finished at nearly 450 °C. The last exothermic effect, which was stronger for the ternary compositions, was observed in the range of 530–600 °C. It could be attributed to the crystallization of TiO_2 (anatase) or other crystalline phases. That peak was accompanied by a very low weight loss (~1%), and the calculated total weight loss for the investigated gels was about 45%.

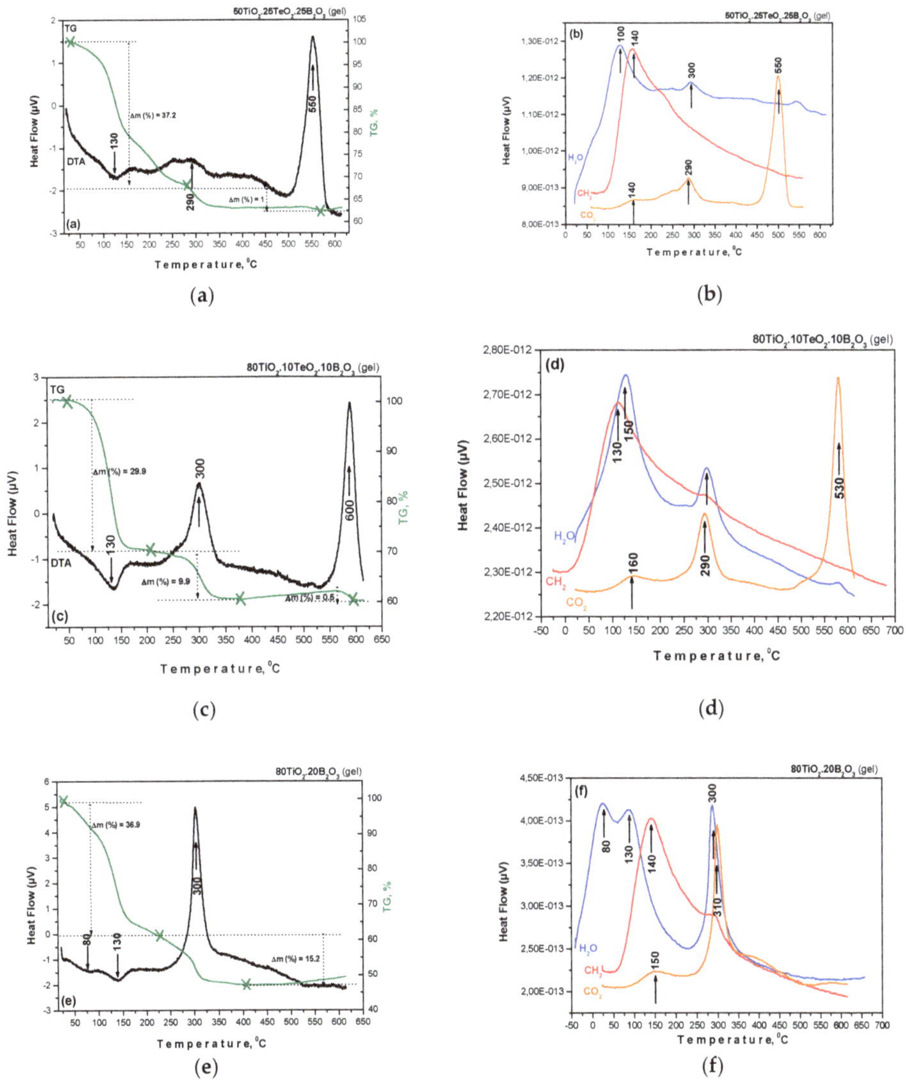

(a) (b)

(c) (d)

(e) (f)

Figure 2. DTA/TG curves (**a**,**c**,**e**) and mass spectra (**b**,**d**,**f**) of the investigated gel compositions.

3.2. SEM Morphology

The morphology of the powders was examined by SEM. Figure 3 shows the micrographs at different magnifications of the powders calcined at 400 °C. As seen from the images, the morphology was a result of the crashing of the monolithic gels during the heating. Most of the particles exhibited plate-like or irregular shapes. A strong trend of agglomeration was also observed, and the average size of the aggregates was above 10 μm. Additional confirmation for the agglomeration of the samples was the BET measurements which showed that the specific surface area of the samples was about 35–45 m^2/g. The microprobe analysis showed no presence of carbon in all samples. Thus, the organic constituents were completely removed at these temperatures.

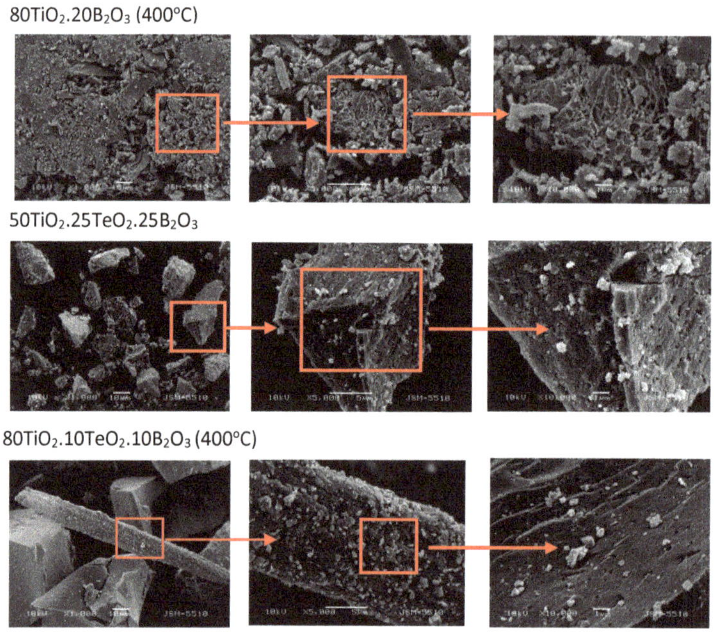

Figure 3. SEM images at different magnifications of the investigated samples heat-treated at 400 °C.

Our results confirmed that the particles obtained by using the sol-gel method exhibited strong agglomeration and less porous structure due to the lack of surfactant used in other synthesis methods which usually helps to reduce agglomeration and the size of the nanoparticles [43].

3.3. Structural Change IR and UV-Vis Spectra

3.3.1. IR Investigations

The IR spectra of the investigated composition system heated at different temperatures are shown in Figure 4. The assignments of the vibration bands were made based on spectral data for the precursors, crystalline phases existing in the system and our previous structural investigations on gels in different binary and ternary TiO_2-containing systems [27–35]. As seen in the figure, at low temperatures (up to 200 °C), all gels exhibited similar behavior in the spectral range of 1600–400 cm^{-1}. Looking at the spectra, several regions can be distinguished: 1500–1200 cm^{-1}, 1200–900 cm^{-1} and 900–400 cm^{-1}. By analogy with our previous paper, the IR bands located between 1500 and 1300 cm^{-1} could be assigned to the bending vibrations of CH_3 and CH_2 groups [41,44]. It is known that bands in the range of 1130–1040 cm^{-1} are related to the characteristic Ti–O–C stretching vibrations that can be used to determine the degree of hydrolysis [41,44,45]. The presence of unhydrolyzed organic groups (bands at 1120, 1080 and 1040 cm^{-1}) and OH groups in the IR spectra of all compositions contributed to the formation of a mixed organic–inorganic amorphous structure. The intensity of absorption bands in the 1120–1040 cm^{-1} region (Ti–O–C stretching vibrations) decreased in the following order: sample A > sample B > sample C. The lowest intensity of these bands was in the spectrum of $80TiO_2.20B_2O_3$ (sample C) as compared to the ternary compositions which indicates that borate units are incorporated in the titanate network and lead to a greater degree of hydrolysis at 200 °C. Obviously, the decrease in the TiO_2 content hinders the hydrolysis–condensation processes as was observed in the spectrum of sample A.

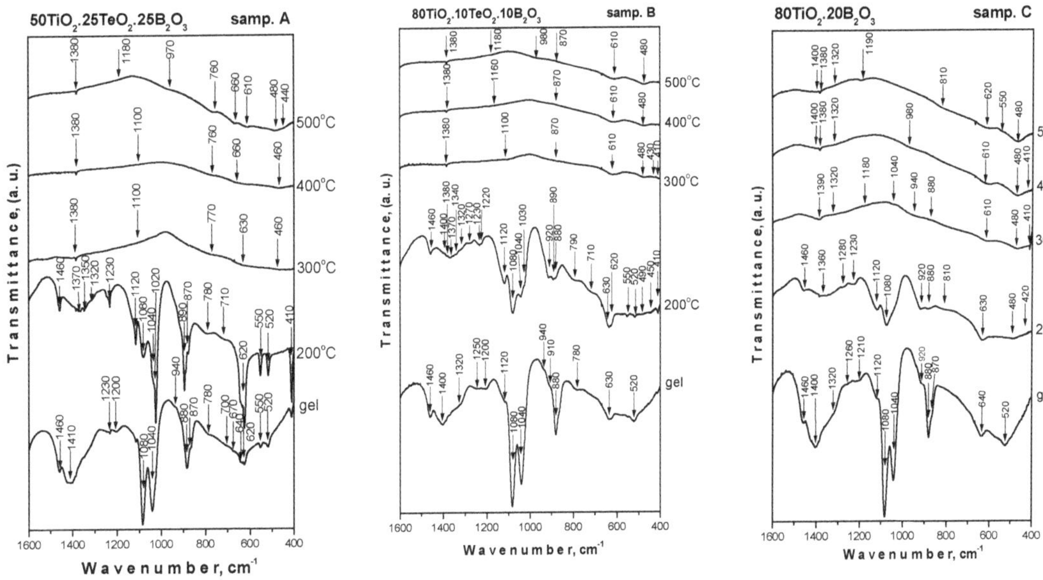

(a) (b) (c)

Figure 4. IR spectra of investigated samples heat-treated at different temperatures. (**a**,**b**) samples A and B; (**c**) sample C.

It has to be noted that in all the gels studied, the TiO_2 content was higher than that of B_2O_3 and TeO_2 content. Therefore, the vibrational modes observed in the range below 700 cm^{-1} were mainly due to the titanate network. The characteristic vibrations of TiO_6 structural units in TiO_2 (anatase) were at 640 and 450 cm^{-1} [20,45,46]. The bands in the range of 700–670 cm^{-1} observed in the spectra of samples A and B could be related to the symmetric TeO_3 units [47], while deformed TeO_4 groups exhibited a strong absorption band at 635 cm^{-1} and a shoulder at 670 cm^{-1}. On the other hand, in the same absorption range (700–600 cm^{-1}), the vibrations of symmetric TeO_6 units from H_6TeO_6 acid were situated [48,49], while bending vibrations of Te–OH were at 1220 and 1130 cm^{-1} [49]. According to the literature data, vibrational modes of different borate units in binary glasses are 1500–1200 cm^{-1} (B–O stretching for trigonal BO_3 units), 1200–850 cm^{-1} (B–O stretching for tetrahedral BO_4 units) and 800–600 cm^{-1} (bending vibrations for various borate segments) [50–52]. The strong absorption bands at 1480 and 1200 cm^{-1} could be assigned to the presence of B–OH bending vibrations [53,54]. Looking at the spectra, it could be seen that the bands above 1000 cm^{-1}, characteristic of the organic groups, were not visible above 300 °C (Figure 4). This could be explained by the transformation of the amorphous phase from organic to inorganic which started at 300 °C. The spectra in the temperature range of 300–500 °C were characterized mainly by bands below 900 cm^{-1}, typical for the inorganic units. These bands are with low intensity and broadened, which is a peculiarity of the disordered systems. In both ternary compositions, the weak bands at 770–760 cm^{-1}, 660 cm^{-1} and 630–610 cm^{-1} can be assigned with the vibrations of TeO_4 units [24]. By analogy with our previous IR investigations, bands in the region 700–400 cm^{-1} are characteristic of the formation of a Ti–O–Ti network [28,55–57]. The overall temperature range (300–500 °C) with the IR spectrum of all samples exhibited absorption bands centered at 1380 and ~1200 cm^{-1} due to the vibrations of BO_3 units [58–60]. The weak bands about 1100–980 cm^{-1} and 880 cm^{-1} could be associated with the vibrations of B–O bonds in BO_4 groups. Based on these results, it could be summarized that at higher temperatures, both tellurium and titanium lead to partial $BO_3 \rightarrow BO_4$ transformations in the investigated ternary compositions (Figure 4). The results obtained correlate well with the XRD data already discussed above as well as with our preceding investigations on the sol-gel obtaining of TiO_2-containing compositions.

3.3.2. UV-Vis Investigations

UV-Vis spectroscopy was used as a tool to obtain useful information for the completeness of the hydrolysis–condensation reactions as well as to accumulate additional structural data for the investigated powders. The UV-Vis spectra of the prepared gels are depicted in Figure 5, and they are compared to those of Ti(IV) butoxide gel as well as pure TiO_2. As seen from the figure, all spectra in the wavelength region from 200 to 800 nm are characterized by two bands 240–260 nm and 310–330 nm, and their UV-Vis spectra are compared to those of TiO_2 obtained by Ti(IV) butoxide. As has been already discussed in our previous papers, the main building units in the unhydrolyzed Ti butoxide are isolated TiO_4 groups with absorption bands in the region 240–260 nm [27,31] due to the ligand-to-metal charge transfer. The polymerization processes transformed the TiO_4 groups into TiO_6 units possessing absorption at higher wavelengths as the charge transfer in TiO_6 groups is about 320–340 nm [27,28,61].

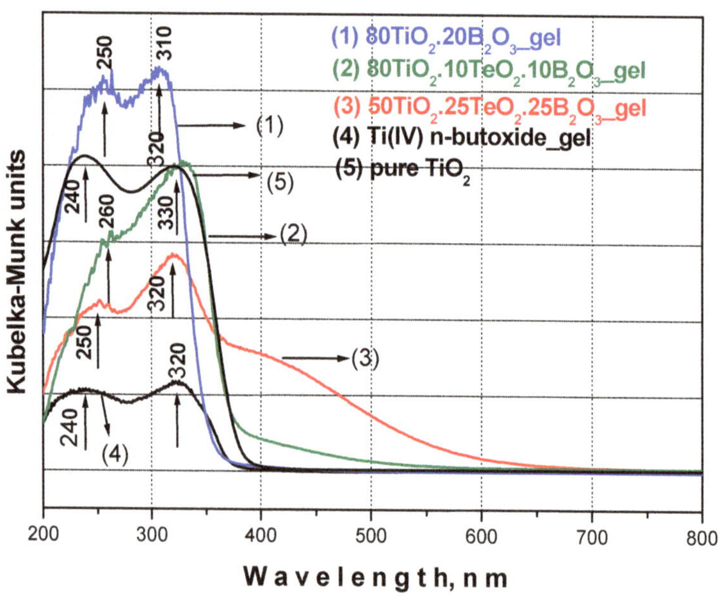

Figure 5. UV-Vis spectra of investigated samples compared with Ti(IV) butoxide gel and pure TiO_2.

Bearing in mind the observed UV-Vis data, it could be generalized that all investigated samples exhibited a stronger absorption band at 330–310 nm, which could be related to the higher number of TiO_6 structural units and consequently more completed hydrolysis–condensation reactions. The other peculiarity is the red shifting toward higher wavelengths. As seen from the figure, sample B ($80TiO_2.10TeO_2.10B_2O_3$) exhibited the highest value of the cut-off (431.09 nm). Based on these results, it could be suggested that sample B possessed good antibacterial and/or photocatalytic properties, which will be discussed below. A stronger absorption above 400 nm was observed only in the spectra of sample A (Figure 5). By analogy with our previous paper [27], this phenomenon could be related to carbon which is responsible for visible light absorption. This is indirect proof for the slow processed hydrolysis–condensation processes in this ternary gel which correspond well to those obtained by IR spectroscopy results. The optical gap values (Table 1) were evaluated from the intercept of the linear portion of each curve hν in the X-axis. The obtained values for the optical band gap of the investigated samples varied depending on composition, and it was affected by the B_2O_3 content. It could be seen that sample A (25 mol% B_2O_3) exhibited a higher Eg value, which confirmed previous investigations obtained by other authors [62,63]. According to the literature data [62–64], the influence of B_2O_3 on the band gap value of TiO_2 is very contradictory. Some authors found that when the boron content is below 10%, it causes a blue shift [18] while other scientific investigations summarized that even the addition of 5% B_2O_3 moved the TiO_2 optical band gap to higher energies [62,63]. Therefore, it could be generalized that more experiments must be performed to clarify the role of boron in TiO_2 optical properties.

Table 1. Cut-off and optical band gap values (Eg) of the investigated gel compositions.

Composition, mol%	Cut-off, nm	Eg, eV
Ti(IV) n-butoxide	389.71	3.18
TiO_2	382.73	3.23
$80TiO_2.10TeO_2.10B_2O_3$ (sample B)	431.09	2.88
$80TiO_2.20B_2O_3$ (sample C)	385.86	3.21
$50TiO_2.25TeO_2.25B_2O_3$ (sample A)	358.19	3.46

3.4. *UV Photocatalytic Properties*

The photocatalytic reaction was performed by using a model aqueous solution of Malachite green (MG) organic dye under UV irradiation (Figure 6a,b). It has to be noted that only heated at 500 °C samples were photocatalytically tested, and the obtained results were compared with those of pure TiO_2 synthesized from Ti(IV) butoxide. As Figure 6a shows, the concentration of MG was reduced with the irradiation time to different degrees depending on the sample composition. Under UV irradiation for 150 min, the maximum degradation efficiency was achieved by the binary sample C (94%), greater than that of pure TiO_2 (60%) and the ternary samples A (11%) and B (9%). The rate constants for the photodegradation of MG were calculated by the slopes of logarithmic plots lnA/Ao (Figure 6b). The highest rate constant (0.0174 min^{-1}) was obtained for the sample $80TiO_2/20B_2O_3$ (C) followed by that of pure TiO_2 (0.0065 min^{-1}). Rate constants for the photodegradation of the ternary samples A ($50TiO_2.25TeO_2.25B_2O_3$) and B ($80TiO_2.10TeO_2.10B_2O_3$) were very low and similar (up to 0.0009 min^{-1}).

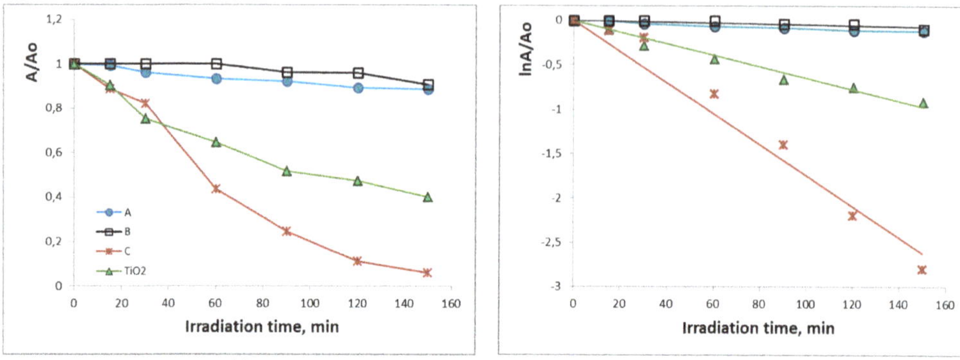

(**a**) (**b**)

Figure 6. Photocatalytic activity of the synthesized samples heated at 500 °C for the degradation of MG under UV illumination: (**a**) change in absorbance as a function of irradiation time; (**b**) plot of lnA/Ao as a function of irradiation time.

Apparently, the highest MG decoloration was reached by the binary $80TiO_2.20B_2O_3$ sample which was higher than that of synthesized pure TiO_2. Both ternary samples exhibited poorer but similar photocatalytic activity compared to the binary one. Up to now, the literature has summarized that the presence of boron promotes the formation of oxygen vacancies and photocatalytic properties [63,65,66]. To the best of our knowledge, the photocatalytic properties of $TiO_2/TeO_2/B_2O_3$ compositions have not been studied so far. Obviously, the addition of TeO_2 did not improve the photocatalytic activity of the synthesized ternary samples. Bearing in mind that the photocatalytic activity depends on many factors [67], one of the possible reasons for the unsatisfying results could be the lower values of the surface area of the investigated samples. It is well known that TiO_2 in the anatase phase is considered to show higher photocatalytic activity than TiO_2 in an amorphous structure [67,68]. On the other hand, the lowest exhibited photocatalytic degradation efficiency of compositions A and B could be related to the XRD results and the presence of rutile as well as other crystalline phases (Figure 1). These results revealed that both the crystalline structure and surface area played a significant role in determining the photocatalytic efficiency of the synthesized powdered materials.

3.5. Antibacterial Properties

The antibacterial ability of the investigated samples calcined at 200 and 400 °C was tested against *E. coli* K12 NBIMCC 407 as a Gram-negative bacterium. The measurement of inhibition zones formed around the materials was the first method for the evaluation of the bactericidal effect of the investigated materials. All photographs of these investigations are given in Figure 7. It was observed that sample A ($50TiO_2.25TeO_2.25B_2O_3$) (Figure 7a,b) showed free growth zones against *E. coli* K12 407 with sizes 18.5 ± 0.25 mm and 36 ± 0.32 mm, at 200 and 400 °C, respectively. Obviously, the powder heat-treated at 400 °C (Figure 7b) inhibited to a great extent the studied strain compared to the sample heat-treated at 200 °C (Figure 7a). In comparison to these results, the other investigated sample B (Figure 7c,d) demonstrated relatively equal size zones for both temperatures, namely 17 ± 0.21 mm and 20 ± 0.27 mm. In this case, the processing temperature of the materials also affects the degree of inhibition of *E. coli* K12 407. The strain showed higher susceptibility to materials prepared at 400 °C (Figure 7d) and lower susceptibility to materials prepared at 200 °C (Figure 7c). In Figure 7e,f can be seen that the last sample C ($80TiO_2.20B_2O_3$) exhibited formation of the smallest zones −11.5 ± 0.25 mm and 9.5 ± 0.19 mm. In contrast to the other samples (A and B), where the larger-size zones were observed, treated at 400 °C, this antibacterial test showed that the opposite tendency was established.

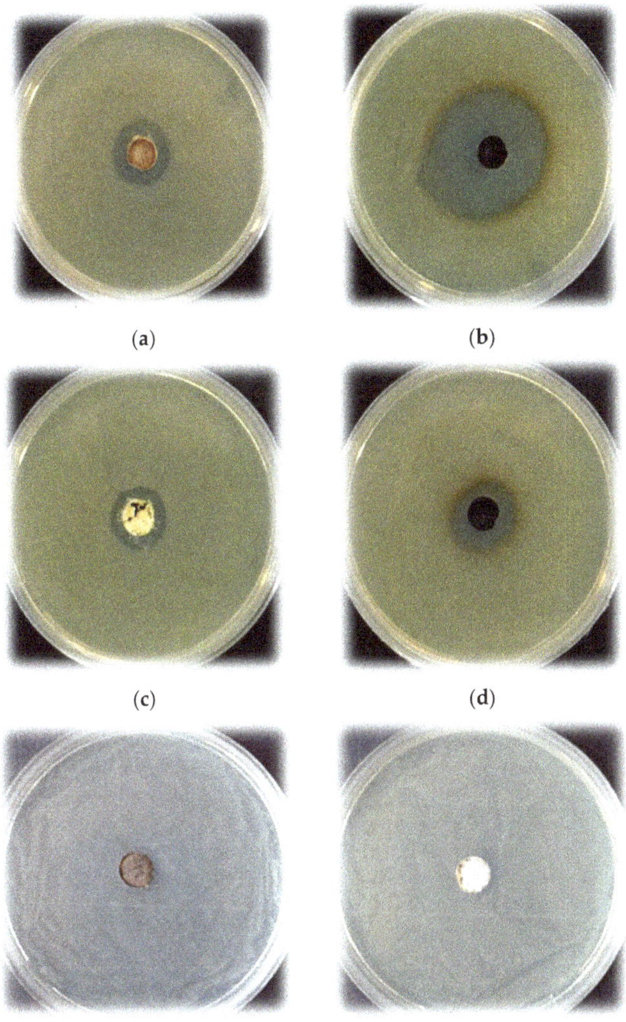

Figure 7. Antibacterial test result of the investigated samples heat-treated at 200 and 400 °C: (**a**,**b**) $50TiO_2.25TeO_2.25B_2O_3$ (sample A); (**c**,**d**) $80TiO_2.10TeO_2.10B_2O_3$ (sample B); and $80TiO_2.20B_2O_3$ (sample C) (**e**,**f**).

The bactericidal effect of the investigated materials was also studied, calculating the percent of cell reduction after exposure of the strain to the materials. Concerning the first ternary sample A heated at 200 °C and 400 °C, the obtained results revealed the growth-inhibiting effect of the strain with 54% and 72% cell reduction, respectively. These values confirmed the above-discussed results from the agar-well diffusion test, namely that higher temperature exerts a stronger inhibitory effect on the strain. Similar to the results from the first antibacterial test where sample B at both temperatures demonstrated relatively equal size zones by using this approach, the same materials inhibit the cell growth with 52% and 56%, respectively. The binary sample C least affected the strain growth. The cell viability after treatment with powder heated at 200 °C was 62% and 70% after treatment with the sample heated at 400 °C.

Generally, in the literature studies, the antibacterial activities of B_2O_3-containing compositions have great potential due to their promising biomedical applications. Moreover, within the last two decades, there has been increased interest in such compositions [4,65]. Aiming to interpret the discussed results, many factors influencing the antibacterial properties, such as specific surface area, morphology, size, etc., should be taken into consideration. Although all investigated samples possessed the average low value of the specific surface area (~35–45 m^2/g), the inhibition of samples A and C against *E. coli* bacteria was higher than that of sample B as was described above. As seen in Figure 7, sample A contained a lower TiO_2 amount (50 mol%), but it exhibited stronger antibacterial activity as compared to the other ternary sample B (80 mol% TiO_2). According to the XRD results (Figure 1) at 400 °C, the presence of an amorphous part along with TiO_2 (anatase) and metallic tellurium was detected in both samples A and B. Obviously, the higher B_2O_3 amount (25 mol%) in sample A is the reason for the increased antibacterial activity. For the binary sample C, it could be suggested that the synergistic effect of TiO_2 and B_2O_3 provided efficient antibacterial activity. By analogy with our previous studies, we suggest that the antibacterial mechanism is realized through the formation of reactive oxygen species which decompose the outer membranes of bacteria and lead to cell death. Additionally, the IR spectroscopy showed that boron exists in the amorphous part of the samples most probably forming Ti–O–B bonds. According to Wang et al. [69], when boron atoms were incorporated into the TiO_2 lattice, it enhanced the generation of the surface oxygen vacancies which improved the antimicrobial activity. The obtained results coincided with the literature studies [16,69–71]. It is well known that antimicrobial activity is an important factor in producing efficient nano-materials for sustainable applications. The obtained results give us reason to continue the investigations in this direction. Future studies will be performed aiming to improve the compositions and their efficiency for environmental applications.

4. Conclusions

Transparent gels containing B_2O_3 were obtained by applying a sol-gel method. They were calcined at different temperatures, and the results showed that above 300 °C, mixed powders of the amorphous phase, Te and TiO_2 (anatase) were obtained. The IR data showed that the presence of H_3BO_3 decreases the rate of hydrolysis. The obtained UV-Vis spectra exhibited a red shift of the cut-off due to the presence of boron and tellurium units. The results from the photocatalytic tests for the degradation of Malachite green dye under UV irradiation revealed that the maximum photoactivity was achieved by the binary sample, whereas the photoactivity of the ternary samples was similar and very low, which could be explained by the effect of the crystalline structure and the low surface area. It was established that samples A and C which contain higher B_2O_3 content (above 20 mol%), exhibited higher antibacterial activity bearing in mind the percent of cell reduction after exposure to the *E. coli* K12 strain. However, more studies are needed to clarify the antimicrobial abilities of B-containing TiO_2 nano-materials which will improve and enlarge their biomedical applications.

Author Contributions: Conceptualization, A.B.-N. and R.I.; methodology, A.B.-N., R.I., A.S. and N.G.; investigation, A.B.-N., V.N. and T.F.; writing—original draft preparation, A.B.-N., R.I., A.S. and N.G. All authors have read and agreed to the published version of the manuscript.

Funding: The authors acknowledge the project Д01-272-"European Network on Materials for Clean Technologies" for providing the opportunity to present the results at the SizeMat4 conference as well as for the financial publication support.

Institutional Review Board Statement: Not applicable.

Informed Consent Statement: Not applicable.

Data Availability Statement: Not applicable.

Acknowledgments: The authors acknowledge the project Д01-272-"European Network on Materials for Clean Technologies" for providing the opportunity to present the results at the SizeMat4 conference as well as for the financial publication support. The authors are also thankful to the Project CoE "National Center of Mechatronics and Clean Technologies" (BG05M2OP001-1.001-0008). Research equipment of the distributed research infrastructure INFRAMAT supported by the Bulgarian Ministry of Education and Science under contract D01-284/17.12.2019 was used.

Conflicts of Interest: The authors declare no conflict of interest.

References

1. Pye, L.D.; Fréchette, V.D.; Kreidl, N.J. *Borate Glasses: Structure, Properties, Applications*; Springer: New York, NY, USA, 1978.
2. Abou Neel, E.A.; Pickup, D.M.; Valappil, S.P.; Newport, R.J.; Knowles, J.C. Bioactive functional materials: A perspective on phosphate-based glasses. *J. Mater. Chem.* **2009**, *19*, 690–701. [CrossRef]
3. Balasubramanian, P.; Büttner, T.; Miguez Pacheco, V.; Boccaccini, A.R. Boron-containing bioactive glasses in bone and soft tissue engineering. *J. Eur. Ceram. Soc.* **2018**, *38*, 855–869. [CrossRef]
4. Lepry, W.C.; Nazhat, S.N. A Review of Phosphate and Borate Sol–gel Glasses for Biomedical Applications. *Adv. NanoBiomed Res.* **2021**, *1*, 2000055. [CrossRef]
5. Yamashita, H.; Yoko, T.; Sakka, T. Preparation of $Li_2B_4O_7$ thin films by Sol–gel method. *J. Mater. Sci. Lett.* **1990**, *9*, 796–798. [CrossRef]
6. Hirano, S.; Yogo, T.; Kikuta, K.; Yamagiwa, K. Preparation of β-BaB_2O_4 Powders and Thin Films by Sol–Gel Method. *J. Amer. Cer. Soc.* **1992**, *75*, 2590–2592. [CrossRef]
7. Nogami, M.; Moriya, Y. Glass formation of the SiO_2-B_2O_3 system by the gel process from metal alkoxides. *J. Non-Cryst. Solids* **1982**, *48*, 359–366. [CrossRef]
8. Tohge, N.; Matsuda, A.; Minami, T. Coating Films of $20B_2O_3.80SiO_2$ by the Sol–gel Method. *J. Am. Ceram. Soc.* **1987**, *70*, 13–15. [CrossRef]
9. Villegas, M.A.; Fernandez Navarro, J.M. Characterization of B_2O_3-SiO_2 glasses prepared via Sol–gel. *J. Mater. Sci.* **1988**, *23*, 2464–2478. [CrossRef]
10. Irwin, A.; Holmgren, J.; Jonas, J. Solid state ^{29}Si and ^{11}B NMR studies of Sol–gel derived borosilicates. *J. Non-Cryst. Solids* **1988**, *101*, 249–254. [CrossRef]
11. Abe, Y.; Gunji, T.; Kimata, K.; Kuramata, M.; Kasgoz, A.; Misono, T. Preparation of polymetalloxanes as a precursor for oxide ceramics. *J. Non-Cryst. Solids* **1990**, *121*, 21–25. [CrossRef]
12. Weinberg, M.C.; Neilson, G.F.; Smith, G.L.; Dunn, B.; Moore, G.S.; Mackenzie, J.D. The preparation and characterization of a lithium borate glass prepared by the gel technique. *J. Mater. Sci.* **1985**, *20*, 1501–1508. [CrossRef]
13. Tohge, N.; Moore, G.S.; Mackenzie, J.D. Part III. Preparation of multicomponent gels and glasses. Structural developments during the gel to glass transition. *J. Non-Cryst. Solids* **1984**, *63*, 95–103. [CrossRef]
14. Ciceo, R.L.; Trandafir, D.L.; Radu, T.; Pota, O.; Simon, V. Synthesis, characterisation and in vitro evaluation of sol–gel derived SiO_2–P_2O_5–CaO–B_2O_3 bioactive system. *Ceram. Int.* **2014**, *40*, 9517–9524. [CrossRef]
15. Gerganova, T.; Ivanova, Y.; Gerganov, T.; Miranda Salvado, I.M.; Fernandes, M.H.V. *The Sol–Gel Process: Uniformity, Polymers and Applications*; Morris, R.E., Ed.; Chapter 13; Nova Science Publishers, Inc.: Hauppauge, NY, USA, 2010.
16. Xue, X.; Wang, Y.; Yang, H. Preparation and characterization of boron-doped titania nano-materials with antibacterial activity. *Appl. Surf. Sci.* **2013**, *264*, 94–99. [CrossRef]
17. Rehman, S.; Ullah, R.; Butt, A.; Gohar, N. Strategies of making TiO_2 and ZnO visible light active. *J. Hazard. Mater.* **2009**, *170*, 560–569. [CrossRef]
18. Jung, K.; Park, S.; Ihm, S. Local structure and photocatalytic activity of B_2O_3–SiO_2/TiO_2 ternary mixed oxides prepared by sol–gel method. *Appl. Catal. B Environ.* **2004**, *51*, 239–245. [CrossRef]
19. Moon, S.C.; Mametsuka, H.; Suzuki, E.; Nakahara, Y. Characterization of titanium-boron binary oxides and their photocatalytic activity for stoichiometric decomposition of water. *Catal. Today* **1998**, *45*, 79–84. [CrossRef]
20. Sakka, S. *Handbook of Sol–Gel Science and Technology: Processing, Characterization and Applications*; Vols. I, II and III.; Kluwer Academic Publishers: Dordrecht, The Netherlands, 2004.
21. Brinker, C.J.; Scherer, G. *Sol–Gel Science: The Physics and Chemistry of Sol–Gel Processing*; Academic Press Inc.: San Diego, CA, USA, 1990.
22. Soulis, M.; Mirgorodsky, A.P.; Merle-Méjean, T.; Masson, O.; Thomas, P.; Udovic, M. The role of modifier's cation valence in structural properties of TeO_2-based glasses. *J. Non-Cryst. Solids* **2008**, *354*, 143–149. [CrossRef]
23. Udovic, M.; Thomas, P.; Mirgorodsky, A.; Durand, O.; Soulis, M.; Masson, O.; Merle-Méjean, T.; Champarnaud-Mesjard, J.C. Thermal characteristics, Raman spectra and structural properties of new tellurite glasses within the Bi_2O_3–TiO_2–TeO_2 system. *J. Solid State Chem.* **2006**, *179*, 3252–3259. [CrossRef]
24. El-Malawany, R. *Tellurite Glasses, Physical Properties and Data*; CRC Press: Boca Raton, FL, USA, 2002.

25. Umar, S.A.; Halimah, M.K.; Azlan, M.N.; Grema, L.U.; Ibrahim, G.G.; Ahmad, A.F.; Hamza, A.M.; Dihom, M.M. Structural, elastic and thermo-physical properties of Er_2O_3 nanoparticles doped bio-silicate borotellurite glasses. *SN Appl. Sci.* **2020**, *2*, 291. [CrossRef]
26. Umar, S.A.; Halimah, M.K.; Chan, K.T.; Amirah, A.A.; Azlan, M.N.; Grema, L.U.; Hamza, A.M.; Ibrahim, G.G. Optical and structural properties of rice husk silicate incorporated borotellurite glasses doped with erbium oxide nanoparticles. *J. Mater. Sci. Mater. Electron.* **2019**, *30*, 18606–18616. [CrossRef]
27. Iordanova, R.; Bachvarova-Nedelcheva, A.; Gegova, R.; Dimitriev, Y. Sol–gel synthesis of composite powders in the TiO_2-TeO_2-SeO_2 system. *J. Sol–Gel Sci. Technol.* **2016**, *79*, 12–28. [CrossRef]
28. Iordanova, R.; Gegova, R.; Bachvarova-Nedelcheva, A.; Dimitriev, Y. Sol–gel synthesis of composites in the ternary TiO_2-TeO_2-B_2O_3 system. *Phys. Chem. Glasses Eur. J. Glass Sci. Technol. B* **2015**, *56*, 128–138. [CrossRef]
29. Bachvarova-Nedelcheva, A.; Iordanova, R.; Gegova, R.; Dimitriev, Y. Sol–gel synthesis and optical characterization of TiO_2/TeO_2 powders. *Bulg. Chem. Commun.* **2016**, *48*, 5–10.
30. Bachvarova-Nedelcheva, A.; Iordanova, R.; Gegova, R.; Dimitriev, Y. Crystallization of gels in the binary TiO_2-M_nO_m (M_nO_m = TeO_2, SeO_2, B_2O_3, ZnO) systems. *Bulg. Chem. Commun.* **2017**, *49*, 110–118.
31. Bachvarova-Nedelcheva, A.; Iordanova, R.; Kostov, K.L.; Gegova, R. Sol–gel powder synthesis in the TiO_2-TeO_2-ZnO system: Structural characterization and properties. *Arab. J. Chem.* **2020**, *13*, 7132–7146. [CrossRef]
32. Bachvarova-Nedelcheva, A.; Iordanova, R.; Naydenov, A.; Stoyanova, A.; Georgieva, N.; Nemska, V.; Foteva, T. Sol–gel obtaining of TiO_2/TeO_2 nanopowders with biocidal and environmental applications. *Catalysts* **2023**, *13*, 257. [CrossRef]
33. Bachvarova-Nedelcheva, A.; Iordanova, R.; Georgieva, N.; Nemska, V.; Stoyanova, A. Photocatalytic and antibacterial assessment of Sol–gel derived TiO_2/TeO_2/ZnO powders. *J. Chem. Technol. Metall.* **2022**, *57*, 589–597.
34. Bachvarova-Nedelcheva, A.; Iordanova, R.; Stoyanova, A.; Georgieva, N.; Angelova, T. Sol–gel synthesis of Se and Te containing TiO_2 nanocomposites with photocatalytic and antibacterial properties. *J. Optoel. Adv. Mater.* **2016**, *18*, 5–9.
35. Shalaby, A.; Bachvarova-Nedelcheva, A.; Iordanova, R.; Dimitriev, Y.; Stoyanova, A.; Hitkova, H.; Ivanova, N. Sol–gel synthesis and properties of nanocomposites in the Ag/TiO_2/ZnO system. *J. Optoel. Adv. Mater.* **2015**, *17*, 248–256.
36. Lecomte, A.; Bamiere, F.; Coste, S.; Thomas, P.; Champarnaud-Mesjard, J.C. Sol–gel processing of TeO_2 thin films from citric acid stabilised tellurium isopropoxide precursor. *J. Europ. Cer. Soc.* **2007**, *27*, 1151–1158. [CrossRef]
37. Weng, L.; Hodgson, S. Sol–gel processing of tellurite materials from tellurium ethoxide precursor. *Mater. Sci. Eng.* **2001**, *87*, 77–82. [CrossRef]
38. Valgas, C.; de Souza, S.M.; Smania, E.F.A., Jr. Screening methods to determine antibacterial activity of natural products. *Braz. J. Microbiol.* **2007**, *38*, 369–380. [CrossRef]
39. Hayakawa, T.; Koyama, H.; Nogami, M.; Thomas, P. Optical properties of TeO_2-TiO_2 thin films doped with Eu^{3+} ions fabricated by Sol–gel processing. *J. Univ. Chem. Technol. Metall.* **2012**, *47*, 381–386.
40. Weng, L.; Hodgson, S. Multicomponent tellurite thin film materials with high refractive index. *Opt. Mater.* **2002**, *19*, 313–317. [CrossRef]
41. Velasco, M.J.; Rubio, F.; Rubio, J.; Oteo, J.L. FT-IR Study of the Hydrolysis and Polymerization of Tetraethyl Orthosilicate and Polydimethyl Siloxane in the Presence of Tetrabutyl Orthotitanate. *Spectr. Lett.* **1999**, *32*, 289–304. [CrossRef]
42. Yordanov, S.I.; Bachvarova-Nedelcheva, A.D.; Iordanova, R.S. Influence of ethylene glycol on the hydrolysis-condensation behavior of Ti(IV) butoxide. *Bulg. Chem. Commun.* **2017**, *49*, 265–270.
43. Dubey, R.S.; Krishnamurthy, K.V.; Singh, S. Experimental studies of TiO_2 nanoparticles synthesized by Sol–gel and solvothermal routes for DSSCs application. *Results Phys.* **2019**, *14*, 102390. [CrossRef]
44. Doeuff, S.; Henry, M.; Sanchez, C.; Livage, J. Hydrolysis of titanium alkoxides: Modification of the molecular precursor by acetic acid. *J. Non-Cryst. Solids* **1987**, *89*, 206–216. [CrossRef]
45. Uzunova-Bujnova, M.; Dimitrov, D.; Radev, D.; Bojinova, A.; Todorovsky, D. Effect of the mechanoactivation on the structure, sorption and photocatalytic properties of titanium dioxide. *Mater. Chem. Phys.* **2008**, *110*, 291–298. [CrossRef]
46. Crisan, M.; Zaharescu, M.; Crisan, D.; Ion, R.; Manolache, M. Vanadium Doped Sol–gel TiO_2 Coatings. *J. Sol–Gel Sci. Technol.* **1998**, *13*, 775–778. [CrossRef]
47. Arnaudov, M.; Dimitrov, V.; Dimitriev, Y.; Markova, L. Infrared-spectral investigation of tellurites. *Mater. Res. Bull.* **1982**, *17*, 1121–1129. [CrossRef]
48. Shashikala, M.N.; Elizabeth, S.; Cary, B.; Bhat, H.L. Raman and infrared spectroscopic studies of the new ferroelectric crystal telluric acid ammonium phosphate. *Curr. Sci.* **1987**, *56*, 861–863.
49. Nakamoto, K. *IR and Raman Spectra of Inorganic Coordination Compounds*, 3rd ed.; John and Wiley and Sons: Hoboken, NJ, USA, 1978; p. 230.
50. Kamitsos, E.; Patsis, A.P.; Karakassides, M.A.; Chryssikos, G.D. Infrared Reflectance Spectra of Lithium Borate Glasses. *J. Non-Cryst. Solids* **1990**, *126*, 52–67. [CrossRef]
51. Varsamis, C.; Kamitsos, E.I.; Chryssikos, G.D. Structure of fast-ion-conducting Ag I-doped borate glasses in bulk and thin film forms. *Phys. Rev. B* **1999**, *60*, 3885–3898. [CrossRef]
52. Kapoutsis, J.A.; Kamitsos, E.I.; Chryssikos, G.D. Synthesis and vibrational investigation of lithium magnesium metaborate glasses. *Phys. Chem. Glasses* **2000**, *41*, 321–324.

53. Soraru, G.D.; Babonneau, F.; Gervais, C.; Dallabona, N. Hybrid $RSiO_{1.5}/B_2O_3$ Gels from Modified Silicon Alkoxides and Boric Acid. *J. Sol–Gel Sci. Technol.* **2000**, *18*, 11–19. [CrossRef]
54. Hsieh, C.W.; Chiang, A.S.; Lee, C.C.; Yang, S.J. Preparation of $TiO_2B_2O_3$ coating by the sol-gel method. *J. Non-Cryst. Solids* **1992**, *144*, 53–62. [CrossRef]
55. Leustic, A.; Babonneau, F.; Livage, J. Structural investigations of the hydrolysis-condensation process ot titanium alkoxides Ti(OR)4 (OR=OPri, OEt) modified by AcAc, 1. Study of the alkoxide modification. *Chem. Mater.* **1989**, *1*, 240–247. [CrossRef]
56. Leustic, A.; Babonneau, F.; Livage, J. Structural investigations of the hydrolysis-condensation process ot titanium alkoxides Ti(OR)4 (OR=OPri, OEt) modified by AcAc, 2. From the modified precursor to the colloids. *Chem. Mater.* **1989**, *1*, 248–252. [CrossRef]
57. Siwinska-Stefanska, K.; Zdarta, J.; Paukszta, D.; Jesionowski, T. The influence of addition of a catalyst and chelating agent on the properties of titanium dioxide synthesized via the Sol–gel method. *J. Sol–Gel Sci. Technol.* **2015**, *75*, 264–278. [CrossRef]
58. Yao, Z.Y.; Möncke, D.; Kamitsos, E.I.; Houizot, P.; Celarie, F.; Rouxel, T.; Wondraczek, L. Structure and mechanical properties of copper-lead and copper-zinc borate glasses. *J. Non-Cryst. Solids* **2016**, *435*, 55–68. [CrossRef]
59. Iordanova, R.; Milanova, M.; Aleksandrov, L.; Khanna, A. Structural study of glasses in the system B_2O_3-Bi_2O_3-La_2O_3-WO_3. *J. Non-Cryst. Solids* **2018**, *481*, 254–259. [CrossRef]
60. Kamitsos, E.I.; Karakassides, M.A.; Chryssikos, G.D. Vibrational spectra of magnesium-sodium-borate glasses. Raman and mid-infrared investigation of the network structure. *J. Phys. Chem.* **1987**, *91*, 1073–1079. [CrossRef]
61. Barlier, V.; Bounor-Legare, V.; Boiteux, G.; Davenas, J. Hydrolysis-condensation reactions of titanium alkoxides in thin films: A study of the steric hindrance effect by X-ray photoelectron spectroscopy. *Appl. Surf. Sci.* **2008**, *254*, 5408–5412. [CrossRef]
62. Saidi, W.; Rasheed, M.; Hfayedh, N.; Girtan, M.; El Maaoui, M.; Megriche, A. Effect of B_2O_3 addition on optical and structural properties of TiO_2 as a new blocking layer for multiple dye sensitive solar cell application (DSSC). *RSC Adv.* **2016**, *6*, 68819–68826. [CrossRef]
63. Zaleska, A.; Sobczak, J.W.; Grabowska, E. and Hupka, Photocatalytic activity of boron-modified TiO_2 under visible light: The effect of boron content, calcination temperature and TiO_2 matrix. *J. Appl. Catal. B* **2008**, *78*, 92–100. [CrossRef]
64. Tijani, J.O.; Momoh, U.O.; Salau, R.B.; Bankole, M.T.; Abdulkareem, A.S.; Roos, W.D. Synthesis and characterization of $Ag_2O/B_2O_3/TiO_2$ ternary nanocomposites for photocatalytic mineralization of local dyeing wastewater under artificial and natural sunlight irradiation. *Environ. Sci. Pollut. Res.* **2019**, *26*, 19942–19967. [CrossRef]
65. Zaleska, A. Doped—TiO_2: A Review. *Recent Pat. Eng.* **2008**, *2*, 157–164. [CrossRef]
66. May-Lozano, M.; López-Medina, R.; Rojas-García, E.; Hernández-Pérez, I.; Martínez-Delgadillo, S.A. Characterization of B-TiO_2 Synthesized under Different Conditions of Hydrolysis. *J. Adv. Oxid. Technol.* **2016**, *19*, 326–337. [CrossRef]
67. Soylu, N.Y.; Soylu, A.; Dikmetas, D.N.; Karbancioglu-Guler, F.; Kucukbayrak, S.; Taygun, M.E. Photocatalytic and Antimicrobial Properties of Electrospun TiO_2–SiO_2–Al_2O_3–ZrO_2–CaO–CeO_2 Ceramic Membranes. *ACS Omega* **2023**, *8*, 10836–10850. [CrossRef] [PubMed]
68. Gu, Q.; Zhao, W.; Yuan, J.; Yao, Y.; Wang, Y.; Wu, W. Adsorption and Photodegradation Behaviors of In-Situ Growth TiO_2 Films with Various Nano-Structures. *Chem. Phys. Lett.* **2019**, *736*, 136804. [CrossRef]
69. Wang, Y.; Xue, X.; Yang, H. Synthesis and Antimicrobial Activity of Boron-doped Titania Nano-materials, Materials and product engineering. *Chin. J. Chem. Eng.* **2014**, *22*, 474–479. [CrossRef]
70. Yang, H.; Wang, Y.; Xue, X. Influences of glycerol as an efficient doping agent on crystal structure and antibacterial activity of B-TiO_2 nano-materials. *Colloids Surf. B Biointerfaces* **2014**, *122*, 701–708. [CrossRef] [PubMed]
71. Wang, Y.Z.; Wu, Y.S.; Xue, X.X.; Yang, H.; Liu, Z.H. Microstructure and antibacterial activity of ions (Ce, Y, or B)-doped Zn-TiO_2: A comparative study. *Mater. Technol.* **2016**, *32*, 310–320. [CrossRef]

materials

Article

Picosecond Pulsed Laser Deposition of Metals and Metal Oxides

Anna Dikovska [1], Genoveva Atanasova [2], Tina Dilova [2,*], Aleksandra Baeva [2], Georgi Avdeev [3], Petar Atanasov [1] and Nikolay Nedyalkov [1]

1 Institute of Electronics, Bulgarian Academy of Sciences, 72 Tsarigradsko Chaussee, 1784 Sofia, Bulgaria; dikovska@ie.bas.bg (A.D.); paatanas@ie.bas.bg (P.A.); nned@ie.bas.bg (N.N.)
2 Institute of General and Inorganic Chemistry, Bulgarian Academy of Sciences, Acad. G. Bonchev Str., bl. 11, 1113 Sofia, Bulgaria; genoveva@svr.igic.bas.bg (G.A.); alexandravbaeva@gmail.com (A.B.)
3 Rostislaw Kaischew Institute of Physical Chemistry, Bulgarian Academy of Sciences, Acad. G. Bonchev Str., bl. 11, 1113 Sofia, Bulgaria; g_avdeev@ipc.bas.bg
* Correspondence: tina@svr.igic.bas.bg

Abstract: In this work, we present the fabrication of thin films/nanostructures of metals and metal oxides using picosecond laser ablation. Two sets of experiments were performed: the depositions were carried out in vacuum and in air at atmospheric pressure. The subjects of investigation were the noble metals Au and Pt and the metal oxides ZnO and TiO_2. We studied and compared the phase composition, microstructure, morphology, and physicochemical state of the as-deposited samples' surfaces in vacuum and in air. It was found that picosecond laser ablation performed in vacuum led to the fabrication of thin films with embedded and differently sized nanoparticles. The implementation of the same process in air at atmospheric pressure resulted in the fabrication of porous nanostructures composed of nanoparticles. The ablation of pure Pt metal in air led to the production of nanoparticles with an oxide shell. In addition, more defects were formed on the metal oxide surface when the samples were deposited in vacuum. Furthermore, the laser ablation process of pure Au metal in a picosecond regime in vacuum and in air was theoretically investigated using molecular dynamics simulation.

Keywords: PLD; picosecond ablation; metals; metal oxides

Citation: Dikovska, A.; Atanasova, G.; Dilova, T.; Baeva, A.; Avdeev, G.; Atanasov, P.; Nedyalkov, N. Picosecond Pulsed Laser Deposition of Metals and Metal Oxides. *Materials* **2023**, *16*, 6364. https://doi.org/10.3390/ma16196364

Academic Editor: Matteo Tonezzer

Received: 2 September 2023
Revised: 16 September 2023
Accepted: 20 September 2023
Published: 22 September 2023

1. Introduction

Recently, femtosecond (*fs*) lasers have become increasingly common and have been intensively studied with respect to their uses in various scientific and industrial fields [1,2]. In particular, applications of *fs* lasers in micro-machining, producing nanoparticles, and forming nanostructures with advanced applications have quickly developed, which have been accompanied by numerous in-depth studies in the field of light–matter interactions [3–5]. However, due to their high price and significant service and maintenance costs on the one hand and the unique working environment requirements on the other, *fs* lasers have currently found limited industrial applicability. At the same time, the particular class of picosecond (*ps*) laser systems that emit pulses with a duration of up to several tens of picoseconds have generated considerable interest for scientists and technologists. The interaction of such pulses with matter retains the key traits of pulses conducted with *fs* laser pulses: a strong thermal effect limitation, lack of a large heat-affected area around the impact area, and realization of non-thermal mechanisms of phase transformations, such as phase explosion and homogeneous melting. Being less complicated and more reliable, these laser systems are considerably less expensive than *fs* laser systems.

Many of the works involving *ps* laser systems that have been reported so far have been stimulated by possibilities for applications, such as cutting, drilling, melting, etc. [6–9]. Thus, they have been primarily focused on modifying the target's morphology with no particular interest in the detailed description of the properties and dynamics of the species that are evaporated and ablated. Furthermore, several studies have reported the use of *ps* lasers for the fabrication of thin films and/or structures and have discussed their possible applications [10]. Huotari et al. described the picosecond laser deposition of a metal oxide nanostructure for high-sensitivity gas sensor applications [11]. Additionally, Kekkonen et al. reported the picosecond pulsed laser deposition of metal-oxide sensing layers with controllable porosity, again with respect to gas sensor applications [12]. Some authors investigated the effects of the picosecond deposition of metal oxides at high pulse repetition rates [13,14]. However, all of these studies have been focused on the properties of the produced films/structures rather than on the mechanism of their fabrication and the dynamics of the ablated species. On the contrary, some authors have only been focused on the theoretical investigation of the mechanisms of ultrashort laser–matter interaction without further interest in the ablated material [15,16].

The aim of this work was to study the material removed from the target when performing picosecond laser ablation. The laser ablation process is the basis of the technology that is used for thin film/nanostructure fabrication, namely, pulsed laser deposition (PLD). Applying *ps* pulses for PLD preserves the features of an ultrashort laser–matter interaction and advances this technology toward its commercial use. This defines the interest in *ps*-PLD from a scientific and research perspective. The subject of investigations in this study were the noble metals Au and Pt and the metal oxides ZnO and TiO_2. The nanoparticles and nanostructures of noble metals such as Au and Pt are of great scientific interest due to the presence of efficient plasmon excitations with resonances in the near UV and visible spectral ranges [17]. A variety of novel applications in nanoelectronics, optical imaging, biomedicine, telecommunications, photovoltaics, photocatalysis, etc., all revolve around field localization and enhancement. ZnO is an important technological material for the effective design of optoelectronic devices, ultraviolet light emitters, photovoltaics, transparent and spin electronics devices, biosensors, etc., due to its useful mechanical, electrical, and optical properties as well as its relatively easy fabrication and low production costs [18]. Among oxides, TiO_2 is a promising semiconductor material because of its high chemical stability, high refractive index, surface functionality, self-cleaning and photocatalytic activity/recycling, antibacterial properties, and biocompatibility [19]. Furthermore, PLD is a flexible technology that allows for the easy fabrication of composite materials, which considerably enriches the possible application of the relevant metal oxides and noble metals.

We studied and compared the phase composition, microstructure, morphology, and physicochemical state of the surface of samples deposited by *ps*-PLD in vacuum and in air. Furthermore, a theoretical investigation of the laser ablation process of Au metal in a picosecond regime in vacuum and in air was implemented using molecular dynamics simulation in order to clarify the mechanism of material decomposition during *ps* pulse irradiation.

2. Materials and Methods

2.1. Experimental

The samples were produced by applying picosecond pulsed laser deposition (PLD). The ablation process was carried out using a picosecond Nd:YAG laser (CNI Laser, PS-A1-1064) operating at a fundamental wavelength of 1064 nm with a pulse duration of 10 ps and a repetition rate of 1 kHz. The laser fluence applied on the target was 0.45 J/cm^2. The targets used for ablation were the noble metals Au and Pt and the metal oxides ZnO and TiO_2. We used quartz and silicon as substrates. Two sets of experiments were performed: the depositions were carried out in vacuum and in air at atmospheric pressure. The experiments in vacuum were conducted at a target–substrate distance of 30 mm and a base pressure of 10^{-4} Torr. In air, the target–substrate distance was kept at 5 mm. All depositions were performed at room substrate temperature for 3 min.

2.2. Theoretical

In order to study the dynamics of a metal target interacting with a *ps* laser pulse, a molecular dynamic (MD) model was developed. It is based on the classical MD approach, whereby the time evolution of an ensemble of atoms can be described by a set of equations of motion for each atom [20]:

$$m_i \frac{\partial^2 r_i}{\partial t^2} = -\nabla_i U(r_1, r_2, \ldots\ldots, r_N) \tag{1}$$

where m_i and r_i are the mass and coordinate of the atom, respectively, and U $(r_1, r_2, \ldots, r_N)$ is the interatomic potential. One of the main advantages of the MD method is that the initial input parameters are only the atoms' positions, the temperature of the system (velocities), and the potential of interaction. No additional consideration and approximations are necessary to obtain the time evolution of the simulated system. In the case presented in the simulation of Au, the simulated system was constructed by $150 \times 1 \times 200$ cubic face-centered unit cells. The initial temperature was 300 K. The interaction between the atoms of the system was considered to be adequately described by the Morse potential, and the specific parameters for Au are taken from [21]. The sets of equations in Equation (1) is solved by the "velocity Verlet" algorithm [20]. Along the lateral direction of the simulation system, periodic boundary conditions are imposed, thus simulating the central part of the laser spot. When the laser pulse duration is shorter than or is in the order of the specific time for energy transfer between the electron system that initially absorbs the laser radiation, the lattice temperature evolution of the system should be described by a two-temperature diffusion model, which consists of one part for the electron system and another part for the lattice [22]. This approach was used in the MD model in order to simulate the interaction with the *ps* laser pulse. The heat diffusion equation was used to predict the temperature of the system, which was set in the atomic system of MD using an appropriate scaling of the atomic velocities. Further details on the simulation model and parameters used can be found in [23].

The MD model was also used to estimate the role of the presence of a high-pressure gas phase in the ablation material dynamics. This case corresponds to ablation in air at atmospheric pressure. Because the real interactions of the ablated species with air, with all of its components, are complex and difficult to describe using an MD model, a simplified interaction is assumed. The main effect of the presence of gas at high pressure on the ablation plume is confinement of the plume close to the target surface [24,25] by decreasing the ablated species' velocities. In the model presented, this effect is accounted for by scaling the ablated atoms' velocities using an appropriate parameter, whose value is estimated based on the data available on the plume velocity distribution in laser ablation in air [26].

2.3. Sample Characterization

Scanning electron microscopy (SEM) was conducted using a LYRA I XMU system (Tescan, Brno, Czech Republic) to study the samples' morphology. Transmission electron microscopy (TEM) and selected area electron diffraction (SAED) images were taken by a JEOL JEM 2100 system (Akishima-Shi, Tokyo) to reveal the samples' microstructure and crystallinity. Some of the samples for TEM analyses were prepared by scratching the sample surface in a drop of distilled water and transferring a small quantity of the material to a Cu grid. Samples for TEM analyses were also prepared using a direct deposition on the TEM Cu grid in the case of ablation of a Au target in order to compare the obtained experimental results with the theoretical ones. In this case, the depositions were performed for a shorter deposition time (1 min) in order to prevent nanoparticles overlapping. X-ray diffraction (XRD) using an Empyrean diffractometer (PANalytical, Malvern, UK) was used to examine the samples' crystalline structure and phase composition. The chemical composition of the samples' surface was examined using X-ray photoelectron spectroscopy (XPS) on an AXIS Supra electron spectrometer (Kratos Analytical Ltd., Manchester, UK).

3. Results

3.1. Experimental Part

3.1.1. Noble Metals—Au and Pt

Figure 1 illustrates the microstructure of the samples obtained by the laser ablation of the Au target. The samples were prepared through a direct deposition on the TEM Cu grid. The TEM image and size distribution of the material deposited in vacuum are presented in Figure 1a. As can be seen, the material ablated from the Au target in vacuum mainly consists of spherically shaped small nanoparticles with sizes in the range of 1–4 nm and a mean diameter of 2 nm. However, larger particles with a size of 40–80 nm are also observed. The SAED image demonstrates that the nanoparticles are crystalline (Figure 1a) and the interplanar distance from the main reflections can be assigned to a family of planes of metallic Au (gold, cubic, ICPDS 96-110-0139).

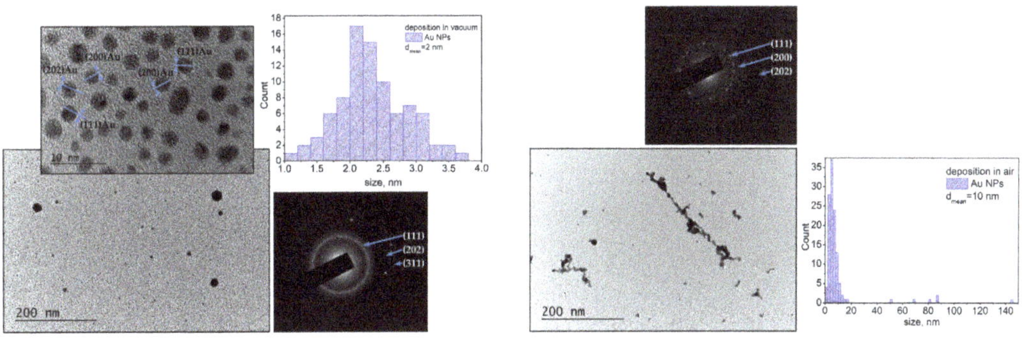

(a) (b)

Figure 1. TEM image, size distribution, and SAED pattern of material ablated from the Au target (**a**) in vacuum and (**b**) in air. The samples were prepared by direct deposition on the TEM Cu grid.

The TEM image and size distribution of the material deposited in air are presented in Figure 1b. The ablated material consists of individual nanometer-scale particles and nanoaggregates with different sizes and shapes. The corresponding size distribution shows that most of the nanoparticles have diameters ranging from 1 nm to 20 nm, with the mean diameter being 10 nm. All nanoparticles and nano-agglomerates are crystalline, which is confirmed by the SAED pattern (Figure 1b).

SEM images of the Au samples produced in vacuum and in air are shown in Figure 2. The sample deposited in vacuum (Figure 2a) displays a flat morphology with a significant amount of droplets, with sizes ranging from 80 nm to 600 nm. An SEM image of the Au sample deposited in air is shown in Figure 2b. The deposition process in air leads to the formation of a fine structure consisting of an ensemble of smaller particles. The XPS analyses of the Au samples that were performed to reveal the physicochemical state of the samples (Figure S1) found no difference in the XPS spectra of the Au samples prepared in vacuum and in air. The presence of Au was confirmed by registering the characteristic Au 4f peak at a binding energy of 84.0 eV, which corresponds to bulk metallic gold (Figure S1).

TEM images of the samples produced by ablation from a Pt target in vacuum and in air are shown in Figure 3. The samples for TEM analyses were prepared by transferring a small quantity of the material that was deposited on a substrate to a TEM grid, as described in the Section 2.3. Figure 3a shows the TEM images and a SAED pattern of the material ablated from the Pt target in vacuum. In the structure formed on the substrate, nanoparticles with sizes in the range of 2–10 nm can be recognized. The SAED pattern indicates that the structure is crystalline, with the patterns being indexed to metallic Pt (platinum, cubic, ICPDS 96-101-1112).

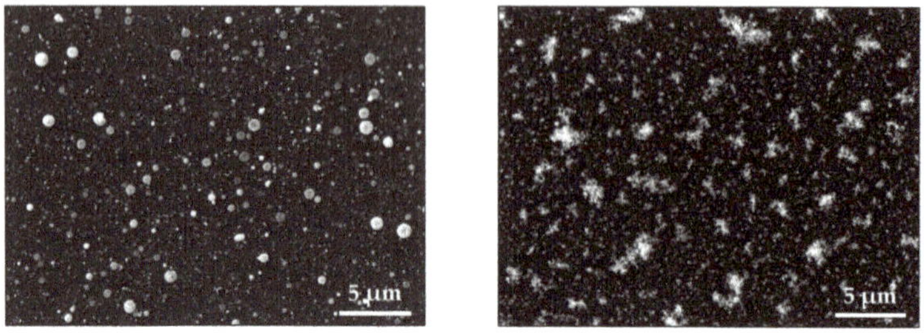

(a) (b)

Figure 2. SEM image of the Au sample produced (**a**) in vacuum and (**b**) in air.

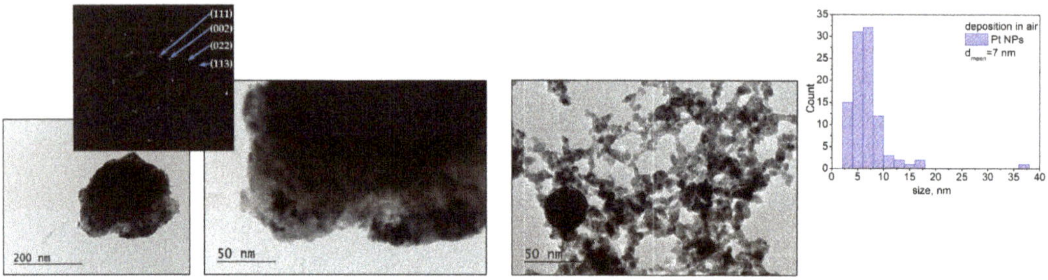

(a) (b)

Figure 3. (**a**) TEM image, corresponding size distribution, and SAED patterns of material ablated from the Pt target in vacuum; (**b**) TEM image and size distribution of material ablated from the Pt target in air. The samples were prepared by transferring a small quantity of the material to a TEM grid.

A TEM image and the size distribution of the material deposited in air are presented in Figure 3b. The nanoparticles and aggregates produced by the ablation of the Pt target in air form a chain-like nanostructure on the substrate. The corresponding SAED pattern shows that the nanostructure is crystalline, and the main reflections can be assigned to a family of planes of metallic Pt.

Figure 4 illustrates the morphology of the Pt samples produced in vacuum and in air. We observe that the Pt sample deposited in vacuum (Figure 4a) has a flat morphology with many droplet formations with sizes in the range of 80–200 nm, while the same deposition performed in air produces a morphology consisting of distinct features, where it is formed by nano-sized particles (Figure 4b).

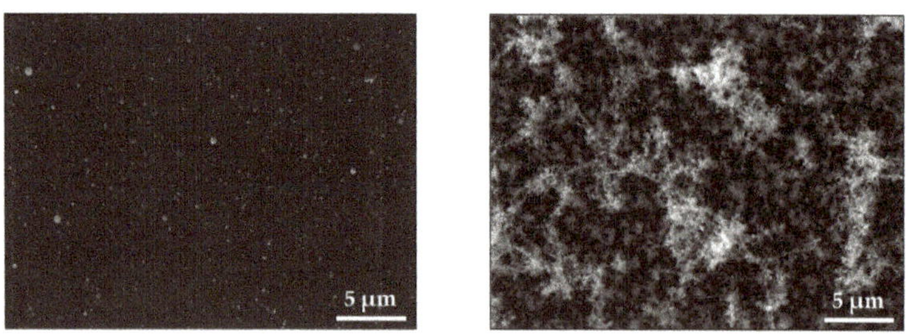

(a) (b)

Figure 4. SEM image of the Pt sample fabricated (**a**) in vacuum and (**b**) in air.

Figure 5 displays high-resolution XPS spectra of Pt 4f taken from the samples deposited from the Pt targets in vacuum and in air. The Pt 4f spectra of both samples were fitted to three major components attributed to metallic Pt^0 at 71.0 eV and oxidized Pt in different oxide states, respectively, Pt^{2+} at 72.4 eV, and Pt^{4+} at 74.3 eV [27]. Metallic Pt^0 is deposited on the surface of the sample in a significantly larger amount in vacuum than in air.

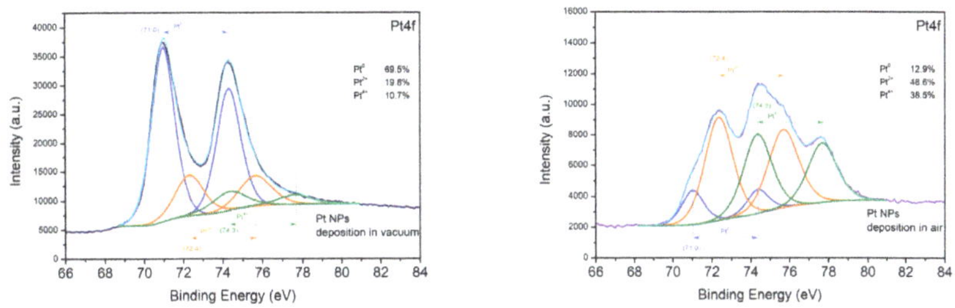

(a) (b)

Figure 5. XPS spectra of the Pt samples deposited (**a**) in vacuum and (**b**) in air.

3.1.2. Metal Oxides—ZnO and TiO_2

The TEM images of the samples produced by ablating a ZnO target are shown in Figure 6. Figure 6a displays the TEM image, corresponding SAED pattern, and high-resolution (HR) TEM of the material deposited on the substrate when the experiments were performed in vacuum. The microstructure of the sample is represented by a polycrystalline structure with embedded nanoparticles with irregular form. The HR TEM image reveals lattice fringe spacings of 2.81 Å, 2.60 Å, and 2.48 Å, which correspond well to the (100), (002), and (101) planes of the ZnO hexagonal wurtzite structure (zinc oxide, hexagonal, ICPDS 96-901-1663). The microstructure of the ZnO sample prepared in air consists of separate nanosized clusters of nanoparticles with irregular shapes and sizes.

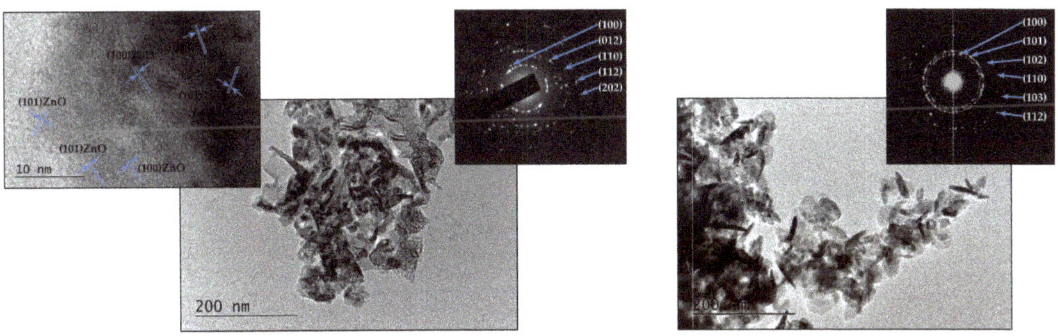

(a) (b)

Figure 6. (**a**) TEM image, corresponding SAED pattern, and HR TEM image of material ablated from ZnO target in vacuum; (**b**) TEM image and SAED pattern of material ablated from ZnO target in air. The samples were prepared by transferring a small quantity of the material to a TEM grid.

Elongated nanoparticles/nanorods with lengths ranging from 25 nm to 65 nm and diameters ranging from 7 to 15 nm are clearly visible in Figure 6b. The SAED pattern demonstrates the crystal nature of the structure (Figure 6b). The phase composition and crystal structure of the sample were also confirmed using the XRD analysis (Figure S2).

The SEM images in Figure 7 illustrate the morphology of the deposited ZnO samples. Figure 7a shows the morphology of a thin ZnO film deposited in vacuum. We observe that the surface is littered by mostly spherical particles/droplets with sizes ranging between 60 and 500 nm. Droplets with irregular shapes are also observed. Figure 7b presents the morphology of the ZnO sample deposited in air. A highly porous structure consisting of aggregated nanoparticles is observed. XPS analyses of the ZnO samples' surface were also performed. No difference was observed in the Zn 2p XPS spectra of the surface of the samples deposited in vacuum and in air.

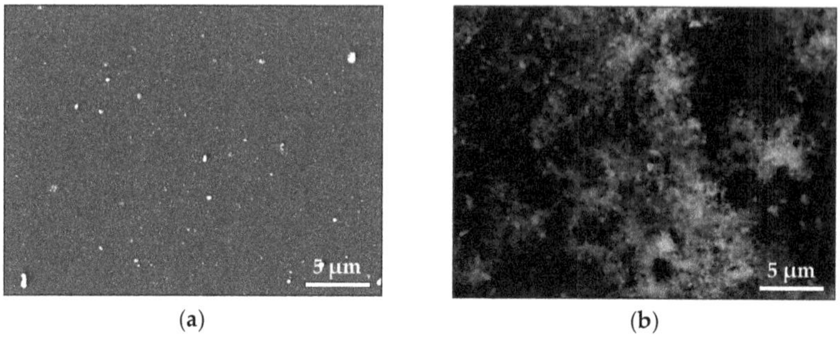

Figure 7. SEM image of material ablated from ZnO target (**a**) in vacuum and (**b**) in air.

We conclude that the Zn atoms are in the Zn^{2+} oxidation state based on the binding energy positions, the width of the Zn 2p peaks, and the spin–orbit splitting of 23.1 eV (Figure S3) [28]. The asymmetric O 1s peak indicates the presence of different oxygen-containing species (Figure S3); it could be fitted by three components that are assigned to lattice O^{2-} ions in the stoichiometric environment of Zn^{2+} ions, oxygen vacancies formed in oxygen-deficient regions of the ZnO lattice, and the presence of adsorbed hydroxyl, carbonate, or O_2 species, respectively [29]. It should be noted that the components associated with the oxygen deficiency state and with the presence of adsorbed species are amplified in the case of deposition in vacuum.

In Figure 8, TEM images are presented of the samples fabricated by ablation of the TiO_2 target in vacuum and in air. As seen in Figure 8a, the morphology of the sample deposited in vacuum has a polycrystalline structure with embedded nanoparticles with sizes ranging from 20 nm to 110 nm and a predominantly spherical shape. The main reflection of the SAED pattern can be identified as non-stoichiometric (Ti_7O_{13}, triclinic, ICPDS 96-100-8197) and anatase (TiO_2, tetragonal, ICPDS 96-101-0943) phases of titanium oxide. Figure 8b shows a TEM image of the sample deposited by ablation of the TiO_2 target in air. The microstructure of the sample consists of spherically shaped big particles with sizes mainly ranging from 60 nm to 130 nm (a mean diameter of approximately 100 nm). Furthermore, nanoparticles with sizes in the range of 2–20 nm (a mean diameter of approximately 7 nm) are clearly distinguished (size distribution in Figure 8b).

The phase composition of the sample identified by XRD analysis (Figure S4) shows that the sample structure is a combination of anatase and rutile phases of TiO_2 (rutile, tetragonal, ICPDS 00-021-1276). SEM images of the TiO_2 samples deposited in vacuum and in air are shown in Figure 9. Figure 9a demonstrates the morphology of the sample deposited in vacuum. One can see that the sample surface is covered by mostly spherical particles that range in size from 100 nm to 1 μm. The morphology of the TiO_2 sample deposited in air (Figure 9b) is a porous structure that is composed of nanoparticles of different sizes. However, separate droplet formations with sizes ranging 80–700 nm are observed across the entire structure.

The high-resolution XPS spectra of the Ti 2p core level of the TiO_2 sample surface are shown in Figure 10. The Ti 2p spectra can be deconvoluted into two close components at 458.7 and 457.3 eV. The Ti $2p_{3/2}$ and Ti $2p_{1/2}$ peaks located at 458.7 eV and 464.0 eV with a separation between the binding energies of 5.7 eV correspond to Ti^{4+} [30]. The Ti $2p_{3/2}$ and Ti $2p_{1/2}$ peaks located at 457.3 eV and 464.0 eV are associated with the presence of Ti^{3+} [30]. The amount of the Ti^{3+} component on the surface of the sample deposited in vacuum is more than twice as high as that on the surface of the sample produced in air (Figure 10a,b). As indicated by the asymmetric O1s peak, different oxygen-containing species are present. The peak can be deconvoluted into components that are attributed to lattice O^{2-} ions in the metal oxides as well as to oxygen vacancies and the presence of species adsorbed on the sample surface. The percentage of the components associated with the oxygen deficiency state and with the presence of adsorbed species is higher in the case of deposition in vacuum.

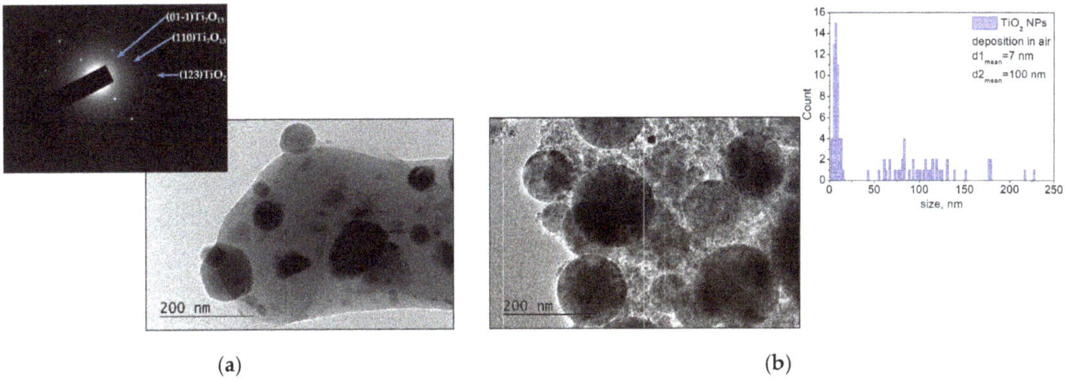

Figure 8. (**a**) TEM image and corresponding SAED pattern of material ablated from the TiO_2 target in vacuum; (**b**) TEM image and size distribution of material ablated from the TiO_2 target in air. The samples were prepared by transferring a small quantity of the material to a TEM grid.

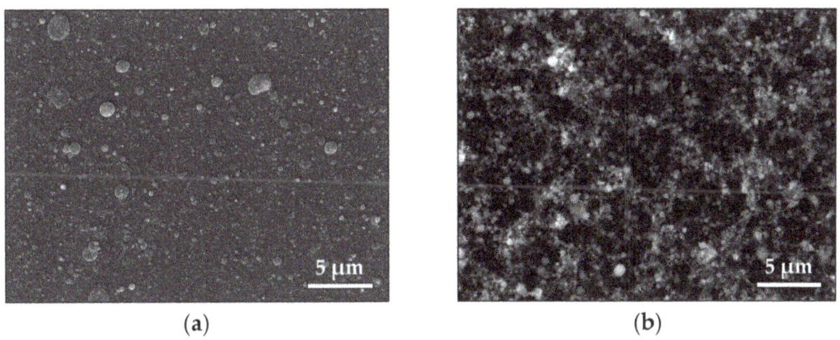

Figure 9. SEM image of material ablated from the TiO_2 target (**a**) in vacuum and (**b**) in air.

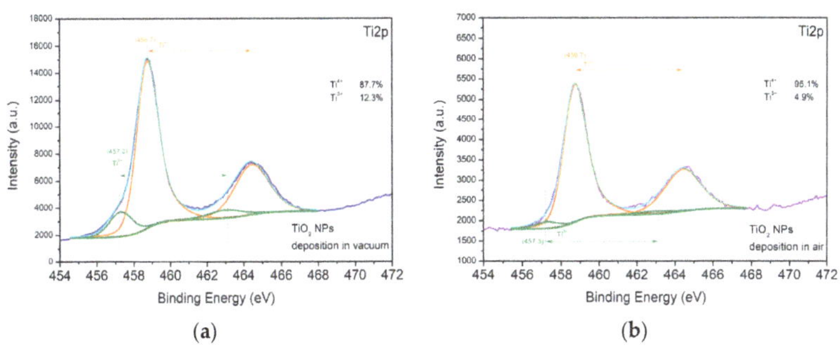

Figure 10. XPS spectra of the TiO_2 samples deposited (**a**) in vacuum and (**b**) in air.

3.2. Theoretical Part

Figure 11 presents the time evolution of the electron and lattice temperatures as calculated by a two-temperature diffusion model for a Au target that is irradiated by a 10 ps laser pulse with a fluence of 0.45 J/cm^2. The two systems reach equilibrium at about 70 ps after the laser pulse onset. The maximum lattice temperature that is reached is 3300 K, i.e., higher than the boiling point of Au (3081 K). Note that as the target is heated, temperature gradients of about 10^{14} K/s are being developed.

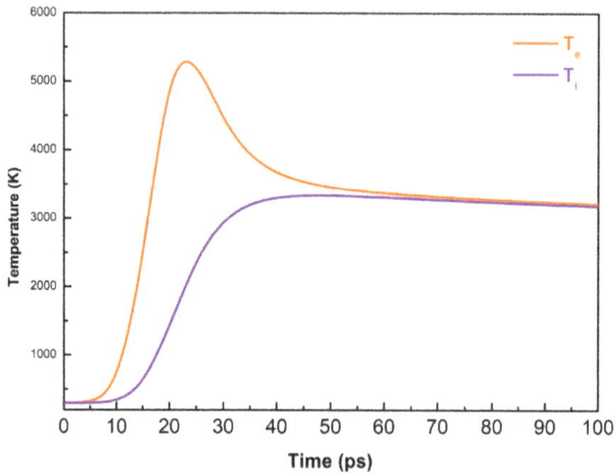

Figure 11. Calculated time evolution of the electron and lattice temperature of Au irradiated by a 10 ps laser pulse with a fluence of 0.45 J/cm^2.

The response of the material irradiated under such conditions is provided in Figure 12. Figure 12a shows a snapshot of the simulated system under the condition 50 ps after the laser pulse onset, when a laser fluence of 0.45 J/cm^2 is applied to the Au target in vacuum. The strong heating of the material results in a rapid expansion; a detailed view shows that the material starts to decompose into single atoms and clusters. The expanded view presented in the figure shows that the clusters that are obtained have irregular shapes. Their size was determined under the assumption that they are spherically shaped with a diameter equal to the mean value of the size along the x and z axes. The size distribution of the clusters shows sizes in the range of 1–7 nm (Figure 12a) with a mean diameter of 4 nm.

Figure 12b is a snapshot of the simulated system 50 ps after the laser pulse onset, when the laser fluence that is applied to the Au target in vacuum is 0.25 J/cm^2. We observe that when a lower fluence is applied to the Au target, its material undergoes similar decomposition. However, the corresponding size distribution demonstrates the formation of bigger clusters, with the mean diameter being estimated as 6 nm (Figure 12b).

Snapshots of Au targets irradiated by a 10 ps laser pulse at a fluence of 0.45 J/cm^2 in air are presented in Figure 12c,d. The development of the simulated system in air was obtained by limiting the velocities of the species obtained in a vacuum. Figure 12c is a snapshot of the simulated system 50 ps after the laser pulse onset. The decomposition of the irradiated material leads to formation of clusters with sizes ranging from 2 nm to 7 nm with a mean diameter of 4 nm (see size distribution in Figure 12c). Figure 12d displays a simulation of the situation 80 ps after the laser pulse onset; the clusters have grown in size, with a mean diameter reaching 6 nm (see size distribution of Figure 12d). The presence of single atoms in the simulated system in air should be noted.

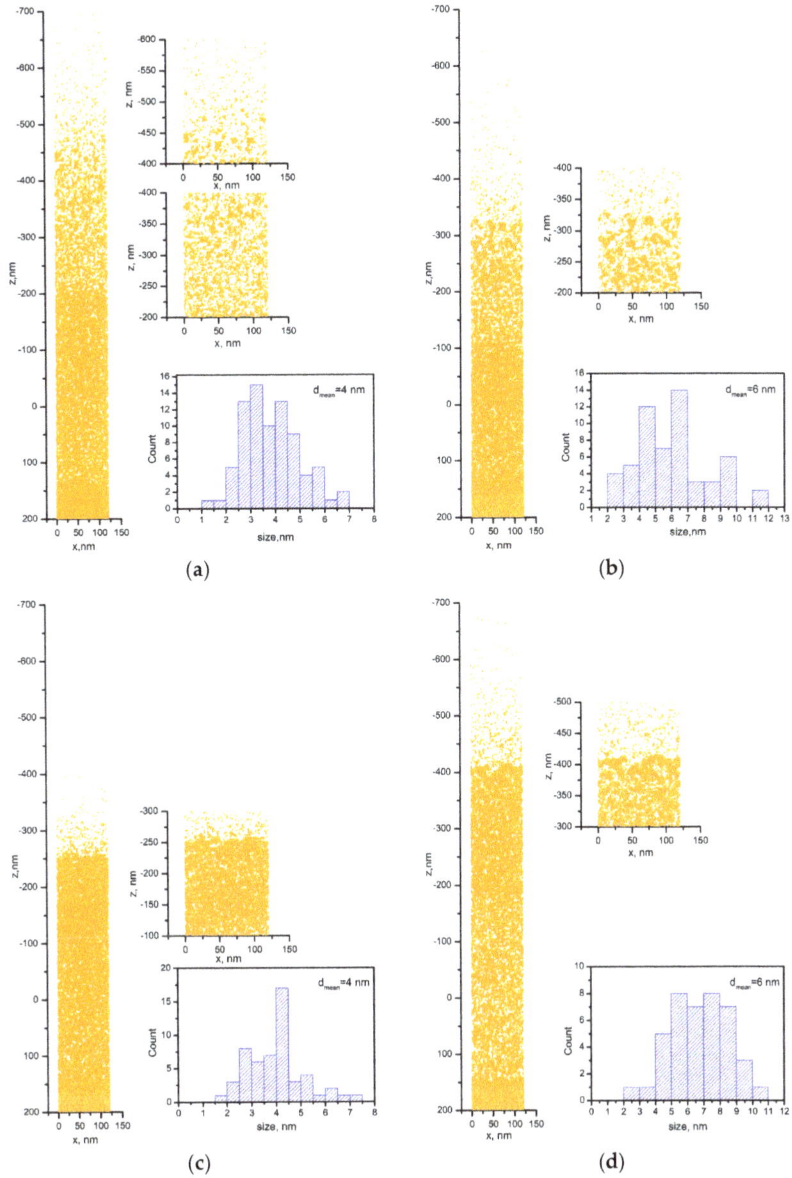

Figure 12. Snapshot of Au target irradiated by a 10 ps laser pulse in vacuum 50 ps after the laser pulse onset with a fluence of (**a**) 0.45 J/cm^2 and (**b**) 0.25 J/cm^2. Snapshot of Au target irradiated by a 10 ps laser pulse in air at a fluence of 0.45 J/cm^2 (**c**) 50 ps and (**d**) 80 ps after the laser pulse onset.

4. Discussion

The interaction of high-power ultrashort laser pulses with metals results in strong heating of the absorbing volume followed by rapid melting of the material. In the case of *fs* pulses, the heating takes place at a nearly constant volume. If the temperature reaches a critical value, the usual relaxation mechanism is phase explosion or homogeneous decomposition of the overheated material into gas and liquid droplets [31]. In the case of picosecond pulses, the material is also rapidly melted, but it may undergo expansion, even during the laser pulse. In such a case, the decomposition mechanism can consist of liquid fragmentation [32], where the fast expansion of the overheated liquid results in the development of density fluctuations within the volume, which further evolve into material decomposition into atoms and liquid droplets.

Based on the above scenario, one can conclude that under the conditions discussed, the decomposition of the material occurs via fragmentation rather than phase explosion, as the maximal temperature that is achieved is significantly lower than the critical temperature for gold (~5000 K) (Figure 11). In the case of metal oxides, other studies [33,34] have demonstrated that the ablation of metal oxides by ultrashort laser pulses also leads to decomposition of the material into nanoparticles. It is also considered that the absorption of the ultrashort laser pulses is realized via multiphoton absorption as the absorbing material is rapidly transformed into a "metallic state" with a high concentration of free electrons. Such effects are difficult to simulate due to a lack of reliable data for the materials at such a state. However, based on the above comments, we can assume that the ablation mechanisms of metal oxides could also be attributed to fragmentation.

To summarize, ablation by *ps* laser pulses leads to the production of nanoparticles regardless of the ambient pressure. As demonstrated by the MD simulation, they originate from an ablation mechanism that leads to a direct decomposition into nanoparticles. However, the morphology of the samples shows a clear difference when the ambient pressure is changed (Figures 2, 4, 7 and 9). It is evident that increasing the pressure from vacuum to atmospheric transforms the thin-film morphology that is normally obtained in vacuum to the highly porous structure formed at atmospheric pressure (Figures 2, 4, 7 and 9). In general, the morphology of the samples deposited in vacuum represents a flat thin film with a significant amount of round-shaped nanoparticles/droplets with sizes depending on the material deposited (Figures 2a, 4a, 7a and 9a). Obviously, the size of these droplets is smaller for the samples prepared by ablation of Pt (compared with Au) and ZnO (compared with TiO_2) targets. We attribute this difference in the size of the droplets to the different ablation thresholds of the materials. Such droplet formations are typical for the classical PLD process performed by nanosecond (*ns*) laser pulses, as their origin is associated with the ejection of molten material due to the overheating of the target [35].

It should be noted that the size of the droplets formed by *ps*-PLD is smaller than that obtained by *ns*-PLD, which is probably related to the limited heat-affected zone when the target is irradiated by ultrashort laser pulses. However, the temperature that is attained on the target surface is considerably higher than the melting point of the material used in this work, as the calculation for the Au target predicts. This is why one should not be surprised to observe particles/droplets on the surface of the sample that is produced by *ps* ablation. It should also be mentioned that the higher the fluence that is applied on the target, the more droplets that will be produced. In this work, we used a fixed laser fluence for the ablation of all target materials independently of their ablation threshold. Furthermore, the microstructure of the samples shows that these particles are embedded in the sample structure (Figures 3a, 6a and 8a). The response of the Au target under a 10 ps irradiation in vacuum consists of decomposition of the material into single atoms and clusters (Figure 12a). Thus, the "building blocks" of the samples deposited in vacuum are single atoms and clusters/nanoparticles that are directly formed at the stage of material decomposition. This way, a thin film with embedded nanoparticles is grown on the substrate via *ps* target ablation.

It should be noted that the experimental results are in good agreement with the results obtained by simulation (Figures 1a and 12a). Note that Figure 1a presents the size distribution of the smallest nanoparticles produced. In the experiments, the laser beam has a Gaussian shape, which means that the real fluence applied to the target will be in a tiny range as the maximum value is 0.45 J/cm^2. In the simulations, for simplicity, a constant value for laser fluence was used; therefore, simulations at a lower laser fluence were also implemented (Figure 12b). The results demonstrate that a broader size distribution should be expected due to the Gaussian shape of the laser beam.

Complex 3D nanostructures are formed when the deposition process is performed in air (Figures 2b, 4b, 7b and 9b). Similar structures were previously obtained and reported when *fs* or *ns* laser pulses were used for ablation at atmospheric pressure [36–39]. These structures are composed of particles with nanometer-scale sizes; the main mechanism of their formation is the fast condensation of the ablated species [35–39]. In air, the response of the Au target under 10 ps of irradiation also leads to decomposition of the material into single atoms and clusters (Figure 12c). The temporal evolution of the simulated system reveals the growing size of the clusters formed (Figure 12d), which is in good agreement with the experimental results (Figure 1b). However, due to the high ambient pressure, the ejected atoms condense and form nanoparticles and/or nanoaggregates. Thus, the "building blocks" of the samples deposited in air are nanoparticles and nanoaggregates (Figures 3b, 6b and 8b).

It is worth noting that, in spite of the samples being deposited in air, no nitrogen inclusions were detected on their surface. However, the Pt nanoparticles were oxidized in air during their formation. When comparing the surface of the metal oxides deposited in vacuum and in air, the former had a higher percentage of components assigned to the oxygen deficiency state and to the presence of adsorbed species. This allows us to conclude that more defects are formed on the metal oxide surface when the samples are deposited in vacuum.

Finally, the method for samples' fabrication is a type of sputtering process. Furthermore, the experiments that were performed were very simple (at room temperature, in vacuum or air). The replication of the obtained results demonstrates that the *ps*-PLD method has good reproducibility. Moreover, the results obtained in this work are in agreement with the results previously reported on PLD performed by using *fs* pulses [40].

5. Conclusions

Laser ablation using *ps* pulses leads to the production of nanoparticles independently of the ambient pressure. Increasing the pressure from vacuum to atmospheric transforms the thin-film morphology normally obtained in vacuum to a highly porous structure that is formed in air at atmospheric pressure. The morphology of the samples deposited in vacuum is represented by a flat thin film with a significant amount of round-shaped nanoparticles/droplets, with the size of these particles depending on the material deposited. The difference in the size of the droplets is related to the difference in the ablation threshold of the materials. Furthermore, the droplets that are formed by *ps*-PLD are smaller than those obtained by the classical *ns*-PLD, which is probably related to the limited heat-affected area on the target that is irradiated by ultrashort laser pulses. When the deposition process is performed in air, complex 3D nanostructures are formed, which are composed of particles with nanometer sizes, with the main mechanism of their formation being the fast condensation of the ablated species. The decomposition mechanism in the case of a 10 ps laser pulse irradiation of the Au target consists of liquid fragmentation, which further evolves into material decomposition into atomic and liquid droplet components.

Supplementary Materials: The following supporting information can be downloaded at: https://www.mdpi.com/article/10.3390/ma16196364/s1, Figure S1: XPS spectra of the Au samples deposited (a) in vacuum and (b) in air; Figure S2: XRD patterns of material ablated from the ZnO target in air; Figure S3: XPS spectra of the ZnO samples deposited (a) in vacuum and (b) in air; Figure S4. XRD patterns of material ablated from the TiO_2 target in air.

Author Contributions: Conceptualization, A.D. and P.A.; methodology, A.D. and N.N.; formal analysis, G.A. (Genoveva Atanasova), G.A. (Georgi Avdeev), T.D. and A.B.; investigation, A.D.; data curation, G.A. (Genoveva Atanasova), T.D., A.B. and G.A. (Georgi Avdeev); writing—original draft preparation, A.D.; writing—review and editing, N.N.; visualization, G.A. (Genoveva Atanasova); project administration, A.D. All authors have read and agreed to the published version of the manuscript.

Funding: The authors wish to thank for the financial support of project KP-06-N37/20, "Formation and physical properties of composite nanostructures of metal oxides and noble metals", under the "Competition for financial support of basic research projects—2019" program of the Bulgarian National Science Fund. The authors acknowledge the TwinTeam project Д01-272 "European Network on Materials for Clean Technologies" for providing the opportunity to present the results at the SizeMat4 conference and for providing the financial publication support.

Institutional Review Board Statement: Not applicable.

Informed Consent Statement: Not applicable.

Data Availability Statement: Not available.

Conflicts of Interest: The authors declare no conflict of interest.

References

1. Sibbett, W.; Lagatsky, A.A.; Brown, C.T.A. The development and application of femtosecond laser systems. *Opt. Express* **2012**, *20*, 6989–7001. [CrossRef]
2. Guo, B.; Sun, J.; Hua, Y.; Zhan, N.; Jia, J.; Chu, K. Femtosecond Laser Micro/Nano-manufacturing: Theories, Measurements, Methods, and Applications. *Nanomanuf. Metrol.* **2020**, *3*, 26–67. [CrossRef]
3. Wang, X.; Yu, H.; Li, P.; Zhang, Y.; Wen, Y.; Qiu, Y.; Liu, Z.; Li, Y.P.; Liu, L. Femtosecond laser-based processing methods and their applications in optical device manufacturing: A review. *Opt. Laser Technol.* **2021**, *135*, 106687. [CrossRef]
4. Vorobyev, A.Y.; Guo, C. Direct femtosecond laser surface nano/microstructuring and its applications. *Laser Photonics Rev.* **2013**, *7*, 385–407. [CrossRef]
5. Andreeva, Y.; Sharma, N.; Rudenko, A.; Mikhailova, J.; Sergeev, M.; Veiko, V.P.; Vocanson, F.; Lefkir, Y.; Destouches, N.; Itina, T.E. Insights into Ultrashort Laser-Driven Au:TiO_2 Nanocomposite Formation. *J. Phys. Chem. C* **2020**, *124*, 10209–10219. [CrossRef]
6. Cheng, J.; Perrie, W.; Sharp, M.; Edwardson, S.P.; Semaltianos, N.G.; Dearden, G.; Watkins, K.G. Single-pulse drilling study on Au, Al and Ti alloy by using a picosecond laser. *Appl. Phys. A* **2009**, *95*, 739–746. [CrossRef]
7. Zhang, H.; Di, J.; Zhou, M.; Yan, Y.; Wang, R. An investigation on the hole quality during picosecond laser helical drilling of stainless steel 304. *Appl. Phys. A* **2015**, *119*, 745–752. [CrossRef]
8. Shestakov, D.; Khairullina, E.; Shishov, A.; Khubezhov, S.; Makarov, S.; Tumkin, I.; Logunov, L. Picosecond laser writing of highly conductive copper micro-contacts from deep eutectic solvents. *Opt. Laser Technol.* **2023**, *167*, 109777. [CrossRef]
9. Spiro, A.; Lowe, M.; Pasmanik, G. Drilling rate of five metals with picosecond laser pulses at 355, 532, and 1064 nm. *Appl. Phys. A* **2012**, *107*, 801–808. [CrossRef]
10. Preißa, E.M.; Krauß, A.; Kekkonen, V.; Barsan, N.; Seidel, H. Characterization of WO_3 thin films prepared by picosecond laser deposition for gas sensing. *Sens. Act. B* **2017**, *248*, 153–159. [CrossRef]
11. Huotari, J.; Kekkonen, V.; Haapalainen, T.; Leidinger, M.; Sauerwald, T.; Puustinen, J.; Liimatainen, J.; Lappalainen, J. Pulsed laser deposition of metal oxide nanostructures for highly sensitive gas sensor applications. *Sens. Act. B Chem.* **2016**, *236*, 978–987. [CrossRef]
12. Kekkonen, V.; Chaudhuri, S.; Clarke, F.; Kaisto, J.; Liimatainen, J.; Pandian, S.K.; Piirto, J.; Siltanen, M.; Zolotukhin, A. Picosecond pulsed laser deposition of metal-oxide sensing layers with controllable porosity for gas sensor applications. *Appl. Phys. A* **2016**, *122*, 233. [CrossRef]
13. Pervolaraki, M.; Mihailescu, C.N.; Luculescu, C.R.; Ionescu, P.; Dracea, M.D.; Pantelica, D.; Giapintzakis, J. Picosecond ultrafast pulsed laser deposition of $SrTiO_3$. *Appl. Surf. Sci.* **2015**, *336*, 278–282. [CrossRef]
14. Salminen, T.; Hahtala, M.; Seppälä, I.; Niemi, T.; Pessa, M. Pulsed laser deposition of yttria-stabilized zirconium dioxide with a high repetition rate picosecond fiber laser. *Appl. Phys. A* **2010**, *98*, 487–490. [CrossRef]
15. Mao, S.S.; Mao, X.; Greif, R.; Russo, R.E. Dynamics of an air breakdown plasma on a solid surface during picosecond laser ablation. *Appl. Phys. Lett.* **2000**, *76*, 31–33. [CrossRef]
16. Mao, S.S.; Mao, X.; Greif, R.; Russo, R.E. Initiation of an early-stage plasma during picosecond laser ablation of solids. *Appl. Phys. Lett.* **2000**, *77*, 2464–2466. [CrossRef]
17. Kravets, V.G.; Kabashin, A.V.; Barnes, W.L.; Grigorenko, A.N. Plasmonic Surface Lattice Resonances: A Review of Properties and Applications. *Chem. Rev.* **2018**, *118*, 5912–5951. [CrossRef]
18. Sharma, D.K.; Shukla, S.; Sharma, K.K.; Kumar, V. A review on ZnO: Fundamental properties and applications. *Mater. Today Proc.* **2022**, *49*, 3028–3035. [CrossRef]
19. Ahmed, R.M.; Hasan, I. A review on properties and applications of TiO_2 and associated nanocomposite materials. *Mater. Today Proc.* **2023**, *81*, 1073–1078. [CrossRef]
20. Allen, M.P.; Tildesley, D.J. *Computer Simulation of Liquids*; Clarendon: Oxford, UK, 1987. [CrossRef]
21. Mohammed, K.; Shukla, M.M.; Milstein, F.; Merz, J.L. Lattice dynamics of face-centered-cubic metals using the ionic Morse potential immersed in the sea of free-electron gas. *Phys. Rev. B* **1984**, *29*, 3117. [CrossRef]
22. Anisimov, S.I.; Kapelovich, B.L.; Perel'man, T.L. Electron emission from metal surfaces exposed to ultrashort laser pulses. *Sov. Phys. JETP* **1974**, *39*, 375–377.

23. Amoruso, S.; Nedyalkov, N.N.; Wang, X.; Ausanio, G.; Bruzzese, R.; Atanasov, P.A. Ultrashort-pulse laser ablation of gold thin film targets: Theory and experiment. *Thin Solid Films* **2014**, *550*, 190–198. [CrossRef]
24. Farid, N.; Harilal, S.S.; Ding, H.; Hassanein, A. Emission features and expansion dynamics of nanosecond laser ablation plumes at different ambient pressures. *J. Appl. Phys.* **2014**, *115*, 033107. [CrossRef]
25. Bai, X.; Ma, Q.; Perrier, M.; Motto-Ros, V.; Sabourdy, D.; Nguyen, L.; Jalocha, A.; Yu, J. Experimental study of laser-induced plasma: Influence of laser fluence and pulse duration. *Spectrochim. Acta Part B* **2013**, *87*, 27–35. [CrossRef]
26. Borchert, H.; Dar´ee, K.; Hugenschmidt, M. Plasma formation during the interaction of picosecond and nanosecond laser pulses with BK7 glass. *J. Phys. D Appl. Phys.* **2005**, *38*, 300–305. [CrossRef]
27. Ihnatiuk, D.; Tossi, C.; Tittonen, I.; Linnik, O. Effect of Synthesis Conditions of Nitrogen and Platinum Co-Doped Titania Films on the Photocatalytic Performance under Simulated Solar Light. *Catalysts* **2020**, *10*, 1074. [CrossRef]
28. Claros, M.; Setka, M.; Jimenez, Y.P.; Vallejos, S. AACVD Synthesis and Characterization of Iron and Copper Oxides Modified ZnO Structured Films. *Nanomaterials* **2020**, *10*, 471. [CrossRef]
29. Tomić, M.; Claros, M.; Gràcia, I.; Figueras, E.; Cané, C.; Vallejos, S. ZnO Structures with Surface Nanoscale Interfaces Formed by Au, Fe_2O_3, or Cu_2O Modifier Nanoparticles: Characterization and Gas Sensing Properties. *Sensors* **2021**, *21*, 4509. [CrossRef]
30. Konatu, R.T.; Domingues, D.D.; França, R.; Alves, A.P.R. XPS Characterization of TiO_2 Nanotubes Growth on the Surface of the $Ti_{15}Zr_{15}Mo$ Alloy for Biomedical Applications. *J. Funct. Biomater.* **2023**, *14*, 353. [CrossRef]
31. Pallotti, D.K.; Ni, X.; Fittipaldi, R.; Wang, X.; Lettieri, S.; Vecchione, A.; Amoruso, S. Laser ablation and deposition of titanium dioxide with ultrashort pulses at 527 nm. *Appl. Phys. B* **2015**, *119*, 445–452. [CrossRef]
32. Sang, L.; Zhang, H.; Ni, X.; Anoop, K.K.; Fittipaldi, R.; Wang, X.; Amoruso, S. Hydrogen-evolving photoanode of TiO_2 nanoparticles film deposited by a femtosecond laser. *Int. J. Hydrogen Energy* **2015**, *40*, 779–785. [CrossRef]
33. Wu, C.; Zhigilei, L.V. Microscopic mechanisms of laser spallation and ablation of metal targets from large-scale molecular dynamics simulations. *Appl. Phys. A* **2014**, *114*, 11–32. [CrossRef]
34. Lorazo, P.; Lewis, L.J.; Meunier, M. Short-Pulse Laser Ablation of Solids: From Phase Explosion to Fragmentation. *Phys. Rev. Lett.* **2003**, *91*, 225502. [CrossRef] [PubMed]
35. Kolasinski, K.W.; Gupta, M.C.; Zhigilei, L.V. Plume and Nanoparticle Formation During Laser Ablation. In *Encyclopedia of Interfacial Chemistry: Surface Science and Electrochemistry*; Wandelt, K., Ed.; Elsevier: Oxford, UK, 2018; Volume 2, pp. 594–603. [CrossRef]
36. Nedyalkov, N.; Nikov, R.; Dikovska, A.; Atanasova, G.; Nakajima, Y.; Terakawa, M. Gold nanostructures deposition by laser ablation in air using nano- and femtosecond laser pulses. *Appl. Phys. A* **2017**, *123*, 306. [CrossRef]
37. Nikov, R.G.; Dikovska, A.O.; Nedyalkov, N.N.; Atanasov, P.A. Fabrication of Au nanostructures by pulsed laser deposition in air. *Proc. SPIE* **2017**, *10226*, 119–125. [CrossRef]
38. Nikov, R.G.; Dikovska, A.O.; Nedyalkov, N.N.; Avdeev, G.V.; Atanasov, P.A. Au nanostructure fabrication by pulsed laser deposition in open air: Influence of the deposition geometry. *Beilstein J. Nanotechnol.* **2017**, *8*, 2438–2445. [CrossRef]
39. Nikov, R.G.; Dikovska, A.O.; Nedyalkov, N.N.; Atanasov, P.A.; Atanasova, G.; Hirsch, D.; Rauschenbach, B. ZnO nanostructures produced by pulsed laser deposition in open air. *Appl. Phys. A* **2017**, *123*, 657. [CrossRef]
40. Bonis, A.D.; Teghil, R. Ultra-Short Pulsed Laser Deposition of Oxides, Borides and Carbides of Transition Elements. *Coatings* **2020**, *10*, 501. [CrossRef]

materials

MDPI

Article

Study on the Dye Removal from Aqueous Solutions by Graphene-Based Adsorbents

Paunka Vassileva *, Vencislav Tumbalev, Diana Kichukova, Dimitrinka Voykova, Daniela Kovacheva and Ivanka Spassova

Institute of General and Inorganic Chemistry, Bulgarian Academy of Sciences, 1113 Sofia, Bulgaria; tumbalev@svr.igic.bas.bg (V.T.); diana123georgieva@gmail.com (D.K.); dimitrinka@svr.igic.bas.bg (D.V.); didka@svr.igic.bas.bg (D.K.); ispasova@svr.igic.bas.bg (I.S.)

* Correspondence: pnovachka@svr.igic.bas.bg

Abstract: In the current investigation, the removal efficiency regarding a cationic dye, methylene blue (MB), from three graphene-based materials was investigated. The materials' characterization process involved instrumental methods such as XRD, XPS, SEM, TEM, FTIR, and nitrogen adsorption at 77 K. The survey examined how various process factors influenced the ability of the studied materials to adsorb cationic dyes. These parameters encompassed contact time, initial dye concentrations, solution pH, and temperature. The adsorption procedure was effectively explained through the application of pseudo-second-order and Langmuir models. The maximum adsorption capacity for the best adsorbent at 293 K was found to be 49.4 mg g^{-1}. In addition, the study also determined the entropy, enthalpy, and Gibbs free energy values associated with the removal of MB and showed that the adsorption of MB is endothermic, feasible, and spontaneous. The results also revealed that the studied materials are suitable adsorbents for the removal of cationic dyes.

Keywords: graphene oxide; adsorption; methylene blue; environment

Citation: Vassileva, P.; Tumbalev, V.; Kichukova, D.; Voykova, D.; Kovacheva, D.; Spassova, I. Study on the Dye Removal from Aqueous Solutions by Graphene-Based Adsorbents. *Materials* **2023**, *16*, 5754. https://doi.org/10.3390/ma16175754

Academic Editor: Sergio Morales-Torres

Received: 1 August 2023
Revised: 16 August 2023
Accepted: 21 August 2023
Published: 22 August 2023

1. Introduction

Water pollution is a vital challenge facing humanity, given that wastewater poses serious risks to both the environment and human health [1]. Within this context, the treatment of dyed wastewater has garnered considerable interest. Dyes represent a significant group of synthetic organic compounds extensively utilized across multiple industries, particularly textiles. These compounds possess intricate aromatic molecular structures, rendering them highly stable and resistant to degradation. Persistence, non-biodegradability, and toxicity are some of the reasons why dyes have a harmful effect on water and soil in the environment [2–4]. Even at low concentrations (less than 1 ppm), their presence in water bodies leads to the abnormal penetration of sunlight and inflicts serious harm on aquatic organisms [5]. Various approaches have been performed for dye-cleaning in water, such as ion exchange, precipitation, osmosis, etc. [6]. However, due to the simplicity of the method and low cost, adsorption is most widely applied.

The basic dye, methylene blue (MB) (tetra-methylthionine chloride), finds extensive applications in the textile sector to dye materials such as cotton, silk, and wood [7,8]. It is also employed in microbiology, diagnostics, and surgery [9]. Although it is not extremely dangerous, it can have harmful effects on human health, causing eye burns, increased heart rate, diarrhea, jaundice, cyanosis, quadriplegia, shock, and tissue necrosis [4,9,10]. Hence, there is a crucial need to conduct research focused on eliminating this dye from water solutions. Methylene blue frequently serves as a representative compound for the removal of cationic dyes and other organic pollutants from aqueous solutions. In response to this challenge, researchers have been striving to create advanced materials capable of efficiently purifying water contaminated with MB.

Numerous studies have extensively examined the process of MB dye adsorption, employing a wide variety of adsorbents such as synthetic and natural polymers, inorganic materials, clays, and others [4–19]. Recently, attention has been directed to the application of carbon-based materials as adsorbents due to their abundance, structure, chemical stability, and suitability for practical application. Among these compounds, graphene-based nanomaterials, comprising graphene, graphene oxide (GO), and reduced graphene oxide (RGO), as newly developed carbon modifications have garnered considerable attention in scientific research, due to their unique structure and remarkable electronic and physicochemical properties [20–22]. Over the past few years, they have been applied as cutting-edge components in various scientific and technological domains, including nanomedicine, electronics, energy storage, power generation, etc. [23,24]. The unique structure of graphene provides the opportunity for all carbon atoms to be at the surface and both sides of the sheets are accessible for pollutant molecules. Graphene oxide is produced by the chemical oxidation process of graphite. This material possesses a specific two-dimensional atomic crystal structure, comprising a single layer of closely packed sp^2-bonded carbon atoms arranged in a honeycomb lattice. Graphene oxide contains various types of functional groups on its surface (carboxyl, carbonyl, alkoxy, epoxide hydroxyl, and other oxygen-based functional groups) [25–28]. These oxygen-containing groups are responsible for the possibility of the surface functionalization of GO, which provides many opportunities for utilizing graphene oxide as a precursor in the synthesis of diverse nanocomposite materials. Reduced graphene oxide (RGO) is obtained by reducing the oxygen content of graphene oxide through thermal, chemical, or other approaches. This reduction process is carried out to enhance the honeycomb hexagonal lattice structure of the material [29]. Extensive research has demonstrated the potential of graphene family nanomaterials as effective nano-adsorbents. Their surface groups are involved in the adsorption processes. The mechanisms of adsorption have been identified as hydrogen bonding, electrostatic forces, π–π stacking, van der Waals forces, etc. [27,28]. Additionally, they act as weak acid cation exchange sites. Therefore, these materials have been widely reported as excellent adsorbents for the removal of diverse pollutants, including dyes, heavy metal ions, and radioactive substances [5,27–32]. Over the last decade, a variety of graphene-based materials, such as GO, RGO, and composite GO materials, have been effectively applied in the removal of methylene blue dye [1,2,5,7,9,27,28,33–40].

Data in the literature show a similar adsorption efficiency for both GO and RGO [41]. Many authors have reported that the hydrophilic character of GO may be a drawback for its practical use in the removal of dyes from aqueous media. They pointed out that GO disperses easily in water, which on the one hand, worsens its adsorption properties, and on the other leads to the difficult process of its separation from water. From this point of view, the RGO is a more appropriate adsorbent for practical use but possesses a lower number of surface functional groups that could embarrass the adsorption process.

For these reasons, the present study applied a new approach for handling the adsorption of MB from aqueous solutions; namely, the use of a monolith with a 3D-RGO hierarchical structure as an adsorbent. Its morphological and chemical uniformity allows for much easier use in practice than powders [42]. Moreover, the 3D-RGO adsorption properties can be further improved by additional surface functionalization.

In this present investigation, a 3D-hierarchical N-functionalized monolith was developed and investigated for its ability to remove the cationic dye methylene blue from water. The 3D-reduced graphene oxide monolith was successfully synthesized by the hydrothermal reduction of graphene oxide in the presence of ammonia. To evaluate its performance, it was compared with that of the starting material graphene oxide, and with that of reduced graphene oxide. Additionally, the study investigated the kinetics, thermodynamics, and desorption potential of methylene blue from the three materials.

2. Materials and Methods

2.1. Materials Preparation

Graphene oxide. GO was synthesized by a Tour method [43] from synthetic graphite (Sigma-Aldrich, Saint Louis, MO, USA, 99% carbon, 50 meshes), with the use of appropriate amounts of concentrated sulfuric acid, phosphoric acid, and potassium permanganate. The suspension was stirred for 1 h in an ice bath, followed by the addition of distilled water. The resulting material was washed to pH 6.5 with water. Then, by sonication of the GO for 1 h using the ultrasonic processor Sonix (Sonics & Materials, Inc., Newtown, CT, USA) (20 KHz, 750 W) the effective exfoliation to GO sheets was performed.

Reduced graphene oxide. RGO was obtained from the synthesized GO by reduction with L-ascorbic acid. Such prepared RGO was dried at 363 K for 12 h and thermally treated for 3 h at 873 K in Ar flow.

Monolithic 3D-RGO. The 3D graphene-based material was prepared by gelation of as-synthesized GO suspension under basic conditions. The aqueous GO suspension with a concentration of 25.5 $mg.mL^{-1}$ was prepared by evaporation. In an autoclave equipped with a Teflon vessel, 8 mL of the GO suspension was mixed with 2 mL of concentrated NH_4OH. The mixture was subjected to hydrothermal treatment at 433 K for 15 h in order to obtain a monolithic material. The preparation is based on a reduction process of GO to create self-assembled hierarchical porous structures, thus forming a macroscopic monolith. The obtained 3D graphene-based material was frozen in freeze-dry equipment the Alpha-Christ Freeze-Dryer (Christ, Osterode am Harz, Germany) for about 18–20 h using liquid nitrogen. The sample was denoted as M-RGO-N.

2.2. Characterization

The phase composition of the materials was studied by powder X-ray diffraction (XRD). The diffractograms were collected within the 5–80° 2θ range on a Bruker D8-Advance Diffractometer (Karlsruhe, Germany) with CuK_{α} radiation and a LynxEye detector. The EVA software package supported by the ICDD-PDF-2(2021) database was applied for phase identification.

The morphology of the samples was studied by combining scanning electron microscopy (SEM) and transmission electron microscopy (TEM). SEM images were obtained on JEOL JSM-6390 (Tokyo, Japan) and TEM was performed on a TEM JEOL 2100 at 200 kV. Standard C/Cu grid was applied for loading the samples on it.

Low-temperature nitrogen adsorption at 77 K on Quantachrome Nova 1200e (Anton Paar Quanta Tech Inc., Boynton Beach, FL, USA) was performed for the determination of the texture parameters. A BET equation was used for the determination of the specific surface area; the total pore volumes and the mean pore diameters were calculated at $p/p_0 \approx 0.99$, according to the Gurvitch rule. The pore-size distributions were made by NLDFT (slit pores, equilibrium model).

The XPS measurements were performed on an AXIS Supra electron spectrometer (Kratos Analytical Ltd., Manchester, UK) using Al K_{α} radiation (1486.6 eV). The photoelectron spectra were analyzed in C1s, O1s, and N1s peak regions.

Fourier-transform infrared spectra (FTIR) of GO, RGO, and M-RGO-N in KBr were collected using a Thermo Nicolet Avatar 360 FTIR spectrometer (Thermo Fisher Scientific, Waltham, MA, USA), with a resolution of 2 cm^{-1}.

2.3. Adsorption Studies

Adsorption tests were conducted in 50 mL conical flasks with a pre-weighed amount of 10 mg of adsorbent mixed with 10 mL of methylene blue dye solution. The mixtures were then subjected to shaking on a rotary shaker for varying periods. The solid particles were separated from the solution through centrifugation. The concentrations of methylene blue were analyzed using a Spekol 11 spectrophotometer (Carl Zeiss Industrielle Messtechnik GmbH, Jena, Germany), measuring the absorbance at λ_{max} = 650 nm. Batch adsorption experiments were conducted to examine the impact of contact time, pH, initial MB concentration, and temperature on the adsorption of methylene blue.

The equilibrium adsorption capacity (Qe, mg/g) was determined using the following equation:

$$Q_e = (C_0 - C) \times V/g \quad (1)$$

Here, C_0 and C_e represent the initial and equilibrium concentrations of MB (mg L^{-1}), respectively. V stands for the solution volume (L), and m corresponds to the mass of the sorbent (g). It is essential to note that each experiment was conducted in triplicate to ensure accuracy and reliability.

Methylene blue (MB) with a chemical formula of $C_{16}H_{18}N_3ClS$ and a molecular weight of 319.85 g mol^{-1} was procured from Merck (Darmstadt, Germany). To prepare a stock solution of MB dye with a concentration of 500 mg L^{-1}, an appropriate quantity of dye powder was dissolved in deionized water (pH = 6.9). Subsequently, all the necessary working solutions with desired concentrations were prepared by diluting the stock solution with deionized water. This process allowed for the accurate preparation of solutions for the experiments. The adsorption behavior of methylene blue (MB) was investigated under various experimental conditions. To explore the impact of contact time, experiments were conducted at pH 6.9 with MB concentration of 50 mg L^{-1}. The influence of the solution's acidity on MB removal efficiency was examined over a pH range of 2.0 to 10.0, using an MB concentration of 50 mg L^{-1}. To understand the effect of the initial MB concentration on adsorption capacity, different concentrations of MB ranging from 10 to 100 mg L^{-1} were tested at a pH of 8.0. Additionally, the influence of temperature was studied at 293, 303, and 333 K while maintaining a dye concentration of 100 mg L^{-1}. Before mixing the suspensions, the pH of the initial solutions was carefully adjusted to the desired level using 0.1 M HCl and 0.1 M NaOH solutions. The desorption of MB was examined in a batch system using four different eluents: 2 M HCl, ethanol, water: ethanol (50:50), and 2 M HCl: ethanol (50:50). A weighed amount of the pre-adsorbed materials was added to 10 mL of each eluent and stirred for 24 h. The eluents were subsequently filtered and analyzed to determine the amount of desorbed methylene blue.

3. Results and Discussion

3.1. Materials Characterization

The XRD patterns of the GO, RGO, and M-RGO-N samples are presented in Figure 1. The pattern of GO comprises two main peaks—the first located at 12.3° 2θ (d = 7.1Å), and the second one at 42.4° 2θ (d = 2.1Å). The first peak is connected with the (002) interplanar distance of the graphite-type cell and reflects the increase of the **c** unit cell parameter perpendicular to graphite sheets. Such an increase is due to the intercalation of various molecular groups between the graphite sheets during the oxidation process. The FWHM of the (002) peak corresponds to a 8 nm size within this direction (≈10 graphene sheets). The second peak corresponds to the (100) plane of graphite. The patterns correspond to those already published in [28,39]. Upon reduction, the (002) peak shifts to the higher angle and for the RGO sample it is situated at 25.8° 2θ (d = 3.4 Å), similar to [34,36]. The FWHM of this peak reveals a 4 nm size within this direction, indicating a preserving of the number of stacked graphene sheets as in GO. As for the M-RGO-N pattern, the (002) peak is at 26.6° 2θ (d = 3.3Å) which corresponds to the reduced graphene oxide. The size within the <002> direction is about 13 nm (≈40 graphene sheets). The results show that the hydrothermal treatment led to the effective reduction of GO to RGO, although with an increased crystallite size.

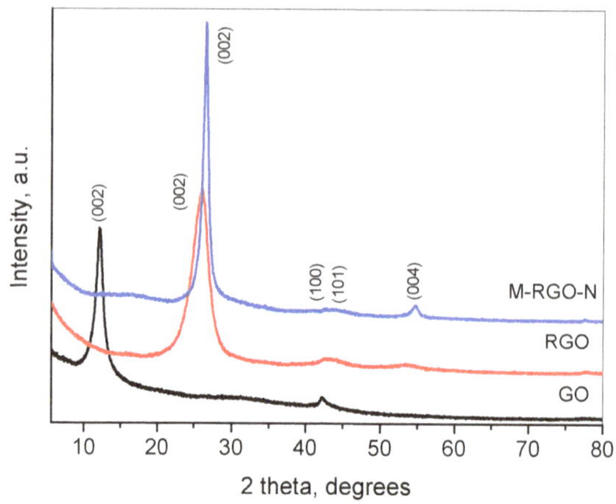

Figure 1. XRD patterns of GO, RGO, and M-RGO-N.

The morphology of GO, RGO, and M-RGO-N combining SEM and TEM images of the corresponding samples is shown in Figure 2.

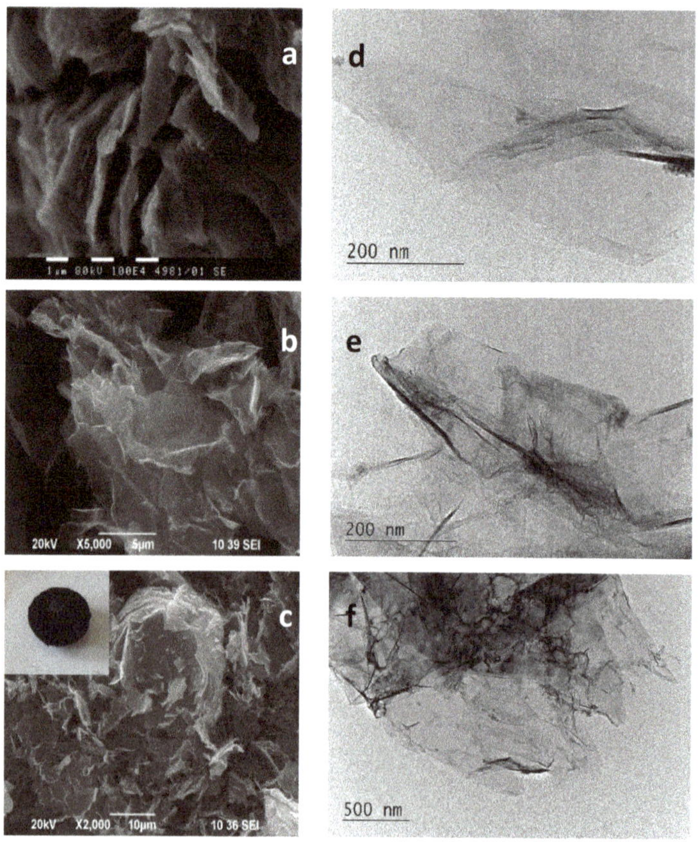

Figure 2. SEM (**left**) images of: (**a**) GO; (**b**) RGO, and (**c**) M-RGO-N and TEM (**right**) images of: (**d**) GO; (**e**) RGO, and (**f**) M-RGO-N.

A sheet-like structure of all materials is clearly observed which reveals successful exfoliation of the initial graphite material. All samples consist of well-preserved graphene nanosheets, slightly rippled, and stacked in an arbitrary way, allowing the exposition of a large number of active sites for adsorption. The difference between them is in the size of the sheets and in the presence of a higher number of wrinkles and ripples in the reduced RGO and M-RGO-N samples. For the latter, the size of the sheets is the smallest and they are more crumpled, plicated, and broken. This fact is a prerequisite for increased specific surface area and pore volume.

The N_2 adsorption–desorption isotherms and the pore-size distribution of GO, RGO, and M-RGO-N are presented in Figure 3a,b, respectively. The texture parameters of the above-mentioned samples are placed in Table 1. GO and RGO exhibit II-type adsorption isotherms, according to the IUPAC classification, which is characteristic of a nonporous or macroporous material. The very narrow hysteresis is of type H3, which implies slit-shaped pores between the graphene sheets [44]. The isotherm of M-RGO-N shows a mixed-type isotherm—type I (attributed to micropores filling at low p/p_0) and type II–pseudo II (connected with mesopores filling at high p/p_0) [44]. The hysteresis shape H3 type reveals the non-rigid nature of the monolith, and the closure point of the adsorption and the desorption branches at about ≈0.4 p/p_0 is consistent with the metastability of the condensate. The observed hysteresis loop is shifted to a higher p/p_0 pressure and is an indication of the appearance of some macropores, as could also be seen from the pore-size distribution (Figure 3b). The pore-size distribution and the presence of a small volume of micropores (V_{mi}, Table 1) are evidence for the formation of a hierarchical pore structure of the hydrothermally obtained M-RGO-N.

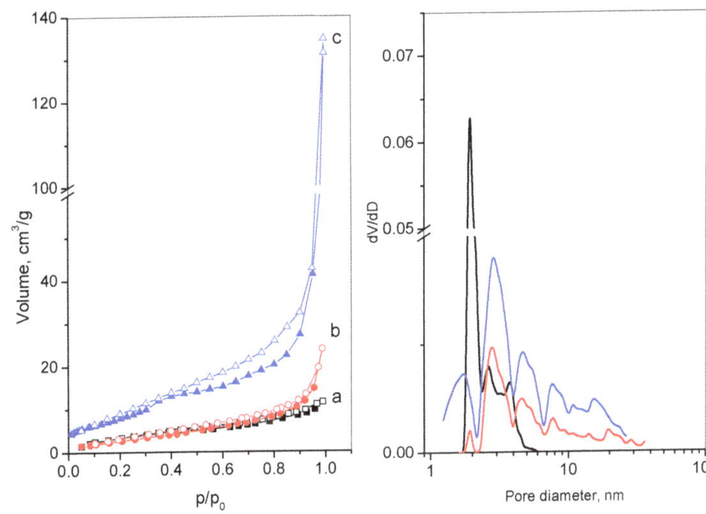

Figure 3. N_2 adsorption–desorption isotherms (**left**) and pore-size distributions (**right**) of: (a) GO—black line; (b) RGO—red line; (c) M-RGO-N—blue line.

Table 1. Texture characteristics of GO, RGO, and M-RGO-N.

Sample	S [1] m^2/g	V [2] cm^3/g	D_{av} [3] nm	V^4_{mi} cm^3/g
GO	14	0.02	5	-
RGO	18	0.05	12	-
M-RGO-N	40	0.18	18	0.02

[1] BET; [2,3] at $p/p_0 \approx 0.99$; [4]—Dubinin–Radushkevich method.

The texture parameters of GO and RGO are similar in their values.

The simultaneous process of GO reduction and N-modification during the hydrothermal synthesis leads to the enhanced specific surface area, total pore volume, and average pore diameter for M-RGO-N and is an indication of more successful exfoliation of the graphite. This confirms the results from XRD, SEM, and TEM. However, although freeze-drying was applied to restrict the effect of the capillary forces during the material's drying, the obtained graphene sheets possess a relatively low surface area compared to the theoretical one, probably due to the incomplete exfoliation or sticking of the graphene sheets.

Figure 4 presents the Fourier-transform infrared spectra of the investigated materials. The GO spectrum (Figure 4a) consists of broadband at the region 3200–3600 cm^{-1} assigned to stretching vibrations of O–H. The bands at 2925 and 2854 cm^{-1} could be due to the C–H stretching vibrations [45]. The next bands at 1730 cm^{-1} and 1574 cm^{-1} are attributed to C=O and C=C in aromatic rings, respectively. Some overlapping bands form a broad band in the region 1400–890 cm^{-1}, which should be a reflectance of the coexisting vibrations of epoxy, carboxyl, and carbonyl groups [46]. This band appears with decreased intensity in the spectrum of RGO, which indicates a successful reduction. The reduction also led to the disappearance of the band at 1730 cm^{-1} (C=O) and the appearance of a band at 1630 cm^{-1}, assigned to the C=C bond. The spectrum of M-RGO-N shows an additional hump at about 3300–3100 cm^{-1} assigned to the stretching vibrations of N–H bonds. The intense band in 1285–1030 cm^{-1} indicates, similar to GO, the presence of oxygen-containing as well as nitrogen-containing groups [47].

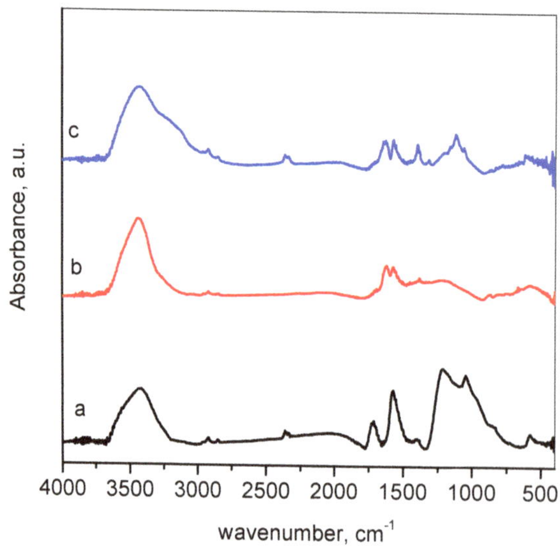

Figure 4. FTIR spectra of: (a) GO—black line; (b) RGO—red line; (c) M-RGO-N—blue line.

XPS was performed in order to investigate the surface of the synthesized adsorbents. Figure 5a–c presents the C1s spectra of GO, RGO, and M-RGO-N, and the N1s spectrum of M-RGO-N, respectively. It is evident, that the main peak in the C1 spectrum of GO is situated at 284.5 eV and could be attributed to the C=C sp^2 graphitic carbon bonds. This XPS peak is wide and asymmetric with shoulders at higher binding energies, suggesting the presence of sp^3 carbon and oxygen-containing functional groups. The XPS C1s spectra for RGO and M-RGO-N are similar, with peaks with binding energy at 284.3 eV, less asymmetric than the peak for GO, which could be related to the reduction of the oxygen-containing groups. The N1s spectrum for M-RGO-N evidences the successful incorporation of nitrogen atoms during the hydrothermal production of the monolith and the surface nitrogen concentration was determined to be 4.5 at.%. The N1s peak is centered at 399.8 eV, but is asymmetric, suggesting a presence of nitrogen atoms in pyridinic, pyrrolic, or amine species [48].

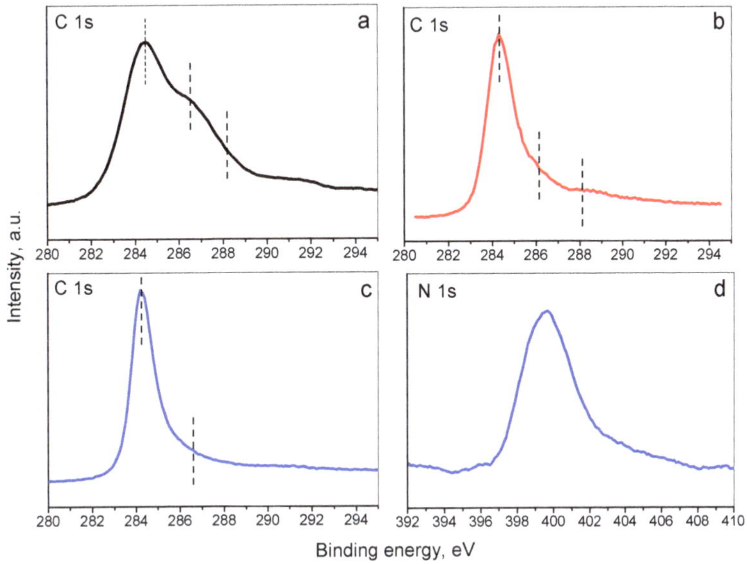

Figure 5. XPS photoelectron spectra of: (**a**) C1s of GO; (**b**) C1s of RGO; (**c**) C1s of M-RGO-N; (**d**) N1s of M-RGO-N.

3.2. Adsorption Studies

3.2.1. Effect of pH on the Adsorption Process

The pH of the aqueous solution affects the adsorption of dyes by changing the surface charge and ionization of the adsorbent as well as the dye. The regulation of electrostatic interactions between the adsorbent and the adsorbate plays a pivotal role in this context. To explore the impact of the initial solution's pH on the adsorption of MB by the investigated materials, experiments were conducted across a pH range from 2.0 to 10.0. The outcomes of these experiments are depicted in Figure 6.

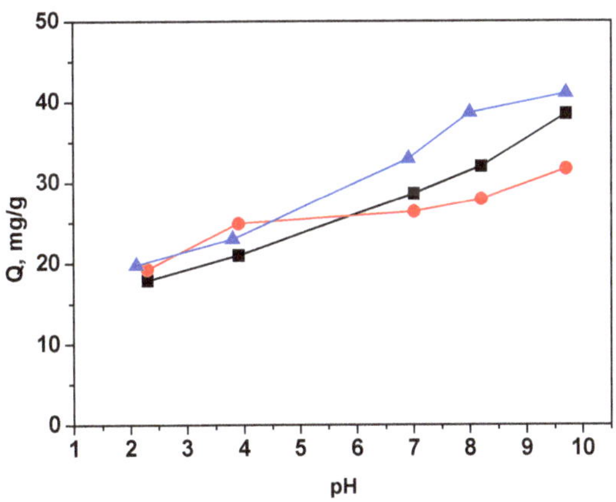

Figure 6. Effect of pH on MB adsorption by GO (black), RGO (red), and M-RGO-N (blue).

It is seen that the adsorption steadily increased with increasing the solution pH on the three sorbents. At low pH values, the surface of the adsorbents was positively charged (due to protonation of COOH, OH, etc. groups). Therefore, repulsion occurs between the surface of the studied samples and the cationic dye molecules, resulting in lower adsorption. At higher pH values, an improvement in the removal capacity of MB is seen. The increase in adsorption with increasing pH is due to the deprotonation of the functional groups present on the surface of the GO-based materials. It generates the negative charges on the surface of the adsorbents that attract the dye molecules, the electrostatic attraction becoming stronger as the pH of a solution increases. Additionally, the presence of liberated H^+ ions can be reduced through a neutralization process, which, in turn, enhances the adsorption of MB [49]. The maximum adsorption performance of MB at pH 10 and higher was attained.

The final pH (pH_f) of the adsorption solution is evaluated to calculate ΔpH (pH_f—pH_i). ΔpH was plotted against pH_i to determine the point of zero charge (PZC) associated with the investigated materials. The pH_{pzc} of GO, RGO, and M-RGO-N was observed to be around 2.8, 3.0, and 4.8, respectively. It is evident that these three materials exhibit a negative charge, even when placed in an acidic environment.

3.2.2. Effect of Contact Time and Kinetic Study

The impact of contact time on the MB removal by GO, RGO, and M-RGO-N was explored throughout a 5 h period. Figure 7a–c shows the adsorbed amount, % removal, and concentration of MB as a function of contact time, respectively. As the agitation time increased, the adsorbed amount and % removal of MB (according to [50]) showed a corresponding increase, reaching its maximum value within 120 min for both GO and M-RGO-N. With RGO, the dye adsorption required less time to reach equilibrium—60 min. As can be seen from Figure 7c, the largest decrease in the initial MB concentration upon reaching equilibrium is achieved with the M-RGO-N material. Nevertheless, these results suggest a good affinity of the three investigated sorbents for MB. All subsequent experiments were conducted with a contact time of 5 h.

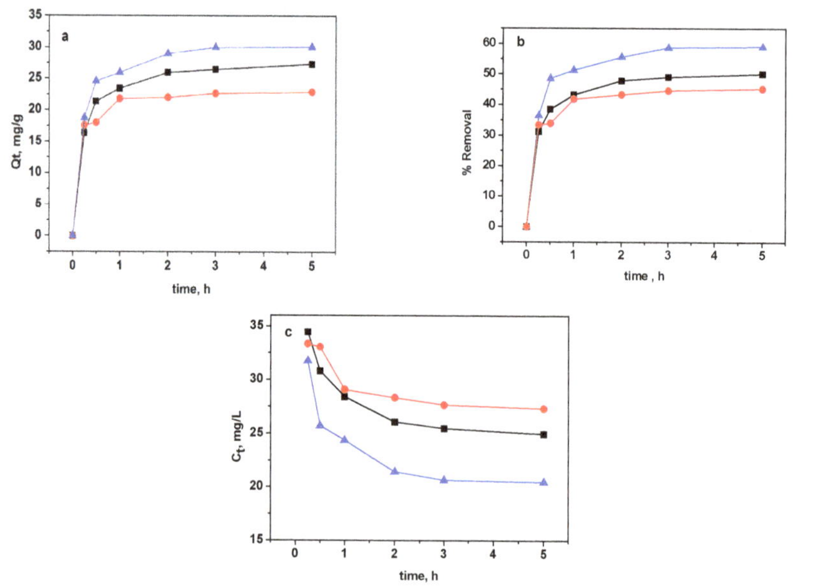

Figure 7. Plots of the adsorbed amount (**a**), % removal (**b**), and concentration (**c**) of MB as a function of time: for GO (black), RGO (red), and M-RGO-N (blue).

The adsorption mechanism can be elucidated using various models. Among them, the commonly employed models for investigating adsorption kinetics include the pseudo-first-order, pseudo-second-order kinetic models, and intraparticle diffusion models.

$$log(Q_e - Q_t) = log(Q_e) - (k_1/2.303)t \tag{2}$$

$$(t/Q_t) = (1/k_2Q_e) + (1/Q_e)t \tag{3}$$

$$Q_t = k_{id}\, t^{1/2} + C \tag{4}$$

Here, Q_t (mg g^{-1}) signifies the adsorbed amount of MB at a specific time t, Q_e (mg g^{-1}) represents the equilibrium adsorption capacity, k_1 (h^{-1}), k_2 (g $mg^{-1}h^{-1}$) and k_{id} (mg $g^{-1}h^{-1/2}$) represent the rate constant of pseudo-first-order adsorption, pseudo-second-order adsorption and intraparticle diffusion constant, respectively. Additionally, the values of k_1 were determined from the slope of the plots $log(Q_e - Q_t)$ versus t (Figure 8a), k_2 values were calculated from the slope of the plots t/Q_t versus t (Figure 8b), and k_{id} values were derived from the slope of the plots Q_t versus $t^{1/2}$ (Figure 8c). The intercepts of the curves from the adsorption models were utilized to determine the essential parameter: equilibrium capacity (Q_e and C). The correlation coefficients (r^2-values), rate constants, and other relevant parameters for the three kinetic models (pseudo-first-order, pseudo-second-order, and intraparticle diffusion) were computed and are presented in Table 2.

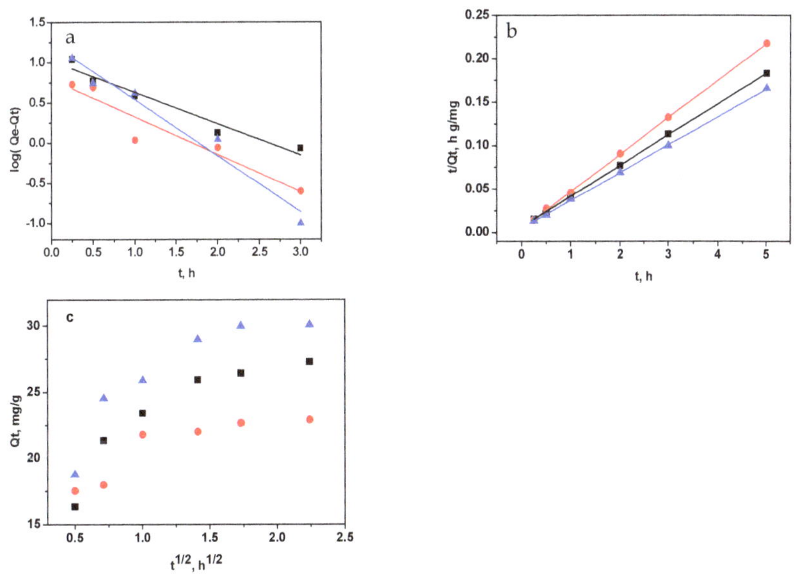

Figure 8. Application of (**a**) pseudo-first-order model, (**b**) pseudo-second-order model, and (**c**) intraparticle diffusion model for the adsorption of MB by GO (black), RGO (red), and M-RGO-N (blue).

The adsorption of methylene blue on the three studied materials exhibits kinetics that closely resemble a second-order process. The correlation coefficients calculated for this model were found to be very close to unity, indicating a strong agreement with the experimental data. Furthermore, the equilibrium capacities estimated from the pseudo-second-order model showed a strong correlation with the experimentally observed Q_e values. The findings indicate that the MB adsorption process can be effectively described by the pseudo-second-order kinetic equation. These observations suggest that the rate of adsorption reaction is affected by two main factors: the concentration of MB in the solution and the abundance of available adsorption sites on the surfaces of the adsorbents [51]. Moreover, the adsorption behavior of methylene blue for all the studied materials is predominantly governed by chemical-controlling processes.

Table 2. Kinetic parameters for MB adsorption.

Sample	Pseudo-First-Order Constants			Pseudo-Second-Order Constants			Intraparticle Diffusion Constants		
	Q_e (mg g^{-1})	k_1 (h^{-1})	r^2	Q_e (mg g^{-1})	k_2 (g $mg^{-1}h^{-1}$)	r^2	k_{id} (mg g^{-1} $h^{-1/2}$)	C (mg g^{-1})	r^2
GO	10.46	0.169	0.9399	28.24	5.531	0.9996	5.650	16.33	0.8695
RGO	6.19	0.203	0.8820	23.41	8.173	0.9997	3.499	16.02	0.8728
M-RGO-N	17.18	0.303	0.9545	31.15	6.275	0.9996	5.912	11.12	0.8808

To assess the significance of diffusion as a controlling process in determining the rate of MB adsorption on GO, RGO, and M-RGO-N, the diffusion kinetic equation proposed by Weber and Morris was also employed. The findings from the kinetic analysis suggest that the adsorption process involves multiple stages. The kinetics plot exhibits two distinct linear regions, implying a complex adsorption mechanism. However, the results clearly indicate that diffusion does not act as the rate-determining step during MB adsorption in this particular case.

3.2.3. Adsorption Isotherms

To investigate the impact of the initial concentration of MB on the adsorption capacities of the studied GO-based materials, a range of MB concentrations from 10 to 100 mg L^{-1} was used at pH 8.0 (as shown in Figure 9a). pH 8 was chosen for the study because at high pH values, the molecular structure of MB can undergo changes due to stepwise demethylation [33]. The adsorption isotherm provides valuable insights into the distribution of adsorbate molecules or ions between the solid adsorbent and the solution at equilibrium. It describes how the quantity of adsorbate on the adsorbent's surface relates to its concentration in the solution, under specific conditions such as temperature and concentration. Different types of model adsorption isotherms help in understanding the adsorption process and characterizing the adsorbent's behavior in terms of its capacity and affinity for the adsorbate. In the present study, MB equilibrium data were analyzed using three widely used isothermal models: the Langmuir, Freundlich, and Dubinin–Radushkevich models, whose equations are formulated based on different principles, as described in [51]. The Langmuir model hypothesizes that adsorption occurs at uniform and homogenous sites within the adsorbent, making it suitable for describing monolayer adsorption processes. On the other hand, the Freundlich model suggests that the adsorbates are taken up on a heterogeneous surface through multi-layer adsorption. The Dubinin–Radushkevich isotherm, in contrast, sheds light on the adsorption mechanism by employing potential theory to explain the adsorption process. The compatibility of the experimental data with a particular equation provides insight into the probable mechanism of adsorption. The linear graphs for the three isotherm models are presented in Figure 9b–d. The calculated values for isotherm constants and correlation coefficients are given in Table 3. The data for the adsorption of MB on the three GO-based materials were well fitted by the Langmuir isotherm model (Figure 9b), as demonstrated by the high correlation coefficients (r^2 greater than 0.996), compared to those for the Freundlich, and Dubinin–Radushkevich (D-R) isotherm models. These findings suggest that the predominant mechanism of MB adsorption involves binding to a relatively uniform surface and the formation of a monolayer.

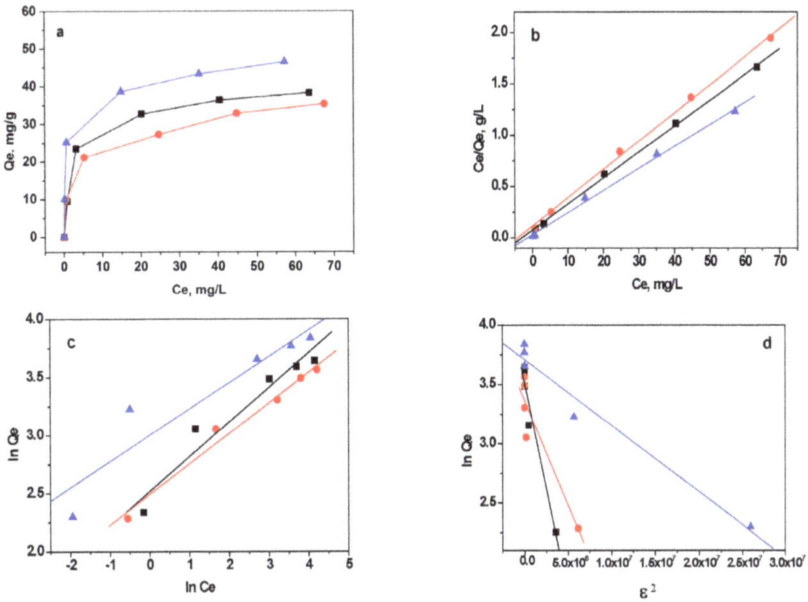

Figure 9. (**a**) Experimental adsorption isotherms and the application of (**b**) Langmuir, (**c**) Freundlich and (**d**) Dubinin–Radushkevich isotherms to the adsorption of MB on GO (black), RGO (red), and M-RGO-N (blue).

Table 3. Isotherm constants and correlation coefficients for MB adsorption.

Sample	Langmuir Parameters [1]			Freundlich Parameters [2]			D-R Parameters [3]		
	Q_0 (mg g^{-1})	K_l (L mg^{-1})	r^2	k_F (mg^{1-n}L^n g^{-1})	n (L mg^{-1})	r^2	Q_m (mg g^{-1})	E (kJ mol^{-1})	r^2
GO	39.81	0.335	0.9984	12.39	3.344	0.9546	32.46	1.20	0.9352
RGO	36.63	0.241	0.9977	12.07	3.802	0.9512	27.99	1.74	0.8227
M-RGO-N	49.38	0.711	0.9958	20.09	4.415	0.9508	33.32	3.78	0.9567

[1] $C_e/Q_e = 1/K_L Q_0 + C_e/Q_0$; [2] $lnQ_e = lnk_F + (1/n)\ lnC_e$; [3] $lnQ_e = lnQ_m - \beta\varepsilon^2$, $\varepsilon = RTln(1 + 1/C_e)$, $E = 1/(-2\beta)^{1/2}$.

GO shows a relatively higher MB adsorption capacity compared to RGO because the surface of GO comprises more oxygen-containing functional groups and therefore, more adsorption sites than in RGO. The M-RGO-N material shows the highest adsorption capacity towards MB. This could be explained by the texture parameters (specific surface areas, pore volumes) and the presence of hierarchical pores, as well as the surface functionalization with nitrogen-containing groups that may play the role of additional active centers for dye adsorption. Thus, the combination of its porous structure and surface functionalization with nitrogen groups makes M-RGO-N a highly efficient adsorbent for MB.

The capacities towards MB obtained in the present study are compared to other adsorbents based on GO reported in the literature (Table 4) [28,34–40,49,52–55]. It could be clearly seen that the maximum Langmuir capacity varies very significantly, depending on the type of material, the conditions of the experiment, and the mechanism of adsorption. It is worth mentioning that the adsorption capacities for MB of the materials in the present study display comparable values to the data presented in the literature, which makes the investigated graphene-based materials appropriate for MB removal from aqueous solutions.

Table 4. Comparison of the values of the adsorption capacities of graphene-based materials with respect to MB reported in the literature.

Adsorbent	Q_{max} (mg/g)	Conditions	Reference/ Year
Monolayer GO	440.5–803.7	pH = 3, 7, 11; 2.35 mg GO; MB range: 0.05–0.475 g/L; 298 K	[28] (2022)
rGO	81	283 K, 10–50 mg/L, 0.01 g adsorbent	[34] (2013)
GO	119.0	293 K, 180 min, pH = 4, 600 mgL^{-1}	[35] (2021)
rGO	64.8 and 80.1	MB 20–50 mgL−1 dosage of 50 mgL^{-1}; pH = 4	[36] (2019)
GO	429.5	5–25 ppm MB, 10 mg adsorbent, 313 K, 15 min, pH 8	[37] (2019)
rGO	121.9	pH = 6, 298 K, 10–100 mg/L MB, 200 mg adsorbent^{-1}	[38] (2020)
GO	311.7	6–100 ppm MB, 0.5 gL^{-1} adsorbent, 293 K, pH = 10	[39] (2021)
GO-CS composites	7.5	MB 10 mg/L, pH = 5, adsorbent 0.143 g/L, 125 min	[40](2022)
Chitosan-Functionalized GO	23.3	MB 10–50 mg/L, 35 min, GO@CS 50 mg/10 mL, pH = 7, 25 °C	[49] (2022)
rGO/TNT	26.3	MB 1–60 mg/L, 25 °C, 0.1 g/10 mL adsorbent, pH 6.8	[52] (2019)
GO/polyaniline rGO/polyaniline	14.2 19.2	MB 10–80 mg/L, 270 min, 0.3–0.4 g adsorbent, pH = 6.5	[53] (2019)
GO	48.8–598.8	298 K, pH = 7, 0.015 g GO, MB 5–600 mg dm^{-3}	[54] (2014)
GO	17.3	MB 10–50 mg/L, 25 °C, 0.22 mg/mL adsorbent	[55] (2011)
GO RGO M-RGO-N	39.8 36.6 49.4	MB 10–100 mg/L, 293 K, pH = 8, 0, 10 mg adsorbent	Present study

The n-values obtained from the Freundlich isotherm model (Figure 9c) for GO, RGO, and M-RGO are 3.34, 3.80, and 4.41, respectively [51]. These values point to a favorable adsorption process for all three materials, as higher n-values signify stronger adsorption. On the other hand, the Dubinin–Radushkevich isotherm model provides insights into the adsorption energy. The calculated mean free energy values for GO, RGO, and M-RGO-N are 1.20, 1.74, and 3.78 kJ mol^{-1}, respectively. These values suggest that the predominant adsorption mechanism is physisorption.

3.2.4. Thermodynamic Studies

The thermodynamic parameters of adsorption from solutions are crucial in providing valuable insights into the type and mechanism of the adsorption process. They offer information about the driving forces that govern the interaction between the adsorbate and the adsorbent. By studying these thermodynamic parameters, researchers can determine the feasibility, spontaneity, and nature of the adsorption process, which is essential for optimizing and understanding the various practical applications of adsorption in different fields. To investigate the effect of temperature on MB removal, experiments were conducted at different temperatures: 293 K, 303 K, and 333 K. As the temperature increased, the adsorbed amounts of MB on each adsorbent also increased. This observation suggests that the adsorption process is characterized as endothermic.

The changes in Gibbs free energy (ΔG^0), enthalpy (ΔH^0), and entropy (ΔS^0) were calculated using the following equations:

$$K_d = Q_e/C_e \tag{5}$$

$$\Delta G^0 = -RT\, lnK_d \tag{6}$$

$$lnK_d = \Delta S^0/R - \Delta H^0/RT \tag{7}$$

In the equations provided, K_d represents the equilibrium constant, R is the gas constant (measured in J mol^{-1}K^{-1}), and T stands for the temperature (in Kelvin). By plotting lnK_d against $1/T$, we can determine the enthalpy change, ΔH^0, from the slope of the graph, and the entropy change, ΔS^0, from its intercept (as shown in Figure 10). The values of the thermodynamic parameters ΔG^0, ΔH^0, and ΔS^0 are compiled and presented in Table 5.

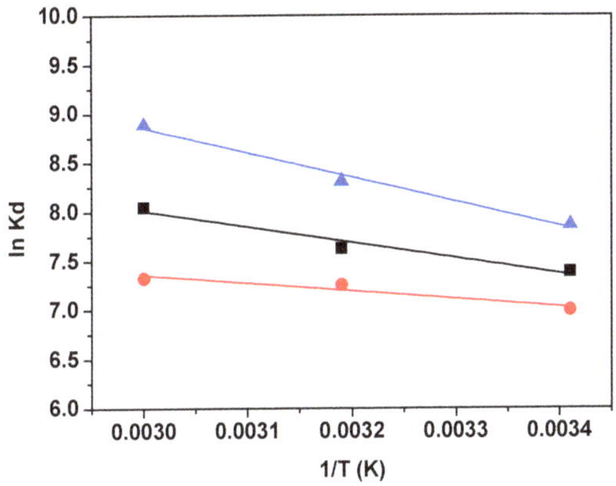

Figure 10. The relationship between the equilibrium constant (K_d) and temperature: GO (black), RGO (red), and M-RGO-N (blue).

Table 5. Values of ΔG^0, ΔH^0, and ΔS^0 at different temperatures for MB adsorption.

Adsorbent	ΔH^0, kJ mol^{-1}	ΔS^0, J mol^{-1} K^{-1}	ΔG^0, kJ mol^{-1}		
			293 K	313 K	333 K
GO	13.72	107.82	−17.87	−18.95	−22.18
RGO	7.28	83.14	−17.05	−17.91	−20.36
M-RGO-N	21.41	138.01	−19.03	−20.41	−24.55

Changes in the standard free energy at different temperatures have negative values, indicating that the adsorption of MB dye on the materials GO, RGO, and M-RGO-N is feasible and spontaneous. The ΔG^0 value for the MB−M-RGO-N system is more negative compared to the other two systems, suggesting more spontaneous adsorption of MB onto M-RGO-N. Generally, ΔG^0 values for physisorption range between 0 and 20 kJ mol^{-1}, while chemisorption is between 80 and 400 kJ mol^{-1} [56]. The calculated ΔG^0 values in the present study reveal that physisorption is the dominating mechanism. For the MB−M-RGO-N system, ΔG^0 values are in the range −19.03–24.55 kJ mol^{-1}, which indicated a more complex adsorption mechanism. On the other hand, the positive values of ΔH^0 (in the range 7.28–21.41 kJ mol^{-1}) confirmed that the adsorption was mainly physical and the fact that the adsorption efficiency increased with the temperature increase. The positive values of the entropy change (ΔS^0) correspond to a decrease in the molecular orderliness at the interface between the solid and liquid phases during the adsorption process, leading to an increase in the degree of freedom of the adsorbed dye molecules. The positive value of ΔS^0 may indicate the irreversibility and stability of the adsorption [38,57]. The observed small positive values of ΔS^0 in the studied systems suggest that there are no significant structural changes in the surfaces of the adsorbents during the adsorption process.

Adsorbent regeneration is a problem of great importance. Therefore, the desorption capability of MB from the studied adsorbents was examined. In the present study, batch desorption experiments were conducted using different solvents, including 2 M HCl, ethanol, water: ethanol (50:50), and 2 M HCl: ethanol (50:50). However, the desorption efficiency was found to be remarkably low. The highest percentage of MB desorption was achieved using a water: ethanol mixture, with values of 18% for GO, 6% for RGO, and 14% for M-RGO-N, respectively. The limited desorption percentage can be attributed to the special structure of the studied materials. Due to the small size of the MB molecule, it easily becomes adsorbed and forms a stable complex with the adsorbent, making the desorption process highly difficult [58]. On the other hand, this outcome could be attributed to a more intricate combined adsorption mechanism, a mixed physisorption–chemisorption process. It is possible that not only electrostatic interactions but also π–π interactions between the aromatic rings of MB and graphene oxide-based materials as well as hydrogen bonding are involved in the adsorption process [31].

4. Conclusions

The removal efficiency of three graphene-based materials for the adsorption of methylene blue from water was studied. The graphene materials—graphene oxide, reduced graphene oxide, and novel 3D-reduced graphene oxide monolith—were successfully prepared via the modified Tour method, reduction with ascorbic acid, and hydrothermal reduction in the presence of ammonia, respectively. The results from various characterization techniques revealed in detail the structural, morphological, and surface characteristics of the adsorbents. The prepared monolith was confirmed to be 3D-hierarchical and nitrogen functionalized. The influence of main process factors (contact time, initial dye concentrations, solution pH, and temperature) on the ability of these materials to adsorb cationic dyes was studied. The adsorption procedure was effectively explained through the application of pseudo-second-order and Langmuir models. The determined entropy, enthalpy, and Gibbs free energy values for the removal of MB showed that the adsorption of MB is feasible, spontaneous, and endothermic. The highest adsorption capacity was found to be exhibited by a 3D-reduced monolith of graphene oxide—49.4 mg g^{-1} at 293 K. The combination of its porous structure and surface functionalization with nitrogen groups makes M-RGO-N an efficient adsorbent for MB; moreover, it can be easily separated from the solution. Nevertheless, the results established that the three studied materials are suitable adsorbents for the removal of cationic dyes.

Author Contributions: Conceptualization, P.V., D.K. (Daniela Kovacheva), and I.S.; methodology, P.V., D.K. (Daniela Kovacheva), and I.S.; formal analysis, D.K. (Diana Kichukova), D.K. (Daniela Kovacheva), P.V., I.S., V.T. and D.V.; investigation, P.V., D.K. (Daniela Kovacheva), and I.S.; data curation, P.V., D.K. (Daniela Kovacheva), and I.S.; writing—original draft preparation, P.V., D.K. (Daniela Kovacheva), and I.S.; writing—review and editing, P.V., D.K. (Daniela Kovacheva), and I.S.; visualization, P.V., D.K. (Daniela Kovacheva), and I.S.; project administration, I.S. All authors have read and agreed to the published version of the manuscript.

Funding: This research was funded by the National Science Fund, Project KP-06-H 59/8.

Institutional Review Board Statement: Not applicable.

Informed Consent Statement: Not applicable.

Data Availability Statement: Data sharing is not applicable to this article.

Acknowledgments: The authors acknowledge the TwinTeam project D01-272 "European Network on Materials for Clean Technologies" for providing the opportunity to present the results at the SizeMat4 conference as well as for the financial publication support. In this investigation, research equipment of the distributed research infrastructure INFRAMAT supported by the Bulgarian Ministry of Education and Science (contract D01-172/28.07.2022) was used.

Conflicts of Interest: The authors declare no conflict of interest.

References

1. Basha, I.K.; El-Monaem, E.M.A.; Khalifa, R.E.; Omer, A.M.; Eltawei, A.S. Sulfonated graphene oxide impregnated cellulose acetate foated beads for adsorption of methylene blue dye: Optimization using response surface methodology. *Sci. Rep.* **2022**, *12*, 9339. [CrossRef] [PubMed]
2. Lellis, B.; Fávaro-Polonio, C.Z.; Pamphile, J.A.; Polonio, J.C. Effects of textile dyes on health and the environment and bioremediation potential of living organisms. *Biotechnol. Res. Innov.* **2019**, *3*, 275–290. [CrossRef]
3. Al-Tohamy, R.; Ali, S.S.; Li, F.; Okasha, K.M.; Mahmoud, Y.A.-G.; Elsamahy, T.; Jiao, H.; Fu, Y.; Sun, J. A critical review on the treatment of dye-containing wastewater: Ecotoxicological and health concerns of textile dyes and possible remediation approaches for environmental safety. *Ecotoxicol. Environ. Saf.* **2022**, *231*, 113160. [CrossRef]

4. Nezhad, A.A.; Alimoradi, M.; Ramezani, M. One-step Preparation of graphene oxide/polypyrrole magnetic nanocomposite and its application in the removal of methylene blue dye from aqueous solution. *Mater. Res. Express* **2018**, *5*, 025508. [CrossRef]
5. Paton-Carrero, A.; Sanchez, P.; Sánchez-Silva, L.; Romero, A. Graphene-based materials behaviour for dyes adsorption. *Mater. Today Commun.* **2022**, *30*, 103033. [CrossRef]
6. Shao, Y.M.; Wang, X.; Kang, Y.; Shu, Y.H.; Sun, Q.Q.; Li, L.S. Application of Mn/MCM-41 as an adsorbent to remove methyl blue from aqueous solution. *J. Colloid Interface Sci.* **2014**, *429*, 25–33. [CrossRef]
7. Zhang, H.; Han, X.; Yang, Z.; Zou, J.; Tang, H. Enhanced Adsorption of Methylene Blue on Graphene Oxide by Tuning the Oxidation Degree of Graphene Oxide. *J. Nanomater. Mol. Nanotechnol.* **2013**, *S1*, 003. [CrossRef]
8. Vassileva, P.; Voykova, D.; Uzunov, I.; Uzunova, S. Methylene blue adsorption by *Triticum monococcum* L. husks based materials. *CR Acad. Bulg. Sci.* **2018**, *71*, 1192–1199.
9. Martis, L.J.; Parushurama, N.; Sangappa, Y. Preparation, characterization, and methylene blue dye adsorption study of silk fibroin–graphene oxide nanocomposites. *Environ. Sci. Adv.* **2022**, *1*, 285–296. [CrossRef]
10. Senthilkumaar, S.; Varadarajan, P.R.; Porkodi, K.; Subbhuraam, C.V. Adsorption of methylene blue onto jute fiber carbon: Kinetics and equilibrium studies. *J. Colloid Interface Sci.* **2005**, *284*, 78–82. [CrossRef]
11. Rafatullah, M.; Sulaiman, O.; Hashim, R.; Ahmad, A. Adsorption of methylene blue on low-cost adsorbents: A review. *J. Hazard. Mater.* **2010**, *177*, 70–80. [CrossRef] [PubMed]
12. Ethaib, S.; Zubaidi, S.L. Removal of Methylene Blue Dye from Aqueous Solution Using Kaolin. *IOP Conf. Ser. Mater. Sci. Eng.* **2020**, *928*, 022030. [CrossRef]
13. Mohrazi, A.; Ghasemi-Fasaei, R. Removal of methylene blue dye from aqueous solution using an efficient chitosan-pectin bio-adsorbent: Kinetics and isotherm studies. *Environ. Monit. Assess.* **2023**, *195*, 339. [CrossRef]
14. Bayomie, O.S.; Kandeel, H.; Shoeib, T.; Yang, H.; Youssef, N.; El-Sayed, M.M.H. Novel approach for effective removal of methylene blue dye from water using fava bean peel waste. *Sci. Rep.* **2020**, *10*, 7824. [CrossRef]
15. Rahman, M.M.; Foisal, J.A.; Ihara, H.; Takafuji, M. Efficient removal of methylene blue dye from an aqueous solution using silica nanoparticle crosslinked acrylamide hybrid hydrogels. *New J. Chem.* **2021**, *45*, 20107–20119. [CrossRef]
16. Benhachem, F.; Attar, T.; Bouabdallah, F. Kinetic study of adsorption methylene blue dye from aqueous solutions using activated carbon. *Chem. Rev. Lett.* **2019**, *2*, 33–39. [CrossRef]
17. Rahman, S.; Ajmal, M.; Muhammad, S. Micron sized anionic poly (methacrylic acid) microgel particles for the adsorptive elimination of cationic water pollutants. *Z. Phys. Chem.* **2023**, *237*, 121–145. [CrossRef]
18. Amjad, S.; Shaukat, S.; Rahman, H.M.A.; Usman, M.; Farooqi, Z.H.; Nazar, M.F. Application of anionic-nonionic mixed micellar system for solubilization of methylene blue dye. *J. Mol. Liq.* **2023**, *369*, 120958. [CrossRef]
19. Naseem, K.; Farooqi, Z.H.; Begum, R.; Ghufran, M.; Muhammad, Z.R.; Najeeb, J.; Irfan, A.; Al-Sehemi, A.G. Poly(N-isopropylmethacrylamide-acrylic acid) microgels as adsorbent for removal of toxic dyes from aqueous medium. *J. Mol. Liq.* **2018**, *268*, 229–238. [CrossRef]
20. Sindona, A.; Pisarra, M.; Gomez, C.V.; Riccardi, P.; Falcone, G.; Bellucci, S. Calibration of the fine-structure constant of graphene by time-dependent density-functional theory. *Phys. Rev. B* **2017**, *96*, 201408. [CrossRef]
21. Coello-Fiallos, D.; Tene, T.; Guayllas, J.L.; Haro, D.; Haro, A.; Gomez, C.V. DFT comparison of structural and electronic properties of graphene and germanene: Monolayer and bilayer systems. *Mater. Today Proc.* **2017**, *4*, 6835–6841. [CrossRef]
22. Abdelhalim, A.O.E.; Semenov, K.N.; Nerukh, D.A.; Murin, I.V.; Maistrenko, D.N.; Molchanov, O.E.; Sharoyko, V.V. Functionalisation of graphene as a tool for developing nanomaterials. *J. Mol. Liq.* **2022**, *348*, 118368. [CrossRef]
23. Ghulam, A.N.; dos Santos, O.A.L.; Hazeem, L.; Pizzorno Backx, B.; Bououdina, M.; Bellucci, S. Graphene Oxide (GO) Materials—Applications and Toxicity on Living Organisms and Environment. *J. Funct. Biomater.* **2022**, *13*, 77. [CrossRef] [PubMed]
24. Sang, M.; Shin, J.; Kim, K.; Yu, K.J. Electronic and Thermal Properties of Graphene and Recent Advances in Graphene Based Electronics Applications. *Nanomaterials* **2019**, *9*, 374. [CrossRef]
25. Zhao, L.; Yang, S.T.; Feng, S.; Ma, Q.; Peng, X.; Wu, D. Preparation and application of carboxylated graphene oxide sponge in dye removal. *Int. J. Environ. Res. Public Health* **2017**, *14*, 1301. [CrossRef] [PubMed]
26. Ali, I.; Basheer, A.A.; Mbianda, X.Y.; Burakov, A.; Galunin, E.; Burakova, I.; Mkrtchyan, E.; Tkachev, A.; Grachev, V. Graphene based adsorbents for remediation of noxious pollutants from wastewater. *Environ. Int.* **2019**, *127*, 160–180. [CrossRef]
27. Rahman, D.A.; Helmy, Q.; Syafila, M.; Gumilar, A. Adsorption of Dyes Using Graphene Oxide-Based Nano-Adsorbent: A Review. *Jurnal Presipitasi Media Komunikasi Pengembangan Teknik Lingkungan* **2022**, *19*, 384–397. [CrossRef]
28. Ederer, J.; Ecorchard, P.; Slušná, M.; Tolasz, J.; Smržová, D.; Lupínková, S.; Janoš, P. A Study of Methylene Blue Dye Interaction and Adsorption by Monolayer Graphene Oxide. *Adsorpt. Sci. Technol.* **2022**, *2022*, 7385541. [CrossRef]
29. Sengupta, I.; Chakraborty, S.; Talukdar, M.; Pal, S.K.; Chakraborty, S. Thermal reduction of graphene oxide: How temperature influences purity. *J. Mater. Res.* **2018**, *33*, 4113–4122. [CrossRef]
30. Adel, M.; Ahmed, M.A.; Elabiad, M.A.; Mohamed, A.A. Removal of heavy metals and dyes from wastewater using graphene oxide-based nanomaterials: A critical review. *Environ. Nanotechnol. Monit. Manag.* **2022**, *18*, 100719. [CrossRef]
31. Khurana, I.; Bharti, A.S.; Khurana, J.M.; Rai, P.K. Removal of Dyes Using Graphene-Based Composites: A Review. *Water Air Soil Pollut.* **2017**, *228*, 180. [CrossRef]
32. Verma, S.; Kim, K.-H. Graphene-based materials for the adsorptive removal of uranium in aqueous solutions. *Environ. Int.* **2022**, *158*, 106944. [CrossRef] [PubMed]

33. Chia, C.H.; Razali, N.F.; Sajab, M.S.; Zakaria, S.; Huang, N.M.; Lim, H.N. Methylene blue adsorption on graphene oxide. *Sains Malays.* **2013**, *42*, 819–826.
34. Chen, L.; Yang, J.; Zeng, X.; Zhang, L.; Yuan, W. Adsorption of methylene blue in water by reduced graphene oxide: Effect of functional groups. *Mater. Express* **2013**, *3*, 281–290. [CrossRef]
35. Chilakapati, R.B.; Kumar, S.H.; Satyanarayana, S.V.; Behara, D.K. Adsorptive removal of methylene blue (MB) and malachite green (MG) dyes from aqueous solutions using graphene oxide (GO). *Z. Phys. Chem.* **2021**, *235*, 1645–1660. [CrossRef]
36. Vinh, L.T.; Khiem, T.N.; Chinh, H.P.; Tuan, P.V.; Tan, V.T. Adsorption capacities of reduced graphene oxide: Effect of reductants. *Mater. Res. Express* **2019**, *6*, 075615. [CrossRef]
37. Soudagar, S.; Akash, S.; Venkat, M.S.; Poiba, V.R.; Vangalapati, M. Adsorption of methylene blue dye on nano graphene oxide-thermodynamics and kinetic studies. *Mater. Today Proc.* **2022**, *59*, 667–672. [CrossRef]
38. Arias, F.A.; Guevara, M.; Tene, T.; Angamarca, P.; Molina, R.; Valarezo, A.; Salguero, O.; Gomez, C.V.; Arias, M.; Caputi, L.S. The Adsorption of Methylene Blue on Eco-Friendly Reduced Graphene Oxide. *Nanomaterials* **2020**, *10*, 681. [CrossRef]
39. Ciğeroğlu, Z.; Haşimoğlu, A.; Özdemir, O.K. Synthesis, characterization and an application of graphene oxide nanopowder: Methylene blue adsorption and comparison between experimental data and literature data. *J. Dispers. Sci. Technol.* **2021**, *42*, 771–783. [CrossRef]
40. Khiam, G.K.; Karri, R.R.; Mubarak, N.M.; Khalid, M.; Walvekar, R.; Abdullah, E.C.; Rahman, M.E. Modelling and optimization for methylene blue adsorption using graphene oxide/chitosan composites via artificial neural network particle swarm optimization. *Mater. Today Chem.* **2022**, *24*, 100946. [CrossRef]
41. Sharma, P.; Hussain, N.; Borah, D.J.; Das, M.R. Kinetics and adsorption behavior of the methyl blue at the graphene oxide/reduced graphene oxide nanosheet–water interface: A comparative study. *J. Chem. Eng. Data* **2013**, *58*, 3477–3488. [CrossRef]
42. Yun, S.; Lee, H.; Lee, W.-E.; Park, H.S. Multiscale textured, ultralight graphene monoliths for enhanced CO_2 and SO_2 adsorption capacity. *Fuel* **2016**, *174*, 36–42. [CrossRef]
43. Marcano, D.C.; Kosynkin, D.V.; Berlin, J.M.; Sinitskii, A.; Sun, Z.; Slesarev, A.; Alemany, L.B.; Lu, W.; Tour, J.M. Improved synthesis of graphene oxide. *ACS Nano* **2010**, *4*, 4806–4814. [CrossRef]
44. Thommes, M.; Kaneko, K.; Neimark, A.; Olivier, J.P.; Rodriguez-Reinoso, F.; Rouquerol, J.; Sing, K.S.W. Physisorption of gases, with special reference to the evaluation of surface area and pore size distribution (IUPAC Technical Report). *Pure Appl. Chem.* **2015**, *87*, 1051–1069. [CrossRef]
45. Rao, G.-S.; Nabipour, H.; Zhang, P.; Wang, X.; Xing, W.; Song, L.; Hu, Y. Lightweight, hydrophobic and recyclable carbon foam derived from lignin–resorcinol–glyoxal resin for oil and solvent spill capture. *J. Mater. Technol.* **2020**, *9*, 4655–4664. [CrossRef]
46. Acik, M.; Lee, G.; Mattevi, C.; Pirkle, A.; Wallace, R.M.; Chhowalla, M.; Cho, K.; Chabal, Y. The role of oxygen during thermal reduction of graphene oxide studied by infrared absorption spectroscopy. *J. Phys. Chem. C* **2011**, *115*, 19761–19781. [CrossRef]
47. Van Khai, T.; Na, H.G.; Kwak, D.S.; Kwon, Y.J.; Ham, H.; Shim, K.B.; Kim, H.W. Comparison study of structural and optical properties of boron-doped and undoped graphene oxide films. *Chem. Eng. J.* **2012**, *211–212*, 369–377. [CrossRef]
48. Long, D.; Li, W.; Ling, L.; Miyawaki, J.; Mochida, I.; Yoon, S.-H. Preparation of Nitrogen-Doped Graphene Sheets by a Combined Chemical and Hydrothermal Reduction of Graphene Oxide. *Langmuir* **2010**, *26*, 16096–16102. [CrossRef]
49. Nayl, A.A.; Abd-Elhamid, A.I.; Arafa, W.A.A.; Ahmed, I.M.; El-Shanshory, A.A.; Abu-Saied, M.A.; Soliman, H.M.A.; Abdelgawad, M.A.; Ali, H.M.; Bräse, S. Chitosan-Functionalized-Graphene Oxide (GO@CS) Beads as an Effective Adsorbent to Remove Cationic Dye from Wastewater. *Polymers* **2022**, *14*, 4236. [CrossRef]
50. Ajmal, M.; Demirci, S.; Siddiq, M.; Aktas, N.; Sahiner, N. Amidoximated poly (acrylonitrile) particles for environmental applications: Removal of heavy metal ions, dyes, and herbicides from water with different sources. *J. Appl. Polym. Sci.* **2016**, *133*, 7. [CrossRef]
51. Vassileva, P.; Uzunov, I.; Voykova, D. Kinetics, equilibrium and thermodynamics of Congo red removal by cationized cellulose obtained from cereals by-product. *Cellul. Chem. Technol.* **2022**, *56*, 1117–1128. [CrossRef]
52. Nguyen, C.H.; Juang, R.-S. Efficient removal of methylene blue dye by a hybrid adsorption–photocatalysis process using reduced graphene oxide/titanate nanotube composites for water reuse. *J. Ind. Eng. Chem.* **2019**, *76*, 296–309. [CrossRef]
53. El-Sharkaway, E.A.; Kamel, R.M.; El-Sherbiny, I.M.; Gharib, S.S. Removal of methylene blue from aqueous solutions using polyaniline/graphene oxide or polyaniline/reduced graphene oxide composites. *Environ. Technol.* **2020**, *41*, 2854–2862. [CrossRef] [PubMed]
54. Yan, H.; Tao, X.; Yang, Z.; Li, K.; Yang, H.; Li, A.; Cheng, R. Effects of the oxidation degree of graphene oxide on the adsorption of methylene blue. *J. Hazard. Mater.* **2014**, *268*, 191–198. [CrossRef] [PubMed]
55. Ramesha, G.K.; Kumara, A.V.; Muralidhara, H.B.; Sampath, S. Graphene and graphene oxide as effective adsorbents toward anionic and cationic dyes. *J. Colloid Interface Sci.* **2011**, *361*, 270–277. [CrossRef]
56. Shikuku, V.O.; Zanella, R.; Kowenje, C.O.; Donato, F.; Bandeira, N.M.G.; Prestes, O.D. Single and binary adsorption of sulfonamide antibiotics onto iron-modified clay: Linear and nonlinear isotherms, kinetics, thermodynamics, and mechanistic studies. *Appl. Water Sci.* **2018**, *8*, 175. [CrossRef]

57. Sahmoune, M.N. Evaluation of thermodynamic parameters for adsorption of heavy metals by green adsorbents. *Environ. Chem. Lett.* **2019**, *17*, 697–704. [CrossRef]
58. Momina; Rafatullah, M.; Ismail, S.; Ahmad, A. Optimization study for the desorption of methylene blue dye from clay based adsorbent coating. *Water* **2019**, *11*, 1304. [CrossRef]

materials

MDPI

Article

Effect of Cu as a Minority Alloying Element on the Corrosion Behaviour of Amorphous and Crystalline Al-Ni-Si Alloy

Vanya Dyakova [1,*], Yoanna Kostova [1], Boriana Tzaneva [2], Hristina Spasova [1] and Daniela Kovacheva [3]

1 Institute of Metal Science, Equipment and Technologies with Hydro- and Aerodynamics Centre "Acad. A. Balevski", Bulgarian Academy of Sciences, Shipchenski Prohod Str., 67, 1574 Sofia, Bulgaria; y_kostova@is.bas.bg (Y.K.); h.spasova@ims.bas.bg (H.S.)
2 Department of Chemistry, Technical University of Sofia, Kliment Ohridski Blvd., 8, 1000 Sofia, Bulgaria; borianatz@tu-sofia.bg
3 Institute of General and Inorganic Chemistry, Bulgarian Academy of Sciences, Acad. Angel Bonchev Str., 11, 1113 Sofia, Bulgaria; didka@svr.igic.bas.bg
* Correspondence: v_diakova@ims.bas.bg

Abstract: The effect of copper as a minority alloying element on the corrosion behaviour of amorphous and crystalline $Al_{74}Ni_{16}Si_{10}$ and $Al_{74}Ni_{15}Si_9Cu_2$ alloys was investigated in this study. Amorphous alloys were produced as rapidly solidified ribbons using the Chill Block Melt Spinning (CBMS) method and subsequently annealed to complete crystallisation. The corrosion rate of alloys was obtained through continuous immersion tests in 3.5% NaCl at 25 °C and 50 °C for 360 h. The electrochemical parameters corrosion current density (J_{corr}) and corrosion potential (E_{corr}) were determined via the potentiodynamic polarisation test. The results showed better corrosion characteristics of amorphous alloys. The addition of 2 at.% copper to the $Al_{74}Ni_{16}Si_{10}$ alloy improved pitting corrosion resistance without significant effect on the corrosion current and potential. In immersion tests at 25 °C, the presence of copper resulted in an increase in the corrosion rate of about 300% for both amorphous and crystalline alloys. At a temperature of 50 °C, this increase is on average 130%. The apparent difference between the results of the two test methods is discussed in terms of the imperfections on the surface of rapidly solidified ribbons. The results of this study will contribute to a more complex understanding of the nature of amorphous alloys and their application.

Keywords: corrosion; amorphous Al-Ni-Si alloys; crystallisation; copper additive

Citation: Dyakova, V.; Kostova, Y.; Tzaneva, B.; Spasova, H.; Kovacheva, D. Effect of Cu as a Minority Alloying Element on the Corrosion Behaviour of Amorphous and Crystalline Al-Ni-Si Alloy. *Materials* **2023**, *16*, 5446. https://doi.org/10.3390/ma16155446

Academic Editor: Amir Mostafaei

Received: 30 June 2023
Revised: 31 July 2023
Accepted: 1 August 2023
Published: 3 August 2023

1. Introduction

The corrosion resistance of amorphous and crystalline alloys is one of the determining factors for their application. The corrosion properties of amorphous alloys are expected to surpass those of their crystalline analogues for several reasons: (i) the formation of a monophasic solid solution with good chemical and structural homogeneity which eliminates the influence of microstructural defects (including grain boundaries) and particles of a second phase; (ii) the addition of rare earth, transition and refractory elements which stimulate the formation of a passive surface layer with these metals; and (iii) the tendency to repassivation which has been observed in amorphous alloys after the occurrence of local corrosion [1,2].

The first evidence of high corrosion resistance of metallic glasses was reported by Naka, Hashimoto and Masumoto in 1974 [3]. Since then, numerous studies have been published on the corrosion behaviour of amorphous alloys based on iron [4–9], zirconium [10–13], magnesium [14–16], copper [17–19], aluminium [20–30] and others [31–33].

Amorphous and nanocrystalline alloys based on aluminium, especially compositions with transition and rare earth metals, show improved mechanical properties and good corrosion resistance [21]. Vitrification of aluminium alloys via rapid solidification is known to be a difficult process and vitrification of only local areas of the specimens has been found in some cases. All the advantages of homogeneous structure could potentially be lost in the presence of crystal areas. Nevertheless, the corrosion behaviour of rapidly solidified aluminium alloys is of interest even in cases when complete vitrification is not achieved [22].

The research on the influence of minority alloying elements usually focuses on their role in improving the glass formation of metallic glasses. The role of minority solutes in corrosion resistance has not been extensively investigated and is not well understood due to the difficulty of glass formation and the impossibility to produce amorphous alloys in the absence of a minority solute [34].

Zhang et al. studied the influence of Ti and Cr as minority alloying elements on the corrosion behaviour of Al-Ni-Y amorphous alloys [23]. They found that the improved corrosion resistance is due to an increased concentration of Ti or Cr in the formed surface passive films. Furthermore, the effect of Ti is stronger due to the higher Ti concentration in the film, leading to higher film resistance. The effect of Ti and Cr as alloying elements on the corrosion behaviour of amorphous alloys Al-Ti-Mg, Al-Cr-Mo and Al-Cr-Mo was also investigated by Akiyama [24]. The influence of alloying additions on the composition and structure of the passive film was studied. The authors [35] believe that the protective ability of the passive film in aluminium–manganese–molybdenum (Al-Mn-Mo) alloys is due to the formation of a less defective and a more compact structure. They explain the role of the alloying additive Mo in both the substrate and the passive film.

Amorphous aluminium alloys containing W and Mo have been reported to show significant corrosion resistance, especially in chloride environments [25].

Yoshioka et al. [26] investigated the effect of alloying on the pitting corrosion resistance of rapidly solidified binary and ternary aluminium alloys containing Mg, Ti, Mn, Cr, Fe, Ni, Cu, Zn, Zr, and Nb. They proved that the rapid melt spinning method is quite effective in refining the pitting potential by several hundred millivolts for all alloys, except those containing Mg, Fe, or Zn.

The influence of the transition metals ratio on the corrosion electrochemical behaviour of Al–Ni–Co–Nd amorphous alloy was investigated by Karfidov et al. [27]. According to the gravimetric and chemical analysis data, the $Al_{86}Ni_4Co_4Nd_6$ alloy was found to demonstrate higher corrosion resistance than $Al_{86}Ni_6Co_2Nd_6$. With the increase in Ni and decrease in Co concentrations, the corrosion potential shifted in an electronegative direction. All amorphous compositions investigated in [27] had a wide passivation area due to the formation of a stable layer of aluminium oxide on the surface, and this determined their high corrosion resistance.

Aburada et al. [28] reported on the positive influence of Ni as a minority alloying element (Ni < 2 at.%) on the corrosion resistance of amorphous $(Al_{75}Cu_{19}Mg_6)_{100-x}Ni_x$ alloys. Nickel was considered superior to Cu for improving the pitting resistance of Al.

Research on the effect of copper on the corrosion behaviour of aluminium alloys is incomplete and the results are conflicting The addition of small amounts of copper (about 1.5%) is known to increase the strength and hardness of commercial Al-based alloys [36,37]. In a NaCl solution, when copper content increases (within the solubility limit of Cu in Al), the current density is reduced [38] and the pitting potential shifts in the positive direction [39]. In contrast, Wang et al. demonstrated that in the case of prolonged salt spray exposure to Al-Si-Mg-Cr alloys, copper deteriorates the corrosion resistance [40]. Deterioration of corrosion resistance is mainly associated with the separation of secondary Cu-rich phases, which have cathode relations concerning the aluminium matrix during the pit initiation [41].

In our previous research, we investigated the influence of Ni on the corrosion behaviour of amorphous alloys $(Al_{74}Cu_{16}Mg_{10})_{100-x}Ni_x$ (x = 1, 2, 3 at.%) [29]. The results showed a significant improvement in the pitting corrosion resistance of all amorphous alloys compared to their crystalline analogues, which was explained by the absence of secondary phases Al_2Cu and Al_2CuMg in the chemically homogenous amorphous structure.

The Influence of minority alloying elements, Zn and Zr, on the corrosion behaviour of rapidly solidified and annealed nanocrysalline $Al_{74}Cu_{16}Mg_{10}$ alloys was studied in [30]. The effect of the amorphous–crystalline structure transformation on the corrosion rate was negative, and the effect was most significant in the Zn-containing alloy. The main reason for the registered enhanced local corrosion in the crystalline alloys was the chemical and structural inhomogeneity due to the presence of the active intermetallic phases, Al_2CuMg, $Al_2(Cu,Zn)$, and Al_3Zr_4, in the aluminium matrix.

Yoshioka et al. [26] studied the corrosion behaviour of conventionally cast and rapidly solidified Al-Ni-Si alloys. They reported that in conventionally cast Al-Ni-Si alloys, pitting corrosion usually attacked the Al_3Ni phase. Rapidly solidified alloys formed an α-Al phase supersaturated with dissolved Ni and Si atoms, and this resulted in a reduced amount of silicon phases and intermetallic compounds, which are usually responsible for preferential corrosion attack.

Gao et al. [42] synthesised and investigated dense Al-based amorphous metallic coating deposited on AA 2024 substrate. They produced the coating using high-velocity air fuel (HVAF) spraying of an $Al_{86}Ni_6Y_{4.5}Co_2La_{1.5}$ alloy which is known to possess high glass-forming ability. The authors found an enhanced corrosion resistance of the coating compared to AA 2024, which provided a sacrificial anode-based cathodic protection for the substrate under a chloride-containing environment.

Most available studies confirm a better corrosion resistance of amorphous glasses compared to their crystalline analogues [39], but in some cases, no improved corrosion resistance has been found in amorphous alloys, despite their homogeneous and defect-free microstructure [43–46]. It is also not yet clear how the level of partial crystallisation affects the corrosion resistance, what is the effect of the minor surface imperfections, what is the influence of the test method and duration, etc. The influence of the material surface, as the interface between the material and the corrosive environment, is an important factor in evaluating the corrosion behaviour of amorphous metals [47]. Despite the lack of boundaries, segregation or other defects in the microstructure of the corrosion process is influenced by both the roughness and the chemical inhomogeneity of metallic glasses. Unfortunately, these topics are hardly addressed in scientific research. The topic of corrosion mechanisms in different types of metallic materials is also discussed in a limited number of scientific studies [48].

In our previous studies, we investigated the effect of copper as a minority alloying element on the glass formation of a nearly eutectic Al-Ni-Si alloy. There is a lack of data in the scientific literature on the effect of small amounts of Cu on the corrosion properties of amorphous Al-based alloys. This study aims to evaluate the influence of 2 at.% copper on the corrosion resistance of amorphous Al-Ni-Si alloys and their crystalline analogues. We believe that this will contribute in enriching the knowledge of the influence of the amorphous–crystalline transition on the properties of Al-Ni-Si and Al-Ni-Si-Cu alloys.

2. Materials and Methods

2.1. Materials

Al-Ni-Si and Al-Ni-Si-Cu alloys were synthesised from pure metals of Al, Ni, Si and Cu in an installation including a resistance electric furnace mounted in a water-cooled pneumovacuum chamber in an argon atmosphere. The alloys were obtained in the form of ribbons using the Chill Block Melt Spinning (CBMS) method. They are about 3–4 mm wide and 26–40 μm thick. To obtain a crystalline structure, the ribbons were annealed at a temperature of 350 °C, which is 100 °C higher than their crystallisation temperatures determined via the DSC analysis. The production processes of the ligatures and the ribbons were described in detail in our previous publications [49].

The composition of the alloy was determined based on our preliminary studies, which showed that a near-eutectic composition is optimal for producing an amorphous Al-Ni-Si alloy. We determined the amount of copper added to be 2 at.% based on research to combine good GFA, mechanical properties and corrosion resistance [50–52].

The chemical composition of the Al-Ni-Si and Al-Ni-Si-Cu ribbons was determined using a HIROX 5500 scanning electron microscope (SEM, HIROX Europe, Limonest, France) with a BRUCKER EXDS system (BRUCKER Co., Germany). The average values of the results obtained from measurements in 15 fields are presented in Table 1. The indices in the designation of the alloys correspond to the content of the elements Al, Ni, Si and Cu in at.%. Based on the obtained results, the designation of the Al-Ni-Si and Al-Ni-Si-Cu alloys is $Al_{74}Ni_{16}Si_{10}$ and $Al_{74}Ni_{15}Si_9Cu_9$, respectively.

Table 1. Chemical composition of the tested Al-Ni-Si and Al-Ni-Si-Cu alloys.

Designation of the Alloy	Al [at.%]	Ni [at.%]	Si [at.%]	Cu [at.%]
Al-Ni-Si	73.50	16.08	10.42	-
Al-Ni-Si-Cu	73.97	14.94	8.76	2.33

2.2. Characterisation Methods

The microstructure of the rapidly solidified and annealed alloys was characterised using X-ray diffraction analysis (XRD) and transmission electron microscopy (TEM) observation and electron diffraction.

XRD analysis was carried out for determining the phase composition of the alloys and corrosion products. A Bruker D8 Advance powder X-ray diffractometer (Karlsruhe, Germany) with Ni-filtered Cu Kα radiation and LynxEye solid-state position-sensitive detector was applied. The PDF-2 (2021) database of the International Center for Data Diffraction (ICDD) and the DiffracPlusEVA software v.4.0 (Bruker AXS 2010-2014, Karlsruhe, Germany) package was used to perform the phase analysis

XRD proved that the microstructure of the rapidly solidified ribbons $Al_{74}Ni_{16}Si_{10}$ and $Al_{74}Ni_{15}Si_9Cu_2$ is completely amorphous. These alloys will be denoted hereafter with the index "am". The type, amount and sizes of the crystalline phases in the annealed samples were determined. Three phases were registered in the crystalline $Al_{74}Ni_{16}Si_{10}$ alloy, Al-fcc, Al_3Ni, and $NiSi_2$, with sizes 88 nm, 71 nm, and 43 nm, respectively; in the crystalline $Al_{74}Ni_{16}Si_9Cu_2$ alloy, four phases were registered, Al-fcc, Al_3Ni, $NiSi_2$, and $Cu_{3.8}Ni$ with sizes 136 nm, 88 nm, 45 nm, and 60 nm, respectively and $Al_{0.4}Cu_{0.6}Ni_3$ traces [53]. Based on the XRD results, after devitrification, we designated the annealed $Al_{74}Ni_{16}Si_{10}$ alloy as a nanocrystalline and the annealed $Al_{74}Ni_{15}Si_9Cu_2$ alloy as an ultra-fine crystalline. Further in this work, we use the indexes "ncr" and "ufcr", respectively, for these two alloys [54].

High-Resolution Transmission Electron Microscope (HRTEM) JEOL JEM 2100, (JEOL Ltd., Tokyo, Japan) at an accelerating voltage of 200 kV in Selected Area Diffraction (SAED) and High Resolution (HRTEM) modes was used for transmission electron microscopy observation of the microstructures of the Al-Ni-Si and Al-Ni-Si-Cu amorphous alloys. The results of the observations and electron diffraction confirmed the amorphous nature of the ribbons obtained.

The corrosion products deposited after the immersion test were observed using SEM HIROX 5500 (HIROX Europe, Limonest, France) and were analysed via XRD.

2.3. Corrosion Test Methods

Two methods for determining the corrosion properties of $Al_{74}Ni_{16}Si_{10}$ and $Al_{74}Ni_{15}Si_9Cu_2$ alloys were applied: gravimetric (immersion) test for testing metals' resistance to general corrosion and electrochemical polarisation test to indicate the propensity to local corrosion.

General corrosion tests were performed through continuous immersion of specimens in 3.5% NaCl for 360 h in a laboratory thermostat at a temperature of 25 °C and for 50 h at a temperature of 50 °C. Before testing, the specimens were degreased in acetone and treated in diluted HNO_3 for one minute. The test temperature of 25 °C is the room temperature normally used in standard immersion tests and in electrochemical polarisation tests according to standards ASTM G31 and ASTM G3.

The following formula was used to determine the corrosion rate through the gravimetric tests: $CR = \frac{W}{S.T}$ [g $m^{-2}h^{-1}$], where W represents the mass loss index in [g] as the difference between the weight of test specimens before the test and the removal of corrosion products after the test, $W = w_1 - w_2$; S is the area of the specimen [m^2]; and T is the time of immersion in the corrosion medium [h]. The mass of the samples was weighed on an analytical balance. All CR results obtained are averages of a minimum of three parallel tests [53].

After the test, the test specimens were rinsed with distilled water repeatedly and the released corrosion products were analysed via XRD.

The general and localised corrosion of the ribbons was studied using the cyclic potentiodynamic polarisation method. Test specimens were degreased in alcohol and immersed in a solution of 3.5% NaCl at a temperature of 25 °C. A conventional tri-electrode cell with a working electrode from the studied amorphous and crystalline tested alloys, a Pt plate counter electrode and a silver chloride reference electrode (Ag/AgCl) was used. For the electrochemical test, the surface on the cooling copper disk side (Figure 1a) of ribbons is insulated with varnish to define the working surface at 0.5 cm^2 on the smoother side (Figure 1b). All potentials in this work reported are relative to the Ag/AgCl electrode. The electrochemical tests were performed with galvanostat-potentiostat Autolab model PGSTAT 204 and computer software NOVA 2.1 (Metrohm Autolab, Netherland, Utrecht).

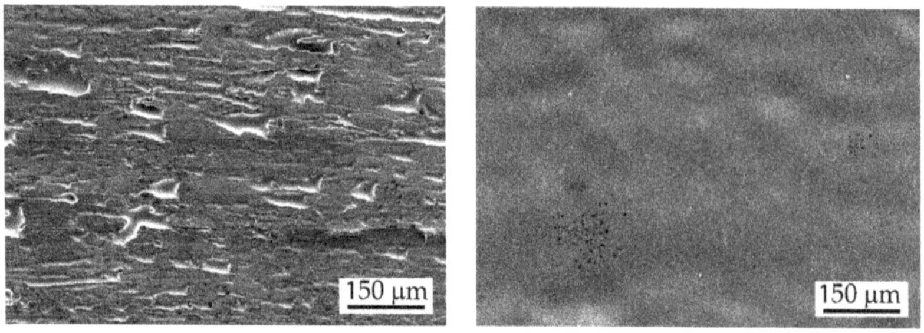

(**a**) (**b**)

Figure 1. Surface morphology of the amorphous $Al_{74}Ni_{15}Si_9Cu_2$-am ribbon, SEM: (**a**) surface on the side of cooling copper disk; and (**b**) top surface.

The specimens were soaked for 10 min in 3.5% NaCl to stabilise the open circuit potential (OCP). Electrochemical impedance spectroscopy (EIS) or potentiodynamic tests were then performed. EIS was carried out at an OSP whit AC potential fluctuations with an amplitude of 10 mV and frequencies from 1 kHz to 0.1 Hz. The cyclic potentiodynamic studies were carried out at a scan rate of 1 mV s^{-1} in an anodic direction from an initial potential of −0.25 V vs. OCP until exceeding the threshold current density of 1 mA cm^{-2}; then, the potential was scanned in a reverse cathodic direction until the two branches of the polarisation curve intersect. For each tested alloy, at least 5 specimens were tested to verify the reproducibility of the results.

3. Results

3.1. Gravimetric Studies

The results of the gravimetric tests for the determination of the corrosion rate CR are presented in Table 2. Based on the averaged values of CR at both test temperatures, the ratios A, B, C, and D were calculated. The ratios $A = CR_{cr25}/CR_{am25}$ and $B = CR_{cr50}/CR_{am50}$ demonstrate the change in corrosion rate for each alloy after the amorphous → crystalline transformation caused by annealing at the corrosion test temperature of 25 °C or 50 °C, respectively. Analogically, the ratios $C=CR_{am50}/CR_{am25}$ and $D=CR_{cr50}/CR_{cr25}$ indicate the influence of the test temperature on the corrosion rate in both states, amorphous and nano-/ultra-fine crystalline.

Table 2. Corrosion rate of $Al_{74}Ni_{16}Si_{10}$ and $Al_{74}Ni_{15}Si_9Cu_2$ alloys in amorphous and crystalline states.

	Corrosion Rate CR [g $m^{-2}h^{-1}$]					
Designation of the Alloy	**CR_{25} T = 25 °C**	**CR_{50} T = 50 °C**	A	B	C	D
$Al_{74}Ni_{16}Si_{10}$ am	0.35×10^{-2}	0.67×10^{-2}	2.9	8.4	1.9	-
$Al_{74}Ni_{16}Si_{10}$ ncr	1.02×10^{-2}	5.63×10^{-2}			-	5.5
$Al_{74}Ni_{15}Si_9Cu_2$ am	0.92×10^{-2}	0.95×10^{-2}	3.3	7.7	1	-
$Al_{74}Ni_{15}Si_9Cu_2$ ufcr	3.07×10^{-2}	7.3×10^{-2}			-	2.4

At a test temperature of 25 °C (Table 2), the amorphous $Al_{74}Ni_{16}Si_{10}$-am alloy has the lowest corrosion rate, CR_{am25}. The corrosion rate CR_{25} of the Cu-containing amorphous alloy $Al_{74}Ni_{15}Si_9Cu_2$-am is about 2.6 times higher compared to the copper-free $Al_{74}Ni_{16}Si_{10}$-am alloy. At a test temperature of 50 °C, the lowest corrosion rate CR_{50} is obtained again for the amorphous $Al_{74}Ni_{16}Si_{10}$-am, but its difference with $Al_{74}Ni_{15}Si_9Cu_2$-am is only about 1.4 times. Nevertheless, the corrosion rates of both amorphous alloys at both temperatures remain low and do not exceed 1×10^{-2} g $m^{-2}h^{-1}$.

Both annealed alloys, in which the structure was transformed from amorphous to crystalline, have significantly higher corrosion rates at 25 °C and especially at 50 °C than their amorphous analogues. The highest CR50 = 7.3×10^{-2} g $m^{-2}h^{-1}$ was measured for the ultra-fine crystalline $Al_{74}Ni_{15}Si_9Cu_2$-ufcr alloy.

It should be noted that at a test temperature of 25 °C, the effect of amorphous → crystalline transformation on the corrosion process is more negative for the $Al_{74}Ni_{15}Si_9Cu_2$ alloy than for $Al_{74}Ni_{16}Si_{10}$; the corrosion rate of $Al_{74}Ni_{15}Si_9Cu_2$-ufcr increases 3.3 times compared to 2.9 times for $Al_{74}Ni_{16}Si_{10}$-ncr. A reverse dependence is observed at 50 °C; the effect is more negative for $Al_{74}Ni_{16}Si_{10}$ (B = 8.4) than for $Al_{74}Ni_{15}Si_9Cu_2$ (B = 7.7). However, taking into account the relatively small deviations (3.3 vs.2.9 for A and 8.4 vs.7.7 for B) and the uncertainty of the tests, it can be assumed that at each of the two test temperatures, the effect of crystallisation is negative and similar for both alloys; the increase in corrosion rate in crystallised alloys at a test temperature of 25 °C is about 3 times and at a test temperature of 50 °C about 8 times.

The increase in test temperature from 25 °C to 50 °C doubled the corrosion rate of the amorphous alloy $Al_{74}Ni_{16}Si_{10}$-am (C = 1.9), but hardly affected the corrosion rate of the amorphous alloy $Al_{74}Ni_{15}Si_9Cu_2$-am (C = 1). The tendency for an increase in the corrosion rate of $Al_{74}Ni_{16}Si_{10}$ with increasing test temperature is observed in the crystalline state also: for $Al_{74}Ni_{16}Si_{10}$-ncr, it is D = 5.5 ad and for $Al_{74}Ni_{15}Si_9Cu_2$-ufcr, it is D = 2.4.

The corrosion products formed during the immersion test on the surface of the tested amorphous and crystalline ribbons at both temperatures were observed using SEM (Figure 2) and analysed via XRD (Figure 3). After a prolonged immersion test in 3.5% NaCl, the surface of the amorphous alloys was affected by surface corrosion with clearly visible caverns oriented along the ribbon length. Corrosion products appear relatively uniformly deposited and tightly adhered on the surface (Figure 2a,c). On annealed alloys, the corrosion products have an uneven and loose structure, especially in the case of the copper-containing alloy (Figure 2b,d).

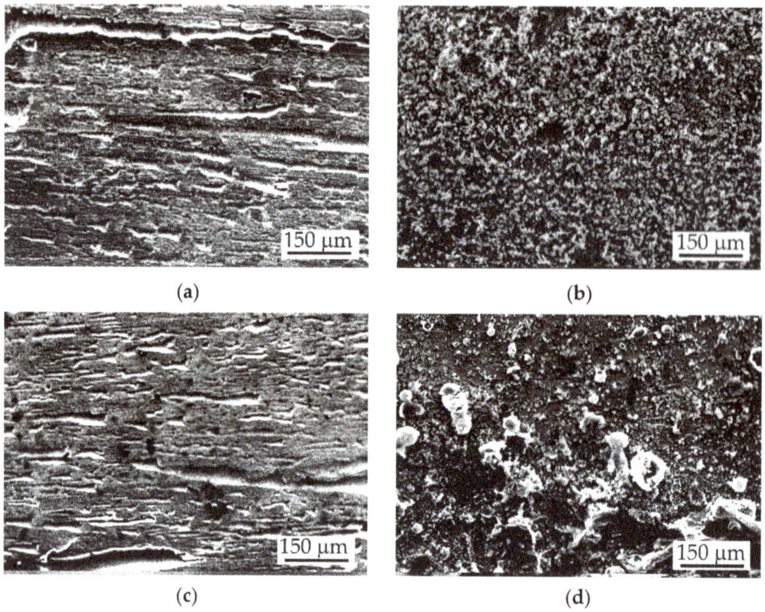

Figure 2. SEM images of Al-Ni-Si and Al-Ni-Si-Cu alloys corroded samples after prolonged immersion test in 3.5% NaCl solution: (**a**) $Al_{74}Ni_{16}Si_{10}$-am; (**b**) $Al_{74}Ni_{16}Si_{10}$-ncr; (**c**) $Al_{74}Ni_{15}Si_9Cu_2$-am; and (**d**) $Al_{74}Ni_{15}Si_9Cu_2$-ufcr.

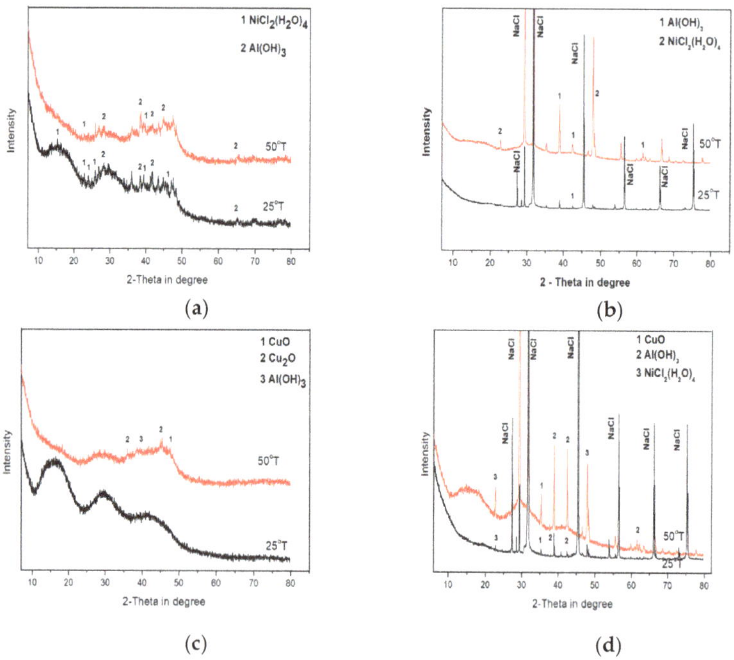

Figure 3. XRD patterns of the corrosion products of $Al_{74}Ni_{16}Si_{10}$ and $Al_{74}Ni_{15}Si_9Cu_2$ alloys after 360 h immersion test in 3.5% NaCl; (**a**) $Al_{74}Ni_{16}Si_{10}$-am; (**b**) $Al_{74}Ni_{16}Si_{10}$-ncr; (**c**) $Al_{74}Ni_{15}Si_9Cu_2$-am; and (**d**) $Al_{74}Ni_{15}Si_9Cu_2$-ufcr.

Three amorphous humps are clearly visible on the XRD patterns of the corrosion products of all the amorphous alloys. These humps are situated at around 15, 28 and 40 degrees 2θ and correspond to the main peaks of AlO(OH)-Bohemite (Figure 3a,c). These humps are of higher intensity and are more pronounced in the pattern of the Cu-containing $Al_{74}Ni_{15}Si_9Cu_2$-am alloy (Figure 3c).

On the XRD pattern of the corrosion products of the amorphous $Al_{74}Ni_{16}Si_{10}$-am alloy (Figure 3a), released at 25 °C and 50 °C, low-intensity peaks of the crystalline phases nickel chloride hydrate $NiCl_2(H_2O)_4$ and $Al(OH)_3$ are observed together with the amorphous humps. This indicates that the released corrosion products at both test temperatures are a mixture of amorphous and crystalline compounds. A similar structure of the corrosion products has been observed by other researchers also [55–58].

The diffraction pattern of the products of the amorphous $Al_{74}Ni_{15}Si_9Cu_2$-am alloy released at 25 °C shows a completely X-ray amorphous structure (Figure 3c). The humps intensity of this alloy, compared with the XRD pattern of the $Al_{74}Ni_{16}Si_{10}$-am alloy, are greater, indicating a greater amount of released amorphous products. Single peaks with a low intensity of the crystal phases CuO and Cu_2O are observed on the XRD pattern at 50 °C (Figure 3c). Based on these features of the diffractogram, we assume that the structure of the released products at both temperatures is mostly X-ray amorphous.

On the XRD patterns of the corrosion products, released at 25 °C from the nanocrystalline $Al_{74}Ni_{16}Si_{10}$-ncr and the ultra-fine crystalline $Al_{74}Ni_{15}Si_9Cu_2$-ufcr alloys, only great intensity NaCl peaks (residues from the 3.5% NaCl solution) are identified. On the XRD patterns at 50 °C of both alloys, amorphous humps of AlO(OH) are visible around 2θ (10–25), (25–35), and (35–48), (Figure 3b,d), and the humps in the copper-containing alloy $Al_{74}Ni_{15}Si_9Cu_2$-ufcr again are more intense. Based on the form of the XRD patterns, we conclude that the microstructure of the released corrosion products from the both the $Al_{74}Ni_{16}Si_{10}$-ncr and $Al_{74}Ni_{15}Si_9Cu_2$-ufcr alloys is a mixture of an X-ray amorphous matrix with singe crystalline phases.

3.2. Electrochemical Corrosion Tests

Figure 4 shows the dependences of the change in the open circuit potential (OCP) in contact with 3.5% NaCl. It can be seen that during the first 20 min, the values of OCP for the amorphous alloys are close, but with longer immersion in the chloride media, the potential of the alloy $Al_{74}Ni_{15}Si_9Cu_2$-am shifts to the negative direction. Negative polarisation is most often associated with accelerating metal dissolution. The values of OCP for the annealed alloys are about 300 mV more negative than those of their amorphous analogues. Moreover, for both the crystalline alloys, the potentials are unstable and continuous fluctuations are observed on the curve. Initially, the amplitude of these oscillations exceeds 50 mV, but gradually decreases to 10 mV.

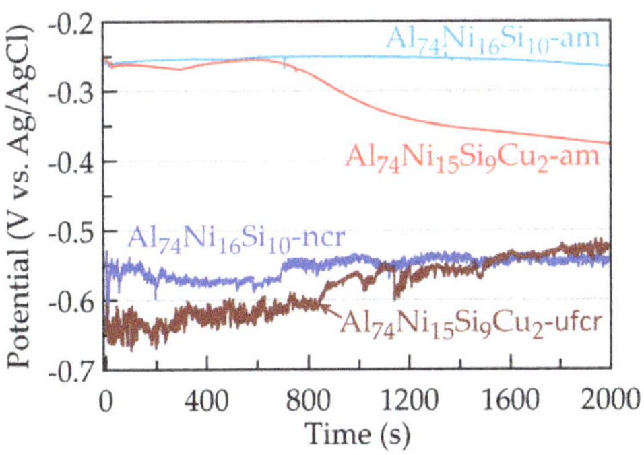

Figure 4. Dependence of open circuit potential on time.

Electrochemical impedance spectroscopy provides additional information on the behaviour of alloys under OCP. The Nyquist plot shows incomplete semicircles typical of passive metal systems. Semicircle response is due to the charge transfer reaction. An increase in semicircle diameter corresponds to an increase in charge transfer and hence the corrosion resistance. At high frequencies, Bode impedance plots are in straight line parallel to the *x*-axis and is a result of the frequency independence of resistance. The intersection of the dependences with the ordinate axis of the impedance module is usually associated with the solution resistance R_S, which dominates in the high-frequency range. In the low-frequency region, the charge transfer process (the electrochemical reaction) dominates. It can be seen from Figure 5b that the amorphous alloys have the highest impedance at 0.1 Hz, while their crystalline analogous are about one order of magnitude lower.

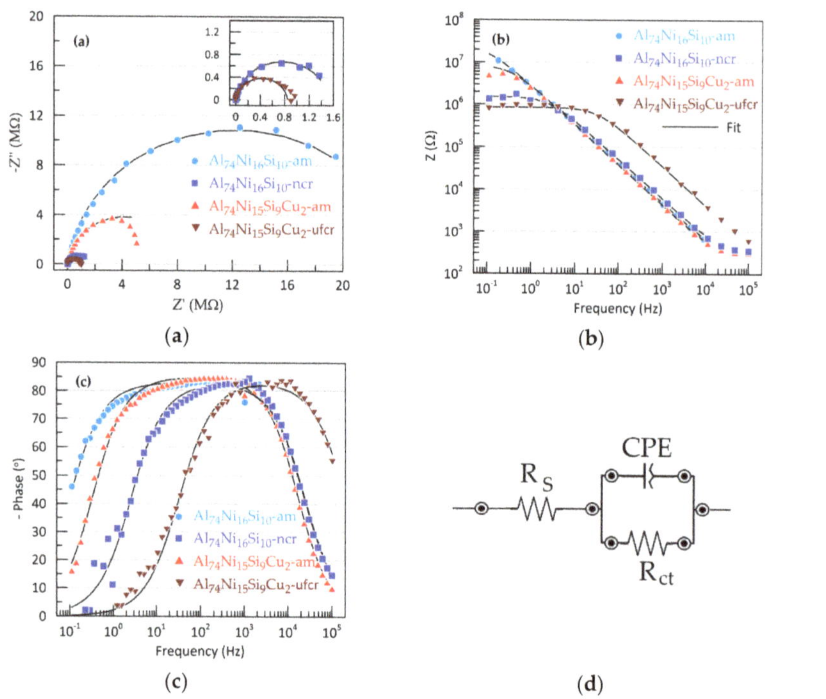

Figure 5. EIS results for Sb-Cu layers: (**a**) Nyquist's plots, (**b**) Bode impedance plots, (**c**) Bode phase angle plots and (**d**) equivalent circuit used to fit EIS data.

These observations are also confirmed by the Bode phase angle plots (Figure 5c), which have a bell shape. At both very high and very low frequencies, the phase angle deviation tends to be 0, indicating the resistive behaviour of the circuit. In the middle frequency range, the dependences for all tested alloys show a wide plateau, not reaching $-90°$. The shift in the Bode magnitude maxima to higher frequencies in crystalline alloys indicates an expansion of the frequency range where the charge transfer of corrosion reaction dominates, and charging of the electric double layer takes place at increasingly higher frequencies.

The equivalent circuit model presented in Figure 5d was used to interpret and analyse the EIS data. This model describes processes at the metal–electrolyte interface and electron transfer. The electrical circuit consists of a series-connected R_s (solution resistance) and a block of parallel R_{ct} (charge transfer resistance) and a constant phase element CPE corresponding to the capacitance of electrochemical double. The CPE is used instead of a pure capacitor since the phase angle in the middle frequencies region does not reach $90°$ (Figure 5c) as well as an imperfect semicircle in the Nyquist plots. The deviation from the ideal capacitor behaviour is expressed through the constant n. However, the recorded deviations are relatively weak, since the value of n is above 0.92.

The resulting fits based on this circuit are represented as black lines on the Nyquist and Bode plots (Figure 5).

The values of the parameters of the equivalent circuit for the tested alloys are summarised in Table 3. The EIS data show that while the R_S values for all tested alloys remain relatively close, the charge transfer resistance and CPE decrease with the addition of copper and with the transition from the amorphous to the crystalline structure.

Table 3. Equivalent electrical circuit parameters for Al-Ni-Si-(Cu) alloys at OCP.

Designation of the Alloy	R_S (Ω)	R_{ct} (MΩ)	CPE (μΩ.s^n)	n
$Al_{74}Ni_{16}Si_{10}$-am	256.0	24.51	0.064	0.924
$Al_{74}Ni_{15}Si_9Cu_2$-am	295.0	8.36	0.059	0.945
$Al_{74}Ni_{16}Si_{10}$-ncr	310.7	1.52	0.051	0.925
$Al_{74}Ni_{15}Si_9Cu_2$-ufcr	248.9	0.85	0.008	0.931

The corrosion behaviour of the Al-Ni-Si-(Cu) alloys was also tested via the cyclic potentiodynamic polarisation method. Typical example of the measurements is presented in Figure 6. Characteristic corrosion electrochemical parameters such as corrosion potential (E_{corr}), corrosion current density (J_{corr}) and pitting potential (E_{pitt}) were determined from the polarisation dependences. The average values of these parameters are presented in Table 4.

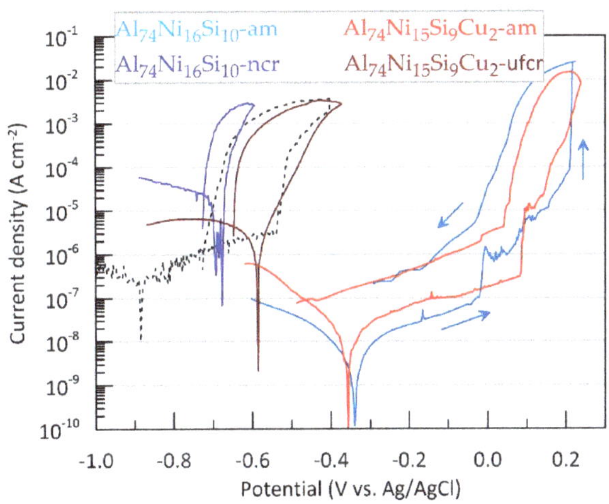

Figure 6. Polarisation dependences of alloys $Al_{74}Ni_{16}Si_{10}$ and $Al_{74}Ni_{15}Si_9Cu_2$ in amorphous and crystalline state and of pure aluminium (black dotted line). Arrows indicate the progress of the potential scan.

The values of E_{corr} and J_{corr} for the two amorphous alloys are close. The more negative value of E_{corr} and the lower value of Jcorr for the copper-containing alloy can be considered insignificant. A similar trend was also reported by Setiady and Soegijono [38]. The more negative average Ecorr values compared to OCP measured immediately before the polarisation tests showed that the low value of cathodic polarisation and the short time of its application nevertheless negatively polarised the alloys.

Table 4. Corrosion parameters extracted from the polarisation dependences of amorphous and crystalline Al-Ni-Si-(Cu) alloys.

Designation of the Alloy	E_{corr} (V)	J_{corr} (μA/cm^2)	E_{pitt} (V)
$Al_{74}Ni_{16}Si_{10}$-am	−0.326 ± 0.043	0.009 ± 0.002	−0.029 ± 0.050
$Al_{74}Ni_{15}Si_9Cu_2$-am	−0.349 ± 0.044	0.005 ± 0.003	0.068 ± 0.046
$Al_{74}Ni_{16}Si_{10}$-ncr	−0.669 ± 0.010	14.60 ± 4.20	-
$Al_{74}Ni_{15}Si_9Cu_2$-ufcr	−0.557 ± 0.034	0.508 ± 0.077	-

The average E_{corr} values for $Al_{74}Ni_{15}Si_9Cu_2$ alloy are about 20 mV more negative and the values for J_{corr} deviate over a wide range of more than one order of magnitude (Table 4). Furthermore, it can be noted that the corrosion rate of the amorphous alloys is two orders of magnitude lower and the corrosion potential is more than 500 mV more positive compared to the corresponding values obtained for high-purity polycrystalline aluminium (Figure 6, black dotted line).

The annealing of the alloys leads to a significant deterioration in the corrosion resistance, manifested both in a deviation of E_{corr} in the negative direction (with an average of about 340 mV for $Al_{74}Ni_{16}Si_{10}$ and 200 mV for $Al_{74}Ni_{15}Si_9Cu_2$) and in an increase in J_{corr} (with an average of about three and two orders of magnitude for $Al_{74}Ni_{16}Si_{10}$ and $Al_{74}Ni_{15}Si_9Cu_2$, respectively). The lowest corrosion resistance is demonstrated by the annealed copper-free alloy (Table 4).

The anodic branch of the polarisation curves of the amorphous alloys shows diffusion-limited anodic process, which is typical for the presence of passive layers (Figure 6). Upon reaching a certain critical potential called the pitting potential (E_{pitt}), the anodic current density increases sharply. The latter is an indication of the development of pitting corrosion, typical for aluminium alloys under anodic polarisation in a chloride media. The registered slight fluctuations on the anodic branch of the polarisation dependence before reaching the E_{pitt} can be associated with the development of metastable pits which are repassivated immediately after their nucleation. When additional anodic polarisation exceeds E_{pitt}, some of the pits grow steadily, resulting in a sharp increase in current.

The pitting potential of $Al_{74}Ni_{15}Si_9Cu_2$-am alloy is more positive by about 100 mV on average compared to $Al_{74}Ni_{16}Si_{10}$-am. This shows that the presence of 2% copper broadens the passive state. These results are similar to the ones reported by Muller and Galvele for aluminium alloys containing less than 5% copper [59]. According to these authors, the $CuAl_2$ phase formation during heating leads to copper depletion of grain boundaries, an increased susceptibility to intercrystalline and pitting corrosion around this phase. These authors also reported Cu enrichment in the interior of the pits.

Scanning the potential in the reverse direction reveals the local corrosion development, characterised by rapid dissolution of limited active areas, which raises the anodic current density to several orders of magnitude higher than that of the forward branch. It is noteworthy that for both amorphous alloys partial repassivation is observed at potentials close to E_{pitt}. After this value, the reverse branch current remains only about an order of magnitude higher than that of the forward branch, and the polarisation curve maintains a relatively small slope until crossing the cathode branch.

In the case of the annealed alloys, the anodic polarisation does not lead to diffusion limitation of the current; they are not passive, but actively dissolve in the chloride medium.

4. Discussions

The initiation and evolution of corrosion processes in metallic materials depend on the chemical composition and the microstructure of the alloys, the chemical composition and temperature of the corrosion environment and on the interaction of these factors.

The results of our immersion tests showed that the corrosion rate of the copper-containing $Al_{74}Ni_{15}Si_9Cu_2$ alloy both in amorphous and crystalline state was higher than that of the corresponding copper-free alloys. The probable reason for this behaviour in crystalline state is the presence of (Ni,Si,Cu)-containing phases of various composition and size in the $Al_{74}Ni_{15}Si_9Cu_2$-ncr alloy [47], but this cannot explain the behaviour of the amorphous alloy.

In our previous research, we confirmed via TEM observations the amorphous microstructures of both $Al_{74}Ni_{16}Si_{10}$ and $Al_{74}Ni_{15}Si_9Cu_2$ rapidly solidified alloys. A typical amorphous structure uniform matrix with well-defined diffraction halo was registered on the diffractograms of both alloys. Small white spots randomly scattered in the dark uniform matrix were observed on the TEM images of $Al_{74}Ni_{15}Si_9Cu_2$ [50]. We performed additional high resolution HRTEM observations and found regions corresponding to the white spots in both alloys, but in much larger number in $Al_{74}Ni_{15}Si_9Cu_2$ (Figure 7). These regions showed a specific striation contrast which made us to believe that they were clusters of atoms of the dissolved alloying elements, probably nickel, silicon and/or copper. No grain boundaries bordered the striated regions indicated that the clusters were formed during rapid solidification but did not succeed to combine into crystal grains. Therefore, we concluded that the microstructures of both rapidly solidified alloys consists of an X-ray amorphous matrix with single clusters of atoms of alloying elements, probably Ni and Si in $Al_{74}Ni_{16}Si_{10}$, and a large number of clusters, probably Ni, Si and/or Cu in $Al_{74}Ni_{15}Si_9Cu_2$. The clusters act as microcathodes to the anodic aluminium matrix and form microgalvanic cells that are potential areas for corrosion development [60]. The larger number of clusters in $Al_{74}Ni_{15}Si_9Cu_2$-am is the reason for the higher corrosion rate of this alloy compared to $Al_{74}Ni_{16}Si_{10}$-am in long-term corrosion tests.

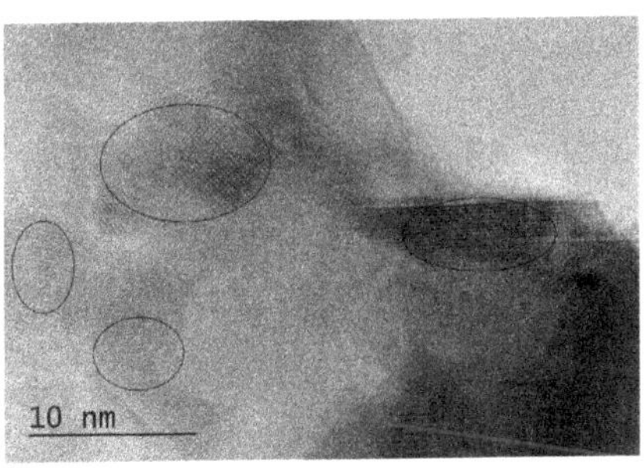

(**a**)

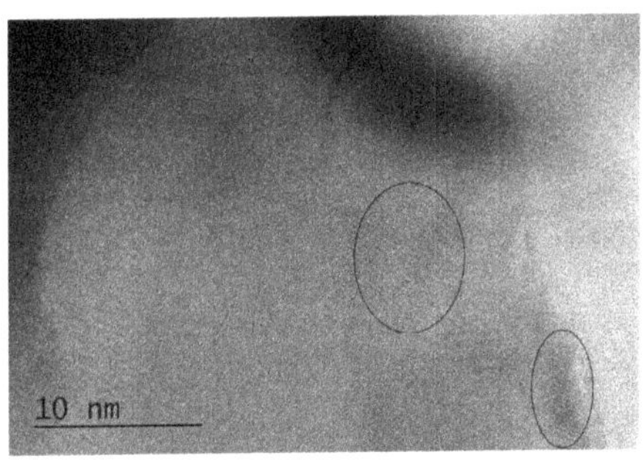

(**b**)

Figure 7. Microstructure of the amorphous $Al_{74}Ni_{16}Si_{10}$–am (**a**) and $Al_{74}Ni_{15}Si_9Cu_2$–am (**b**) alloys, HRTEM. Amorphous matrix with clusters of Ni, Si/Cu atoms.

When comparing the results obtained via gravimetric and electrochemical tests at 25 °C, some discrepancies between the two methods appear. The immersion test results show a lower corrosion rate CR_{25} of $Al_{74}Ni_{16}Si_{10}$-ncr compared to $Al_{74}Ni_{15}Si_9Cu_2$-ufcr, while the potentiodynamic polarization test shows a negligible difference in corrosion potential. A probable reason for this discrepancy can be found in the morphological features of the rapidly solidified ribbons and the specificity of the used corrosion test methods.

The two sides of the ribbons obtained via the CBMS method have different roughness (Figure 1). The surface which has been in contact with the surface of the cooling copper disk is rather rough (Figure 1a)—there are hollows (the darker areas) and ridges (the lighter areas) on it. These places are additional potential sites for corrosion initiation. During the gravimetric immersion tests, the rough side of the samples was not isolated in order to ensure a larger active surface area. After the enhanced corrosion initiation occurred due to surface roughness, the cathodic areas on the rough surface are increasingly revealed during the prolonged immersion in the corrosive environment (360 h), and this accelerates the general dissolution of the anodic Al-matrix.

We can conclude that the absence of a significant influence of copper on the electrochemical test results of the amorphous alloys is a result of both the smoother surface and the short immersion time in the chloride medium. Over time, the differences between the two sides of the ribbons decrease and the negative effect of the copper-rich cathodic phases accelerates the corrosion process. This is clearly observed from both the OCP dependences and the SEM observations. After annealing, the corrosion resistance of both alloys decreases due to an increase in surface heterogeneity.

The prolonged immersion tests involve the rough side also and this results in higher calculated values of the corrosion rate. Keeping in mind the significantly more positive potential of copper, it can be expected that the participation of copper, in spite of the smaller content of Cu than Ni in the alloy, will have a stronger negative effect on the corrosion of microgalvanic cells during long-term treatment compared to those based on nickel.

Summarizing the results of both the corrosion tests and the TEM and SEM observations, we can conclude that in the amorphous Al-Ni-Si and Al-Ni-Si-Cu alloys in 3.5% NaCl solution, general surface corrosion is mainly observed, which is accelerated by the anodic dissolution of the aluminium matrix caused by the microgalvanic cathodes (polymetallic clusters). Corrosion damage is visibly better expressed in the copper-containing amorphous alloy. Due to the greater total area of the cathodic phases in crystalline alloys, corrosion develops primarily through a galvanic mechanism.

5. Conclusions

The addition of 2 at.% copper to the amorphous $Al_{74}Ni_{16}Si_{10}$ base alloy improves the pitting corrosion resistance of the amorphous $Al_{74}Ni_{15}Si_{9}Cu_{2}$, shifting the pitting potential by about 100 mV in the positive direction, but does not significantly affect the corrosion current and potential in the duration when short-term polarisation tests were performed. However, in long-term immersion tests, the copper-containing alloy $Al_{74}Ni_{15}Si_{9}Cu_{2}$ demonstrates 2.6 times lower general corrosion resistance both in amorphous and crystalline states and at both temperatures.

The amorphous $Al_{74}Ni_{16}Si_{10}$ and $Al_{74}Ni_{15}Si_{9}Cu_{2}$ alloys have better corrosion resistance compared to their crystalline analogues. After crystallisation, the corrosion rate of both the alloys increases by about 3 times at a test temperature of 25 °C and by about 8 times at a temperature of 50 °C. The corrosion rates of the crystal alloys $Al_{74}Ni_{16}Si_{10}$ and $Al_{74}Ni_{15}Si_{9}Cu_{2}$ increase 5 and 2 times at test temperatures 25 °C and 50 °C, respectively.

The thin film of corrosion products released from the tested amorphous and crystallised copper-containing $Al_{74}Ni_{15}Si_{9}Cu_{2}$ alloys is completely X-ray amorphous, and the film released from the copper-free alloy $Al_{74}Ni_{16}Si_{10}$ is mostly X-ray amorphous with single peaks of crystalline phases.

Author Contributions: Conceptualisation, V.D. and Y.K.; methodology, V.D., Y.K. and B.T.; validation, Y.K., B.T. and D.K.; formal analysis, Y.K., H.S. and D.K.; investigation Y.K., B.T., H.S. and D.K.; writing—original draft preparation, V.D. and B.T.; writing—review and editing V.D., B.T. and D.K.; visualisation, Y.K. and B.T.; project administration, V.D. All authors have read and agreed to the published version of the manuscript.

Funding: This research was funded by a contract with the Bulgarian National Scientific Found №KP-06-H37/13 of 6 December 2019.

Institutional Review Board Statement: Not applicable.

Informed Consent Statement: Not applicable.

Data Availability Statement: Not applicable.

Acknowledgments: The authors are indebted to their colleagues, Jordan Georgiev and Ivan Penkov, (IMSETHAC-BAS) for their help and support in the preparation of the alloys.

Conflicts of Interest: The authors declare no conflict of interest.

References

1. Russew, K.; Stojanova, L. *Glassy Metals*; Springer: Berlin/Heidelberg, Germany, 2016. [CrossRef]
2. Wang, D.; Ding, H.; Ma, Y.; Gong, P.; Wang, X. Research Progress on Corrosion Resistance of Metallic Glasses. *J. Chin. Soc. Corros. Prot.* **2021**, *41*, 277–288. Available online: https://www.jcscp.org/EN/10.11902/1005.4537.2020.089 (accessed on 18 June 2023).
3. Naka, M.; Hashimoto, K.; Masumoto, T. Corrosion resistivity of amorphous Fe alloys containing Chromium. *J. Jpn. Inst. Met.* **1974**, *38*, 835–841. [CrossRef]
4. Souzaa, A.C.; Ribeiroa, D.V.; Kiminam, C.S. Corrosion resistance of Fe-Cr-based amorphous alloys: An overview. *J. Non-Crystall. Sol.* **2016**, *442*, 56–66. [CrossRef]
5. Pang, S.J.; Zhang, T.; Asami, K.; Inoue, A. Synthesis of Fe–Cr–Mo–C–B–P bulk metallic glasses with high corrosion resistance. *Acta Mater.* **2002**, *50*, 489–494. [CrossRef]
6. Kiminami, C.S.; Souza, C.A.C.; Bonavina, L.F.; Bolfarini, C.; Botta, W.J. Partial crystallization and corrosion resistance of amorphous Fe-Cr-M-B (M = Mo, Nb) alloys. *J. Non-Cryst. Solids* **2010**, *356*, 2651–2657. [CrossRef]
7. Mariano, N.A.; Souza, C.A.C.; May, J.E.; Kuri, S.E. Influence of Nb content on the corrosion resistance and saturation magnetic density of FeCuNbSiB alloys. *Mater. Sci. Eng. A* **2003**, *354*, 1–5. [CrossRef]
8. Botta, W.J.; Berger, J.E.; Kiminami, C.S.; Roche, V.; Nagueira, R.P.; Bolfarini, C. Corrosion resistance of Fe-based amorphous alloys. *J. Alloys Compd.* **2014**, *586*, 105–110. [CrossRef]
9. Hertsyk, O.M.; Kovbuz, M.O.; Hula, T.H. Corrosion Resistance of Modified Amorphous Alloys Based on Iron in Sulfuric Acid. *Mater. Sci.* **2021**, *56*, 755–763. [CrossRef]
10. Pang, S.; Hang, T.Z.; Asami, K.; Inoue, A. Corrosion behaviour of Zr-(Nb)Al-Ni-Cu glassy alloys. *Mater. Trans.* **2001**, *42*, 376–379. [CrossRef]
11. Mudali, U.K.; Scudino, S.; Kühn, U.; Eckert, J.; Gebert, A. Polarisation behaviour of the $Zr_{57}Ti_8Nb_{2.5}Cu_{13.9}Ni_{11.1}Al_{7.5}$ alloy in different microstructural states in acid solutions. *Scr. Mater.* **2004**, *50*, 1379–1384. [CrossRef]
12. Raw, V.R.; Kuhn, U.; Wolff, U.; Schneider, F.; Eckert, J.; Reiche, R.; Gebert, A. Corrosion behaviour of Zr-based bulk glass-forming alloys containing Nb or Ti. *Mater. Lett.* **2002**, *57*, 173–177. [CrossRef]
13. Wans, D.P.; Wang, L.; Wang, J.O. Relationship between amorphous structure and corrosion behaviour in a Zr-Ni metallic glass. *Corros. Sci.* **2012**, *59*, 88–95. [CrossRef]
14. Chunming, W.; Yang, S.; Youwen, Y.; Da, Z.; Xiongwei, L.; Shuping, P.; Cijun, S. Amorphous magnesium alloy with high corrosion resistance fabricated by laser powder bed fusion. *J. Alloys Compd.* **2022**, *897*, 163247. [CrossRef]
15. Liming, X.; Xingwan, L.; Kang, S.; Rao, F.; Gang, W. Corrosion behaviour in magnesium-Based Alloys for Biomedical Applications. *Materials* **2022**, *15*, 2613. [CrossRef]
16. Esmaily, M.; Svensson, J.E.; Fajardo, S.N.; Birbilis, G.S.; Frankel, S.; Virtanen, R.; Arrabal, T.S.; Johansson, L.G. Fundamentals and advances in magnesium alloy corrosion. *Prog. Mater. Sci.* **2017**, *89*, 92–193. [CrossRef]
17. Bala, H.; Szymura, S. Acid corrosion of amorphous and crystalline Cu-Zr alloys. *Appl. Surf. Sci.* **1989**, *35*, 41–51. [CrossRef]
18. Li, C.; Wang, Y.; Geng, H. Crystallization Behaviour and Corrosion Resistance of $Cu_{50}Zr_{40}Ag_{10}$ Amorphous Alloy. *J. Inorg. Organomet. Polym.* **2011**, *21*, 919–924. [CrossRef]
19. Xie, C.; Milosev, I.; Renner, F.; Kokalj, A.; Bruna, P.; Crespo, D. Corrosion resistance of crystalline and amorphous CuZr alloys in NaCl aqueus environment and effect of corrosion inhibitors. *J. Alloys Compd.* **2021**, *879*, 160464. [CrossRef]
20. Inoue, A. Amorphous, nanoquasicrystalline and nanocrystalline alloys in Al-based systems. *Prog. Mater. Sci.* **1998**, *43*, 365–520. [CrossRef]
21. Louzguine-Luzgin, D.V. Aluminium-base amorphous and nanocrystalline materials. *Met. Sci. Heat. Treat.* **2012**, *53*, 472–477. [CrossRef]
22. Sweitzer, J.; Shiflet, G.; Scully, J.R. Localized corrosion of Al90Fe5Gd5 and Al87Ni8. 7Y4. 3 alloys in the amorphous, nanocrystalline and crystalline states: Resistance to micrometer-scale pit formation. *Electrochim. Acta* **2003**, *48*, 1223–1234. [CrossRef]
23. Zhang, L.M.; Ma, A.L.; Hu, H.X.; Zheng, B.J.; Yang, B.J., Wang, J.Q. Effect of Microalloying with Ti or Cr on the Corrosion Behavoir of Al-Ni-Y Amorphous Alloys. *Corrosion* **2018**, *74*, 66–74. [CrossRef] [PubMed]
24. Akiyama, E.; Habazaki, H.; Kawashima, A.; Asami, K.; Hashimoto, K. Corrosion-resistant amorphous aluminium alloys and structure of passive films. *Mater. Sci. Eng. A* **1997**, *226–228*, 920–924. [CrossRef]
25. Principe, E.L.; Shaw, B.A.; Davis, G.D. Role of Oxide/Metal Interface in Corrosion Resistance: Al-W and Al-Mo Systems. *Corrosion* **2003**, *59*, 295–313. [CrossRef]
26. Yoshioka, H.; Yoshida, S.; Kawashima, A.; Asami, K.; Hashimoto, K. The pitting corrosion behaviour of rapidly solidified aluminium alloys. *Corros. Sci.* **1986**, *26*, 809–812. [CrossRef]
27. Karfidov, E.A.; Rusanov, B.A.; Sidorov, V.E.; Nikitina, E.V.; Janichkovic, D.; Svec, S.R.P. Corrosion-electrochemical behaviour of Al–Ni-Co-Nd. *Amorph. Alloys Melts* **2022**, *2*, 189–195. [CrossRef]
28. Aburada, T.; Unlu, N.; Fitz-Gerald, J.M.; Shiflet, G.J.; Scully, J.R. Effect of Ni as a minority alloying element on the corrosion behaviour in Al–Cu–Mg–(Ni) metallic glasses. *Scr. Mater.* **2008**, *58*, 623–626. [CrossRef]
29. Dyakova, V.; Kostova, Y.; Spasova, H. Influence of Ni as Minority Alloying Element on the Corrosion Behaviour of Amorphous Al-Cu-Mg Alloys in Chloride Solution. In Proceedings of the 14th International Scientific and Practical Conference "Environment. Technology. Resources", Rezekne, Latvia, 15–16 June 2023; Available online: http://journals.rta.lv/index.php/ETR/article/view/7201/5993 (accessed on 13 June 2023).

30. Dyakova, V.L.; Tzaneva, B.R.; Kostova, Y.G. Influence of Zn as Minority Alloying Element on the Uniform and Local Corrosion of Amorphous Rapidly Solidified AlCuMg(Zn) Ribbons. In Proceedings of the AIP Conference, TehSys, Plovdiv, Bulgaria, 27 May 2021. [CrossRef]
31. Hara, M.; Hashimoto, K.; Masumoto, T. Anodic characteristics of amorphous palladium-iridium-phosphorus alloys in a hot concentrated sodium chloride solution. *J. Non-Cryst. Solids* **1983**, *54*, 85–100. [CrossRef]
32. Chen, N.; Qin, C.L.; Xie, G.Q.; Louzguine-Luzgin, D.V.; Inoue, A. Investigation of a ductile and corrosion-resistant $Pd_{79}Au_{1.5}Ag_3Si_{16.5}$ bulk metallic glass. *J. Mater. Res.* **2010**, *25*, 1943–1949. [CrossRef]
33. Hara, M.; Hashimoto, K.; Masumoto, T. Anodic characteristics of amorphous ternary palladium-phosphorus alloys containing ruthenium, rhodium, iridium, or platinum in a hot concentrated sodium chloride solution. *J. Appl. Electrochem.* **1983**, *13*, 295–306. [CrossRef]
34. Chen, N.; Martin, L.; Luzguine-Luzgin, D.; Inoe, A. Role of Alloying Additions in Glass Formation and Properties of Bulk Metallic Glasses. *Materaials* **2010**, *3*, 5320. [CrossRef]
35. Chen, J.; Xiao, J.; Wang, W.; Zhao, J.; Michel, F.; Deng, C.; Cai, W. Effects of alloying concentration on the aqueous corrosion and passivation of aluminium-manganese-molybdenum concentrated alloys. *Corros. Sci.* **2022**, *198*, 110–137. [CrossRef]
36. Shabestari, S.G.; Moemmeni, H. Effect of copper and solidification conditions on the microstructure and mechanical properties of Al–Si–Mg alloys. *J. Mater. Procesing Technol.* **2004**, *153–154*, 193–198. [CrossRef]
37. Zeren, M. Effect of copper and silicon content on mechanical properties in Al–Cu–Si–Mg alloys. *J. Mater. Procesing Technol.* **2005**, *169*, 292–298. [CrossRef]
38. Setiady, A.; Soegijono, B. The Effect of Copper on Corrosion Resistance and Structure of Al-8% wt Si Alloys for Engine Block Application. In Proceedings of the IOP Conference Series: Materials Science and Engineering, 694, 1st International Symposium on Advances and Innovations in Mechanical Engineering, Jakarta, Indonesia, 9–10 October 2019. [CrossRef]
39. Bakare, M.; Voisey, K.; Chokethawai, K.; McCartney, D. Corrosion behaviour of crystalline and amorphous forms of the glass forming alloy $Fe_{43}Cr_{16}Mo_{16}C_{15}B_{10}$. *J. Alloys Compd.* **2012**, *527*, 210–218. [CrossRef]
40. Wang, Z.; Dong, L.; Hu, B.; Chen, B. The Effect of Cu Addition on Corrosion Resistance of Al-Si-Mg-Cr Alloy. *Metals* **2023**, *13*, 795. [CrossRef]
41. Chen, H.; Zhang, X.; Li, Y. Effect of the Secondary Phase on the Corrosion of Al-Zn-Mg-Cu Alloy. *Int. J. Metall. Met. Phys.* **2020**, *5*, 058. [CrossRef]
42. Gao, M.; Lu, W.; Yang, B.; Zhang, S.; Wang, J. High corrosion and wear resistance of Al-based amorphous metallic coating synthesized by HVAF spraying. *J. Alloys Compd.* **2018**, *735*, 1363–1373. [CrossRef]
43. Coimrao, D.D.; Zepon, G.; Koga, G.Y. Corrosion properties of amorphous, partially, and fully crystallized $Fe_{68}Cr_8Mo_4Nb_4B_{16}$ alloy. *J. Alloys Compd.* **2020**, *826*, 154123. [CrossRef]
44. Mehmood, M.; Zhang, B.P.; Akiyama, E.; Habazaki, H.; Kawashima, A.; Asami, K.; Hashimoto, K. Experimental evidence for the critical size of heterogeneity areas for pitting corrosion of Cr-Zr alloys in 6 M HCl. *Corros. Sci.* **1998**, *40*, 1–17. [CrossRef]
45. Schroeder, V.; Gilbert, C.J.; Ritchie, R.O. Comparison of the corrosion behaviour of a bulk amorphous metal, $Zr_{41.2}Ti_{13.8}Cu_{12.5}Ni_{10}Be_{22.5}$, with its crystallized form. *Scr. Mater.* **1998**, *38*, 1481–1485. [CrossRef]
46. Li, W.H.; Chana, K.C.; Xiaa, L.; Liuc, L.; He, Y.Z. Thermodynamic, corrosion and mechanical properties of Zr-based bulk metallic glasses in relation to heterogeneous structures. *Mater. Sci. Eng. A* **2012**, *534*, 157–162. [CrossRef]
47. Safavi, M.S.; Mehdad, F.; Ahadzadeh, I. Feasible strategies for promoting the mechano-corrosion performance of Ni-Co based coatings: Which one is better? *Surf. Coat. Technol.* **2021**, *420*, 127337. [CrossRef]
48. Miao, Y.; Tiancheng, C.; Dapeng, Z.; Rongzhi, L.; Jibin, P.; Changsheng, L. Improved oxidation and hot corrosion resistance of the NiSiAlY alloy at 750 °C. *Mater. Today Commun.* **2021**, *29*, 102939. [CrossRef]
49. Dyakova, V.; Stefanov, G.; Penkov, I.; Kovacheva, D.; Marinkov, N.; Mourdjeva, Y.; Gyurov, S. Influence of Zn on Glass Forming Ability and Crystallization Behaviour of Rapidly Solidified Al-Cu-Mg (Zn) alloys. *J. Chem. Technol. Metall.* **2020**, *57*, 622–630. Available online: https://journal.uctm.edu/node/j2022-3/23_21-64_br_3_pp_622-630.pdf (accessed on 18 June 2023).
50. Perepezko, J.H.; Hebert, R.J. Amorphous aluminium alloys—Synthesis and stability. *JOM* **2002**, *54*, 34–39. [CrossRef]
51. Egami, T.; Waseda, Y. Atomic size effect on the formability of metallic glasses. *J. Non-Cryst. Solids* **1984**, *64*, 113–134. [CrossRef]
52. Xue-Kui, X.; Magdalena, T.S.; Yan-Hui, L.; Wei-Hua, W.; Yue, W. Structural changes induced by microalloying in $Cu_{46}Zr_{47-x}Al_7Gd_x$ metallic glasses. *Scr. Mater.* **2009**, *61*, 967–969. [CrossRef]
53. Dyakova, V.; Spasova, H.; Kostova, Y.; Mourdjeva, Y.; Stefanov, G. Effect of Cu as Minority Alloying Element on Glass Forming Ability and Crystallization Behaviour of Rapidly Solidified Al-Si-Ni Ribbons. In Proceedings of the 14th International Scientific and Practical Conference Environment. Technologies. Resources, Rezekne, Latvia, 15–16 June 2023; Available online: http://journals.rta.lv/index.php/ETR/article/view/7200/5992 (accessed on 13 June 2023).
54. Dyakova, V.L.; Stefanov, G.N.; Kovacheva, D.G.; Mourdjeva, Y.S.; Marinkov, N.E.; Penkov, I.G.; Georgiev, J.S. Influence of Zr and Zn as Minority Alloying Elements on Glass Forming Ability and Crystallization Behaviour of Rapidly Solidified AlCuMg ribbons. In Proceedings of the AIP Conference, TehSys, Plovdiv, Bulgaria, 27 May 2021; Available online: https://aip.scitation.org/doi/abs/10.1063/5.0090746 (accessed on 1 September 2022).
55. Birringer, R.; Gleiter, H.; Klein, P.; Marquardt, P. Nanocrystalline materials an approach to a novel solid structure with gas-like disorder? *Phys. Lett. A* **1984**, *102*, 365–369. [CrossRef]

56. *ASTM G31-21*; Standard Guide for Laboratory Immersion Corrosion Testing of Metals. American Technical Standard: Washington, DC, USA, 1 January 2021.
57. Ram, L.K.; Park, J.M.; Soo Jee, S.; Kim, S.Y.; Kim, S.J.; Eun-Sung, L.; Kim, W.T.; Gebert, A.; Eckert, J.; Kim, D.H. Effect of thermal stability of the amorphous substrate on the amorphous oxide growth on Zr–Al–(Cu,Ni) metallic glass surfaces. *Corr. Sci.* **2013**, *73*, 1–6. [CrossRef]
58. Hsu, P.; Bates, T. Formation of X-ray amorphous and crystalline aluminium hydroxides. *Mineral. Mag. J. Mineral. Soc.* **1964**, *33*, 749–768. [CrossRef]
59. Muller, I.L.; Galvele, J.R. Pitting potential of high purity binary aluminium alloys—I. Al-Cu alloys. Pitting and intergranular corrosion. *Corros. Sci.* **1977**, *17*, 179. [CrossRef]
60. Tao, M.J. Surface Composition and Corrosion Behavior of an AlCu Alloy. Ph.D. Thesis, Sorbonne University Pierre and Marie Curie Campus, Paris, France, 2016.

materials

MDPI

Article

Impact of Beam Deflection Geometry on the Surface Architecture and Mechanical Properties of Electron-Beam-Modified TC4 Titanium Alloy

Maria Ormanova [1,*], Borislav Stoyanov [2], Nikolay Nedyalkov [1] and Stefan Valkov [1,3]

1 Academician Emil Djakov Institute of Electronics–Bulgarian Academy of Sciences, 72 Tsarigradsko Chaussee Blvd, 1784 Sofia, Bulgaria; nned@ie.bas.bg (N.N.); stsvalkov@gmail.com (S.V.)
2 Department of Industrial Desing and Textile Engineering, Technical University of Gabrovo, 4 H. Dimitar Srt, 5300 Gabrovo, Bulgaria; stoyanov_b@mail.bg
3 Department of Mathematics, Informatics and Natural Sciences, Technical University of Gabrovo, 4 H. Dimitar Srt, 5300 Gabrovo, Bulgaria
* Correspondence: m.ormanova@ie.bas.bg

Abstract: This paper aims to investigate the impact of beam deflection geometry on the structure, surface architecture, and friction coefficient of electron-beam-modified TC4 titanium alloys. During the experiments, the electron beam was deflected in the form of different scanning geometries, namely linear, circular, and matrix. The structure of the treated specimens was investigated in terms of their phase composition by employing X-ray diffraction experiments. The microstructure was studied by scanning electron microscopy (SEM). The surface architecture was examined by atomic force microscopy (AFM). The friction coefficient was studied by a mechanical wear test. It was found that the linear and circular deflection geometries lead to a transformation of the phase composition, from double-phase $\alpha + \beta$ to α' martensitic structure. The application of a linear manner of scanning leads to a residual amount of beta phase. The use of a matrix does not tend to structural changes on the surface of the TC4 alloy. In the case of linear geometry, the thickness of the modified zone is more than 800 μm while, in the case of EBSM using circular scanning, the thickness is about 160 μm. The electron-beam surface modification leads to a decrease in the surface roughness to about 27 nm in EBSM with linear deflection geometry and 31 nm in circular deflection geometry, compared to that of the pure TC4 substrate (about 160 nm). The electron-beam surface modification of the TC4 alloy leads to a decrease in the coefficient of friction (COF), with the lowest COF values obtained in the case of linear deflection geometry (0.32). The results obtained in this study show that beam deflection geometry has a significant effect on the surface roughness and friction coefficient of the treated surfaces. It was found that the application of a linear manner of scanning leads to the formation of a surface with the lowest roughness and friction coefficient.

Keywords: electron-beam modification; TC4 titanium alloy; beam deflection geometry; phase composition; structure; surface architecture; coefficient of friction

Citation: Ormanova, M.; Stoyanov, B.; Nedyalkov, N.; Valkov, S. Impact of Beam Deflection Geometry on the Surface Architecture and Mechanical Properties of Electron-Beam-Modified TC4 Titanium Alloy. *Materials* **2023**, *16*, 5237. https://doi.org/10.3390/ma16155237

Academic Editor: Pan Gong

Received: 22 June 2023
Revised: 19 July 2023
Accepted: 24 July 2023
Published: 26 July 2023

1. Introduction

Titanium-based alloys have been widely used for the production of orthopedic implants and prostheses [1–3], as well as for orthodontic archwires and dental instruments [4,5], due to their significant number of exceptional properties (good corrosion resistance, better bio-compatibility, and higher specific strength). However, when placing implants in the human body, a number of adverse reactions may occur (fracture of the implant, deterioration of adhesion, release of metal ions, etc.), leading to a new intervention or the course of acute inflammatory processes [6]. Therefore, the Ti-based alloys used in implantology have to possess suitable functional properties, such as a low Young's modulus, low coefficient of friction, etc. The modulus of elasticity has to be close to that of human bones. [7,8]. In this case, stresses at the implant–bone interface could be significantly reduced and the longevity of placed implants increased.

The improvement of the surface properties of implant materials could be achieved by different surface modification methods [9–12]. In general, methods for surface modification of implant materials can be classified as physical, chemical, and mechanical [13–15]. These methods and techniques can be used both individually [16–18] and in combination [19–21]. It is essential to choose a suitable technique for the modification of surfaces of implants depending on their specific applications [22–24].

Surface treatments of metals and alloys by a flux of accelerated electrons comprise a commonly used technology used to improve the performance characteristics of materials [23,25–27]. These techniques are considered very reproducible, low cost, and have very low process times [28]. The mechanism of electron-beam surface modification (EBSM) is based on the transformation of the kinetic energy of electrons into heat, which is accompanied by a distribution of heat from the surface to the bulk [28–30]. High heating and cooling rates lead to phase and structural changes that significantly improve the surface properties of metals and alloys [31,32]. The electron-beam surface treatment can be performed via different scanning trajectories of the electron beam, namely linear, circular, matrix, etc. [33]. In this way, precise control of the heating and cooling rates and, therefore, the obtained structure and functional properties of the material can be achieved.

The authors of [32] have discussed the effect of low-energy high-current pulsed electron-beam (LEHCPEB) surface treatment on the structure and properties of a near-α Ti alloy. The results showed that the LEHCPEB process leads to phase and structural transformations, leading to a significant enhancement in the surface hardness and corrosion properties of the titanium alloy.

Gao [34] has studied the effect of electron-beam surface treatment on the structure and properties of titanium alloys. The work-pieces were modified in a pulsed mode and experiments were realized under different pulsed energy densities, durations of pulses, and numbers of pulses. It was found that the hardness was improved due to the higher number of dislocations caused by heat stresses. Similarly, Gao [35] investigated the pulsed electron-beam modification of TA15 titanium alloy where the kinetic energy was in the range of 10–40 keV and the results showed that the microstructure was significantly refined, leading to an increase in the microhardness. In Refs. [36,37] the electron-beam technique was used to design a specific surface (cp-Ti and Ti6Al4V alloy) to control the effect of contact guidance and limit microfouling. The results showed a change in crystallographic structure and microstructure; moreover, a significant reduction in bacterial adhesion was observed in the EB-structured samples. The relationship between the microstructure of titanium alloys and antiadhesive activity against bacteria opens a new strategy for the development of innovative antiadhesive/antibacterial metallic materials [36,37]. Our previous investigations [38] were based on the study of the change in the phase composition and microstructure of Ti6Al4V alloy surfaces after electron-beam surface treatment. During the experiment, the electron-beam current was in the range of 20 mA to 25 mA, and a circular manner of scanning was applied. The results showed a change in the phase composition and microstructure of the surface of the treated samples. A typical martensitic transformation for the electron-beam-processed titanium alloys was observed. The structural changes that occurred had a significant impact on the improvement of the surface hardness of the titanium alloy samples.

It is obvious that the use of electron-beam surface modification is very suitable for the improvement of the surface properties of a number of metals and alloys. However, investigations on surface modification by a more complex manner of scanning, as well as those on the influence of the beam deflection geometry on the functional properties of Ti-based materials, are currently lacking in the scientific literature. As already mentioned, the use of different manners of scanning, defined by the geometry of beam deflection, leads to different thermal distributions from the surface to the bulk and, therefore, different thermal cycling gradients during the electron-beam surface modification procedure. This leads to different structures and corresponding functional properties. Thus, the aim of the present work is to study the impact of beam deflection geometry on the surface architecture and mechanical properties of an electron-beam-modified TC4 titanium alloy. The presented results here provide new information regarding structure and property modifications associated with electron-beam processing when different geometries of beam deflection are considered. This study could be a basis for the design and development of advanced techniques for material surface structuring with tunable properties.

2. Materials and Methods

The TC4 titanium alloy samples used in this study had a cylindrical shape with diameter of 30 mm and thickness of 4 mm. The chemical composition of the samples is shown in Table 1.

Table 1. Chemical composition of TC4 titanium alloy.

Element	Al	V	Fe	Co	Mo	Pd	Hf	Ti
wt. %	5.80	4.67	0.16	0.09	0.07	0.02	0.14	Bal.

The surfaces of the samples were mechanically polished (grinding papers of 320 to 1000) and ultrasonically cleaned in water. The surface modification was conducted by means of a scanning electron beam using an Evobeam Cube 400 machine (Evobeam GmbH, Nieder-Olm, Germany). The following technological conditions were applied: accelerating voltage—60 kV; electron-beam current—25 mA; scanning frequency—1 kHz; speed of sample's motion—5 mm/s. The beam deflection geometry was in the form of a line (sample 1), a circle (sample 2), and a matrix (sample 3). All experiments were realized under the same technological conditions with an oscillation amplitude of 16 mm. The selection of the discussed technological conditions was based on systematical optimization and the parameters presented above were considered as the most representative. It was established that for much lower values of electron-beam current, as well as for higher values of speed of movement and scanning frequency, the surface temperature became insufficient and a modification of the structure and properties was not achieved. At significantly higher values of beam current, as well as lower values of speed of movement and scanning frequency, very strong modifications of the surface topography occurred. Also, the melting processes became predominant which was not considered a desirable effect. The geometries of the beam deflection are shown in Figure 1. These are available in most commercial electron-beam systems and their use expresses a variety of conditions wherein different heating dynamics can be realized. Several experiments were realized under these technological conditions and the corresponding results were identical, as their deviation was in the range of 10%, confirming their reproducibility.

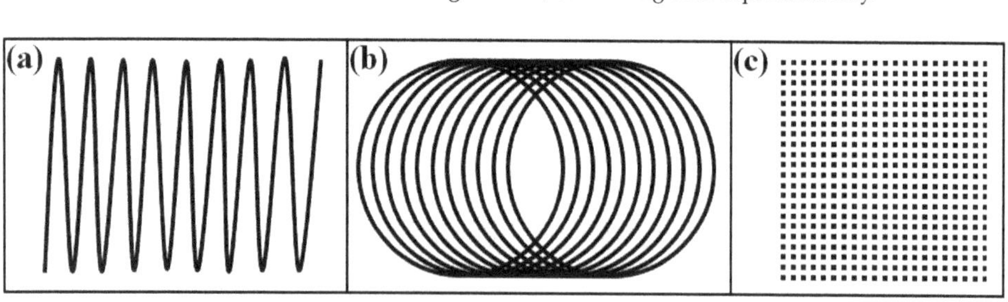

Figure 1. Schematical representation of the beam deflection geometries used: (**a**) line; (**b**) circle; (**c**) matrix.

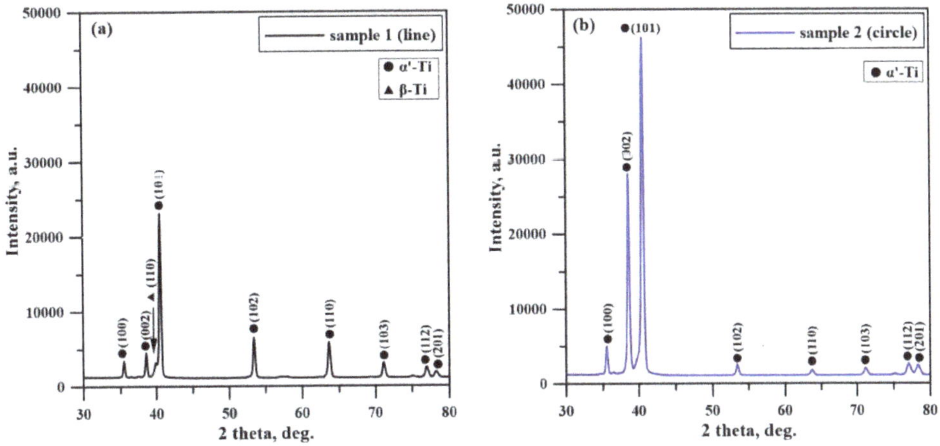

Figure 6. XRD patterns of the electron-beam-modified TC4 alloy: (**a**) sample 1; (**b**) sample 2.

Figure 7 presents the 3D AFM micrographs of the surface architecture of the pure TC4 titanium alloy (Figure 7a), namely sample 1, modified with a linear geometry (Figure 7b), and sample 2, modified with a circular geometry (Figure 7c). The modification process using a scanning electron beam has a significant influence on the surface topography and architecture of the examined samples. The untreated titanium sample exhibits a rough surface (Figure 7a). This indicates residual roughness despite the initial mechanical polishing with different grinding papers. On the other hand, sample 1 possesses a wave-like topography, with relatively homogeneously distributed surface formations (Figure 7b). During the EBSM process with a scanning electron beam, the molten material from the peaks flows down and fills the valleys. Thus, an almost flat surface is formed. Considering the surface architecture of the specimen treated with circular-geometry beam deflection, some peak-like formations on the flattened surface can be seen. In this case, the material stays in a molten state for a longer time due to the overlap of the beam trajectory, compared to the case of linear geometry. The process of the formation of the peaks may be related to the evaporation of the TC4 material and subsequent condensation of the vapors, despite the low surface temperature during treatment using the circular manner of scanning, which could be the reason for the formation of the discussed peak-like formations [42].

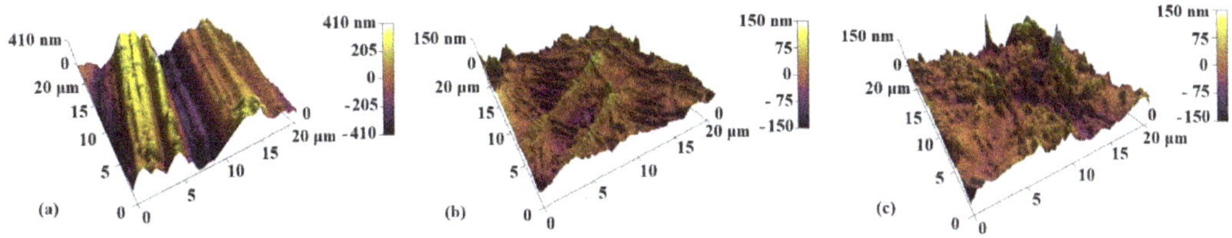

Figure 7. Three-dimensional (3D) AFM micrographs of the surface architecture of the samples: (**a**) base (pure TC4 alloy); (**b**) sample 1 (line); (**c**) sample 2 (circle).

The results for the surface roughness of the studied samples are summarized in Table 2. Electron-beam surface modification can significantly affect the resultant surface architecture in two ways [43]. In the case of very high roughness, EBSM results in a decrease in the surface roughness as the molten peaks fill the valleys. On the other hand, the roughness may increase due to the formation of craters on the surface of the modified samples. In our particular case, the presence of residual roughness (Figure 7a) even after mechanical polishing is a prerequisite for a reduction in surface roughness after the EBSM process. This fact is confirmed by the drastic decrease in the discussed parameter to about 27 nm in EBSM with linear deflection geometry (sample 1) and to 31 nm in EBSM with circular deflection geometry (sample 2), compared to that of the initial TC4 substrate (about 166 nm). The formation of the discussed peak-like formations in the case of the circular manner of scanning is likely the reason for the slightly higher surface roughness than that of the sample treated with linear-geometry beam deflection.

Table 2. Surface roughness and coefficient of friction of the TC4 substrate and electron-beam-modified TC4 samples.

Samples	Surface Roughness, nm	Coefficient of Friction (COF)
substrate (pure TC4)	165.62	0.392 ± 0.008
sample 1 (line)	26.70	0.324 ± 0.009
sample 2 (circle)	31.41	0.344 ± 0.010

It should be noted that the results obtained on surface roughness are of significant importance in the field of modern biomedicine, mostly in implant manufacturing. It is known from the literature that the surface nanostructuring of biomaterials has a direct impact on the cellular response by increasing protein adsorption, bone cell migration, and osseointegration [44,45]. On the other hand, a higher roughness corresponds to a larger contact area, and this can stimulate bone regeneration and binding to bone tissue [46].

The coefficient of friction (COF) of the pure TC4 alloy and the electron-beam-modified titanium samples was investigated by a mechanical wear test (ball-on-flat). The obtained results are shown in Figure 8 and are summarized in Table 2. The average value of the COF of the TC4 substrate is about 0.39, the COF of sample 1 (modified with linear deflection geometry) is about 0.32, and the COF of sample 2 (EBSM with circular deflection geometry) is about 0.34. The results show that surface modification with a continuous electron beam leads to a reduction in the coefficient of friction, with the lowest COF values obtained in the case of linear deflection geometry (0.32). Therefore, the friction coefficient of the modified surfaces is improved due to two different reasons. As already mentioned, the application of the electron-beam surface modification procedure leads to a transformation from $\alpha + \beta$ to α' martensitic structure in both considered cases [33,38,39]. The latter is characterized by much better wear properties, which could be a reason for the decrease in the COF. On the other hand, the surface roughness also has a significant influence on the friction properties [47]. As was already established, the application of surface modification using a scanning electron beam tends to a decrease in the surface roughness and, therefore, to the friction coefficient. When the investigated surface is smooth, the main wear mechanism is abrasive wear, leading to relatively low values of the friction coefficient [48]. In the case of linear deflection geometry, the surface roughness is the lowest (about 27 nm), which corresponds to a low coefficient of friction (0.32), as a result of abrasive wear. The coefficient of friction of sample 2 (0.34) modified with a circular deflection geometry is also lower than that of the untreated specimen (0.39). This means that the wear mechanism in this case is also abrasive. In general, the COF of the electron-beam-modified TC4 alloy is lower than that of the initial TC4 alloy, indicating that EBSM can contribute to a reduction in the friction coefficient and improvement of the wear properties. This makes EBSM technology promising for enhancing the performance of implant materials, where lower friction coefficients and improved wear properties are of major importance. The authors of [49] have studied the coefficient of friction of different coatings (TiN, TiCN, and TiAlN) deposited on Ti6Al4V alloy for biomedical applications, and the results showed that the COF is in the range of 0.5 to 0.7 and concluded that the surface roughness is a major factor influencing COF values. The results obtained in this study are comparable with those published by the authors of [23], wherein Ti substrates were modified by the formation of surface alloys in a system of Ti–Ta.

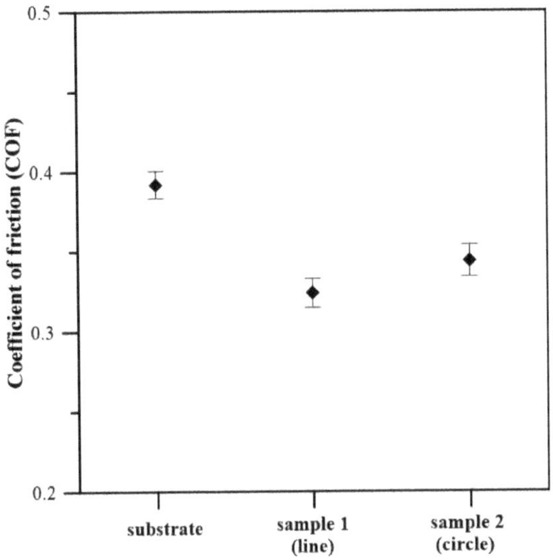

Figure 8. The average coefficient of friction of the TC4 substrate and electron-beam-modified TC4 samples.

In this study, results of electron-beam surface modification of TC4 alloy are presented. The impact of beam deflection geometry during the EBSM process on the phase composition, microstructure, surface architecture, and coefficient of friction of the modified samples was investigated. The results show that the discussed parameter (beam deflection geometry) influences the surface topography and roughness of the specimens.

This study concerns the electron-beam surface modification of TC4 titanium alloy which was realized using a deflection of an electron beam in the form of a line, circle, and matrix. The results show that the application of a linear manner of scanning leads to the deepest modified zone, while the use of a circular manner of scanning tends to the formation of a modified zone with a significantly smaller thickness. It should be noted that the rate of modification using linear and circular approaches of scanning is different due to the different dimensions of both figures. In the case of the circular geometry, the modified zone is thinner because of the larger dimension of the scanning figure, which results in a lower deposited energy density. The application of a beam deflection in the form of a matrix does not tend to the formation of a modified area using the present technological conditions. However, the question remains open regarding the influence of technological conditions and parameters on the structure and properties of electron-beam-modified metals and alloys using beam deflection in the form of a matrix. This is of major importance since it will provide new knowledge on the structure formation and corresponding functional properties of electron-beam surface modification technologies employing more complex manners of scanning.

The results for the microstructure obtained by both approaches of electron-beam treatment (i.e., using linear and circular manners of scanning) show that a fine martensitic structure was obtained in the two cases. During the experiments, the kinetic energy of the electrons is transferred into heat and a thermal distribution from the surface to the bulk is formed, where the cooling rate is very high (10^4–10^5 K/s), leading to the formation of significantly finer structure. As already mentioned, the temperature in the case of the linear manner of scanning is higher than that of the circular one, leading to a deeper modified zone. Considering the treatment using a circular manner of scanning, the decrease in the amplitude of scanning (i.e., the decrease in the diameter of the circle) leads to a rise in the surface temperature and a deeper treated zone. At the same time, the use of a circular approach of electron-beam modification leads to overlap of the trajectory of the beam (see Figure 1), leading to an increase in the maximal temperature and a lower cooling rate in comparison with the linear approach of surface modification and, therefore, coarser microstructure and worse functional properties are obtained. These statements are in agreement with those published in Ref. [33], where it is stated that a linear manner of scanning is much more appropriate for surface modification based on martensitic transformation. It should, however, be mentioned that careful choice of the processing conditions and the beam scanning geometry may result in a wide variety of material surface modifications since, as it is mentioned above, these define the temperature evolution of material which is the crucial parameter for phase transformations.

It was demonstrated the surface roughness is significantly lower in the case of the electron-beam modified TC4 alloy with linear deflection geometry than the initial TC4 substrate. It reached values of 26.70 nm, which is about six times lower in comparison with the untreated titanium alloy. The surface roughness values under circular-geometry beam deflection are 31.41 nm, which is about five times lower than that of the untreated alloy. As mentioned above, the surface roughness of biomaterials on macro-, micro-, and nano-scales play an important role in the successful acceptance of implants by the human body [44]. Surface nanostructuring is relevant to cellular response and osseointegration. On the other hand, an increase in surface roughness corresponds to a larger contact area, which in turn can stimulate bone regeneration and bonding of bone tissue to the implant [45]. The roughness of the surface significantly influences the friction coefficient of materials, where higher values lead to deterioration in the wear properties. This is also of major importance for the application of titanium and its alloys for implant manufacturing and modern biomedicine. It is also known that friction occurs between inserted implants and human bone, which can lead to the release of a large number of metal ions [50]. This is an unwanted effect as it contributes to implant failure and the occurrence of inflammatory processes in the human body. Therefore, precise control and improvement of the friction coefficient (COF) are very important to prevent these side effects. The results obtained in this work for COF show that electron-beam surface modification led to its reduction; this effect is more pronounced in the EBSM procedure with linear deflection geometry. The results obtained in this study completely correlate with these statements. The specimen characterized by the lowest surface roughness exhibited the lowest friction coefficient. However, the contact area in this case is the lowest which is expected to result in deteriorations of cell adhesion, bone regeneration, and bonding of bone tissue to the implant. Therefore, the technological conditions of the surface modification of TC4 alloys using scanning electron beams should be optimized to obtain surface architecture and topography with appropriate nano-roughness for cell adhesion and bonding of bone tissue to implants.

The possibility of employing different electron-beam geometries through the beam deflection system during the process of electron-beam surface modification allows for superb control over the process of cooling atop the surface of the specimens [28]. As mentioned, the use of different beam deflection geometries is accompanied by different cooling rates, and the resulting structure and properties of the modified materials can be very well controlled [33]. This is one of the main advantages of EBSM technology in comparison with other methods for surface modification. For that reason, the correct choice of the technological conditions of the EBSM process is very important and directly depends on the requirements for the materials and their application in the various branches of industry.

4. Conclusions

The impact of beam deflection geometry on the surface architecture and mechanical properties of electron-beam-modified TC4 titanium alloy have been investigated. During the EBSM process, the beam deflection geometry was in the form of a line and a circle. It was found that for linear and circular deflection geometries, the phase composition is in the form of an α' martensitic structure. The thicknesses of the treated zones via both manners of scanning significantly differ from each other. In the case of linear geometry, it is more than 800 μm, while in the case of EBSM using circular scanning, it is 162 μm. The effect is based on different temperature evolutions due to the difference in the deposited energy density at fixed geometrical parameters. Electron-beam surface modification leads to a decrease in surface roughness to about 27 nm in EBSM with linear deflection geometry and 31 nm in circular deflection geometry, compared to that of the pure TC4 substrate (about 166 nm). The results obtained for the friction coefficient exhibit that electron-beam surface modification of the TC4 alloy leads to a decrease in the COF, with the lowest COF values obtained in the case of linear deflection geometry (0.32). The results obtained in this paper on the surface architecture and coefficient of friction would find application in the production of orthopedic and dental implants. It also provides information about the crucial role of electron-beam scanning geometry regarding surface modifications and can be used in the design of processing techniques where optimal efficiency (processed area per time) and desired surface properties can be obtained.

Author Contributions: Conceptualization, M.O. and S.V.; methodology, M.O., B.S., N.N. and S.V.; formal analysis, M.O., B.S. and S.V.; investigation, M.O., B.S., and S.V.; writing—original draft preparation, M.O. and S.V.; writing—review and editing, M.O., S.V., and N.N.; visualization, M.O.; project administration, S.V. All authors have read and agreed to the published version of the manuscript.

Funding: This work was supported by the Project of OP "Science and Education for Smart Growth", "Creation and development of centres of competence"—BG05M2OP001−1.002−0023−C01 "INTELLIGENT MECHATRONICS, ECO- AND ENERGY-SAVING SYSTEMS AND TECHNOLOGIES" (IMEEST).

Institutional Review Board Statement: Not applicable.

Informed Consent Statement: Not applicable.

Data Availability Statement: Not applicable.

Conflicts of Interest: The authors declare no conflict of interest.

References

1. Kaur, M.; Singh, K. Review on titanium and titanium based alloys as biomaterials for orthopaedic applications. *Mater. Sci. Eng. C* **2019**, *102*, 844–862. [CrossRef] [PubMed]
2. Chen, L.-Y.; Cui, Y.-W.; Zhang, L.-C. Recent development in beta titanium alloys for biomedical applications. *Metals* **2020**, *10*, 1139. [CrossRef]
3. Kupur, A.; Dhatrak, P.; Khasnis, N. Surface modification techniques of titanium and titanium alloys for biomedical dental applications: A review. *Mater. Today Proc.* **2021**, *39*, 84–90.
4. Stoyanova-Ivanova, A.; Cherneva, S.; Petrunov, V.; Petrova, V.; Ilievska, I.; Mikli, V.; Iankov, R. Investigation of mechanical and physiochemical properties of clinically retrieved titanium-niobium orthodontic archwires. *Acta Bioeng. Biomech.* **2020**, *22*, 31–39. [CrossRef] [PubMed]
5. Stoyanova-Ivanova, A.; Ilievska, I.; Petrova, V.; Gueorgieva, M.; Petrov, V.; Andreeva, L.; Zaleski, A.; Mikli, V. Physicochemical characterization on clinically retrieved TriTanium orthodontic archwires. *Bulg. Chem. Comm.* **2018**, *50*, 73–79.
6. Rahyussalim, A.J.; Marsetio, A.F.; Saleh, I.; Kurniawati, T.; Whulanza, Y. The needs of current implant technology in orthopaedic prosthesis biomaterials application to reduce prosthesis failure rate. *J. Nanomater.* **2016**, *2016*, 5386924. [CrossRef]
7. Liang, S. Review of the design of titanium alloys with low elastic modulus as implant materials. *Adv. Eng. Mater.* **2020**, *22*, 2022555. [CrossRef]
8. Tamayo, J.A.; Riascos, M.; Vargas, C.A.; Baena, L.M. Additive manufacturing of Ti6Al4V alloy via electron beam melting for the development of implants for the biomedical industry. *Heliyon* **2021**, *7*, e06892. [CrossRef]
9. Panin, A.V.; Kazachenok, M.S.; Borodovitsina, O.M.; Sinyakova, E.L.; Ivanov, Y.F.; Leontieva-Smirnova, M.V. Surface modification of structural materials by low-energy high-current pulsed electron beam treatment. *AIP Conf. Proc.* **2014**, *1623*, 467–470.
10. Zhang, L.-C.; Chen, L.-Y.; Wang, L. Surface modification of titanium and titanium alloys: Technologies, developments, and future interests. *Adv. Eng. Mater.* **2020**, *22*, 1901258. [CrossRef]
11. Valkov, S.; Parshorov, S.; Andreeva, A.; Rabadzhiyska, S.; Nikolova, M.; Bezdushnyi, R.; Petrov, P. Influence of beam power on the surface architecture and corrosion behavior of electron-beam treated Co-Cr-Mo alloys. *Nucl. Instrum. Methods Phys. Res. Sect. B* **2021**, *494–495*, 46–52. [CrossRef]
12. Zhang, L.-C.; Chen, L.-Y. A review on biomedical titanium alloys: Recent progress and prospect. *Adv. Eng. Mater.* **2019**, *21*, 1801215. [CrossRef]

13. Bacela, J.J.; Kielan-Grabowska, Z.; Borak, B.; Sobieszczanska, B.; Walczuk, U.; Kawala, B.; Ziety, A.; Detyna, J.; Sarul, M. Antiadherent and antibacterial properties of TiO_2-coated and TiO_2:Ag-coated stainless steel orthodontic wires against *S. mutans* bacteria. *Acta Bioeng. Biomech.* **2022**, *24*, 107–118. [CrossRef]
14. Liu, Y.; Rath, B.; Tingart, M.; Eschweiler, J. Role of implants surface modification in osseointegration: A systematic review. *J. Biomed. Mater. Res.* **2020**, *108A*, 470–484. [CrossRef] [PubMed]
15. Sasikumar, Y.; Indira, K.; Rajendran, N. Surface modification methods for titanium and its alloys and their corrosion behavior in biological environment: A review. *J. Bio-Tribo-Corros.* **2019**, *5*, 36. [CrossRef]
16. Catauro, M.; Barrino, F.; Poggetto, G.D.; Milazzo, M.; Blanco, I.; Ciprioti, S.V. Structure, drug absorption, bioactive and antibacterial properties of sol-gel SiO_2/ZrO_2 materials. *Ceram. Int.* **2020**, *46*, 29459–29465. [CrossRef]
17. Garcia, E.; Miranzo, P.; Sainz, M.A. Thermally sprayed wollastonite and wollastonite-diopside compositions as new modulated bioactive coatings for metal implants. *Ceram. Int.* **2018**, *44*, 12896–12904. [CrossRef]
18. Hakakzadeh, M.; Jafarian, H.R.; Seyedein, S.H.; Eivani, A.R.; Park, N.; Heidarzadeh, A. Production of Ti-CNTs surface nanocomposites for biomedical applications by friction stir processing: Microstructure and mechanical properties. *Mater. Lett.* **2021**, *300*, 120138. [CrossRef]
19. Nikolova, M.; Ormanova, M.; Nikolova, V.; Apostolova, M.D. Electrochemical, tribological and biocompatible performance of electron beam modified and coated Ti6Al4V Alloy. *Int. J. Mol. Sci.* **2021**, *22*, 6369. [CrossRef]
20. Samanta, A.; Rane, R.; Kundu, B.; Chanda, D.K.; Ghosh, J.; Bysakh, S.; Jhala, G.; Joseph, A.; Mukherjee, S.; Das, M.; et al. Bio-tribological response of duplex surface engineered SS316L for hipimplant application. *Appl. Surf. Sci.* **2020**, *507*, 145009. [CrossRef]
21. Rabadzhiyska, S.; Kotlarski, G.; Shipochka, M.; Rafailov, P.; Ormanova, M.; Strijkova, V.; Dimcheva, N.; Valkov, S. Duplex Surface modification of 304-L SS substrates by an electron-beam treatment and subsequent deposition of diamond-like carbon coatings. *Coatings* **2022**, *12*, 401. [CrossRef]
22. Kirmanidou, Y.; Sidira, M.; Drosou, M.-E.; Bennani, V.; Bakopoulou, A.; Tsouknidas, A.; Michailidis, N.; Michalakis, K. New Ti-alloys and surface modifications to improve the mechanical properties and the biological response to orthopedic and dental implants: A review. *BioMed Res. Int.* **2016**, *2016*, 2908570. [CrossRef]
23. Valkov, S.; Dechev, D.; Ivanov, N.; Bezdushnyi, R.; Ormanova, M.; Petrov, P. Influence of beam power on Young's modulus and friction coefficient of Ti–Ta alloys formed by electron-beam surface alloying. *Metals* **2021**, *11*, 1246. [CrossRef]
24. Behzadi, P.; Badr, M.; Zakeri, A. Duplex surface modification of pure Ti via thermal oxidation and gas nitriding: Preparation and electrochemical studies. *Ceram. Int.* **2022**, *48*, 34374–34381. [CrossRef]
25. Nevskii, S.; Sarychev, V.; Konovalov, S.; Granovskii, A.; Gromov, V. Formation mechanism of micro- and nanocrystalline surface layers in titanium and aluminum alloys in electron beam irradiation. *Metals* **2020**, *10*, 1399. [CrossRef]
26. Liang, S.X.; Liu, K.Y.; Yin, L.X.; Huang, G.W.; Shi, Y.D.; Zheng, L.Y.; Xing, Z.G. Review of major technologies improving surface performances of Ti alloys for implant biomaterials. *J. Vac. Sci. Technol. A* **2022**, *40*, 030801. [CrossRef]
27. Geng, Y.; Chen, X.; Konovalov, S.; Panchenko, I.; Ivanov, Y.; Deev, V.; Prusov, E. Ultrafast microstructure modification by pulsed electron beam to enhance surface performance. *Surf. Coat. Technol.* **2022**, *434*, 128226. [CrossRef]
28. Valkov, S.; Ormanova, M.; Petrov, P. Surface manufacturing of materials by high energy fluxes. In *Advanced Surface Engineering Research*; Chowdhury, M.A., Ed.; IntechOpen: London, UK, 2018.
29. Ramskogler, C.; Warchimika, F.; Mostofi, S.; Weinberg, A.; Sommitsch, C. Innovative surface modification of Ti6Al4V alloy by electron beam technique for biomedical application. *Mater. Sci. Eng. C* **2017**, *78*, 105–113. [CrossRef] [PubMed]
30. Angelov, V.; Ormanova, M.; Kaisheva, D.; Lazarova, R.; Dimitrova, R.; Petrov, P. Selective electron beam surface alloying of aluminum with TiCN nanoparticles. *Nucl. Instr. Meth. Phys. Res. Sec. B* **2019**, *440*, 88–94. [CrossRef]
31. Petrov, P. Optimization of carbon steel electron-beam hardening. *J. Phys. Conf. Ser.* **2010**, *223*, 012029. [CrossRef]
32. Zhang, X.D.; Zou, J.X.; Weber, S.; Hao, S.Z.; Dong, C.; Grosdidier, T. Microstructure and property modifications in a near α Ti alloy induced by pulsed electron beam surface treatment. *Surf. Coat. Technol.* **2011**, *206*, 295–304. [CrossRef]
33. Valkov, S.; Ormanova, M.; Petrov, P. Electron-beam surface treatment of metals and alloys: Techniques and Trends. *Metals* **2020**, *10*, 1219. [CrossRef]
34. Gao, Y.-K. Surface modification of TA2 pure titanium by low energy high current pulsed electron beam treatments. *Appl. Surf. Sci.* **2011**, *257*, 7455–7460. [CrossRef]
35. Gao, Y.-K. Influence of pulsed electron beam treatment on microstructure and properties of TA15 titanium alloy. *Appl. Surf. Sci.* **2013**, *264*, 633–635. [CrossRef]
36. Ferraris, S.; Warchomicka, F.; Ramskogler, C.; Tortello, M.; Cochis, A.; Scalia, A.; Gautier di Confiengo, G.; Keckes, J.; Rimondini, L.; Spriano, S. Surface structuring by electron beam for improved soft tissues adhesion and reduced bacterial contamination on Ti-grade 2. *J. Mater. Process. Technol.* **2019**, *266*, 518–529. [CrossRef]
37. Ferraris, S.; Warchomicka, F.; Iranshahi, F.; Rimondini, L.; Cochis, A.; Spriano, S. Electron beam structuring of Ti6Al4V: New insights on the metal surface properties influencing the bacterial adhesion. *Materials* **2020**, *13*, 409. [CrossRef]
38. Ormanova, M.; Nikolova, M.; Angelov, V.; Petrov, P. Investigation of the microstructure and the thermal processes in a Ti6Al4V alloy surface-modified by scanning electron beam. *J. Phys. Conf. Ser.* **2020**, *1492*, 012065. [CrossRef]

39. Petrov, P.; Dechev, D.; Ivanov, N.; Hikov, T.; Valkov, S.; Nikolova, M.; Yankov, E.; Parshorov, S.; Bezdushnyi, R.; Andreeva, A. Study of the influence of electron beam treatment of Ti5Al4V substrate on the mechanical properties and surface topography of multilayer TiN/TiO_2 coatings. *Vacuum* **2018**, *154*, 264–271. [CrossRef]
40. Panin, V.E.; Panin, A.V.; Perevalova, O.B.; Shugurov, A.R. Mesoscopic structural states at the nanoscale in surface layers of titanium and its alloy Ti-6Al-4V in ultrasonic and electron beam treatment. *Phys. Mesomech.* **2019**, *22*, 345–354. [CrossRef]
41. Nikolova, M.; Yankov, E.; Nikolov, D.; Tonev, D.; Zaharieva, V.; Ormanova, M.; Valkov, S.; Hikov, T.; Petrov, P. Influence of the Chemical Composition, Structure and Heat Treatment on the Topography and Mechanical Properties of Ti5Al4V and Ti6Al4V after Electron Beam Surface Treatment. *IOP Conf. Ser. Mater. Sci. Eng.* **2018**, *416*, 012031. [CrossRef]
42. Zhang, K.; Zou, J.; Grosdidier, T.; Gey, N.; Weber, S.; Yang, D.; Dong, C. Mechanisms of structural evolutions associated with the high current pulsed electron beam treatment of a NiTi shape memory alloy. *J. Vac. Sci. Technol. A* **2007**, *25*, 28–36. [CrossRef]
43. Hu, J.J.; Zhang, G.B.; Xu, H.B.; Chen, Y.F. Microstructure characteristics and properties of 40Cr steel treated by high current pulsed electron beam. *Mater. Technol.* **2012**, *27*, 301–303. [CrossRef]
44. Bonfante, E.A.; Jimbo, R.; Witek, L.; Tovar, N.; Neiva, R.; Torroni, A.; Coelho, P.G. Biomaterial and biomechanical considerations to prevent risks in implant therapy. *Periodontology 2000* **2019**, *81*, 139–151. [CrossRef] [PubMed]
45. Sirdeshmukh, N.; Dongre, G. Laser micro & nano surface texturing for enhancing osseointegration and antimicrobial effect of biomaterials: A review. *Mater. Today Proc.* **2021**, *44*, 2348–2355.
46. Agarwal, R.; Garcia, A.J. Biomaterial strategies for engineering implants for enhanced osseointegration and bone repair. *Adv. Drug Deliv. Rev.* **2015**, *94*, 53–62. [CrossRef]
47. Sedlacek, M.; Podgornik, B.; Vizintin, J. Influence of surface preparation on roughness parameters, friction and wear. *Wear* **2009**, *266*, 482–487. [CrossRef]
48. Kang, J.; Wang, M.; Yue, W.; Fu, Z.; Zhu, L.; She, D.; Wang, C. Tribological Behavior of Titanium Alloy Treated by Nitriding and Surface Texturing Composite Technology. *Materials* **2019**, *12*, 301. [CrossRef]
49. Danisman, S.; Adabas, D.; Teber, M. The Effect of Coatings on the Wear Behavior of Ti6Al4V Alloy Used in Biomedical Applications. *IOP Conf. Ser. Mater. Sci. Eng.* **2018**, *295*, 012044. [CrossRef]
50. Chen, Q.; Thouas, G.A. Metallic implant biomaterials. *Mater. Sci. Eng. R* **2015**, *87*, 1–57. [CrossRef]

MDPI
St. Alban-Anlage 66
4052 Basel
Switzerland
www.mdpi.com

Materials Editorial Office
E-mail: materials@mdpi.com
www.mdpi.com/journal/materials

www.ingramcontent.com/pod-product-compliance
Lightning Source LLC
Chambersburg PA
CBHW061603220926
51850CB00005B/407